Enchanted Looms

Conscious networks in brains and computers

The title of this book was inspired by a passage in Charles Sherrington's *Man on his Nature*. When that famous physiologist died in 1952, the prospects for a scientific explanation of consciousness seemed remote. *Enchanted Looms* shows how the situation has changed dramatically, and provides what is probably the most wide-ranging account of the phenomenon ever written.

Rodney Cotterill bridges the gap between the bottom-up approach to understanding consciousness, anchored in the brain's biochemistry, anatomy and physiology, and the top-down strategy, which concerns itself with behaviour and the nervous system's interaction with the environment. Equally at home describing the intricacies of neural networks, the methods of monitoring brain activity and relevant aspects of psychology, the author argues that an explanation of consciousness is now at hand, and extends the discussion to include intelligence and creativity. Those who believe that their conscious selves control their bodies will be shocked to learn that even our spur-of-the-moment decisions are made unconsciously – that we are merely captive audiences of our own muscular acts. The remarkable thing is that the underlying deterministic mechanism requires that we be conscious. Evolution's boldest trick greatly increased our competitive edge, by giving the body a conscious servant. The fascinating thing is that this ultimately opened the flood gates of our culture and our technology, including – soon – consciousness in computers.

This beautifully written and illustrated book will be valued be scientists and general readers alike for its easy access to one of science's last great challenges. It will change forever our view of consciousness, and our concept of the human being.

Born near Jamaica Inn in Cornwall, Rodney Cotterill spent his early years on the Isle of Wight, and he was educated at University College London, Yale and Cambridge. Originally a physicist, he gradually extended his interests to the biological domain and he now specializes in biophysics at the Danish Technical University, where he has been a professor for the past three decades. Prior to that, he spent five years at Argonne National Laboratory, near Chicago, and he has had two briefer periods as visiting professor at the University of Tokyo. He has been studying the human brain since the mid-1980s, and has concentrated particularly on the neural correlates of consciousness and intelligence. He is also interested in the possibility of simulating consciousness with a computer; he holds a patent in this area, and is currently involved in an industrial collaboration which aims to introduce artificial consciousness to the Internet. *Enchanted Looms* is Professor Cotterill's third popularization, his previous books being *No Ghost in the Machine* (1989) and *The Cambridge Guide to the Material World* (1985), which one reviewer referred to as an instant classic. He is a fellow of the Royal Danish Academy of Sciences and Letters and also of the Danish Academy of Technical Science. When not researching, teaching or writing, he enjoys sailing and choral singing.

Enchanted Looms

Conscious networks in brains and computers

Rodney Cotterill

PUBLISHED BY THE PRESS SYNDICATE OF THE UNIVERSITY OF CAMBRIDGE
The Pitt Building, Trumpington Street, Cambridge CB2 1RP, United Kingdom

CAMBRIDGE UNIVERSITY PRESS
The Edinburgh, Cambridge CB2 2RU, UK http:/www.cup.cam.ac.uk
40 West 20th Street, New York, NY 10011–4211, USA http://www.cup.org
10 Stamford Road, Oakleigh, Melbourne 3166, Australia

First published 1998

Printed and bound in Great Britain by Biddles Ltd, Guildford and King's Lynn

Typeset in Quadraat 9.5/12pt, in QuarkXPress™

A catalogue record for this book is available from the British Library

Library of Congress Cataloguing in Publication data

Cotterill, Rodney, 1933–
Enchanted looms : conscious networks in brain and computers /
Rodney Cotterill.
p. cm.
Includes bibliographical references and index.
ISBN 0 521 62435 5 (hardbound)
1. Neural networks (Computer science) 2. Neural networks
(Neurobiology) I. Title.
QA76.87.C685 1998
006.3'2–dc21 98–20548 CIP3

ISBN 0 521 62435 5 hardback

for all who care for the mentally infirm

Should we continue to watch the scheme, we should observe after a time an impressive change which suddenly accrues. In the great head-end which has been mostly darkness spring up myriads of twinkling stationary lights and myriad trains of moving lights of different directions. It is as though activity from one of those local places which continued restless in the darkened main-mass suddenly spread far and wide and invaded all. The great topmost sheet of the mass, that where hardly a light twinkled or moved, now becomes a sparkling field of rhythmic flashing points with trains of travelling sparks hurrying hither and thither. The brain is waking and with it the mind is returning. It is as if the Milky Way entered upon some cosmic dance. Swiftly the head-mass becomes an enchanted loom where millions of flashing shuttles weave a dissolving pattern, always a meaningful pattern though never an abiding one; a shifting harmony of subpatterns. Now as the waking body rouses, subpatterns of this great harmony of activity stretch down into the unlit tracks of the stalk-piece of the scheme. Strings of flashing and travelling sparks engage the length of it. This means that the body is up and rises to meet its waking day.

Charles Scott Sherrington (*Man on his Nature*)

Contents

Preface

One cannot lightly borrow a phrase from someone else's famous literary work and use it as the title for one's own book. The idea of the brain's neural networks functioning like an enchanted loom has long held a magnetic fascination for me, as it no doubt has for others. But appropriating it for a title brings a heavy responsibility; there is the obvious danger that one could be perceived to be sullying something precious. Charles Sherrington was one of the towering figures of twentieth century physiology, and if I say that I have sensed his presence while writing this book, let me hasten to add that this does not imply that I would anticipate his approval. On the contrary, the writing has been accompanied by an increasing feeling of misgiving.

For a start, I could not hope to write to Sherrington's own exacting standards. Even more seriously, my reading in the subject has led me to a mechanistic view which I suspect Sherrington would not have wanted to share. His spirit continues to be with us, not the least because so many of his scientific descendants are still living. Benjamin Libet, for example, whose work is so prominent in this book, was an associate of John Eccles, and Eccles himself was one of Sherrington's students. Eccles espoused a form of dualism, as is well known, and my most recent contact with Libet left me with the impression that he entertains the idea of free will. These beliefs are shared by many, of course; indeed, they appear to enjoy the support of common sense.

Looking back on our species' earlier attempts to explain the objects and phenomena in Nature, it must be admitted that we have an indifferent track record, when it comes to popular opinion. Common sense has often proved to be an unreliable servant in such matters. We once believed the Earth to be flat, with edges that threatened the too-adventurous voyager. Later, when its true shape was becoming appreciated, we still perceived the Earth as the centre of the Universe, around which the other celestial bodies dutifully rotated.

As our myths gave way to more rational interpretations, anthropomorphic images were replaced by less romantic concepts. Helios had to defer to a thermonuclear reaction, and Thor's hammer and anvil saw themselves superseded by electrical discharges within and between clouds of water vapour. A major contribution to the gradual decline in the supernatural view has been our increasing acceptance of things and agencies that lie beyond the scope of the unaided senses. Organisms and forces that we cannot see have become an unavoidable part of our daily lives.

Even our perception of ourselves and our origins has undergone radical change. Vague ideas about the mixing of blood have long since been replaced by the statistical principles first elucidated by Mendel, and many had reconciled themselves to Darwin's view of evolution well before John Scopes stood trial in Tennessee in 1925. The biological revolution, bolstered by the advent of molecular biology, has now progressed so far that molecular cartographers are well along on an expedition that will chart our genetic territory in its entirety, down to the last base pair.

Along with such issues as the origin of the Universe, the nature of time, and the relationship between the four fundamental forces, consciousness is fast becoming one of the last major bastions of mystery. And because it strikes even closer to home than those three other puzzles, there are many who feel that it ought to be out of bounds. The researcher is all too aware that the scientific boon can become the environmental bane – the researching paragon, the researching pariah. Our minds are the human dimension of the environment, of course, and there must be many who regard the type of work discussed in this book as a form of spiritual pollution.

There has indeed been a sharp rise in the public's awareness of research on the brain. This has come about through the collective effects of a number of factors, not the least of which has been the increase in television coverage of the subject. Just as important has been the emergence of the new term 'neural network'. The successes scored by the computational strategies associated with this term have been winning them considerable financial support, from both the public and the private sectors, and this has naturally provided them with much visibility. Given that research funding has otherwise been meager of late, this development has been a most welcome one.

There has been a disadvantage to these advances, however, for it has led to much confusion about what is actually being achieved. The difficulty stems from the term itself, because it can easily mislead people into thinking that great strides are being made in understanding real biological brains. Unfortunately, that has not been the case, and some of those working in this area have not always taken sufficient care to draw the distinction between computational strategies inspired by the brain and the actual neural networks that we have in our heads.

I have written this book in an effort to help clarify this matter. The text makes what I believe to be the necessary distinction between the different areas of activity which could claim a place under the neural network banner. I hope that the divisions thus advocated will not be perceived as being too pedantically delineated. The book has been set down in this manner because my own experiences in recent years have led me to believe that the misunderstanding is both real and deep.

This approach, which I have not encountered elsewhere, will hopefully help to elucidate these matters. The brain-inspired strategies are sufficiently exciting in their own right, and they have no need of the cheap plaudits that have sometimes come from those who mistakenly believed that such endeavours were contributing to the understanding of the real brain. And by pointing out the difficulties that still confront those who, contrariwise, are trying to understand the genuine article, I hope that this will lead to a better appreciation of their uphill struggle.

The key issue addressed in the book is, of course, consciousness. I must confess that the starkly reductionist picture of this phenomenon that emerged during my evaluation of a large corpus of research data has come as a considerable shock. When I started to write this book, almost a decade ago, I was still reasonably sure that Nature had earmarked a volitional role for consciousness. The strong indications are that this is not the case, and I must admit to being disturbed by such an idea. I cannot help wishing that Sherrington was still alive today, so that we could all benefit from his own views as to what the enchanted loom appears to be doing for us, and indeed *to* us.

Rodney Cotterill

Acknowledgements

I am grateful for the help and encouragement I have received from many people during the writing of this book. Particular thanks are due to Roger Carpenter, who generously devoted much time to reading the manuscript, and who made many valuable suggestions for its improvement. I would also like to thank Roger for all the productive discussions we have enjoyed during these past several years; his input has been the single most important contribution to the entire enterprise.

It is good to have this opportunity of thanking several other people, namely Erik Fransén, Anton Lethin, Franz Mechsner, John Skoyles and Gerd Sommerhoff, for extensive discussions and correspondence; these interactions have had a large impact on my thinking. Similarly, it is a pleasure to have this chance of acknowledging my indebtedness to past and present members of the Danish Technical University's brain science discussion group, and especially to Claes Hougård, Morten Kringelbach, Bjørn Nielsen, Claus Nielsen and Lis Nielsen.

Arthur Pece, Andreas Engel, Peter Földiák, Hans Liljenström and Christian Balkenius all read drafts of the early chapters, and their help was invaluable in getting the book under way. On the technical side, Ove Broo Sørensen skillfully produced beautiful illustrations from my primitive sketches; his contribution to the final product has been a major one.

Finally, I would like to express my thanks to the many others with whom I have had discussions and correspondence. Their number is considerable, and it is my prayer that no such colleague has been inadvertently omitted from the following list: Peter Århem, Bernard Baars, Christen Bak, Rick Bennett, Margaret Boden, Joseph Bogen, Jim Bower, Claus Bundesen, John Clark, David Comings, Francis Crick, John Duncan, Rolf Eckmiller, Örjan Ekeberg, Walter Freeman, Chris Frith, Uta Frith, Jeffrey Gray, Richard Gregory, Sten Grillner, Stephen Grossberg, Hermann Haken, Francesca Happé, Mitra Hartmann, Detlef Heck, J. Allan Hobson, R. Peter Hobson, James Houk, Philip Johnson-Laird, Gareth Jones, Brian Josephson, Christof Koch, Bent Kofeod, Teuvo Kohonen, Richard Lane, Anders Lansner, Benjamin Libet, Ingemar Lindahl, Peter Matthews, Björn Merker, Chris Miall, Pradeep Mutalik, Jim Newman, Terence Picton, Edmund Rolls, Steven Rose, Knud Særmark, Alwyn Scott, Hitomi Shibuya, Wolf Singer, Mircea Steriade, Donald Stuss, John Taylor, Anne Treisman, Max Velmans, Max Westby and Semir Zeki.

1 Introduction

What! Out of senseless Nothing to provoke,
A conscious Something to resent the Yoke!
Edward Fitzgerald (*Omar Khayyam*)

'The sum of all the parts of Such –
Of each laboratory scene –
Is such.' While science means this much
And means no more, why, let it mean!
But were the science-men to find
Some animating principle
Which gave synthetic Such a mind
Vital, though metaphysical –
To Such, such an event, I think,
Would cause unscientific pain:
Science, appalled by thought, would shrink
To its component parts again.
Robert Graves (*Synthetic Such*)

A forbearing machine

In recent months, I have been periodically guilty of what would normally be looked upon as bad behaviour, namely the making of audible comments during games of chess. Seated across the board from my commendably silent opponent, I have been voicing remarks about my adversary's moves, and indulging in outbursts of self-approbation whenever it appeared that I was gaining the upper hand.

Had these games been played in a chess club, I would no doubt have been asked to leave the premises, and even though they actually took place in my own home, the conduct nevertheless seems reprehensible. So why have my taunts been borne with such forbearance? Why, in the face of such provocation, has my rival continued to display mute equanimity? For the simple reason that it is a machine: a computer, which happens to have a chess program in its memory.

It is a sign of the times that I must qualify these statements, by adding that my computer is equipped with neither a speech synthesizer nor a contraption for receiving and analysing the human voice. Both types of device are now conceivable attachments to such machines. But even though it is thus not out of the question for a chess-playing computer to be able to give as good as it receives, verbally, the fact remains that it would be unlikely to react in the way that would be quite normal for a human being; it would probably not refuse to play with a partner prone to such boorish manners.

But a machine could be built which would *physically* respond in just that fashion. Robots exist which are capable of walking, and although this represents a task that is far from trivial, it would be possible, in principle, to furnish such a device with a means of receiving my words, of evaluating their meaning, of perceiving my undesirability as a chess opponent, and of simply leaving the board and walking away. In so doing, it would be reacting in what could broadly be called a human manner. But most people would shy from the idea of calling it a pseudo-person; they would be unlikely to accord the robot human status despite anything that was done to make it more sophisticated.

What is the origin of our scepticism, when confronted with this issue? Why are we so reluctant to allow the possibility that a machine could think, in much the same way that we do? What is the quality that is so quintessentially human that it defies replication in any man-made machine, however complex? I believe that most people would opt for *consciousness*, (or possibly *self*-consciousness[a]). This is the capacity that seems to belong so uniquely to *Homo sapiens*, and there are some who are loathe to credit even the other animals with it, let alone machines. There are others, however, who feel that it will one day be possible to build conscious devices, and some have already striven to give computers what is generally referred to as *artificial intelligence*.

Is that a realistic quest, or is it doomed to fail, because of an inability to grasp some fundamental difference between the human mind and anything that could be fabricated by technology? This will be one of the themes of the present book, and I will be seeking to put the reader in a better position to judge the issue. To judge it, that is, not from the standpoint of traditional artificial intelligence (for that venerable field of technology has already been the subject of considerable discussion), but rather through an appreciation of what are known as *neural networks*.[b] And part of my job will be to explain just what is meant by that expression, indeed, for it would be easy to be confused by the coupling of a biological word with another that so smacks of technology.

Let me state immediately, therefore, that the term neural network is commonly applied in a variety of contexts. These span a spectrum that stretches from the neural circuitry we have in our brains to computers in which some of the components are wired up in a manner reminiscent of the way the brain's nerve cells are interconnected. In particular, and when it is preceded by the qualifier *artificial*, the term is used to describe certain computational strategies that are inspired by the brain's structure. Such strategies have scored a number of successes in recent years, and I will describe some particularly prominent examples. The discussion of artificial neural networks will also serve to illustrate certain principles that these versions have bor-

a / Self-consciousness in this context refers to the ability to examine one's thoughts, rather than to shyness.

b / A commonly encountered alternative term is *parallel distributed systems*. The neural enterprise, both in real brains and as a computational strategy, is broadly referred to by the terms *connectionism*, *computational neuroscience* and *neural computation*.

rowed from the brain itself. It must be emphasized, however, that there are fundamental differences between neural computer hardware and computational strategies, on the one hand, and the genuine article on the other. I will be drawing a careful distinction between the real and the artificial in this book, but I will also be venturing to judge whether the dividing line is ever likely to be breached.

Research on the real brain has of course being going on for centuries. There are reports on the effects of damage to that organ which date from the second millennium BC. Attempts at theoretically modelling the brain are on the point of being able to celebrate their hundredth jubilee, whereas computational strategies employing artificial neural networks arrived on the scientific stage only a few decades ago. Brain-inspired computer hardware is of even more recent vintage; at the time this book is being written, the marketing of such devices is still in its infancy.

The present book's title was taken from a passage in Charles Sherrington's *Man on his Nature*. Sherrington's contributions to the study of the brain were made during a remarkably productive scientific career which stretched over almost seventy years. The revised edition of his seminal book was published in 1952, a year before his death.[1] At that time, the systematic study of the brain's neural networks was barely under way, and it is not surprising that even a man of Sherrington's intellectual stature should have looked upon the brain's workings as something almost magical. But the view of the brain to which he and his contemporaries had been contributing allowed precious little room for the mysterious; with each new advance, the picture was becoming increasingly mechanistic. One can imagine Sherrington and his colleagues speculating about the possibility of constructing a machine capable of thought.

The idea had already been in the air for some time, even in his day. Primitive robots had been built, and there was also the stimulation from literature. A famous step along that road was taken in the story by Mary Wollstonecraft Shelley, published in 1818, in which a certain Dr Frankenstein fabricates a creature from parts of dead humans, and galvanizes the completed body to life by subjecting it to a jolt of electricity. That novel gave birth to an entirely new genre, and the idea that intelligent machines can be created is now so common in fiction that it almost leaves us blasé.[c]

Because we nevertheless draw a sharp distinction between fact and fantasy, confrontation with an actual robot, particularly one of the advanced type, tends to intrigue us. I recall seeing examples at the international fair Expo '85, held at Tsukuba, near Tokyo. Because the organizers wished to give those attending a taste of things to come, this exhibition was well stocked with all manner of clever technological achievements, but the real crowd

c / Another human trait that has recently been observed in clever machines is their susceptibility to 'infection', as typified by the notorious viruses that now plague so many home and institutional computers.

Figure 1.1
This sketch of the author was made by a portrait-drawing robot constructed by a research team at the National Panasonic Division of the Matsushita Electrical Industrial Company. The robot scanned the scene with a television camera, and its image processing unit evaluated the variations between the lighted and shaded parts of the original image. It then extracted the contours in order to form the elements of the line diagram. The associated image processing was carried out by a 16-bit microcomputer in about ten seconds. The early part of the human visual system is known to carry out a similar extraction of line elements from the visual scene, and it has also been shown that there are neurons deeper in the visual pathway which respond to juxtapositions of lines. As has now been established experimentally, there are neurons lying even deeper in the primate brain which exclusively react to a highly specific object, such as a familiar face.

attractors were the robots. And it was easy to locate them, because long queues formed outside the pavilions where they were being put through their paces. At one of these, I had the eerie experience of sitting in front of a portrait-drawing robot constructed by a research team at the National Panasonic Division of the Matsushita Electrical Industrial Company.

Once I had settled in the chair, it moved toward me and gazed at me with a video camera reminiscent of a Cyclopean eye. Then, after a brief pause, which in itself was a little unnerving because one wondered what it was up to, the machine started to make a drawing of me. But the impressive thing was that this was an outline sketch, with all the highlights and shadings normally present in a photographic image removed (see figure 1.1). This is no mean accomplishment for a machine. The image processing unit has to gauge the variations between the lighted and shaded parts of the original image, and it must then extract the contours in order to form the elements of the line diagram. (As will be discussed in a later chapter, it is believed that similar extraction occurs during the early part of the processing of visual information by our own brains.) In this artistic robot, the job of processing the information was carried out by a 16-bit microcomputer in about ten seconds.[d]

Another of the exhibits was still more impressive, in retrospect at least, even though the robot in question looked singularly cumbersome. The ungainly impression stemmed from the fact that this device was attempting to do something that seemed rather commonplace, namely mount a staircase. On closer inspection, however, one began to realize why it held the crowd so enthralled. It turned out that the staircase had been deliberately constructed with uneven step heights, and the machine would soon have toppled over if it had not been able to allow for this lack of uniformity. It too was a one-eyed monster, and the thing that most fascinated the large group of spectators was the way it would pause and seem to cogitate each time it had inspected a new step.

As impressed as we undoubtedly are by such feats of technical élan, we reserve our strongest reactions for those cases in which there is even the suggestion of a personality within the machine. Mere mechanical prowess, however complex, pales in comparison with the implication that such a device even has a mind of its own. This idea too is quite common in literature, of course, a striking example being that of the computer called HAL in Arthur C. Clarke's novel *2001: A Space Odyssey*.[e] Its job was ostensibly to control such things as the navigation and life-support systems in a space ship, and it was equipped with a speech synthesizer, to ease communication with the crew. But although this enabled it to exchange English words with the humans on board, that in itself was not particularly unusual; as noted earlier, a speech capability is now an optional extra for some personal computers. What made

d / The most important pieces of computer jargon are defined and explained in the glossary, as are the key biological terms.
e / Other prominent examples include the robots *R2D2* and *C3PO* in the *Star Wars* series.

HAL different was that it began to make decisions based on what could be called its own motives. And where these were in conflict with those of the crew, it gave priority to its own interests.

We have reached a key issue, for what could be more characteristic of a thinking machine than that it use its thoughts to further its own aims.[f] Is it not this, above all, that is the hallmark of the intelligent agency? But just how are a machine's aims to be defined, and how would one judge the intelligence of something which nevertheless has different prerequisites than that of a human being? These are just some of the points that will have to be discussed in this book.

Intelligence and consciousness – a first look

The concept of *intelligence* is so familiar that we feel we know exactly what is meant by the word. When pressed for a definition, however, we are likely to discover that it is easier to be aware of intelligence than to say just what it is. Our surprise at being thus stumped is likely to be compounded by a frustration stemming from the knowledge that some people not only recognize intelligence in their fellow humans but even manage to quantify it. The existence of intelligence tests, and the general acceptance of the idea of an IQ, tell us that this is the case.[2] Why then does delineation of the concept cause us so much trouble?

We need feel no embarrassment at this perceived inadequacy, for the fact is that even the professional measurers of intelligence do not agree on just what constitutes this obviously desirable characteristic. The manner in which a person's IQ should be gauged is in fact the subject of no small amount of dispute. There are many, indeed, who feel that the use of a single number to denote such a complicated concept is ludicrous. The more cautious of those active in this difficult branch of science identify several different facets of intelligence, and maintain that it must therefore be looked upon as a composite endowment.

With an anxious eye on the machines that may one day usurp us, we might find it easier to say what intelligence is *not*. The insatiable appetite that the typical computer displays for handling large amounts of data, and our feeling that these machines pose no immediate threat, suggests that mere manipulation of numbers and symbols cannot be accepted as a criterion of intelligence. We would be looking for some evidence of what could be called *wisdom*. This, surely, is what puts even the less-educated rustic well above anything that can be presently produced by the computer manufacturer. What is it that gives a person common sense, however, if it is not the ability to take note of the relationships between cause and effect, and act accordingly? And are these not reducible to the very manipulations of symbols that we might be tempted to denigrate when we see them performed by a computer?

f / The word *aims* should not be interpreted in a narrow, egoistic sense; I intend it to include altruistic goals.

Perhaps the most telling aspect of what we call intelligence is the capacity for reacting to novel situations. But the suitability of an individual's response would always have to be judged against the yardstick of what has gone before, even though the conditions that have to be coped with might be unique and thus without precedent. In short, the defining features of intelligence are the faculty for learning from experience, and the ability to apply acquired knowledge to fresh circumstances.

This brief definition circumscribes the concept, but it is insufficiently detailed to serve as a means of ascertaining the presence or absence of intelligence. Because such detection lies at the heart of what this book attempts to accomplish, we will have to delve deeper into the ramifications of sagacity and construct a set of practical rules by which it can be recognized. This is going to be forced upon us by the ultimate need to say whether or not thinking machines will ever be fabricated, but the demand is not a new one. On the contrary, it is the same requirement that burdens those who attempt to measure intelligence in our species. And the fact that such measurements aim at finding a specific level amongst a continuous spectrum of levels endorses the view that intelligence can only be adequately evaluated by taking many factors into account simultaneously. Intelligence, in other words, is not an all-or-nothing thing; each of us possesses intelligence to a certain degree. Let us take a closer look at some of its constituent factors, and let us use the word *agency* to cover both human and machine.

As we have already seen, the intelligent agency must be able to learn from experience, and it must be able to use that experience in making decisions. And the latter will, if they are to be more than mere caprice, involve foreseeing the possible repercussions of given courses of action. Such anticipation is impressive, because it unavoidably entails the ability both to imagine a variety of scenarios and to discern between consequences that are similar but not identical. The handling of novel situations thus implies categorization, association and generalization.

Turning to the question of how intelligence is actually evaluated, we find a situation that is far from ideal. For there are a variety of measures, each reflecting the emphases of their advocates. Indeed, they have sometimes also reflected the interests and exigencies that raised the need for intelligence testing in the first place. A case in point occurred in Paris in 1905, when the school authorities asked Alfred Binet to supply them with criteria for weeding out children with intelligence levels so low that they were unlikely to benefit from formal education.[3] To his credit, Binet devised a series of tests, thirty in all, which were unbiased by any education that the candidate had already received. And he rationalized his observations by defining a person's *mental age* as that of the majority of subjects who displayed the same level of ability as that of the one in question. He went on to recommend use of an intelligence quotient, IQ for short, which is the ratio of the mental age to the *biological age*, expressed as a percentage. Precocious children, well in advance of their years, thus revealed themselves as having IQs much greater than 100, while their poorly equipped counterparts had IQs considerably below the hundred mark.

Even this brief formal definition exposes a limitation in the context of our discussion, because it could not be applied to a machine. What, for example, are we to use for the biological age of a computer? The time that has elapsed since it emerged from the production line? The quite defensible use of biological age when determining the IQ of a human reminds us of the important fact that organisms evolve. But where is the counterpart of this process in a machine? In the case of a computer that possesses a program modifiable by experience, it might be justifiable to equate biological age with the number of opportunities that the program has been given of changing itself, but machine intelligence would have to be judged in a more general manner.

This was a problem that intrigued Alan Turing in the late 1940s, and he came up with a novel solution which he called the imitation game.[4] It can be couched in other terms, but let us consider it in the version originally put forward by Turing himself. The game is played by three individuals: a woman, a man, and an interrogator. The latter may be of either sex, and he or she is located in a room separate from that occupied by the other two. All communication proceeds exclusively by a teleprinter, to avoid the divulging of identity by tone of voice. (Had he been alive today, Turing would probably have replaced the latter by interlinked computer terminals.) The point of the game is that the man must attempt to hoodwink the interrogator into believing that he is a woman, while the real woman merely responds in such a way as to not reveal her gender explicitly. To that end, the two of them are even permitted to lie. Turing believed that a machine can be credited with intelligence if, having replaced the man in this game, it does just as well in fooling the interrogator as to its identity. The candidate machine would obviously have to be rather sophisticated, because it would need to be able to handle language well enough not to give the game away.[g]

Like several other of Turing's inventions, the machine in this game was merely part of an intellectual exercise; he did not actually build such a device. And there is an aspect of this thought experiment that is not well-defined, namely the amount of time that should be allowed to elapse before an assessment is made of the machine's performance; prolonged indefinitely, the test would almost certainly lead to the machine being unmasked. The point that Turing was trying to make was that verbal interaction with a real human serves as a reliable indicator of intelligence if it can be maintained for a reasonable period of time. We should note, too, that the ground rules in this game were not chosen arbitrarily. Turing could have settled on any of a wide variety of attributes, against which the machine's performance was to be appraised, but he settled on something that has almost endless ramifications, namely the difference between the human sexes. The myriad facets of this

g / I have heard remarks to the effect that it was fortunate that Turing did not ask the masquerader to feign being a psychiatrist, since every question would then have been convincingly parried with a *but what do YOU think*.

difference, which permeate the very fabric of human culture, are all potential pitfalls for the aspiring machine.

In contrast, any test that has well-defined rules, however intricate, will readily fall within the capabilities of a sufficiently sophisticated device, so such a task will not be able to expose the machine for what it is. This is an important point, in the present context, because the critical step in training an artificial neural network is invariably the selection of appropriate representations for the input and the output. And yet the discovery of suitable representations is tantamount to exposing the fact that the things being symbolized indeed *have* their rationale. When we later discuss examples of artificial neural networks in action, therefore, we will have to appreciate that the very tractability of a given training assignment might undermine the claim that the network functions intelligently. Turing knew perfectly well what he was doing when he chose the male–female schism as the basis for his experiment. How, for example, would one go about training an artificial network to emulate the proverbial woman's wiles?

No neural network has yet been able to match human behaviour for a sustained period. It seems prudent, therefore, to settle for the lesser aim of demonstrating that current artificial neural networks possess some rudimentary *aspects* of intelligence, such as learning from experience, even though they are not capable of displaying all the characteristics of the human variety. We will see that this achievement is impressive enough in its own right. The inventors of these strategies and machines have no cause to feel sheepish about what they have accomplished. The same thing applies, most emphatically, to the advances secured in recent years by those who study the real brain. Progress has been made across a broad front, contributions having been made by researchers in many disciplines.

We saw earlier that Turing's concern for the need to define intelligence led him to devise an imaginary experiment. The question of consciousness has naturally also given rise to much conjecture, one philosophical (and solipsistic) favourite being the difficulty of proving that one's fellow human beings actually possess it. For example, I know of no method whereby I could demonstrate to the reader of this book that it has not merely been written by an unconscious machine. Similarly, I would be at a loss to devise some testing procedure by which I could unequivocally establish the presence of consciousness in another person. The best that one can settle for appears to be an argument based on similarity. According to this prescription, which is in keeping with Ludwig Wittgenstein's philosophy, an individual who is aware both of his own consciousness and of his general resemblance to other people simply ascribes consciousness to them too. This is expedient, though it unfortunately proves nothing.

The difficulty of establishing the presence of consciousness in another individual is certainly not just a philosophical inconvenience. If that person is a patient on the operating table, and the surgeon wishes to verify *lack* of consciousness before getting down to work with his instruments, the issue acquires an obviously serious dimension. John Kulli and Christof Koch quote from a number of case reports in which patients actually retained consciousness during surgery.[5] These unfortunate people were all pharmacologically paralysed while under anaesthesia, but they were conscious. Their various accounts of the experience make for chilling reading. As one of them put it:

The feeling of helplessness was terrifying. I tried to let the staff know I was conscious but I couldn't move even a finger nor eyelid. It was like being held in a vice and gradually I realized that I was in a situation from which there was no way out. I began to feel that breathing was impossible and I just resigned myself to dying.

Such failures of anaesthesia are not the fault of the surgery staff. Their guidelines are clear enough, and the routine tests of the effectiveness of what they administer usually produce an unequivocal result. The awful thing is that in some cases these chemicals simply do not have the desired effect.

Given the fact that identification of consciousness can sometimes be a very serious matter, it comes as quite a disappointment to find that even those who spend much time thinking about the question are still uncertain as to what consciousness actually is. A particularly disheartening note is struck in a dictionary entry by Stuart Sutherland.[6] It ends with the sentences:

Consciousness is a fascinating but elusive phenomenon; it is impossible to specify what it is, what it does or why it evolved. Nothing worth reading has been written about it.

Others do not throw in the towel quite as readily as this, but their attempts at a definition frequently betray an air of exasperation over the issue. Daniel Dennett, for example, seems overwhelmed by the plenitude of the phenomenon.[7] He has stated that:

Consciousness is both the most obvious and the most mysterious feature of our minds. On the one hand, what could be more certain or manifest to each of us than that he or she is a subject of experience, an enjoyer of perceptions and sensations, a sufferer of pain, an entertainer of ideas, and a conscious deliberator? On the other hand, what in the world can consciousness be? How can physical bodies in the physical world contain such a phenomenon?

These typical examples convey the nonplussed feeling experienced by so many writers on the subject.

There are two points that can be made immediately about Dennett's views. One is that it is not unusual for physical bodies to display phenomena that appear to be non-physical. The conduction, by solid objects, of both heat and electricity used to seem quite mysterious, and these properties were earlier taken to be the manifestations of ethereal processes. Now that such conductivities have been given physical explanations, however, they appear rather commonplace. And this example is a relevant one, because the transmission of signals to and from the brain, and also within the brain itself, is known to be mediated by a somewhat similar (electrochemical) type of conduction. The other question that Dennett's opening remarks raise concerns the implicit assumption that consciousness is a product of the mind. One could ask whether the reverse is not closer to the truth.

This is not to say that Dennett is alone in adopting his position. The *Oxford English Dictionary* defines *mind* as being *the seat of consciousness, thought, volition and feeling*. The trouble is that the entry for *consciousness* merely sees this as being *the totality of a person's thoughts and feelings*. Between them, therefore, the two definitions seem to squeeze consciousness into a subsidiary position, the implied main connection being between mind, on the one hand, and thoughts and feelings on the other.

This sort of uncertainty, over the most basic aspects of the matter, suggests that our misunderstandings might be of a fundamental nature. The solution to the consciousness problem might be right under our noses. Our failure to realize this may be due to the fact that we have cluttered up the picture with too many unnecessary complications. Dennett's inventory of things mediated by consciousness underlines the richness of mental processes, but trying to cope simultaneously with all these, not to mention the many other manifestations of the brain's activities, might be counterproductive. Perhaps we should rather aim at paring the issue down to its bare essentials.

One could start by considering a new-born baby. It seems reasonable to accord consciousness to the wakeful neonate. But at its tender age, it must have a mind that is rather blank. During its first few moments, indeed, dangled upside-down by its ankles and bawling at the top of its lungs, its every response would seem to be the product of instinct rather than thought. My belief that this is the case draws strength from the lack of variation in infants' responses in this situation; one characteristic of the mind, on the other hand, is that it imbues its possessor with the ability to react in a *non*-standard manner.

I have chosen this initial illustration in order to emphasize the need to separate consciousness from its products. Even at this early point, I am thus advocating a somewhat different view from the one expressed by Dennett. I believe that the passage cited above *does* have things back to front, indeed, and that *it is the mind that is the product of consciousness*. I believe, moreover, that it is the sheer abundance of experience mediated by consciousness that fools us into misunderstanding the nature of this fundamental attribute.

In this book, I will probe the question of how consciousness arises from the brain's anatomy and physiology. A description of the workings of the brain's neural machinery will thus be an essential preliminary. Discussion of such details is unavoidable, if the arguments are to be more than mere hand waving. And an appreciation of these essentials will enable the reader to judge the merits of a number of suggestions. There will be conjecture that it might be possible to monitor consciousness externally, for example, and there will be speculation about the prospect of accounting for the biological determinants of intelligence, which enable an individual to reap the benefit of prior experience. I will also be discussing the physical mechanism underlying thought itself.

These deliberations on human consciousness and thought will lead naturally to considerations of their possible counterparts in computers. And in discussing those matters, I will be adopting a position which many, before reading the necessary preamble, would find surprising. Similarly, the issue of animal thought will have to be squarely addressed, as will the question of whether one can nevertheless salvage a unique niche for our own species.

In view of the mysterious position still occupied by consciousness, some of the claims will admittedly sound lofty. But I believe that the understanding of the brain in general, and of the factors which lead to consciousness and thought in particular, are now at the sort of threshold that genetics found itself on just a few decades ago. It is not long since that subject too was shrouded in a mist, one caused to some extent by the inadequacies of our terminology. We used to speak loosely of mixed blood, for example, when discussing the genetic endowment of progeny. The advent of Mendelian systematics, however, and ultimately of molecular biology, blew away the clouds, and we saw just how logically mechanical inheritance actually is. The belief on which this book is based is that we are not far from witnessing a similar revolution in the understanding of consciousness and the human mind.

2 The ultimate black box

What made this brain of mine, do you think? Not the need to move my limbs; for a rat with half my brain moves as well as I.
George Bernard Shaw *(Man and Superman)*

To move things is all mankind can do, and for such the sole executant is muscle, whether in whispering a syllable or in felling a forest.
Charles Scott Sherrington *(The Linacre Lecture, 1924*[1]*)*

The apparent independence of life behind the eyes

Most people experience occasional brief bouts of insomnia. Some unfortunate individuals are habitual sufferers from this affliction, and they require no special provocation to cheat them of a night's slumber. The luckier of us fail to lose consciousness[a] only when we go to bed with something on our mind. Although the condition is familiar, it is worth contemplating because it underscores something important about the imagination.

One lies there in the dark, eyes shut, and under normal circumstances the level of sound is negligible, so the visual and auditory systems are hardly being used. The olfactory (smell) and gustatory (taste) systems are likewise not very active and, if one lies motionless, there will be little by way of tactile (touch) stimulation apart from the steady pressure of the mattress and covering against the body. But although the great majority of the receptor cells that serve the senses are thus not vigorously dispatching signals along their particular nerve fibres, this does not mean that the brain is idling in neutral gear. On the contrary, the problem that is keeping one awake will be giving rise to intense neural activity.

The fascinating thing about all this cerebral commotion is that it manages to conjure up what is normally achieved by the very senses that we

a / In chapter 9, I will be advocating differentiation between consciousness and mere wakefulness. This distinction is without significance for the present discussion.

just characterized as being almost dormant. As an example of the sort of thing that disturbs normal repose, let us take conflict with another person. One goes over the most recent encounter with that individual, re-enacting what gave rise to the friction. One *hears* the agitated dialogue, *sees* the enraged face, and perhaps even *smells* the scents peculiar to the location at which the dispute took place. If the scene involved the consumption of food – perhaps one had complained about something served at a restaurant – the actual *taste* of the offending fare might be recalled. And in the unlikely event that the protest had given rise to a scuffle, one might *feel* the effects of physical contact.

The seeming reality of such internal dramatizations, and our equally sure knowledge that the senses are not directly participating in them, lead to the conclusion that the imagination enjoys a certain independence from the peripheral machinery of the nervous system. In certain cases, indeed, this independence appears to have been incontrovertible, because the sense being invoked by the act of imagination was no longer functional. Beethoven, for example, continued to compose despite his deafness, writing down the scores of music that could have existed only in the private recesses his mind.

Examples which bring out the vividness of such inner representations are commonplace.[b] My favourite story on this subject concerns a girl who was asked why she had expressed a preference for radio vis-à-vis television. Her emphatic reply was that the pictures are much better on radio! In a similar vein, the British novelist Catherine Cookson has described how she routinely conceives her plots sitting in a darkened room, and she likens the process to being in a cinema.

The imagination thus appears to have a life of its own. Everything that it is capable of bringing forth is undeniably related to what the senses mediate, but it can work perfectly well without them being 'switched on'. It is as if the senses are merely the environment's conduits to the mind's inner sanctum, sight providing a window, hearing a funnel, and so on. This seems to be so obviously the case, indeed, that the presence of an ephemeral something-or-other, a *ghost in the machine* as Gilbert Ryle called it,[2] might appear to be unchallengeable.

The mind–body issue is not the only one to have given birth to a perceived hybrid between the physical and the non-physical. There have been several other celebrated cases, all of which have long since bitten the philosophical dust. A now defunct theory of combustion, for example, postulated that all flammable substances are composed of an ethereal agency called

b / There are simple demonstrations that the sceptical reader ought to carry out at this point. Close your eyes and imagine eating your favourite food. Within a minute or two, you will be salivating profusely. As an encore, imagine taking a fresh lemon, cutting it into quarters, and biting into one of them. Unless you happen to have false teeth, this should make you wince. Turning to another of the senses, try to imagine scraping your fingernails along a blackboard or a piece of galvanized steel.

phlogiston, which escaped during burning, and *calx* (or ash), which was left behind. Reunification of phlogiston and calx, according to this idea, would have been sufficient to reinstate the original substance. Experiments involving careful weighing, carried out by Antoine Lavoisier, revealed the fallacy in this line of reasoning. Yet Lavoisier himself was the chief architect of an equally erroneous theory of heat conduction by solids. He attributed this to the similarly incorporeal *caloric*, which was imagined to transport agitation from one region of a body to another, while leaving the material itself unchanged. That idea was demolished when Benjamin Thompson (later Count Rumford) perceived the true origin of the heating that accompanies mechanical work.[c]

Phlogiston's counterpart in life-related substances, *vis viva* (or vital force), proved to be more tenacious, and its demise was brought about by experimental happenstance, during what had been conceived as a routine piece of chemistry. Using the time-honoured strategy of thermally inducing parts of chemical compounds to exchange places, Friedreich Wöhler had heated a mixture of ammonium chloride and potassium cyanate, expecting to obtain a precipitate of ammonium cyanate. To his amazement, the concoction produced crystals of urea. This compound was well known at the time, but it had only ever been observed in biological contexts; it is a component of vertebrate urine, and an adult human normally excretes about 30 grams of it daily. By this serendipitous discovery, Wöhler had shown that there is no difference between the chemistry of organic and inorganic compounds. Just as importantly for our discussion here, he had unwittingly demonstrated that the matter present in living organisms need contain no spiritual additive.

Excretion is one thing, however, and thought quite another. It is not surprising that Wöhler's discovery did not signal the immediate end of the ghost in the machine. If this ghost has been a good survivor, it is because it has been deft at changing disguises. Until medieval times, it acted through *animal spirits*, a term derived from the Latin words *anima* and *spirare*, the latter meaning to breathe. Later, when these spirits had become discredited, the ghost was forced to ally itself with something even more nebulous. And in so doing it actually strengthened its case, as a brief digression will show. So let us consider some of the mileposts during that earlier period.

It was Galen of Bergama, one of the giants of Greek anatomy,[d] who first focussed attention on the brain's four ventricles.[3] These are hollow regions that lie within the brain's tissue. According to his ideas, nutrients processed in the gut give rise to *natural spirits* in the liver, and these are trans-

c / Rumford's original observations were made during the boring of cannon. Oddly enough, he married Lavoisier's widow, and when that relationship turned sour he taunted her with the flaws in his predecessor's reasoning.

d / It was he who squeezed us all into just four temperamental pigeonholes: *sanguine*, *choleric*, *phlegmatic* and *melancholic*. He believed these to stem from the four bodily fluids: *blood*, *green bile*, *lymph* and *black bile*, respectively.

formed into *vital spirits* in the heart. The latter were postulated to turn into animal spirits in the *rete mirabile*,[4] a system of blood vessels at the base of the brain. Galen believed that the animal spirits, also known as psychic pneuma, were then deposited in the ventricles.

More than a thousand years later, these cerebral chambers still held sway, and the Carthusian monk Gregor Reisch apportioned the main faculties between them. There are four ventricles, but the two that lie to the front of the head are symmetrically arranged, and they were collectively allotted to the *sensus communis*. The latter sounds as if it might be a reference to prudence, but the frontal ventricles were merely regarded as a confluence of the routes taken by information streaming inward from the eyes, nose and tongue. In the Reisch scheme, the remaining ventricles were the seats of judgement, memory, fantasy and thought, together with what is our main concern here, namely imagination.

It has wisely been said of the scientific endeavour that it is derivative and fragmentary; all scientists, however prominent, function within the intellectual environment furnished by their peers, and few get much ahead of the pack. When Leonardo da Vinci turned his attention to the brain, therefore, it is not surprising that he became preoccupied by the ventricles.[5] To his credit, he devised a way of establishing their shape. This was no mean task, because the softness of the surrounding matter ruled out simple sectioning. His ingenious solution involved injecting molten wax into the ventricles, waiting for it to set, and thereby obtaining casts. The irony was, of course, that subsequent extraction of the four molds required the scraping away of something altogether more significant, namely the brain's tissue.

The nature of the interaction between the mind and the animal spirits was naturally the object of much conjecture. A picture produced by the Paracelsian philosopher Robert Fludd was essentially a circuit diagram that embodied the current wisdom.[6] It depicted the familiar arrangement of the ventricles and their interconnecting passages, one of the latter extending down to the rest of the body via a route that lies suggestively close to the spinal cord. But in attempting to portray the mind, Fludd could do no better than to show this as occupying a series of orbits deployed around the surface of the head, these being joined to the ventricles by corridors that were no less tangible than those which linked up the ventricles (see figure 2.1).

René Descartes sought to fill in the details of the mind–body interaction by accounting for the mind's control over the body's extremities. He correctly guessed that reflexes must be mediated by a system of conduits, and his sketches of the paths taken by signals travelling from the limbs to the brain show a resemblance to nerve fibres. As an advocate of animal spirits, however, he naturally thought in terms of a hydraulic mechanism. The conduits in his diagrams were thus tubes, which were assigned the dual task of conveying messages from the skin to the brain (in the case of tactile sensations) and from the brain to the muscles. Descartes believed that a muscle moves through the inflation caused by inflow of animal spirits, and he conceived an ingenious arrangement of one-way valves to account for the difference

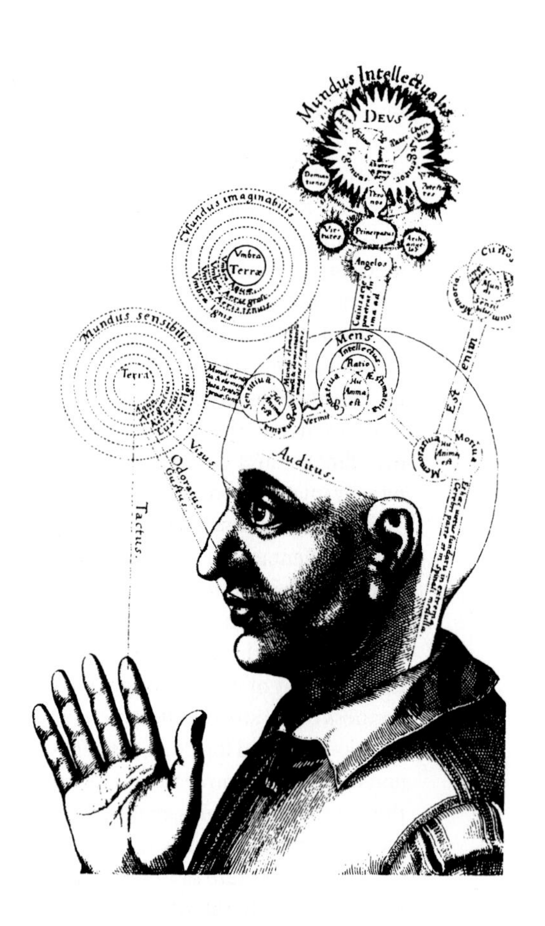

Figure 2.1
This fanciful circuit diagram by the Paracelsian alchemist and mystic, Robert Fludd, attempts to rationalize the mind–brain dualism championed by René Descartes. At the time it was drawn, the general belief was that the working parts of the brain were the ventricles, rather than the surrounding grey matter. This reflected the primacy accorded to the animal spirits, which were assumed to need the intricate system of tunnels shown in the figure. The nebulous and apparently independent nature of the mind led Fludd to position the centres of thought outside the head, but the tunnels leading to them are just as tangible as the inter-ventricle conduits.

between expansion and contraction. Concerning the manner in which the mind exercises control over this system of pipelines, and in contrast to the type of picture epitomized by the work of Fludd, Descartes did not merely opt for further tubes. On the contrary, he looked for a point where the system of passages might come to a focus, and the best candidate seemed to be the pineal body, because this structure lies in the central plane of the brain and it appeared to be the only structure that does not occur as a left–right pair.

These views failed to pass the test of experimental scrutiny. It was noted that the body of a decapitated frog retains the capacity for movement. Its limbs can be made to twitch as a response to pinching and other forms of irritation (including, as Luigi Galvani was subsequently to demonstrate, electrical stimulation). This observation was difficult to reconcile with the Cartesian tubes, which would have been cut through in the beheading of the animal. An even more telling counter to the hydraulic concept was put forward by Jan Swammerdam. He placed a frog's thigh muscle in a sealed flask equipped with a device for detecting changes in volume of the enclosed air, and he demonstrated that the induced contraction of the muscle was accompanied by no such change. Some of his contemporaries favoured an even more direct approach; they simply cut up frogs while these were submerged under water, and finding no signs of effervescence concluded that the pneumatic theory had gone flat!

Natural science was not stumped for a replacement. This was already to hand in the pattern of nerve fibres that had been intensely studied since the time of Leonardo and Andreas Vesalius (the latter's book *De Humani Corporis Fabrica – On the Fabric of the Human Body* – published in 1543, included an exquisitely detailed chart of the nervous system[7]). It has long since been established that such fibres are the real conveyors of signals in the nervous system, and they will figure prominently in the pages of this book. The point to be made here, however, is that this elucidation did not automatically lead to the demise of dualism. On the contrary, it seems almost to have strengthened it, electromagnetic fields being even more ethereal than gases. It is true that gases are invisible (unless they are cooled to below their liquefaction temperatures), but at least one can directly feel the effect of their motion, as blowing on one's hand readily demonstrates.

The signals transported along nerve fibres are electrical (actually, electrochemical), on the other hand, and the body has no means of directly sensing the type of electromagnetic wave which could exert an influence on them. We nevertheless readily accept that the space around us is full of invisible waves, those of radio and television in particular, whose signals can be picked up by the appropriate receiver. And such detection is made possible by the waves' influence on the electrons in the receiver's antenna. It is thus not particularly far-fetched to suppose that there could be a similarly invisible agency interacting directly with the nervous system. On the contrary, one could claim that modern technology has predisposed us to the notion. A neo-Fluddian diagram could be drawn with the orbs and tunnels replaced by arrows used to indicate the orientation of something akin to an electromagnetic field. It would probably enjoy a considerable following.

Dualism, then, has proved to be a remarkably durable idea. It continues to draw strength from what seems to be patently obvious, namely the independence of the imagination from the body. Let us take a closer look at imagination, however, and ask whether this independence hasn't been somewhat overrated.

The suspicious accuracy of the imagination

One of the most intriguing qualities of thought is its fluidity. The mind effortlessly roams from one thing to another, and mental processes seem totally devoid of retarding friction; thought is like a mental equivalent of quicksilver. When we concentrate on one particular thing, the focus can be maintained, but even then there is a natural span of attention, which can be extended only with a certain amount of effort. Thought processes, unlike those occurring in the primary senses, are still barely amenable to direct experimental investigation. Should we gather that this endorses the dualistic separation of mind and matter? We are about to discover that such a conclusion might be rather premature.

We can begin by noting some of the more salient features of sensation, and by making particular reference to the visual system. This is distributed amongst a number of different areas of the cerebral cortex, and various sub-cortical components, and these have been extensively investigated in recent years. The visual areas form a sort of patchwork quilt toward the rear of the head, on either side of the major left–right division. It is also clear that some of these are input regions, because they are the ones that receive connections (albeit indirectly) from the retina. Less obvious is the way in which the various analytical tasks are apportioned amongst the areas, and it is presently not certain whether these are best regarded as forming a hierarchy or merely a confederacy. It *has* been established, on the other hand, that some of the areas participate in the discernment of form and colour, while others contribute to the detection of motion. There are even areas which appear to be involved with several of these tasks simultaneously.

In a manner which is also far from clear at present, the suitably processed visual information is then passed on to other areas in which it can be associated with information emanating from other senses. We do not yet have a cellular-level picture of how useful associations are stored as *short-term memory*, and neither do we know how relevant items are subsequently transferred to *long-term memory*. There are indications that this involves a structure known as the *hippocampus*, however, because damage to that brain component leads to a loss in the ability to lay down new memories, even though neither existing memories nor the capacity for recalling them appear to be impaired.

The big question is, of course, where in the brain imagined events take place. What is the cortical site of these internal representations? In the case of vision, for example, does the mental conjuring up of the face of an absent friend activate the same *neurons* (i.e. nerve cells) as those that fired off their impulses last time one saw the friend in the flesh? There is now evidence, admittedly indirect, that this is indeed the case, and it was supplied by a clever investigation carried out by Stephen Kosslyn.[8] His work was remarkably inexpensive, because his 'apparatus' was simply the minds of his experimental subjects. And the appeal of the study lies in the fact that we can easily substantiate its main findings, with the aid of a stopwatch and a little contemplation.

To begin with, we may note that the visual field of view, that is to say the solid angle subtended at each eye by all that we can see without moving the head or eyes, is constant. This means that what we are able to observe, at any instant, depends upon the distance to the scene we are looking at. In a forest, for example, you will be able to see only one tree if it lies a few centimetres from your nose, whereas you will be able to take in many trees simultaneously if they are several metres away. And from a distance of a few kilometres you will probably be able to see the entire forest. Kosslyn began by noting that mental imagery is subject to the same general constraints; the angular spread of a familiar object, imagined at a particular distance, is about the same as that observed with the eyes open. Trying to imagine specific scenes, he also found that there is a decrease in resolution as one progresses from the centre of gaze out toward the peripheral regions, just as when the eyes are open.[e] Then again, Kosslyn found that it is difficult to imagine having access to a specific level of detail if this is being imagined to lie at too great a distance. The reader can carry out the same test by closing the eyes and trying to imagine someone several paces away holding up a newspaper; you will not be able to imagine reading it, because the print will not be visible. But if you imagine the person slowly walking toward you, first the headlines, and ultimately even the fine print, will become discernable. This is just as it would be if the scene was actually being observed with your eyes open.

Kosslyn's other main investigation was even more revealing. It involved exploration of an imaginary landscape, and he enlisted the help of several experimental subjects. Each of these was first asked to draw the map of an idealized island, the sort of thing that one might see as an illustration in Robert Louis Stevenson's *Treasure Island*. Essential features to appear on this map included the beach, a hut, a coconut tree and buried treasure (suitably marked with an ×), the whole being appropriately embellished with a minor amount of other detail, such as rocks and shrubs. The drawings completed and thoroughly assimilated, Kosslyn then asked the subjects to close their eyes and imagine exploring the island. If this little journey was to start on the beach, for example, the subject might then be asked to proceed to the hut, and thence to the coconut tree, and finally to the treasure. The actual route was chosen by Kosslyn, a different one for every repetition of the test, and the subject was asked to press a button when each feature was reached. From the times it took for the subject to mentally travel between the various points on the map, it was apparent that the mental map was just as reliable an indicator of distance as the physical map originally drawn by the subject.

As simple as these investigations undoubtedly were, their importance could hardly be exaggerated. They strongly suggest that some of the parts of the brain which are used to construct the visual images we see with our eyes open are also involved in the formation of mental images when our eyes are shut. But what is the mechanism by which such remembered images are regenerated? The impetus must come from the brain's association regions, the places where there is a record of what goes with what. But the actual nature of the regeneration process has remained elusive; I will have much to say on that issue in the latter chapters of this book.

There was an earlier experiment which also deserves mention in the context of Kosslyn's findings. It was reported by Cheves Perky in 1910, and it too was remarkably cheap to carry out.[9] Perky placed a subject in front of a translucent screen, onto which pictures could be projected from the rear, the brightness of projection being controlled by a rheostat (a voltage-control device). A favoured example was a picture of a tomato, and the subject was

e / This too is easily checked; while staring straight ahead, slowly move a page of typescript – the current page of this book, for example – from the edge of your visual field in toward the direction of your gaze. You will not be able to read the text until it is almost directly ahead of you.

periodically asked to report whether it had become visible, as the intensity was gradually increased from an initially zero value. Ultimately, of course, the subject indicated that the tomato had indeed appeared.

Perky then switched off the projector and asked the subject, while still staring intently at the blank screen, to try *imagining* the tomato at the location its image had previously occupied. Without divulging the fact, she switched the projector on again and gradually increased its brightness, taking note of the point at which the subject realized what was happening. In all cases, the intensity had achieved a value well above that at which the image had previously been reported as becoming visible. In other words, the subjects experienced great difficulty in distinguishing between an imagined picture and the real thing. A striking augmentation of this result has emerged in recent clinical observations of people suffering from *achromatopsia*, which is caused by damage to a specific area of the cerebral cortex known as V4;[f] they have not only lost their ability to perceive colour, they cannot even recall colours seen before the damage occurred.[g] There could hardly be stronger evidence than this for the most intimate of relationships between the mechanism of seeing and that of imagining visual scenes. In the face of such results, the imagination begins to appear rather less independent than one might have supposed.

We are constantly availing ourselves of this accuracy of the imagination, albeit unwittingly. Consider, for example, the simple act of fastening a clasp at the rear of one's neck, as when putting on a locket. This would be nigh on impossible if one were not able to summon up a mental image of the relevant parts of the ornament. However, it is important to distinguish between such an act and, say, that of walking without looking at one's feet. The former requires conscious attention (before it becomes a habit, at least), whereas the latter merely involves reflexes, after one has learned to walk that is. The difference between conscious and unconscious acts will have the profoundest of consequences for the central theme of this book.

We have now taken a brief initial look at an apparent dichotomy, first establishing the case for independence of the imagination and then undermining this by demonstrating how inextricably tied the imagination must be to the sensory machinery that mediates it. A satisfactory resolution of the issue would require that we understand how the brain works. And such an understanding would have to embrace several different levels of operation, spanning the range from individual nerve cells to the collective functioning that underlies cognition and thought itself. This is sometimes loosely referred to as opening *the mind's black box*.[10] Before proceeding farther, therefore, we would be well advised to contemplate the nature of black boxes.

f / A description of the spatial arrangement of cortical areas, together with an account of how various faculties are distributed amongst them, will be given in the next chapter.

g / The fact that Beethoven retained his ability to imagine music, after the onset of his deafness, suggests that there was something wrong with his ears rather than with his auditory cortex.

Elements of black boxology

The citations at the beginning of this chapter, written many decades ago by two of the twentieth century's great thinkers, appear to advocate diametrically opposing views. Shaw's statement was an appeal against what he saw as denigration of the brain, and he made his point by implying that movement is one of the baser bodily functions. What he had in mind, by way of contrast, were such higher cerebral capacities as thought and imagination, the very things that have been occupying us in this chapter. Sherrington nevertheless made a valid point, because even speech, a human prerogative, is merely the product of muscular movement, the muscles being primarily those that control the diaphragm, jaw, lips, tongue and larynx.

Sherrington's conviction provides us with a timely reminder that the brain is simply a stimulus-response device. It is just an example of what is popularly referred to as a black box, that broad concept naturally covering things that are neither black nor box-like. And he penetrated to the heart of the matter when he perceived the unifying characteristic in all our responses. We might go about things in a more sophisticated manner, but we have it in common with other animals that our goals are all accomplished by muscles. We use various combinations of them when feeding, fighting, fleeing and procreating. And except in those cases where the movement is part of a reflex, the muscular output is determined by processes going on in the brain: mechanisms that produce an appropriate response for a given sensory input. The challenge is to obtain a clear picture of those intervening events, the mental operations that so preoccupied Shaw. Their elucidation is one of the chief aims of this book, of course, and an adequate explanation of brain function will have to include an account of the special case in which the output is merely imagined. It was no doubt the existence of that particular sub-class which gave rise to the Shavian misgivings.

During much of the first half of the twentieth century, when *behaviourism* was the dominant theme of psychology, attempts at elucidating the mechanism underlying the input–output relationship were regarded as pointless. Introspection on the nature of thought and consciousness was not in vogue; there was nothing to be gained from removing the black box's lid. The movement's prime motivator, John Watson, believed that such mental processes as feelings and intentions do not determine what a person does. One of his followers, Clark Hull, saw psychology as resembling physics in its pre-atomic days. At that time, all that one needed to know about the behaviour of a piece of material was the relationships between such variables as stress and strain, pressure and volume, voltage and current, and so on. Knowledge of a material's internal structure was believed to be irrelevant. Hull actually attempted to emulate physics by writing equations that similarly quantified behaviour, purely through the observed correlations between stimulus and response. Behaviourism found its most ardent champion in B(urrhus) F. Skinner. He believed that behaviour is determined by the history of reinforcements that an animal has been subjected to, and that feelings are just the consequences of actions. Like other behaviourists, he dismissed consciousness as simply an epiphenomenon.

If I believed such things of our black box, I would not have written this book. But mere faith that the box contains something important is not enough, of course; we will have to examine its contents and try to understand how they function. As a first tentative step on what is going to be a long journey, we will find it useful to consider how other black boxes go about linking outputs to inputs. I suggest that we do this by contemplating three types of system: mechanical, electrical (electronic) and neural. Moreover, because evolution has probably produced brains that are rather efficient, a good starting position would be to assume that things are not more complicated than they have to be. Let us start, therefore, by considering some really elementary black boxes. Mechanical examples have the advantage that they are readily envisaged, and they tend to be the easiest to understand. They are occasionally encountered in quizzes and usually take the form of diagrams involving levers, rods, pulleys, strings, cog-wheels, weights and the like. One is asked to indicate which of several alternative mechanisms will correctly match a response to a stimulus: the lowering of a weight, for instance, when a lever is raised. Such illustrative pictures can be dispensed with here; we can use our imaginations, apparently independent but suspiciously accurate as we have seen them to be.

It might seem far-fetched to draw comparisons between mechanical gadgetry and living tissue, but we should bear in mind our ultimate need to determine how far along the human route machines are likely to progress, and we will not be in a position to do that unless we really understand the mechanical variety. With a sufficiently large investment of time and materials in its fabrication, a mechanical device could be a considerable improvement on the type of dummy one associates with the name of Marie Tussaud. Apart from being given a flexible exterior, with the texture, feel and smell of real skin, it could be equipped with mechanical, light and audio sensors at the appropriate positions. These would make the dummy capable of moving and reacting to a variety of external commands.

It would be easy to dismiss conjecture about such a contraption, and classify it as belonging to the domain of *thought experiments*, in which one considers in principle things that could not be made in practice. Such rejection would be premature, because it misses an important question: when does a growing child realize that its parents are something more than mere devices? After all, our definition has granted the device all that is required to meet an infant's demands. By what process does the growing toddler latch on to the fact that there is more to the human box than meets the eye? Do children intuitively apply something akin to the Turing test that we discussed in the previous chapter? Are there individuals who never manage the breakthrough? The sad answer is that there might well be such people, namely those who suffer from autism.

It is clear that our discussion still has a long way to go before reaching the degree of sophistication required of a brain-like machine. There are several readily observable features of brain function that even a minimally qualified device, mechanical or electronic, is going to have to emulate if it wishes to be taken seriously as a candidate surrogate. Turning to the issue of time, for instance, we know that real brains often need a period of cogitation before coming up with a suitable response to a stimulus. It would of course be a simple matter to build much longer delays into the working of our box, but this would certainly have to be more than mere delay for delay's sake. A realistic delay would be one made inevitable by the need to accomplish things *en route* to the ultimate response, things somehow related to the response itself.

The 'lifelike' dummy described above was capable of handling multiple inputs. This introduces the further issue of sequence in that attainment of a particular response might be contingent upon presentation of inputs in the correct order. The mechanical world is full of such devices. An automobile will function in the desired manner only if one operates the various pedals and levers in the prescribed succession. A computer, likewise, needs to be directed through the correct sequence of steps in order to produce a particular output, and this is the job of the program. We too possess the equipment for handling multiple inputs in that we can avail ourselves of our five senses, and it is also true of our systems that sequence is usually significant.[h] It is not difficult, for instance, to think of situations in which it is important to know whether a sound precedes or follows a visually or tactually observed movement.

In the automobile, it is of course the internal system of pistons, crank-shafts, cog-wheels and levers which reward the correctly sequenced inputs with the desired locomotion. In ourselves, there are neural counterparts which similarly accomplish the appropriate muscular movements, some of which also produce locomotion indeed. There is something else, however, namely the seemingly paradoxical fact that correct functioning of all these machines, whether mechanical, electronic or living tissue, invariably involves ignoring much of the input. In the case of ourselves, the senses are perpetually being bombarded by all manner of signals that are extraneous to our immediate needs. When we visually observe a particular figure of interest, for example, we exploit our natural ability to separate it from the background.

Mechanical and electronic machines are similarly burdened, of course, each in their own particular way. This is why their designers are at pains to counteract the influence of noise, whether it be of the mechanical type exemplified by the automobile driven over a bumpy road or the electromagnetic variety epitomized by the radio listener's ear glued to the loudspeaker in an attempt to hear a transmission over the background crackle. Such situations are frequently quantified by what is logically termed the *signal-to-noise ratio*, but figure–background separation would seem to involve more than this. Although the background is subsidiary to what is being

h / Multiplicity of inputs is not merely related to our having several senses, of course. A numerically far greater multiplicity arises from the fact that each sense is served by a large number of receptor cells. Each of our retinas, for example, comprises well over a hundred million light-sensitive cells. The manner in which the brain extracts meaning from so many simultaneous inputs will naturally be one of our main concerns in later chapters.

concentrated on, by definition indeed, it is nevertheless not without significance. A bull in a meadow is one thing, for example, while a bull in the matador's arena is something quite different. And a bull in a china shop is something else again.

This need to discard or relegate some (and usually most) of the stimuli that impinge upon our senses is served by the brain's *attention mechanism*. Contemplation of the latter will be one of our chief preoccupations in later chapters, and we will be considering the contributions that various disciplines have recently been making to the problem. Discussion of attention will be inextricably tied to the related enigma of consciousness itself. But we will not be in a position to contemplate these deep matters until much more of the relevant psychology, physiology and anatomy has been reviewed. Let us instead briefly touch upon something related to the question of noise.

When an engineer is confronted with the need make a machine robust against externally imposed variations in its operating conditions, he will often resort to what is known as the *predictor–corrector strategy*. He provides the machine with a device (it can be mechanical or electrical, or a combination of both) which is essentially a model of the machine's ideal behaviour. The machine also has a means of monitoring the instantaneous deviations from the ideal, and these are used to make automatic adjustments. The current shortcoming of this approach is that it is still limited to cases in which a rather simple behavioural model suffices, these cases being referred to by the term *linear* (which is to say imbued with simple proportionality). We will return to this important concept of internal modelling later, both when we consider muscular movements and later still when we look at consciousness itself. I am going to emphasize that it has an important counterpart in the functioning of the real brain. We will see that the brain is special in being able to construct its own model, through its capacity for something that we now ought to add to our black box: *memory*.[i]

It would not be difficult to modify a mechanical version of our black box so as to give it a primitive form of memory. We could, for example, equip it with a ratchet wheel activated by a lever on the input side. For the sake of argument, let us take this wheel to possess ten teeth, just one of these having a form which enables it to trigger a suitable output response. Immediately after the most recent response, it will take ten further activations of the input lever to produce another one. If we press the lever only three times, say, the device will *remember* that it would take seven more lever activations to produce a response. This is certainly a paltry memory, but nevertheless one which could easily have out-performed me on my first birthday. And it would require only the addition of a few more wheels with various numbers of teeth, some of the

i / This addition, if it is accompanied by no other complicating factors, turns the box into what is known as a *finite-state machine*. Such a device has internal states, and both the next internal state and the next output are determined solely by the current internal state and the current input.

wheels interlocking, to produce a memory machine of considerable sophistication, one capable of carrying out quite involved calculations.

This is precisely what Charles Babbage proposed in 1822, as a means of calculating mathematical tables, and he called it a *Difference Engine*. He later conceived the idea of what he referred to as *Analytical Engines*, which were essentially the forerunners of modern computers. Although Babbage did not actually construct these machines, his detailed plans survived and they have now been demonstrated to be functional.[11] In the late 1930s, John Vincent Atanasoff constructed the first electronic counterpart of these mechanical gadgets, namely the *electronic digital computer*.[12]

A simple modification of our ten-toothed ratchet wheel would replace the single specially formed tooth with a cam.[j] And we can easily arrange things so that the output is triggered not only by that tenth tooth but also by the ninth and the eighth. Starting from immediately after the most recent output, we would then have an ambiguous situation in which different inputs, namely different numbers of activations of the input lever, produce the same output. As before, scaling the machine up to the multiple-wheel level would give us a machine with an arbitrary degree of ambiguity-related complexity.

Such devices are nevertheless far less mysterious than those belonging to another class of machines. Let us start with a simple analogy, as usual, and consider the deformation of a thin plastic sheet, a credit card for example. If we hold such a thing between the thumb and the index finger and press with sufficient force, it will suddenly be deformed into an arc. A slight easing of the force allows the card to revert to its straight configuration, and subsequently increasing the force again ultimately gives the arc. Viewed from a given direction, this arc may be either concave or convex, and we find that it is not always easy to predict which alternative configuration the card will adopt. Such deformations have come under intense scrutiny in recent years,[k] because they are one of the phenomena that have been shown to display what is known as *deterministic chaos*. This term describes situations that possess an intrinsic ambiguity of outcome for a given and *fully defined* set of circumstances.

Although there are now electronic versions of both types of ambiguous machine, we need not consider them here. Let us instead pass directly to their neural counterparts. A neural machine in which different inputs can produce the same output has already been touched upon. It is just such an arrangement that permits a figure to be separated from the background. Taking the example given earlier, it is that which enables us to utter the word *bull*, as a primary response at least, irrespective of the animal's surroundings. The actual neural circuitry underlying such a possibility is something which we will have to go into later, when we consider neural networks. For the time being, therefore, let us simply note that the possibility of ambiguity arises from the fact that each neuron is able to interact with many thousands of other

j / In this context, a cam can be regarded as a wider tooth.

k / The mechanical engineer refers to such an element as an Euler strut.

neurons. It is invariably the case that signals from many other neurons converge on a given neuron, as they travel in what we may call the forward direction, and it is easy to imagine a situation in which several different combinations of input signal give one and the same onward-going output signal.

Although it is certainly also the case that a given neuron simultaneously sends signals to numerous other neurons in the down-stream direction, such *divergence* cannot alone be the origin of the above behaviour; a neural circuit with such divergence will not automatically match our deformed credit card by producing intrinsically ambiguous outputs. Neurons, or groups of neurons, that really were capable of such behaviour would obviously be agents of a fundamental indeterminacy in the functioning of the brain. This type of situation has been given serious consideration by a number of people, and we will later have to weigh their arguments. It does not demand too much of the imagination to see that this type of ambiguity could be tied to the issue of *free will*.

It might seem that the brain acts in this manner when its visual domain is confronted by the sort of ambiguous picture that is shown in figure 2.2. Although this particular example was not the first of its kind,[1] it is nevertheless a well-known one. It was devised by the cartoonist W. E. Hill, and it appeared in an article by Edwin Boring[13] in 1930. It can be interpreted as showing either an elderly woman seen at almost full face or a younger woman seen in profile from her left side; the lips of the former provide a neck-band for her young counterpart. There is a complication here, however, because the orientation of one's eyes does not remain fixed as the picture is examined. The centre of gaze scans the various features, in a manner that depends on which of the two interpretations is gaining the ascendancy at any particular moment. In effect, this means that the visual system is presenting itself with *different* inputs, and the issue of ambiguity of response lcses its bite. The example is important, however, because it brings in the question of *selective attention*. We will be investigating the mechanism of attention in chapter 8.

We have not finished adding degrees of complexity to our inventory of devices; there are still other types of black box to consider. But before moving on to them, we must prepare ourselves for a small conceptual leap. Apart from our brief diversion with the credit card, we have been considering examples of machines in which the workings are housed in what could loosely be called a container. The point has arrived at which we must throw off this restriction and contemplate metaphorical black boxes. There are things formerly regarded as mysterious which never saw the inside of a cabinet, an automobile engine compartment or a cranium.

In the heyday of alchemy, what we now call chemistry was a just such a black box. The interactions between substances appeared deeply puzzling, and this is why so many greedy patrons were gullible to the alchemists' propa-

1 / We will be considering the classic examples in chapter 8, and subsequently.

Figure 2.2
This famous picture, drawn by the cartoonist W. E. Hill, and reproduced in the 6 November 1915 number of the magazine *Puck*, was cited by Edwin Boring in a brief article published in 1930. It can be interpreted in two different ways. It can be taken to show either an elderly woman, seen at almost full face, or a younger woman seen in profile from her left side. The old woman's lips serve as the younger woman's neck-band; her left eye is the younger woman's left ear; her right eye-lash is the younger woman's left eye-lash, while her nose is the other's jaw. Although this is a familiar example of an ambiguous figure, it does not require the visual system to function in an ambiguous manner. The attention mechanism accords primacy to different aspects of the figure, depending on which interpretation is currently being given priority.

ganda. Even when reliable identification of the components of materials was turning chemistry into a science, it was still not known why these could be induced to change their affiliations. In many cases, when compound AB was made to interact with compound CD, it could be demonstrated that there had been a rearrangement, producing, say, AC and BD. Later, when the existence of atoms had been established, it became clear that it is the various inter-atomic interactions, together with the prevailing physical conditions, which determine the direction in which a chemical reaction will proceed (actually through a quantity known as the free energy, but we need not go into such details here).

How are we to regard such interactions, now that the logically mechanical nature of chemistry has been revealed to us? In the above example, for instance, would it be correct to say that the atoms of A *remember* that they prefer atoms of C rather than atoms of B, and that they will shift their attentions to the former, given the opportunity? I believe that most people would laugh at this suggestion. And I suspect that the response would be much the same if the reaction was made arbitrarily more complicated. It could, for example, involve a chain of reactions in which AB, CD, EF and GH gradually became BC, DE, FG and HA, a sort of atomic-level musical chairs.

Now the point is that if we shift the scene of this cycle of reactions from the test tube to the extremity of a nerve cell, make AB, CD and so on represent the appropriate chemical compounds, and augment the set of physical conditions with the relevant nerve impulses, we have a situation in which a small part of our nervous system is participating in the learning of something.

It cannot be denied that the order in which the events occur is of paramount importance, but we have already seen that the same is true of quite simple machines. Memory, certainly a senior servant of our apparently independent life behind the eyes, is thus revealed as being mechanical and tied to systematic chemical events taking place in the brain's network of nerve cells. We can fill in the actual details later, but the fact must already be emphasized that the mechanism of memory needs to invoke nothing ineffable.

The significance of timing must not be underestimated when considering chemical reactions. This can be illustrated with a rather gross example. Consider the composite system of an explosive and water, the former being a suitable mixture of chemicals such as that used in a firework. If the water is added after the explosive mixture has been ignited, it may merely serve to douse the glowing cartridge. If it is added prior to ignition, on the other hand, it could give us a damp squib. But the actual timing is not really critical in the latter example, given that it would require a fair amount of care and patience to dry out a wet firework. In the chemical reactions involved in the nervous system's memory processes, however, the timing must be far more precise if they are to be effective. In the conditioning of a certain type of water snail that we will be considering in chapter 5, for example, half a second either way can make all the difference, and there are memory processes taking place in our own cerebral cortices that require a temporal precision in the range of a twentieth of a second.

While in this chemical vein, let us take things just one step farther in the direction of complication. A common acquaintance of those studying elementary science is the *semipermeable membrane*. One learns that it prevents the chemical reunion of some of the components in chemical reactions, thereby giving free play to others. Such membranes were put to good use in the cells that were the forerunners of the modern electrochemical battery. It is noteworthy that the same words, *membrane*, *cell* and *electrochemical* crop up frequently in descriptions of processes in the nervous system, and the membranes involved are indeed semipermeable. The compartmentalization that they represent provides Nature with a means of controlling cascades of chemical reactions – reactions far more recondite than the simple chain mentioned earlier – reactions in which some components modify the interactions between others – reactions in which other components modify the modifiers.

We are approaching the end of our roll-call of black boxes, and we should now contemplate the one in which something is going on internally in the absence of either an input or an output (recent, at least). It is this type of box, no doubt, that provided the basis for the Shavian point of view, and it is the example that most boldly militates against the Sherrington wisdom. Referring back to that opening of this chapter, indeed, we recognize it as the receptacle that was being tormented by those nocturnal reflections: it is the input-less output-less brain of the insomniac. We can absorb the myriad impressions of our environs, as conveyed by our senses, and yet not react; we can sit motionless in thought and suddenly be galvanized into action; and we can simply sit and contemplate, oblivious of everything going on about us, reorganize our impressions of the world, and for the time being at least take no action at all.

Although this review of black boxes could hardly claim to have been exhaustive, it has at least suggested a certain continuity as things progress from the obviously simple to the manifestly more complex. Steadily increasing the demands placed on the performance of our box, we have encountered no radical changes in what could be called the physics of its behaviour. And there has been no indication that the processes underlying brain function involve principles not previously encountered in the scientific enterprise.

That we were wise in embarking on this excursion in the first place has now become apparent, I feel, for there has been no box that could not be realized in a mechanical or electronic version, rather than in living tissue. But the question arises as to whether any one box could possibly show the flexibility we see in the brain. For the latter functions to an order of complexity that directly follows from the task it is being set, at any instant. Even the most confident engineer would wince at the prospect of having to copy such flexible automation. I suspect that this will continue to be true, no matter how much we learn about the brain in the coming decades.

A given output can result from a given input in many different ways. This is what provides the challenge in the quizzes mentioned in the beginning of this section. In the case of the brain, we have sufficiently detailed characterizations of the inputs and outputs. We now wish to know how the brain goes about the task of linking the one to the other. In this quest, we are fortunate in being able to draw on the results of many different types of investigative endeavour. We can even employ different types of overall strategy. In what is known as the *bottom-up approach*, for example, one attempts to synthesize a multi-cellular picture from one's knowledge of how individual nerve cells behave. This route can take advantage of its access to the considerable corpus of anatomical and physiological detail that has accumulated over recent decades.

The *top-down approach*, on the other hand, takes as its starting point observations on the behaviour of the complete individual, under an impressively broad range of circumstances. And it uses these to draw conclusions about the nature of the underlying mechanisms, often employing ingenious pieces of deduction. Putting it in the terms we have been using in this section, this approach involves observations of how *different* inputs and outputs are paired for a given black box, and the analysis leads to conclusions about what lies hidden under the box's lid. Fields which have contributed to this second broad path include those of psychology, cognitive science, psychophysics and computer science.

It is my belief that a proper appreciation of the recent impressive successes scored by the top-down approach can best be obtained through prior consideration of what the bottom-up route has been able to tell us. In what follows, therefore, I will start with the brain's anatomy and physiology.

3 Under the lid

Tell me, where is fancy bred,
Or in the heart, or in the head?
William Shakespeare (*The Merchant of Venice*,III,ii.)

A long-standing curiosity

The discussion of black boxes in the previous chapter produced an important conclusion, namely that finding a mechanism which links a specific output to a given input is not especially difficult. On the contrary, it is invariably the case that many mechanisms can be conjured up which will accomplish the task. It has long been clear that the best way of finding out which mechanisms actually link sensory inputs to muscular outputs, and thereby underlie brain function, would be to make a direct inspection of what our heads contain. Attempts to do this have a venerable history.

The study of brain structure and function began well before it was generally accepted that the brain is the seat of the mind.[a] The documented history of the subject stretches back almost four thousand years, the first written records that have been discovered dating from around 1850 BC. These are Ancient Egyptian papyri, and there are indications that some of them are actually copies of much earlier records. The papyrus discovered by Edwin Smith in a junk shop in Luxor is particularly valuable because of its attention to clinical detail.[1] The manuscript, which Smith found in 1862, describes the case histories of forty-eight injuries, and it lists the symptoms, diagnosis and treatment for each of them. The author appears to have been particularly surprised by

a / The ancient view that the mind resides in the heart (or even in the viscera) is attributable to the influence that emotion has on the pulse (or abdominal comfort).

the fact that a lesion to the head could have a strong influence on regions of the body quite remote from that site. The hieroglyphs disclose the ancient physician's astonishment at observing paralysis of the lower limbs in a patient who had suffered damage to a part of the brain now known as the motor cortex (see figure 3.1).

Medical science was indebted to Ancient Greek civilization for the next major advance, and Hippocrates was especially influential in according the brain its proper status. *From the brain, and the brain only*, he wrote, *arise our pleasures, joys, laughter and jests, as well as our sorrows, pains, griefs and tears.* These deductions were also based on observations of patients with brain injuries, a source of information which has continued to serve brain science right up to the present time. Hippocrates appears to have been the first to realize that the left side of the brain controls the right side of the body, and vice-versa.

During the Renaissance, when other bonds on human thought were being loosened, anatomical dissections of the brain were still not officially sanctioned, and the foremost investigators of the time, Leonardo da Vinci and Andreas Vesalius, had to carry out their work clandestinely. We have already considered Leonardo's resourceful determination of the shape of the brain's four ventricles. Vesalius was more concerned with the brain's outer form, and his charting of the *gyri* (elevations) and *sulci* (depressions) that are visible on the surface of the *cerebral cortex* was remarkable even by today's exacting standards. But he failed to attach any significance to what he observed. On the contrary, he likened these features to *the random clouds drawn by schoolboys.*

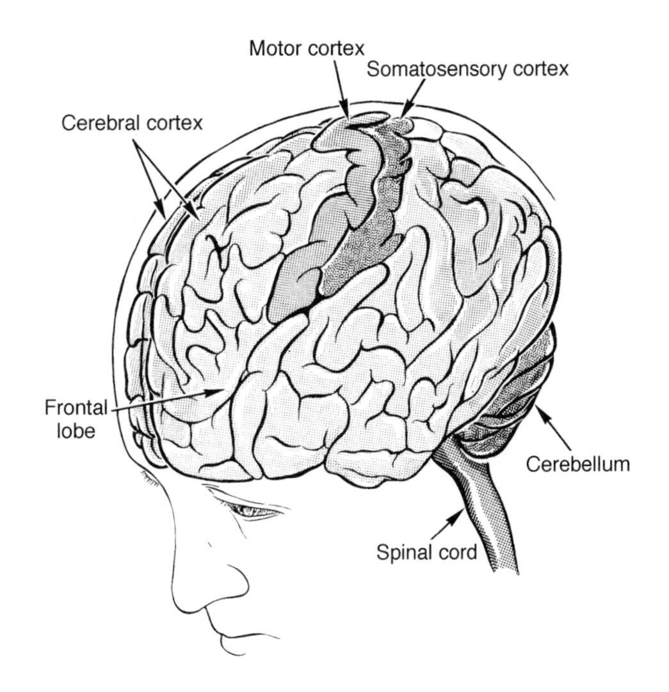

Figure 3.1
The largest part of the brain is the cerebral cortex, which is encased by the dura matter and the pia-arachnoid meninges, just under the skull. It is the main part of the forebrain, and it comprises over fifty contiguous areas. Groups of these are known as lobes, prominent examples being the occipital lobe, which is concerned with vision; the temporal lobe, which comprises the auditory cortex, the limbic cortex and the temporal association cortex; the parietal lobe, which is also instrumental in forming associations; and the frontal lobe, which appears to direct various aspects of thought, and participate in the planning of actions. Also clearly visible in this figure are the motor cortex, which governs muscular movement, and the somatosensory cortex, which receives tactile signals from the different regions of the body.

About a century later, when Descartes was introducing the concept of dualism,[2] his almost unknown contemporary, Niels Steensen, was carrying out admirably detailed anatomical dissections which led him to a profound conclusion.[3] While other anatomists were barely getting to grips with the brain's *grey matter*, Steensen was correctly appreciating the importance of the underlying *white matter*, via which the different areas of the cerebral cortex are connected. In fact, he suggested that it is the white matter which mediates *thought*, a view which will receive strong endorsement in the pages of this book.

Another hundred years were to pass before the significance of nerve fibres was recognized. Many contributed to this awakening, none more than Etienne de Condillac and Julien Offray de la Mettrie.[4] De Condillac, in his *Traité des Sensations*, published in 1754, asserted that the properties of the soul originate from the information that is transmitted from the senses to the brain, via the nerve fibres. De la Mettrie was even more specific in contending that the brain is the origin of all the body's nerve threads. In his *L'Homme machine*, which appeared in 1748, he maintained that this arrangement gave the brain dominion over all bodily processes, however sophisticated or trivial.

Attempts to understand the specific mechanisms underlying the various senses were under way at this time. The astronomer Johannes Kepler, who was of course familiar with the functioning of lenses, put forward a theory of vision in which the eye lens forms an image on the nerves contained in the retina. Thomas Willis proposed an equally imaginative concept of hearing, in which sound was transformed into a mechanical effect on the spring-shaped cochlea of the inner ear. Both of these hypotheses had their merits, but they addressed only the initial stages of the production of sensation. If de la Mettrie was right in according primacy to the brain itself, rather than to its peripheral nerve pathways, the key to the senses would probably be found in the grey matter. About a hundred years were to pass before evidence supporting this idea became available. Then, within the space of a couple of decades, so much testimony accumulated that the issue was established beyond doubt.

The first sign came from the horrible accident that befell the affable and congenial young American railway engineer, Phineas Gage, in 1848. In connection with the blasting of a boulder that lay in the path of a projected branch line, he had bored a hole and filled it with gunpowder. Rather unwisely, Gage allowed himself to be distracted at a critical moment and he proceeded to tap this charge down with a metal rod, even though his assistant had not yet poured in the protective sand. When he inadvertently created a spark, the resulting explosion shot the rod into his skull, just below the left eye. The shaft passed through the poor man's frontal lobe, and it emerged from the top of his head, subsequently to land some fifty metres away. It is a miracle that Gage was not killed instantly, and his survival was a mixed blessing at best, because the terrible injury to the front part of his cerebral cortex had transformed the former likeable personality into one distinguished by unreliability, unbridled aggressiveness, and a proclivity for profanities.[5] This sad case history had provided a strong indication that the brain's frontal region plays a major role in determining character and temperament. (It is noteworthy that some of Gage's colleagues found him more carefree after the accident.)

It had not succeeded in pinpointing the origin of a specific faculty, however. That was achieved just over a decade later. In 1861, Simon Aubertin examined a patient who had made an abortive attempt to commit suicide. The unhappy man had tried to shoot himself in the head, but had merely succeeded in blasting away part of his skull, thereby laying bare part of his brain. Aubertin used the blade of a spatula to apply a light pressure to various regions of the exposed cerebral cortex, simultaneously making clinical observations of the man's behaviour and abilities. It transpired that there is an area towards the front of the left hemisphere which, when thus pressed, causes speech to be suddenly interrupted; removal of the force permits the flow of words to be restored.

Pierre-Paul Broca, a colleague of Aubertin, was soon presented with the opportunity of following up this work. A patient in a local asylum had lost the power of speech about twenty years earlier, an affliction known as *aphasia*. His real name was Leborgne, but he was known as Tan because this was the only syllable that the unfortunate fellow could utter. Tan died shortly after Aubertin had presented his findings to the Société d'Anthropologie in Paris, and Broca's autopsy of the man revealed a lesion near the middle of the left anterior lobe.[6] This area is now named after Broca, in recognition of his subsequent examinations of many patients who suffered from the same deficiency.

Other cortical areas were subsequently identified by autopsy, and this approach was augmented by direct stimulation of the exposed cortex. The first to carry out such investigations were Eduard Hitzig and Gustav Fritsch, in 1870, their initial work being carried out on an unanaesthetized dog.[7] The brain itself possesses no pain centres, so the dog did not suffer, and the lack of an anaesthetic ensured that no potentially enlightening effects would be masked. Hitzig and Fritsch observed no bodily response to the majority of regions that they stimulated, but they found a narrow strip which, when provoked, caused one of the dog's feet to twitch. The region followed the path of one of the cortex's gyri, this running almost vertically downwards from the crown of the head. It was soon established that a nearby point on the same strip, which became known as the *motor cortex*, had control over one of the legs.

In general, it was found that adjacent points on the motor cortex are related to adjacent parts of the body. We can thus look upon this part of the brain as providing a *map* of the body's surface. Albert Grunbaum and Charles Sherrington later showed that a second strip which runs parallel to the motor cortex, now known as the *somatosensory cortex*, is the actual destination of the nerve fibres running in from the various tactile receptors. The mapping in this other strip is virtually identical to that found in the motor cortex (see figures 3.2 and 3.3). In both cases, as established by Edgar Adrian, it is quite distorted, the best represented areas clearly giving the greatest sensitivity. In the

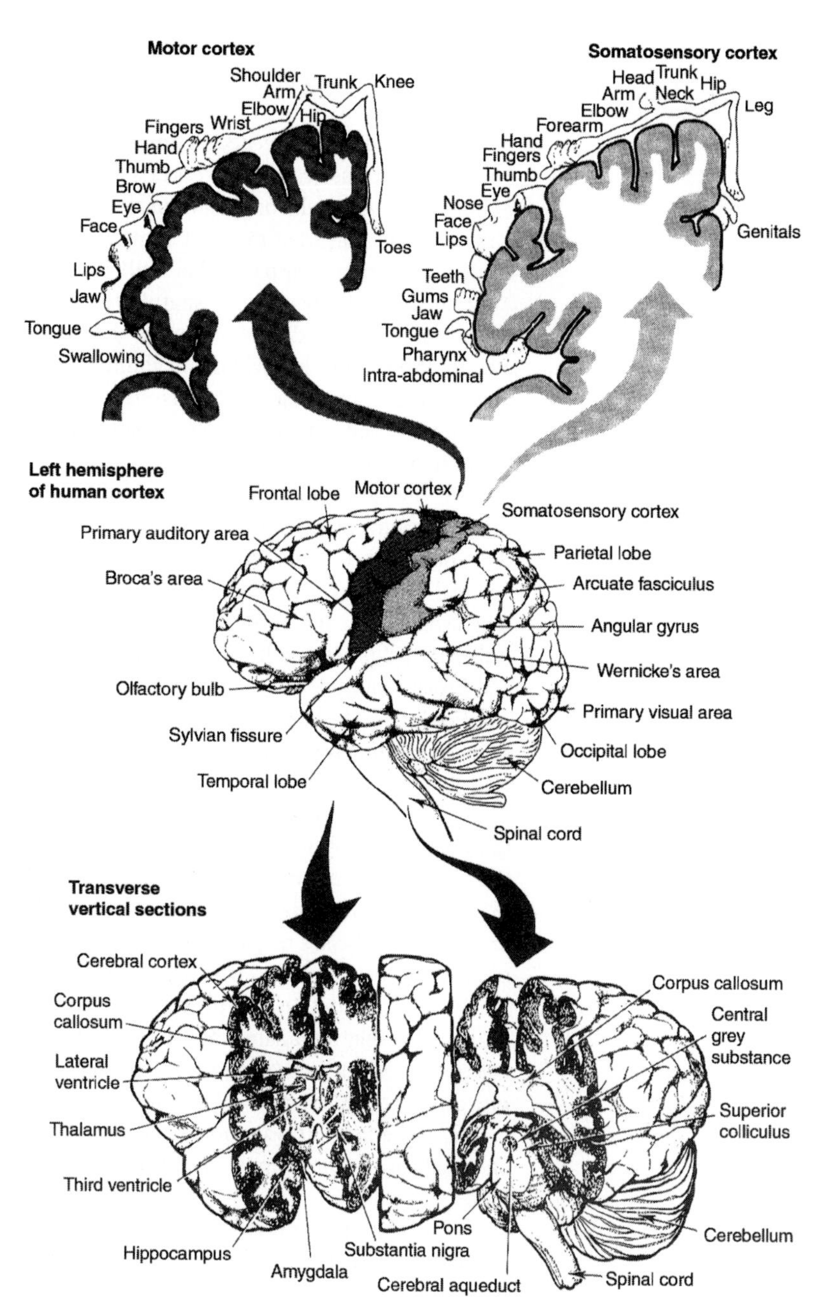

Figure 3.2
Observations of patients with brain damage to specific cortical areas have revealed which faculties are associated with those locations. Some of the grosser divisions are indicated in the central part of the diagram, which shows the left side of the cortex. More detailed investigations have been carried out by direct stimulation of the cortex, and by recordings made with the aid of electrodes. This approach has shown that the surface of the body is mapped, with considerable distortion, in both the motor and somatosensory cortices, as indicated in the upper part of the figure. (Similarly, several of the visual areas comprise maps of the visual field, the distortion in that case favouring the centrally located foveal region.) The coronal sections shown in the lower part of the figure reveal some of the subcortical structures.

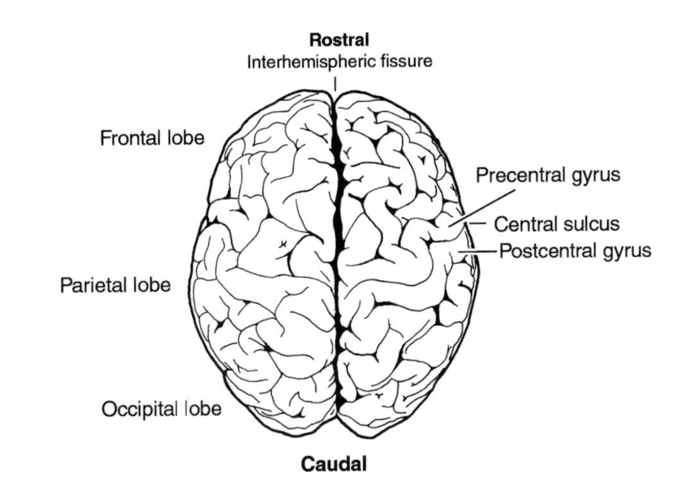

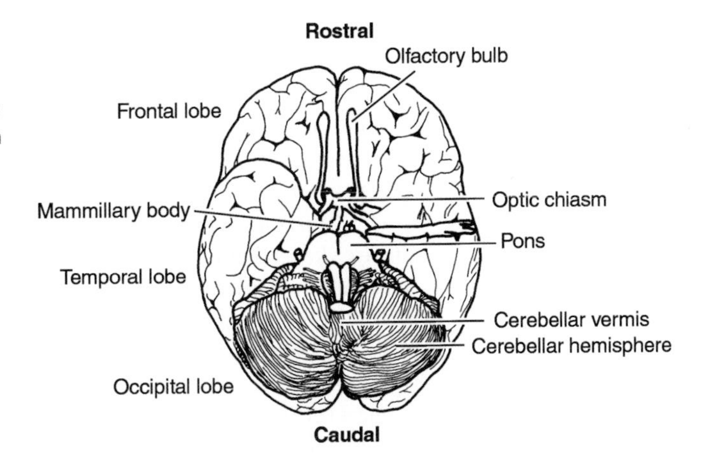

Figure 3.3
The cerebral cortex is a slab of neuronal tissue having a thickness of about three millimetres, and a surface area that is too large to be smoothly accommodated within the endocranial space. It is consequently convoluted into the pattern of gyri (elevations) and sulci (depressions) that are clearly visible in this plan view of the cerebral surface. Although there are minor variations from person to person, the diagram is a reliable guide to the cortical surface's main features. The cerebral tissue is commonly known as the grey matter, a name that derives from the darkening caused by the presence of the neuronal cell bodies. The lower figure shows the view from below, the rostral part of the left temporal lobe (which would have lain to the right) has been cut away, to reveal the underlying left parietal lobe.

mouse, the whiskers are the best represented part, while in the pig it is the snout.[8]

There are still areas of the cortex which have not been identified with any specific function, and it seems likely that they are connected with less understood processes, particularly with the very important process of association. Of the other major regions that have been pinpointed, mention should be made of the *occipital lobe*, located to the rear, which serves sight; the *temporal lobe*, just to the rear of the ear, which amongst other things serves our hearing; the *parietal lobe*, located between the occipital and the somatosensory lobes; and *Wernicke's area*, named after its discoverer Carl Wernicke.[9] This lies around the middle of the triangle formed by the parietal, occipital and temporal lobes, and it provides us with an understanding of speech (as opposed to articulation, which, as we have already seen, is governed by *Broca's area*).

Clinical observations on humans with brain damage are made

possible by what could be called Nature's experiments, and they have been augmented by deliberate lesions to human tissue. This type of approach has been used with experimental animals even more frequently and they have revealed a great deal. The injury to Phineas Gage indicated that the *frontal lobe* makes an important contribution to the formation of personality, as we saw earlier, but this is not a well-defined attribute. Leonardo Bianchi investigated the same cortical region in monkeys by observing the effects of systematic *ablation*, that is to say removal of cortical tissue. He noted that the ablated animals lost what he called their *psychical tone*, by which he meant their capacity for *serializing* and synthesizing groups of actions.[10] Bianchi also found such deficits as lack of initiative, friendship, gratitude, jealousy, maternal and protective impulses, dominance and authority, and a sense of personal dignity. Deliberate ablation of part of the human frontal cortex has in fact been carried out, in connection with the surgical technique known as *lobotomy*, and this has exposed that the region governs the planning and sequencing of motor functions.

The above preliminary list mentioned three of the five senses, namely sight, hearing and touch. The remaining two are taste and smell. The area which serves the latter is the *pyriform cortex*, in the medial temporal lobe, and this in turn projects to what is known as the *entorhinal cortex*. Finally, the ultimate goal of gustatory (taste) information is two *taste areas* of the cortex, located in the *post-central gyrus*.

Just one other cortex-related structure ought to be mentioned in this concise sketch, because of its great importance. This is the *corpus callosum*, a thick bundle of nerve fibres that actually joins the two halves of the cerebral cortex[11] (see figure 3.2). If this main link between the two hemispheres is severed, a number of pathologies arise, some of them rather bizarre because they are suggestive of a splitting of consciousness. One of the byproducts of such surgical severance has been the discovery that the hemispheres are not symmetrical regarding the faculties over which they have dominion. The right side appears to imbue us with creative powers, a feeling for spatial relationships, an appreciation of art and music, and intuition. The left side of the cortex is normally dominant for analysis, logic, abstraction, coordination, a sense of time and, as we have already seen, both speech and the understanding of language.[12]

This brief roll-call of cortical areas has provided only a very crude picture of the actual situation. There is now a large amount of data on the manner in which the larger areas are further divided, but it will be more appropriate to consider these later, in the context of the impairments that result from their damage or removal.

When the brain is used as an idealized motif, the region illustrated is invariably that which would be visible if the skull and underlying protective layers were removed, namely the cerebral cortex. The cerebral hemispheres, with their convoluted walnut-like surfaces, are indeed frequently depicted as if they were the entire brain, and one might gain the impression that the cortex connects directly to the spinal cord. It is important to bear in mind that the cortex is only one part of the brain. It is admittedly the largest, but when exposed by removal of a section of the cranium, it conceals other components, which it encloses under its convoluted mass. There are many such substructures, each with its own function, and there is considerable communication between them. We will shortly direct our attention to these members, which are minor only in respect of their size. Before getting down to that job, however, a few words are necessary about that convoluted appearance of the cortex.

Those corrugations visible on the outer surface are a consequence of the fact that the most recent evolutionary stages of development have presented the cortex with a packing problem. Each cerebral hemisphere is actually a sheet having a thickness that varies between 3 and 5 millimetres, and an area in the vicinity of 1000 square centimetres. It is not difficult to calculate that the surface area, just under the surface of the skull, which would be available to each cerebral hemisphere is about 250 square centimetres, so the cortical sheet is too large by a factor of approximately four. It is for this reason that each sheet has to be folded into its *gyri* and *sulci*.

Structures and strategies

Given the multiplicity of structures in the nervous system, keeping track of their various locations is obviously going to require a frame of reference. The two-dimensional maps in a geographical atlas, with their North–South and East–West coordinates, reflect the two-dimensional nature of the Earth's surface. We are three-dimensional, so our coordinates must be referred to three such systems.[b] One of these naturally distinguishes between left and right. The other two relate respectively to positions along the major axis of the body and in the remaining transverse direction, what we might loosely think of as up and down, and front and back. Both these latter systems are most clearly illustrated by reference to lower vertebrates, such as the lizard, because the central nervous systems in these creatures are deployed along a line stretching from head, denoted by the adjective *rostral*, to tail, denoted by *caudal*. Similarly, the remaining transverse directions are toward the belly (*ventral*, or alternatively *inferior*) and the back (*dorsal*, or *superior*). The various terms are derived from Latin equivalents. Because there is a bending over in our central nervous system, just beyond the top of the spinal cord, rostral above that point refers to the direction of the nose, caudal is toward the back of the head, ventral toward the jaw, and dorsal toward the scalp. Having found our bearings, we can now get down to business, with a lightning tour through the points of major interest.[13]

The nervous system has two parts: the *central nervous system* and the *peripheral nervous system*, the latter being further apportioned into *somatic* and *autonomic* divisions.[14] The job of the somatic division is to monitor *skeletal muscle* and limb position, and to keep the body informed about environmental

b / Definitions of sectional planes will be given later, as the need arises.

conditions, while the autonomic division drives the *viscera*, the *exocrine glands* and what are referred to as the *smooth muscles*. The autonomic division is further separated into the *sympathetic*, *parasympathetic* and *enteric* systems, which respectively serve the body's response to stress, the conservation of the body's resources, and control of the smooth muscles in the abdomen.

The gross divisions of the system are already apparent in the embryo, the elongated form of the neural tube comprising the needle-shaped *spinal cord* and, progressing rostrally from the latter, the successive bulbous swellings known as the *hindbrain*, the *midbrain* and the *forebrain* (see figure 3.4). During subsequent development, the forebrain sub-divides into the *telencephalon* and the *diencephalon*, while the hindbrain separates into the *metencephalon* and the *myelencephalon*. The midbrain remains a single unit as it

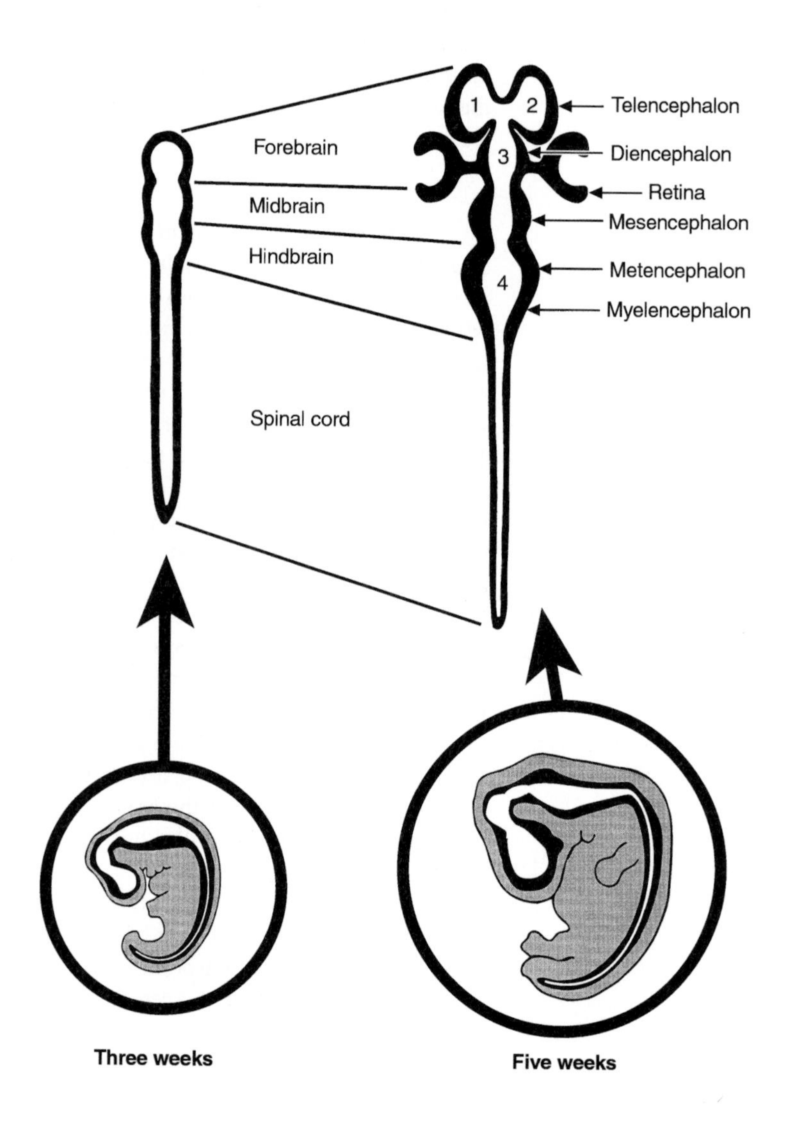

Figure 3.4
A rudimentary nervous system has already been established by the time an embryo is five weeks old. At the three-week stage, the neural tube has developed into the immature spinal cord and the forebrain, midbrain and hindbrain. The cord describes a gentle arc down the length of the embryonic body, and there is flexure between the midbrain and the hindbrain. A dramatic change is visible just two weeks later. There has been marked growth in the lateral direction, and the forebrain displays a clear left–right separation. This growth is particularly pronounced in the forebrain; at its rostral (telencephalon) end, this division is already well on its way to producing the two cerebral hemispheres, while at the caudal (diencephalon) end the new tissue on either side is ultimately destined to become a retina. The region of the diencephalon to which these latter extensions are attached will later develop into the thalamus, amongst other things, so it is not surprising that nerve fibres from the retina project to the thalamus, *en route* to the cortex, in the mature human. The transverse growth is also responsible for the fact that the most rostral of the ventricles divides into a symmetrical pair of these cavities.

develops into the *mesencephalon*. Around the same time, the entire structure is seen to surround a similarly elongated cavity that is destined to become the *ventricles*, their interconnecting passages, and the *lumbar cistern*; they are the collective repository of the *cerebrospinal fluid*.

The cerebral cortex is the main feature of the telencephalon (or endbrain), which also contains the *amygdaloid complex*, the *hippocampal formation*, the *septum*, the *basal ganglia*, and the *olfactory bulb*, all (except the latter) of which lie ventrally to and enclosed by the cortical hemispheres (see figures 3.3 and 3.5). The *amygdala* gets its name from the Latin word *amygdale*, which means almond, an allusion to its shape. The amygdala has been physiologically linked to the control of emotions, to various defence mechanisms, and to feeding and reproductive behaviours. The *hippocampus* also derives its name from Latin, this time for the seahorse. There is now a substantial body of data which indicates that the hippocampus plays a major role in the actual storage of one type of memory, though it is not actually the repository. The term septum comes from the Latin word for fence or hedge, a reference to the fact that this brain structure provides the wall that lies between two of the ventricles. As we have already noted, the latter were looked upon by the ancient anatomists as the chambers which contained the brain's animal spirits. The term *ganglion* comes from the Greek word for knot, and it is applied to regions of the brain that serve as focal points for nerve fibres. The basal ganglia participate in the control of movement and posture, and malfunction of these structures has been implicated in the tremor, involuntary movements and muscular rigidity seen in patients suffering from Huntington's chorea and Parkinson's disease.

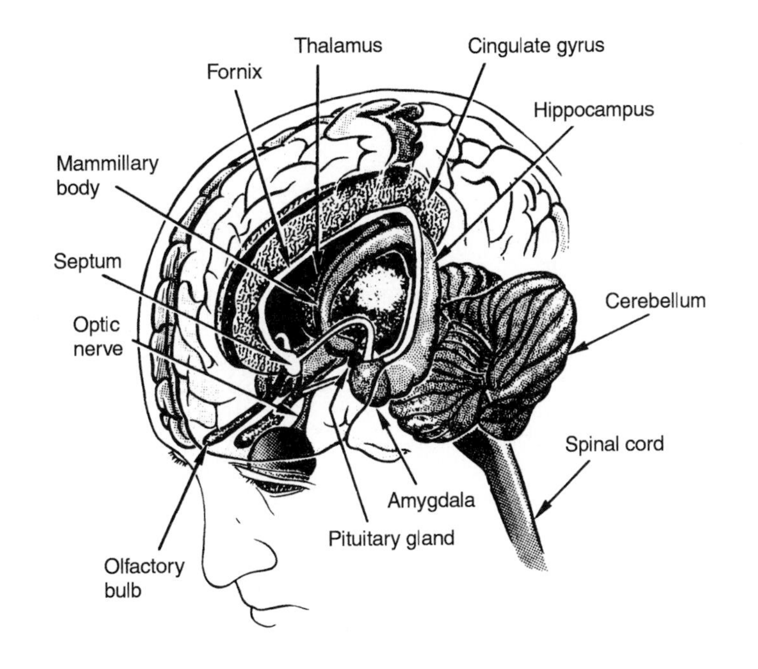

Figure 3.5
The diencephalon lies ventral to, and inside of, the telencephalon, and together with the midbrain governs the more overt aspects of behaviour and the emotions. One of its most prominent members is the thalamus. This structure receives information from the various senses (except that of smell) and sends signals onward to the corresponding areas of the cerebral cortex. But it is more than just a relay station, because it is also the target of the corticofugal fibres which convey reverse-directed information from the cortex. This figure could be misleading in that it makes the hippocampus appear like an almost independent worm-like structure. In reality, it is merely an internal edge of the cortex. Like the thalamus, it both receives signals from, and returns signals to the cortex. The amygdala is the structure that appears to be most directly involved in the orchestration of emotions.

The diencephalon (or betweenbrain) mainly consists of the *thalamus*, the *hypothalamus*, the *optic nerves and tracts*, and the *cerebral peduncles*, the latter actually serving as a connection between the midbrain and the cerebral hemispheres. The thalamus is one of the brain's chief relay stations; it receives nerve signals from the various senses, sends signals to different regions of the cerebral cortex, and receives other signals from those same areas. It is the latter feature that gives the thalamus its control function. The hypothalamus can appropriately be regarded as a sort of lesser thalamus, because it serves those bodily functions which are not under our conscious control. The hypothalamus plays a key role in the automatic control of body temperature, eating, drinking, hormonal balance, a variety of metabolic processes, sexual drive, and general emotional tone. The main parts of the cerebral peduncles are large bundles of nerve fibres that come from the cerebral cortex and project onward to the upper region of the spinal cord.

The mesencephalon (or midbrain) is the smallest part of what is known as the *brainstem*. It includes regions that play key roles in controlling eye movement, and it also has relay nuclei implicated in vision and audition.

The metencephalon (or afterbrain) and myelencephalon (or medullary brain) are actually continuous with the upper part of the spinal cord. The main parts of the former are the *pons* and the *cerebellum*, while the latter's chief component is the *medulla oblongata* (see figure 3.6). The *pontine reticular formation* is involved in the generation of the emotions and in the experience of pain. The pons takes its name from the Latin word for bridge, a reference to the physical appearance of this band of nerve fibres, which lies around the front of the spinal cord's upper end. The medulla oblongata lies

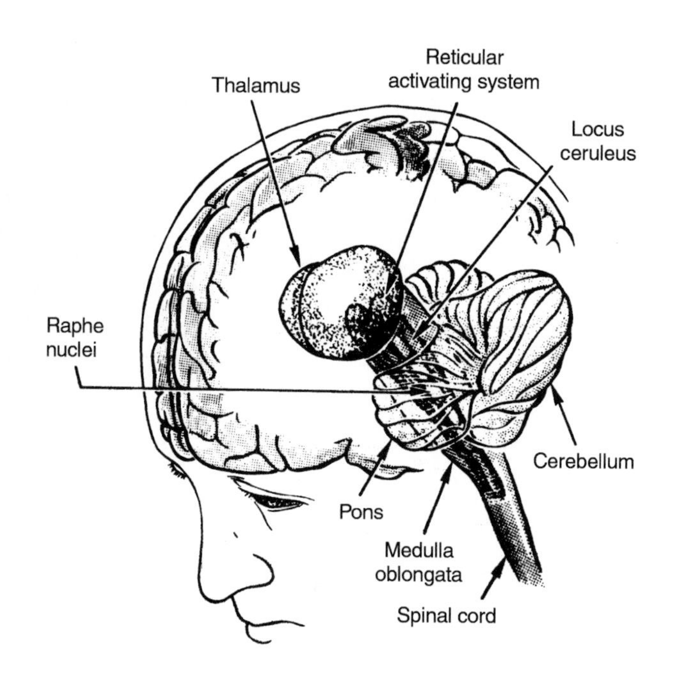

Figure 3.6
The pons and the cerebellum are the most prominent structures in the metencephalon of the hindbrain. They wrap around the top of the brainstem on the ventral and dorsal sides, respectively. The cerebellum is implicated in the fine tuning of motor functions, and evidence is now accumulating for an additional role in cognition. The pons is a distribution point for fibres involved in the control of movement and posture. The reticular activating system exercises control over attention mechanisms, and it is also involved in regulation of the sleep–wake cycle, as is the locus ceruleus. The superior colliculus, which is located on the dorsal side of the brainstem, quite close to the locus ceruleus (see figure 3.7), receives information from the retinas and also from the cortex. It controls the direction of gaze.

underneath the pons. The Latin word *medius* means middle, and oblongata simply refers to this structure's extended shape parallel to the rostral–caudal axis. The medulla oblongata plays a key role in spontaneous respiration, and the maintenance of blood pressure and heart rhythm. These are all vital functions, of course, and the criterion for brain death is that all nervous activity has ceased in this region (and simultaneously in the hypothalamus).

By far the largest part of the metencephalon is the *cerebellum*, which simply means little cerebrum (see figures 3.5, 3.6 and 3.7). It wraps around the rear of the spinal cord's uppermost extremity, rather as the pons wraps around the front, but it is much larger than the pons. Its function is frequently said to be one of the better documented of the brain's parts; it coordinates the contraction and relaxation of our muscles, and thus orchestrates all our movements, large and small. Those things which it helps to put into memory can subsequently be invoked without conscious intervention; they are the things that have become *second nature*. Although we still do not know how the cerebellum accomplishes this, there are strong hints that timing and order are important. I can drum my fingers rapidly in one direction, starting with the little finger and working sequentially forward to the index finger. Using a stopwatch, and repeating the process many times to improve the accuracy of the measurement, I find that I can execute the entire four-finger roll in about 250 milliseconds. This is easy in one direction but not in the other. The reverse roll takes me well over a second, and I have to think about the placement of each finger in turn. There has recently been some discussion as to whether the cerebellum is also implicated in the learning of things not related to movement. Evidence is now rapidly accumulating of its involvement in cognition.

Given the fact that muscular function is the nervous system's only externally directed product, we ought to consider its chief characteristics. It is the output of our black box, of course, and it reverses the process used by the sensory systems. They involve the transformation of physical energy into the electrochemical energy of neural signals; the motor system transforms neural-signal energy into physical energy.[c] And just as the senses are served by sophisticated mechanisms for analysing the various inputs, so are the *motor systems* served by finely tuned mechanisms that plan, coordinate and execute *muscular movements*. The latter are loosely divided into three classes. *Voluntary movements* are purposeful and goal-directed. They can be initiated in response to external stimuli, and they can also be generated internally as a product of thought. These movements can be learned, and they are the ones that the cerebellum helps to become second nature. *Reflex responses* are epitomized by the knee jerk, the cough, and the protective involuntary withdrawal from such adverse stimuli as heat, cold, intense light (the pupil contracts) and sharp objects.[15] Typical examples of *rhythmic motor patterns* are walking, running, jumping, throwing and chewing. This class of movement combines aspects of

c / This is not to say that *all* muscular energy is derived from nerve signals; the latter merely provide the trigger for the subsequent processes.

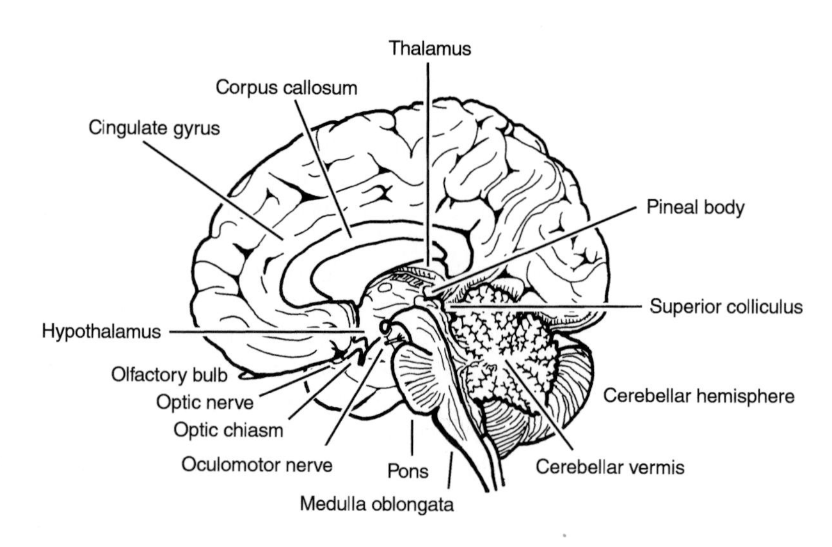

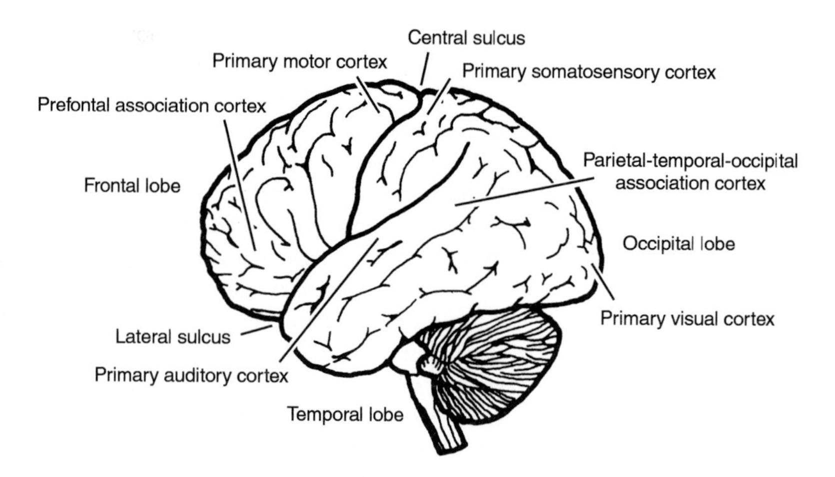

Figure 3.7
The spatial relationships between the pons, cerebellum and medulla oblongata are clearly revealed in this medial view of the right side of the brain. This view also shows the position of the superior colliculus which dictates the direction of gaze. It receives input both from the retinas and from the primary visual area of the cerebral cortex. Two of its layers of nerve cells comprise maps of visual and tactile space, respectively, and these are in mutual registry. The lower part of the figure shows a lateral view of the brain's left side.

both voluntary action and reflex response. The former are involved only in the initiation of the pattern, the actual movements being reflex-like. A good illustration of the overlap between these various types of movement is provided by the smile.

The reciprocal similarity between our input and output systems becomes even more striking when we realize that both involve multiple components. The senses have to extract an essence from the myriad inputs that simultaneously impinge upon them. Our movements, conversely, require the orchestration of many groups of muscles. This is true even of those motions that appear simple. The connection is seen in its most intimate form when we contemplate the signalling that underlies the control of muscular movement. Some of these signals are emitted by sensors located within the muscles themselves. The control processes are obviously one of the cardinal accomplishments of the nervous system, and we should consider how they achieve their ends.

We can illustrate the point with a little experiment. Let your arms dangle loosely and then close your eyes. Note that you can easily bring up either of your hands and touch the tip of your nose with any finger you choose. The movements needed to do this require the coordination of several sets of muscles, but your arm, hand and finger nevertheless describe a smooth arc, even though they do not have sight to guide them. What is the origin of this guidance? The remarkable answer is that the supervision is mediated by an *internal model*. For complex motor tasks indeed, such as the catching of a ball, the nervous system must provide models for both the ball's trajectory and the necessary muscular movements. And the comprehensive nature of the latter model can be appreciated when we note that it has to recruit the involvement of many different opposing pairs of muscles, at the appropriate time and with the appropriate amount of tension. Some of the system's internal models must be present at birth. A foal is up on its legs within minutes of birth, initially lurching about, but soon walking quite confidently. It is difficult to believe that the post-natal requirement, for this task and in this species, is anything more than a bit of fine adjustment. We require more time, however, and concert pianists are certainly made, not born.

One could have misgivings about these internal models, because we would seem to need a very large number of them. That ball that needs to be caught, for example; couldn't it come at us via many different trajectories? Do we have to have a model for every conceivable flight path? Before we tackle that issue, there is another point which is even more puzzling. We can illustrate it with another small experiment. Firstly, I would like you to poke your left index finger into your right ear. This is not a particularly unusual thing to do. We have only two hands, after all, and fingers are often poked into ears. And I would be prepared to bet that you followed my instructions by simply sweeping your hand up across your chest, thereby taking the shortest route. Now I would like you to repeat the performance, but this time *by passing the hand around the back of your neck*. This is an awkward thing to do, but you probably achieved the goal with movements that were nevertheless fairly fluid. Have you ever made that particular set of arm, hand and finger movements before? I very much doubt it, and it is difficult to believe that there has been a model of the process sequestered somewhere in your nervous system all these years, waiting for you to read this book.

This is not the way our systems work. The internal models are modular and hierarchical, with components for gross movement and facilities for making corrections. We also have a *model of the space around us*, including the unseen regions behind us; in the awkward version of our ear-poking exercise, the hand did not have to traverse alien territory. The regulatory mechanism involves the constant monitoring of the states of all the necessary muscles, streams of progress reports being sent to the relevant control centre, in the form of nerve signals. The structures responsible for the sensing are *receptors*. In skeletal muscle, there are two types of these: *muscle spindles* and *Golgi tendon organs*, which respectively have the jobs of reporting on the extent of stretch and the amount of tension (see figure 3.8). The controlling strategy is not unlike that commonly encountered in engineering. When engine speed must be kept constant, for example, one often has a device which compares actual speed with desired speed, and makes adjustments accordingly. In our own systems, information on the counterparts of these two variables, that is to say on the actual and desired states of a muscle, is provided by the *feedback signal* and *reference signal*, respectively.[16]

The overall strategy employed by our motor systems incorporates two main factors: *anticipation* and *compensation*. The latter is achieved in much the same way as in engineering, by *negative feedback*. This *closed-loop mechanism* is relatively slow, however, and it would not by itself enable us to catch a ball. The requisite signalling from the receptors to the brain (transmitting information about the current state), and from the brain back to the muscles (transmitting orders for corrective measures), would simply consume too much time. Anticipation is served by *feed-forward control*, an *open-loop mechanism* in which adjustment signals are dispatched from the brain to the relevant muscles before events occur that would change the states of those muscles. In

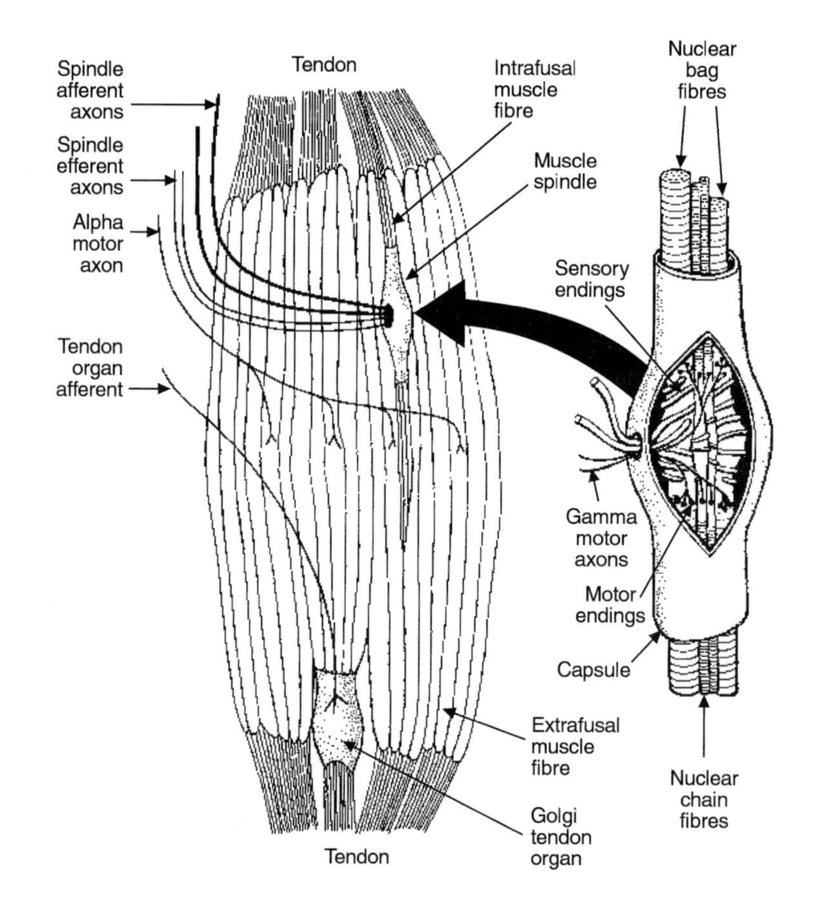

Figure 3.8
Control of skeletal muscle function requires (proprioceptive) information about the instantaneous length of a muscle and about the force it is currently generating. This is supplied by two types of receptor, which act in mutually complementary fashion. The muscle spindles, which are actually composites of the intrafusal fibres and the nuclear fibres (chain and bag types) lie parallel to the extrafusal fibres of the skeletal muscles, while the Golgi tendon organs are in series with the latter. Both types of receptor are innervated by efferent nerve fibres, but the reactions of these to a change in the length of the muscle differ radically. Those that serve the spindles increase their signalling rate when the muscle is stretched, while contraction causes a decrease. The fibres that innervate the tendon organ act oppositely. In humans, precision of control is aided by separation of the innervating fibres, the skeletal muscles being activated by the alpha motor neurons and the spindles by the gamma motor neurons (the latter being further divided into static and dynamic groups).

this second type of control, the result of a current movement is used to reset the system's characteristics, so that the next movement will be more accurate. It is clearly the faster of the two mechanisms.

The most intriguing type of muscular mechanism is that which actually ministers to the senses, because in such cases the output of our black box is directly coupled to its input, and vice-versa. One such system serves vision when either the object of attention or the head, or indeed both, are in motion. No less than five distinct control systems are involved, but they also have quite a lot to take care of. There is a mechanism known as the *smooth pursuit system* which permits the eyes to track the target by keeping its image on the central part of the retina (called the *fovea*), and there are two others which collaborate to continuously compensate for head movement. They are the *vestibulo-ocular reflex* and the *optokinetic reflex*. Smooth pursuit is served by the cerebral cortex, the cerebellum and the pons, while the vestibulo-ocular reflex is steered from the brainstem, with control input also stemming from the cerebellum.[17]

Then again, there is the *vergence movement system* which adjusts the angle between the directions in which the two eyes are pointed, so as to permit focussing on things at different distances. It is coordinated by the midbrain. Yet another, the *saccadic system*, actually steers the eye so that the image of something potentially interesting falls on the fovea. The horizontal component of this latter mechanism is under the control of a structure known as the *pre-pontine reticular formation*. The entire mechanism is a direct participant in attention, the decision as to what might be interesting being made by the post-retinal regions of the visual system itself. All these mechanisms are reflexes. The tennis professional who had to make a conscious effort to keep his eye on the ball would win little prize money.

This section has been pivotal because it has dealt with the chief output of our black box. It is for this reason that I have felt obliged to discuss things in considerable detail. When we later come to evaluate the various ideas that have been put forward regarding the mechanism of attention, and even consciousness itself, we will obviously have to give precedence to arguments that appear to make quantitative sense. A theory gains credence if it gets the numbers right. It is thus fitting that we take our temporary leave of the muscular domain by considering some of the characteristic times. Response to a visual cue may require several hundred milliseconds, but it is clearly the product of a number of individual events, each of which consumes much less time. When the eyes are focussed on an object which suddenly moves out of range of the fovea, eye position is maintained for about 200 milliseconds, whereafter it is rapidly shifted so as to catch up with the object. The speed of this shift, known as a *saccade*, is very high – almost one degree of angle per millisecond. The *latency* (or time delay) of the vestibulo-ocular reflex has been timed at 14 milliseconds, between head and eye movements. The latencies of the various stages of visual processing are rather longer than this. Notification of an instantaneous change of illumination falling on the retina does not reach the output side of that structure until about the 30 millisecond stage, for example, and the relevant regions of the cerebral cortex are not getting the message until almost twice that time has elapsed. Given these numbers, and what was stated earlier, it is clear that the vestibulo-ocular reflex involves open-loop control.

Maps and missions

Around the beginning of this century, anatomists exploited the newly developed technique called *cyto-architectonics* to produce detailed charts of the cerebral cortex. Microscopical observations of the appearance of nerve cells and nerve fibres provided a new way of classifying the various cortical regions. They also revealed that brain tissue has become increasingly differentiated during evolution; advanced species show greater variation in cell form. Amongst the pioneers of this exciting new field were Vladimir Betz, Korbinian Brodmann, Walter Campbell, Theodore Meynert, and Cecile and Oskar Vogt.[18] Brodmann's map of the cortex comprises 52 functionally distinct areas. Area 1 constitutes the *somatosensory map*, for example, while area 17 is the *primary visual cortex*. The *primary auditory cortex* is area 41 on the scheme. Two other locations that we have encountered earlier, namely those associated with the names of Broca and Wernicke, are areas 45 and 22 respectively (see figure 3.9).

Although such anatomical studies clearly established the geography of the cortex, they were unable to reveal what contributions the various areas make to sensory processing. The fact that there are 52 of them, in the Brodmann chart, indicates an underlying complexity. If these were equally divided amongst the senses, the latter would have ten apiece. Why so many? What can several areas achieve that a single area cannot? The areas are *not* equally divided, in fact. In our fairly close relative, the *old world monkey*, over 40 per cent of the cortex is devoted to vision. This would seem to exacerbate the problem of understanding the multiplicity. It also offers a clue, however, because the sheer intricacy of what can be handled by our sight shows that it can be regarded as the senior sense. It appears to be the case, therefore, that a sense will be served by more areas the more things it has to cope with. And vision certainly has much to keep track of, with shape, colour, angular position, proximity and velocity all having potential significance.

The first major strides toward elucidation of this issue had actually been made in the 1960s, through the work of David Hubel and Torsten Wiesel[19] on the primary visual area, V1 (Brodmann's area 17). To appreciate what these pioneers achieved, it will be necessary to consider things at the cellular level, however, so we must postpone discussion of their work until we have had a chance of considering how *nerve cells* function, in the following chapter. The most important subsequent breakthrough was made in the 1970s, through studies of two species of monkey; the *owl monkey* by John Allman and Jon Kaas, and the *macaque monkey* by Semir Zeki.[20] It was they who discovered that vision is served by many areas, and they set about the task of determining what function is managed by which region.

Figure 3.9
Around the beginning of the twentieth century, the new technique of cyto-architectonics was used extensively to chart the arrangement of functionally distinct cortical areas. This microscopic method relied on differences in the appearance of cells and fibres, and its leading practitioners were Vladimir Betz, Korbinian Brodmann, Walter Campbell, Theodore Meynert, and Cecile and Oskar Vogt. The figures reproduced here are two of Brodmann's charts of the cortex. They show the outer surface of the left cerebral hemisphere (above) and the inner (medial) surface of the right hemisphere looking in the outward direction (below). In the Brodmann scheme, areas 17 and 18 are the primary and secondary visual areas respectively. Area 1 is the somatosensory cortex, and area 4 is the motor cortex.

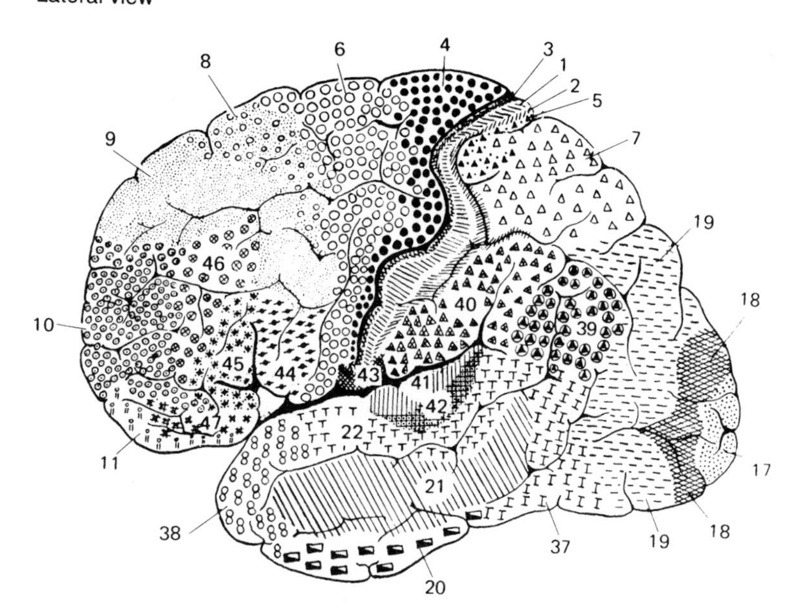

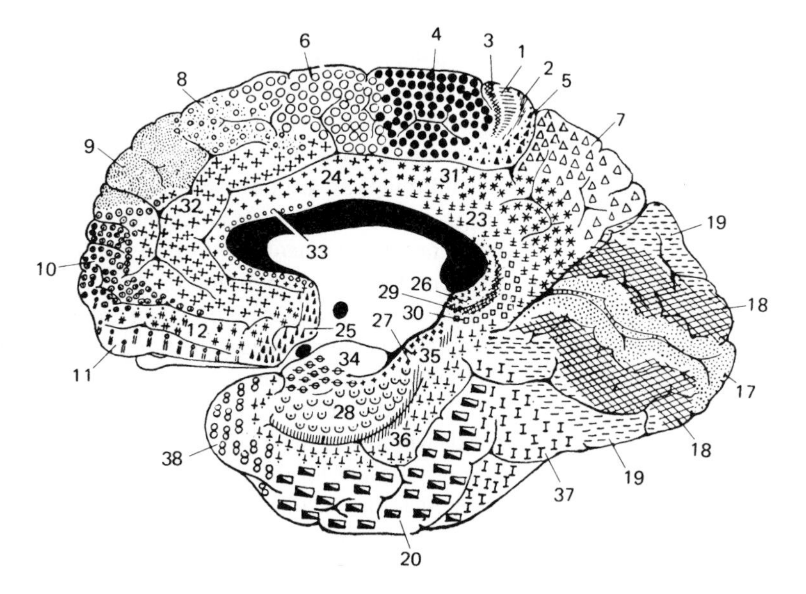

The general approach was the same as that employed by Hubel and Wiesel: to confront monkeys with various visual stimuli while probing the different cortical areas with electrodes, the animals being suitably immobilized with their eyes open. Zeki found an area, now labelled V5, in which all the cells are sensitive to motion but not to colour. The majority of cells in another area, V4, were found to be selective for colour (that is to say, different cells responded to different ranges of wavelength), while two other areas, V3 and V3A, displayed discrimination of form but not colour.

The main input path to the multi-area visual cortex runs from the retinas, via parts of the thalamus known as the *lateral geniculate nuclei*, to V1 and V2. This means that those two areas function as stations *en route* for the various attributes of a visual scene that will later be handled by the higher-numbered areas. The manner in which they accomplish this has now been determined. It could have been that even the smallest part of either V1 or V2 contains different cells variously displaying all the above specializations, but such homogeneity is not what is observed. As David Hubel and Margaret Livingstone were able to demonstrate, the arrangement is quite heterogeneous, with the colour-sensitive cells clustered into well-defined patches (known as *blobs*) that are separated by colour-indifferent regions.[21] A different arrangement in V2 similarly looks after the interests of the various aspects of the image, but we need not go into details here. The main thing to note is the overall strategy of *fragmentation*. Different aspects of the retinal image are channelled into different paths, like letters being sorted according to their post or zip code.

The advent of external imaging techniques has provided the possibility of repeating this type of investigation, non-invasively, in humans. And the advantage of such explorations in our own species is that the investigator can interact with the subject, to a degree that is not possible with other animals. The segregation into many visual areas leads to a similar dispersion of pathologies. This is to our advantage, because reasonably localized lesions tend to produce circumscribed impairments rather than total blindness. The deficits can nevertheless be quite distressing for the victim of one of these afflictions. Damage to V4 produces *achromatopsia*, for example. Sufferers of this infirmity can see only shades of grey, though their ability to perceive shape and motion remains undiminished. Conversely, a lesion to V5 denies detection of movement to those sustaining such damage. The perceptual world of a patient with *akinetopsia*, as this handicap is called, is singularly bizarre. Things are seen clearly when they are stationary, but they disappear every time they move. The result bears a certain resemblance to viewing scenes under stroboscopic illumination.

The fact that several visual areas serve its discernment may be the reason why loss of shape perception has not been encountered clinically. The areas responsible for this faculty collectively form a ring that encircles V1 and V2, so their destruction would probably also demolish those primary areas. The result would be total blindness. Partial damage to the shape-detection apparatus is seen in some patients, however, and these people find it difficult to identify things unless they are in motion. Confronted with the need to recognize a stationary object, those with this impairment will often ease the task by moving their heads from side to side. In so doing, they are probably exploiting a still-intact V3, because that area is activated by both form and motion.

These recent advances in what could be called the cartography of vision have been emulated by those who investigate *language deficits*. In considering a token sample of their achievements, however, we must bear in mind that there is more to language than the mere reception and utterance of sounds. An understanding of the written word naturally implicates vision, but even the understanding of spoken language involves more than just audition. The defining characteristic of any language is its ability to extract the essence from situations, objects, people, actions and attributes, and represent them by written and spoken symbols. It puts specific labels on concepts having general validity. But the formation of a concept might activate one or more patches in various areas of the cortex. Colour, for example, is represented in V2 and V4, as we learned earlier. When we intone the word *blue*, for example, these same areas are called upon, and they act in conjunction with another area which is responsible for assembling the word from its constituent phonemes. The word *conjunction* has special significance in this context, because it implies association mediated by yet another cortical area.

That this is so is endorsed by cases in which the association clearly malfunctions. Antonio and Hanna Damasio have described the plight of patients said to suffer from the defect known as *colour anomia*.[22] These people correctly perceive colour, and their ability to use colour also remains intact; given a sketch of a country scene, and a selection of crayons, they will correctly colour the sky blue, the grass green, and so on. But they are unable to *name* colours. The cortical regions that mediate the link between sensation of colour and the naming of colour apparently fail to do their job in these patients. These regions include an area lying between the temporal and occipital lobes, part of the motor cortex, a region of the frontal cortex and another lying between the parietal lobe and the somatosensory cortex.

Some patients have a rather different type of trouble with names. They are adept at using some common nouns, whereas others which seem no more complex give them serious trouble. The site of this deficit appears to be centred on the temporal lobe, and it stretches across a strip that lies roughly parallel and ventral to the Sylvian fissure, which is the cerebral cortex's most prominent sulcus. As in the case of colour anomia, this impairment affects just one aspect of identification. Sufferers can correctly associate objects (for example) with their function and physical appearance, but they simply cannot put names to them. Failure of noun retrieval can obviously have a devastating effect on the use of language, but verbs are just as important to sentence construction. Verb deficits have recently been the object of investigations carried out by the Damasios, by Caramazza and Gabriele Miceli, and by Rita Berndt, and a link has been established between such words and the *lateral and inferior dorsal frontal regions*. There is also evidence that implicates a patch in the parietal region.

The evidence for separation of functions, and their apportionment to different cortical areas, appears to be rather convincing, but it would seem to raise a major problem for the brain. If the various attributes of a visual scene are processed in different regions, how does that sense nevertheless integrate them into a single perceptual impression. I tend to think of this as the tartan tie problem: a person wearing such a tie moves across my field of view – how does my visual system keep track of all the different colours, always holding them in perfect mutual registry? What prevents the colours from spilling over the outlines, as they deliberately do in the paintings of Raoul Dufy? Why don't blues and yellows sometimes accidentally overlap to give green? As Kevan Martin recently noted, the post-fall state of Humpty Dumpty confronted all the King's Men with a non-trivial challenge.[23]

This is known as the *binding problem*, and it is more general than the above example might suggest. Let us illustrate it by carrying out another brief experiment. Hold your hands out in front of you and clap them together. Your observation of this simple act involved three of your senses; you saw your hands come together; you felt them touch; and you heard the consequence of the collision. There was no doubt in your mind that these visual, tactual and auditory sensations were all related to the same event. The sensations were mediated by different processes in the respective receptors, however. By what agency were they perceived as belonging together? The answer seems obvious, of course: the sensations were received at the same time. But this simultaneity surely refers to arrival at the receptors.[d] To be interpreted as being coincidental, the signals generated by those receptors would presumably have to converge, subsequently, and be suitably timed. No evidence has yet emerged for the presence of a *central clock* in the brain, however. An alternative would be the mutual exchange of signals between the different sensory modalities, instead of their transmission to a sort of neutral adjudicator, but this still leaves the question of how simultaneity is actually to be gauged. Charles Sherrington made the point succinctly when he wrote:[24] '... Pure conjunction in time without necessarily cerebral conjunction in space lies at the root of the solution of the problem of the unity of the mind'.

We are not going to be able to weigh this issue properly without knowledge of the underlying neural circuitry, descriptions of which are the kernel of the following four chapters. Before passing on to that, however, there are still things that should be considered at the macroscopic level of the entire system. And by way of introducing this level, let us take a look at what can now be done to reveal brain activity from outside the head.

The view from without

Of all recent advances in neuroscience, none would have amazed the ancients more than external observation of processes related to thought. As we have noted, some earlier civilizations did not appreciate that the brain is the seat of the mind, and even after this fact had established itself, thought was regarded as something ethereal. As we have also seen, thought was conceptually linked to the motion of animal spirits in the ventricles, rather than to anything occurring in the brain's tissue. One can imagine the amazement with which natural philosophers of other centuries would have observed the external monitoring

d / Strictly speaking, they did *not* arrive at the same time, of course, because light and sound travel at different speeds, but over a short distance this difference is negligible.

of thought-related processes that is now a routine laboratory procedure. Physiologists can study the activity levels in the various parts of the brain, under a given set of circumstances, and the observations are remarkably similar from subject to subject. The technique is now so well underpinned that what is seen on the investigator's screen can be taken as a real indication of the patterns of activity in the subject's head.

One of these techniques could be called a composite development, and we should consider its technical details. It is based on something originally exploited by Louis Sokoloff, namely that a certain modified form of glucose (known as *2-deoxyglucose*) cannot be fully metabolized. All cells, including nerve cells, can absorb molecules of this substance, but they lack the biochemical machinery required to break it down into smaller components, a process which would liberate useful energy. The local concentration of this mutant glucose thus builds up, the excess being highest in the cells that are trying to consume the greatest amount of energy. These latter will naturally be the cells that are currently most active. Sokoloff's trick was to attach a radioactive label to the molecules of this special form of sugar, so that subsequent detection of the radioactivity would reveal which brain regions had been most active at the time of the study.

The one considerable disadvantage with the original implementation of this method was that the radioactivity could only be detected by biopsy of the surgically recovered tissue. This usually meant that the experimental subject had to be sacrificed, and that naturally ruled out investigations of human beings. It was for this reason that modification of the technique, in the hands of David Ingvar and Niels Lassen, proved to be so significant. They replaced Sokoloff's source of localized radioactivity by one that emitted positrons, these having the merit that they mutually annihilate with electrons (actually within a few million-millionths of a second), thereby liberating what are known as gamma rays. These rays are so penetrating that they can pass right through the brain's tissue and the surrounding cranium, thus permitting detection by instruments located outside the head.[25]

The technique requires a track to be kept of the direction taken by each emerging gamma ray, so that its route can be retraced to reveal the ray's place of origin. This reconstruction is facilitated by the fact that two gamma rays are liberated during each annihilation event; retracing of the two paths fixes the point at which the positron was annihilated. The outcome is pictures on a monitoring screen that show which parts of the brain are active when the subject is exposed to a well controlled set of circumstances. Clear differences between the brain activity patterns are observed for different stimuli, and typical motifs have now been charted for a number of standard situations, both active and passive (see figure 3.10).

Positron emission tomography (PET for short), as this technique is called, was not the first procedure developed for externally detecting brain activity. This priority must be accorded to *electroencephalography* (EEG), which involves using suitably sensitive measuring circuitry for discerning the weak voltages that brain activity generates on the surface of a person's scalp[26] (see

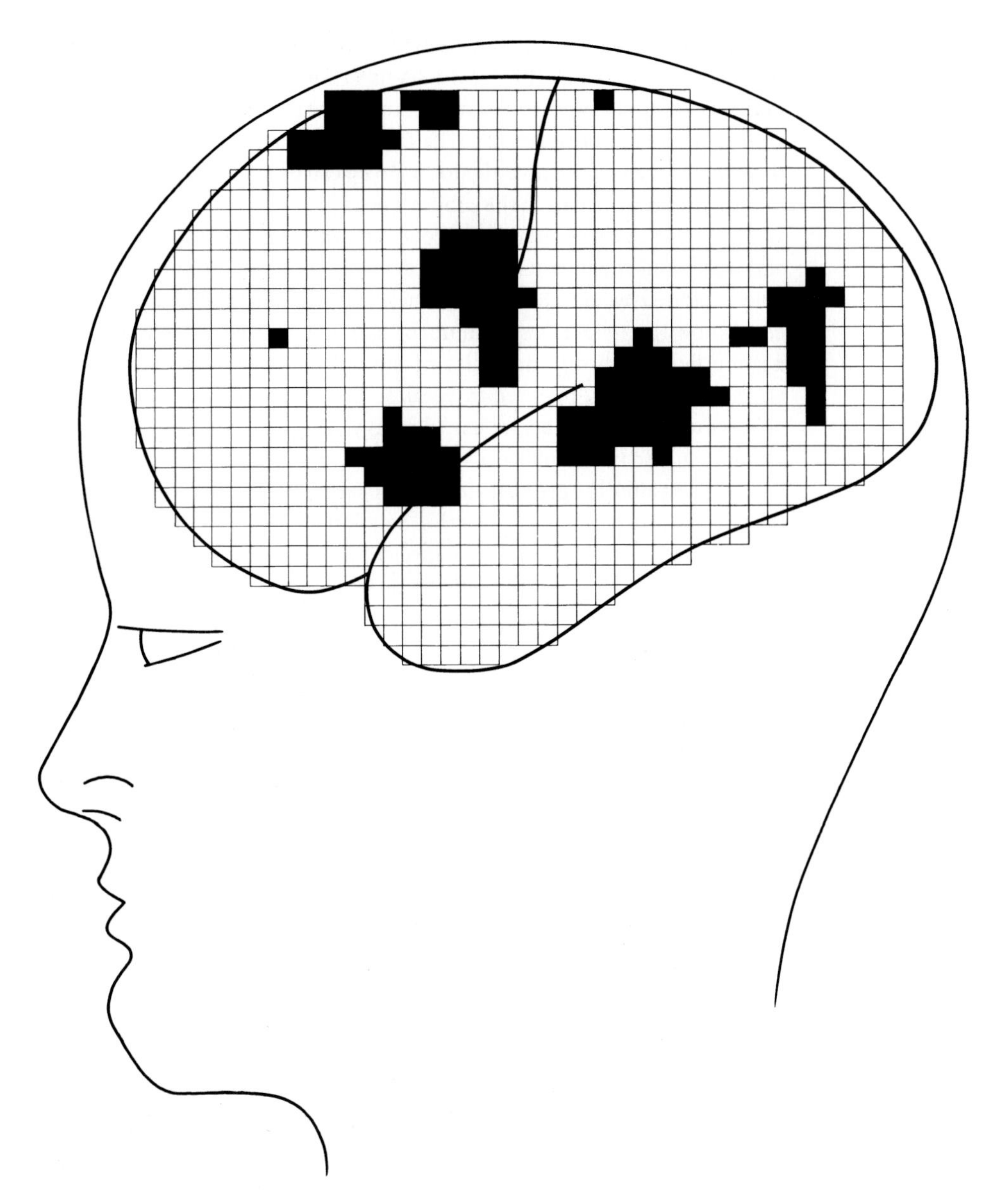

Figure 3.10
Positron emission tomography (PET) is a non-invasive technique which permits the activity levels in various regions of the brain to be monitored externally. A mutant form of glucose, which cannot be fully metabolized, is preferentially absorbed by the most active areas and tends to accumulate in them. The glucose molecules are labelled with an isotope which emits positrons. These mutually annihilate with electrons (within less than a nanosecond) and the resultant two gamma rays travel in straight lines, but in different directions, easily passing through both neural tissue and the cranium. They can readily be detected upon emerging from the subject's head. This enables the experimenter to get a 'fix' on their point of origin, and the end result is an indication of which areas are most active under a given set of circumstances. In the example shown here, the subject was reading aloud, and the most active regions have been coloured black.

figure 3.11). But the EEG is not, at its present state of development at least, a particularly sharp tool. Although *magnetoencephalography* (MEG), its magnetic counterpart, appears to hold more promise in this respect, it still cannot compete with the positron method when it comes to pinpointing where the action is at a given moment. Conversely, these techniques far surpass PET in the time domain. EEG and MEG traces can be readily obtained with temporal resolutions in the millisecond range, a time scale which is well beyond the reach of PET at present.

It must be emphasized, however, that mere detection of brain

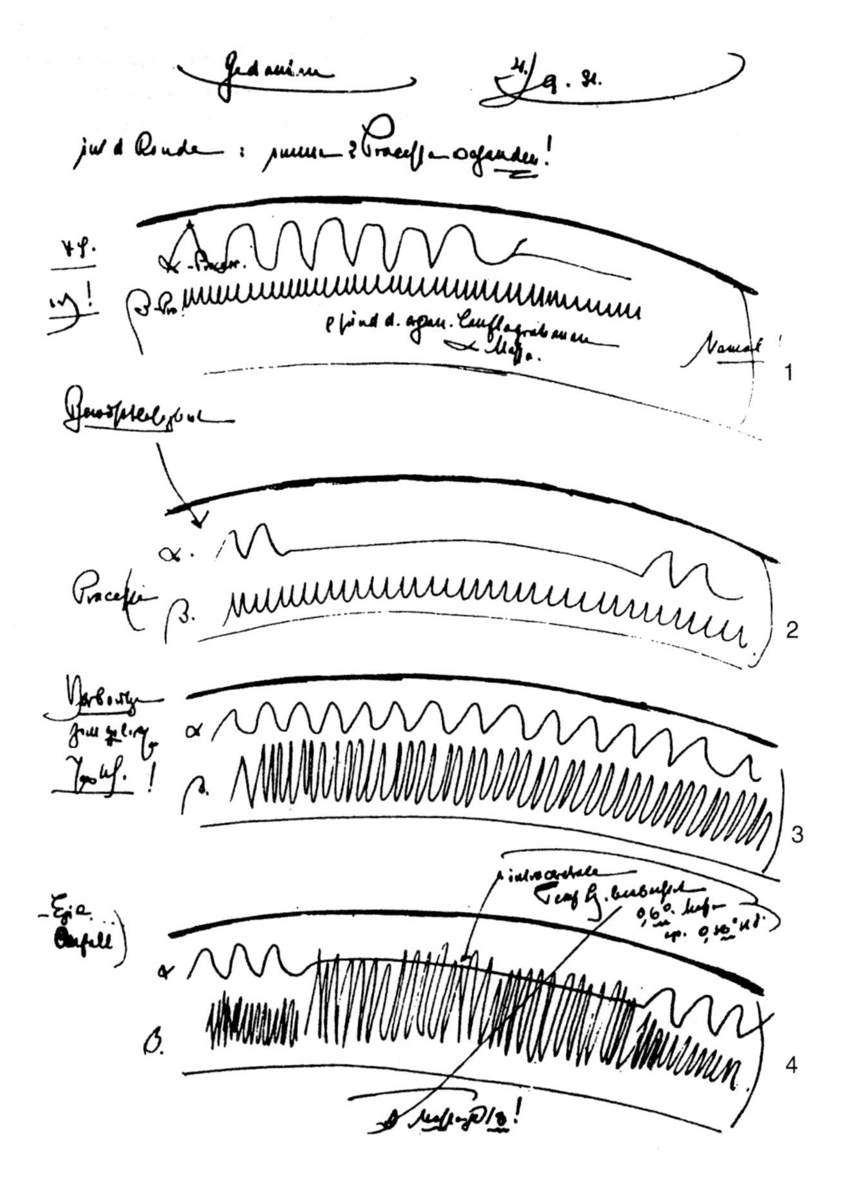

Figure 3.11
This facsimile page from Hans Berger's laboratory notebook records his observations from a day's work on the electroencephalogram of one of his subjects. His handwriting is a mixture of normal German and a type of shorthand. The heading reads as follows. **Thoughts 21.9.31. In the cortex: Always 2 processes present!** The English translation of the four numbered entries read: **1. Psi Phi. Psychophysical, Alpha-process. Nutrition! Beta-process. That is the organ. Conflagration of Mosso. Normal! 2. Unconsciousness. Process Alpha. Beta. 3. Preparation for epileptic seizure. Aura! Alpha. Beta. 4. Epileptic seizure. Alpha. Beta. Intercerebral temperature increase measured 0.6 degrees, Mosso 0.36 degrees in the human. According to Mosso, not always, however!** (The person referred to in the final section was Angelo Mosso (1846–1910), who made the first noteworthy attempt to investigate blood circulation in the brain, in 1881.)

activity, by any of these methods, is still a far cry from exposure of the actual thoughts that give rise to it. These involve the interactions between the brain's myriad nerve cells, and the ultimate goal of relating the dynamics of such interactions to the underlying thought processes is still far from being realized.[e] The positron method is now capable of spatial resolution down to a few millimetres, if a fluorine isotope is used, but it must be borne in mind that the body of a nerve cell has a diameter about a hundred times smaller than this. It is seems unlikely that such precision of observation will fall within the grasp of the positron technique in the near future.

Mention should be made here of a technique that may soon supersede PET, not the least because the latter involves the less desireable measure of admitting radioactivity to the brain. It is a variant of *magnetic resonance imaging* (*MRI*), which is already well established, and which provides impressive pictures of the brain's gyri and sulci. The new possibility comes from exploiting the fact that the magnetic properties of the haemoglobin molecule (which is of course present in blood) depend upon whether or not it has an attached oxygen molecule. The deoxygenated form has the ability to locally distort a magnetic field, and the effect is greatly amplified by the proximity of water molecules. Because the state of the haemoglobin is determined by the local energy consumption, *BOLD* (*blood oxygen level dependent*) *imaging* produces pictures in much the same way as PET. The resolution is now typically of the order of a millimetre, but one-tenth of this is already achievable in ideal cases.[27] The signals grow to usable intensities within 5 to 10 seconds after a suitable stimulus. These characteristics already make the technique very significant, and any considerable increase in the temporal resolution would be a major milepost in the study of the brain.

Despite the great promise of these new techniques, much use is still made of direct measurement of neuronal activity by *electrodes*. This has been a staple of neurophysiology for well over half a century. It is now well appreciated that merely gauging the voltage fluctuations in a single nerve cell tells one very little about the mental processes which such activity is mediating. Because thoughts must involve the simultaneous, or near simultaneous, stimulation of many cells, the experimenter is confronted with a major challenge. And although some headway has been made in this direction, with *multiple-electrode measurements* becoming increasingly common, we are still a long way from being able to observe thoughts in all their multicellular splendour.[28] Because of this, the brain scientist is still forced to depend on other, less direct, approaches. Amongst these are *psychophysics*, which seeks to quantify sensation and mathematically model the relationship between brain and mind, and *cognitive science*. We will be considering some of the victories scored by these approaches in later chapters. Our immediate concern, however, should be with what has been discovered concerning the apportioning of tasks to the various regions of the cortex.

e / I will nevertheless be attempting to do precisely that in the latter chapters of this book.

The logic of emotion

Although our *emotions* are generally regarded as something that should be kept under control, they nevertheless have a positive side. Amongst other things, the artistic dimensions of our culture would be barren without them. Possession of the capacity for emotion is even used as a test to distinguish the human from the mere machine. Earlier in this book, I speculated on the differences between people and chess-playing computers, and the tell-tale behavioural distinctions cited there are directly attributable to our having emotions.

We have already seen that the emotions are mediated by various structures in the midbrain, particularly those belonging to the *limbic system*. These parts of the brain evolved long before the cerebral cortex. The hindbrain, with its control over mental qualities that are still more basic than the emotions, developed even earlier. It might seem that our drives and emotions are just remnants of earlier mental attributes that have outgrown their usefulness; the former are manifestly primitive, while the latter are more refined but still on a lower plane than thought. The cortex having finally been produced by evolution, drives and emotions might appear to be mere vestiges of the past. This was the attitude championed by Charles Darwin,[29] in his *The Expression of the Emotions in Man and the Animals*, published in 1872. He had naturally gained insight into the evolution of species by making observations on other outdated bodily structures, such as the *coccyx*, which is our degenerate tail, the *appendix*, which has long since ceased to serve a useful purpose in our alimentary canal, and the *goose-pimple reaction* that raises no fur because we no longer have any fur to raise. It was natural for Darwin to think along the same lines when considering the emotions. The question remains, however, as to whether he was correct.

An alternative view would see the emotions as having retained whatever utility they had, and it would see their advantage as having been augmented by the more refined responses afforded by the cortex's thought processes. In such a view, retention of emotions during evolution would not merely be perceived as a propitious accident; it would be regarded as vital, and as serving the organism's vested interests.

What might be the advantages of possessing the capacity for emotion. Emotion's principal function appears to be a dual one; it serves both to concentrate our attention,[f] under a given set of external circumstances, and to assist the brain in defining the organism's immediate priorities. At any instant, many different sensory stimuli are vying for our concern. The various parts of the cortex are constantly receiving signals, processing them, and sending the result both onward to other cortical areas and (somewhat surprisingly, but very importantly) back to areas from which the original signals were transmitted. In a manner which we will later have to consider in detail, the

f / Samuel Johnson put the idea succinctly when he wrote, in *Letter to Boswell*: 'Depend upon it, Sir, when a man knows he is to be hanged in a fortnight, it concentrates his mind wonderfully'.

cortex thereby detects the *correlations* through which we can extract meaning from our ever-changing environment. The emotions, far from being cumbersome and unwanted passengers, actually serve to orchestrate the brain's handling of these kaleidoscopic inputs.

Philip Johnson-Laird and Keith Oatley have developed what they call a *cognitive theory of the emotions*, with which they have attempted to explain how we evaluate events in relation to our goals.[30] They believe that we judge circumstances in terms of five categories: the achievement of minor goals in solving problems as they arise; the loss of a goal; the frustration of a plan or goal by some person or circumstance; a conflict of goals, including conflict with the goal of self-preservation; and the perception that something or someone is noxious. Although a precise label cannot be put on the emotional adjuncts of these divisions, the simple words that come closest to doing the job are *happiness*, *sadness*, *anger*, *fear* and *disgust*.

Johnson-Laird and Oatley emphasize that their scheme is not restrictive, because although they identify just *five basic emotions*, these are augmented by an indefinitely large number of *specific emotions*. The latter are each composed of the basic emotions and information about the circumstances that create them. *Love*, for example, is basically the happiness generated by one's interactions with another person. *Jealousy* is basically anger provoked by the possibility of being alienated from a loved one, by the intervention of a third person. *Despair* is basically sadness, induced by one's inability to rectify an adverse situation, and so on. Johnson-Laird and Oatley have examined the *semantics* of almost 600 English words having emotional connotations, and they have shown that all of them can be rationalized in this manner.

It is interesting to see emotions being accounted for in this way, but the ultimate aim must be to link the various emotional categories with specific biological processes, and this is clearly a far more difficult task. Some progress can be made, however, by considering what we already know about components of the nervous system that display a differential reaction to circumstances. The sympathetic and parasympathetic nerves are complementary in that they produce physiological effects that are in mutual opposition. Let us consider an example that exposes the way in which they are paired. You have eaten a satisfying meal and you start to doze off. Your *parasympathetic nerves* slow down your heart rate and step up your digestive activity. Then you start to smell smoke, and open your eyes to discover that flames are visible through the kitchen door! Your *sympathetic nerves* take over, increasing your heart rate; diverting blood away from your digestive system so as to increase the flow through the muscles and brain; expanding your lungs to provide more oxygen; activating your sweat glands to promote the cooling that will be necessary because of the anticipated exertions; dilating your pupils to give you more light. This rather extreme example clearly shows that emotions can bring all manner of useful physiological reactions into play. Happiness was replaced by fear, and this switch from one basic emotion to another resulted in physical changes that were obviously for the good of the organism.

The most profound of the hypotheses put forward in the work of

Johnson-Laird and Oatley concerned the connection between the emotions and consciousness. In our conscious awareness of our emotions, these researchers believe, the latter effectively become *communications to ourselves*. The structures that might be involved lie in the hindbrain, and they are the *brainstem reticular formation*, the *raphe nuclei*, and the *locus ceruleus*. Together, these components control our various states of *arousal* and *attention*; they are effectively a collective *gain control*. They are indeed also responsible for dictating when we sleep.

There is a delicately poised interplay between these three centres, and this enables the brain to be prodded into the various states. This is the case only if nothing has happened to upset the balance, however. That is why our sleeping habits are normally so regular. The brainstem reticular formation controls the level of arousal, while the raphe nuclei cause sleep by effectively blocking the influence of the reticular formation. That the locus ceruleus also gets into the act can be seen from the fact that damage to this structure causes its owner to spend a disproportionately large amount of time in slumber.

These various facts are reconcilable with the idea that the emotions, with their neural processes occurring in the midbrain, transmit at least part of their activity downwards into the hindbrain. Looked at in this way, the emotions are seen to be subservient to our most primitive urges, which, after all, are the most likely directors of our conscious attention. They are the most reliable arbiters, moreover, when it comes to setting our priorities. Further support for these ideas comes from two different types of observation: that a number of drugs with proved potency for the brain structures in question have a marked influence on the emotions, in the absence of any changes in the external circumstances; and that different emotions produce different constellations of physiological effects in the relevant midbrain components.

Just as important as these essentially instantaneous effects are the long-term consequences of emotional experience. There is plenty of evidence to support the idea that the emotions have a strong influence on the acquisition of memories. By pure coincidence, I happen to be writing the first draft of these words on the fiftieth anniversary of the outbreak of the Second World War, and earlier in the day I was listening to the BBC's programme commemorating that sad occasion. The transmission included a recording of the then British Prime Minister Neville Chamberlain's broadcast to the nation, and when I heard the measured cadence of those fateful words – '*I must tell you now that no such undertaking has been received, and that consequently* ***this country is now at war***' – they immediately conjured up all the details of a scene that is indelibly imprinted in my memory. I see again my mother and her two sisters, in the kitchen of my maternal grandmother, surrounded by my brothers and cousins; I can hear the three sisters bursting into uncontrollable sobbing; and I can even smell the food that my grandmother continued to prepare with remarkable equanimity – she had been through it all once before, and knew that life must nevertheless go on. Turning to a more recent event of the same type, those of us who are old enough can remember precisely where we were and just what we were doing when we heard of the assassination of President John F. Kennedy.

Hierarchies and lowerarchies

When considering the functions of the brain, it is quite natural to think in terms of a *hierarchy*. It seems reasonable to assume that faculties became gradually more sophisticated as the brain's components evolved. It is only the possessors of cerebral cortices, moreover, that appear to be capable of the sort of advanced cogitation that our own species prides itself on. Even the layout of the various cortical areas seems to necessitate the hierarchical view, for the primary areas send out nerve fibres to what are generally referred to as the higher areas. The primary visual cortex, for example, sends such conduits to the higher visual regions, via the cortex's white matter. And when we conceptualize brain function, our diagrams almost inevitably reflect a hierarchical way of thinking. They are inclined to involve pyramids, with broad bases to indicate relatively rudimentary faculties, and with the more refined and complex attributes only emerging as one travels upwards towards the pinnacle.

The trouble with such a picture is that one is ultimately confronted with the need to put a label on the pyramid's uppermost point. There must be many who would be quite content to write in the word *soul* at that zenith, and accept that this term is inevitably nebulous; they would be happy to let the top of the pyramid be lost in the clouds. But it is possible that the various cortical areas form no such upwardly pointing pattern, either anatomical or conceptual. As an alternative, I believe that one could look upon the brain as what could rather be called a *lowerarchy*.[31] This idea may be anathema to some, because it could conjure up the spectre of the lowest common denominator; the idea seems to smack of baser instincts, and these are invariably belittled. Why, then, should one be prepared to denigrate what Plato called *the divinest part of us*? What scientific reward could justify such a jaundiced view of so sophisticated a piece of anatomical circuitry? An immediate reward would be that one is thereby relieved of deciding which part of the cerebral cortex is to be accorded primacy. That elusive top of the pyramid would be replaced by something rather more tangible: the top of the spinal cord!

The brain's only function is to contribute to the body's survival and reproduction, and each stage of evolution merely furthered the organism's goals. Considered in this light, the limbic system, and ultimately the cerebral cortex, become embellishments that simply serve the same general cause. And each of these two elaborations represented far more than just an increase in the amount of participating tissue; they both opened up new strategic and tactical possibilities for the brain. Thus the limbic system provided the potentialities of the emotions, and the cortex supplied the additional power inherent in thought. But although these extra faculties widened the brain's repertoire of responses to incoming sensations, they were nevertheless subservient to the fundamental goals.

This idea is not in conflict with the generally accepted criterion for evolutionary development. Anything which improves interaction with the environment will enhance the organism's chance of survival. The sprouting of each new structure is thus not seen as a transfer of control away from the basic instincts, towards more lofty ideals. It is rather to be looked upon as consolidation of the same ancient goal, through the acquisition of a widened inventory of operational choices.

It is important not to confuse hierarchy in anatomical structure and the type of hierarchy that arises in rational thought. The undeniable existence of the latter does not prove that the former really is present in the brain. The hierarchies that occur in our logic have only recently become a feature of human thought, whereas the cortical machinery that makes such thought possible evolved thousands of years earlier. The two things are quite independent; a brain structured as a lowerarchy would be perfectly capable of hierarchical thought.

To reiterate, then, although basic urges are usually disparaged, we should remember that high-minded pursuits are of little consequence if there are not going to be any tomorrows. And tomorrows will come automatically if an animal succeeds in the competitive treadmill that Darwin and Wallace first perceived when they examined the dynamics of ecosystems. The lowerarchy concept sees a well-developed forebrain as giving its possessor a decisive advantage, not because it adds higher steps to some intellectual pyramid but rather because it provides the basic priorities with a more sensitive antenna. *Thinking is a bodily function.*

4 Games neurons play

The functional superiority of the human brain is intimately linked up with the prodigious abundance and unaccustomed wealth of forms of the so-called neurons with short axons.
Santiago Ramón y Cajal (*Histologie du systeme nerveux*)

The workhorses of the nervous system

The key property of the nervous system is its ability to conduct signals, and the conduits that make such signaling possible are the *nerve fibres*.[1] These mediate the interactions between the various parts of the system, of which the brain is the main structure. The brain and its attendant components are composed of *nerve cells*, which are referred to as *neurons*, and the mutual interactions between these cells are fundamental to the functioning of the system. It is thus appropriate that we now take a closer look at the neuron, and ask how it acquires its remarkable properties.

Given that all biological tissue is composed of cells, one might guess that this also applies to nerve fibres, with the cells aggregated so as to produce elongation in one direction. This is not the case. Nerve fibres are actually composed of numerous threads, lying parallel to one another, an individual thread being just *part* of a neuron. It was Robert Remak who first demonstrated this, using one of the then recently developed achromatic compound microscopes, and he also discovered that a thread is frequently sheathed by a sort of insulating material now known as *myelin*. Jan Purkinje's observations, made independently at about the same time, allowed him to add the important fact that neurons have essentially the same form throughout the nervous system.

In 1839, shortly after these advances, Theodor Schwann put forward what became known as the *cell theory*, which indeed saw the individual threads as extensions of neurons.[2] The theory's picture of the brain's white matter was that this consists *only* of nerve threads, grouped together to form fibres, while the grey matter is composed of both threads and the bodies of neurons, these central regions being referred to as *soma*. (The word soma comes from the Greek term for body.) This theory is now supported by massive documentation.

A significant milepost was reached in 1873, when Camillo Golgi described a chemical method for rendering entire neurons visible in the optical microscope.[3] His technique involved a silver stain, and by suitably adjusting the conditions, he found that he could selectively colour just a small fraction of the cells and leave their neighbours invisible. This was very important. The neurons in the brain are so tightly packed that a microscopic picture showing all of them would be confusing. Such pictures are easily taken with a more modern type of instrument known as a scanning electron microscope, and what they show bears a striking resemblance to a can of writhing worms. It is possible to observe the detailed shape of a neuron in a Golgi stained specimen of brain tissue, and it has a form not unlike a bush that has been pulled out of the ground; the soma appears as a short bulbous region in the centre, out of which spring branch-like extensions on one side, and root-like extensions on the other. These physical extensions are now known as *processes*, and one should note the somewhat unusual use of that word in this context. Processes are also (less commonly) referred to as *neurites*. The optical microscope naturally permitted determinations of sizes, and the somatic region of neurons was found to vary from the roughly 80 micrometre diameters of *Purkinje cells* of the cerebellum to approximately one-tenth of this for the *granule cells* in the same part of the brain.

The comparison to a bush is appropriate because just as roots and branches serve the plant in different ways so do the two categories of process play different roles in the functioning of a neuron. But the roles are reversed in this case, for it is the branch-like processes, called *dendrites*, which carry information toward the soma, while the root-like processes, known as the *axon*, the *axon collaterals* and the *axon branches*, convey signals away from that central region. Two other conventions should be mentioned. In order to differentiate between the directions in which they convey information, dendrites and axons are referred to as being *afferent* and *efferent* respectively. The same two terms are encountered on the much larger scale of whole nerve fibres (which, as we have already seen, are composed of large numbers of processes); thus fibres which carry signals from the body's sensory receptors toward the brain are called afferent fibres, and those which carry signals from the brain toward such structures as muscles and glands are referred to as efferent fibres.[a]

From microscopical observations, it is relatively easy to measure the distance between neighbouring neurons, and from this it is just as straightforward to calculate the approximate number of such cells in the brain. The answer is roughly a hundred thousand million. Each neuron has upwards of ten thousand processes, and the question arises as to the purpose of these

a / It should be noted that an efferent nerve fibre which carries signals from the brain to a muscle is *not* said to be afferent to the muscle; it is still referred to as being efferent at that location.

extensions. A strong hint was provided by the work of Santiago Ramón y Cajal (see figure 4.1). His painstaking observations of processes showed them to make occasional close encounters with those of surrounding cells, even though these were not physical attachments. Ramón y Cajal also noticed that the terminal region of an axon often makes one of these near contacts with a dendrite of one of its neighbouring nerve cells. In 1897, Charles Sherrington named such junctions *synapses*, and they were subsequently found to play a major role in the interactions between neurons (see figure 4.2). We have still to discuss the properties of an individual neuron, however, for it is not yet clear how it manages to convey signals. That it must have a rather special structure was already clear in the time of Ramón y Cajal, and it was one of his contemporaries, Heinrich von Waldeyer-Hartz, who coined the name neuron,

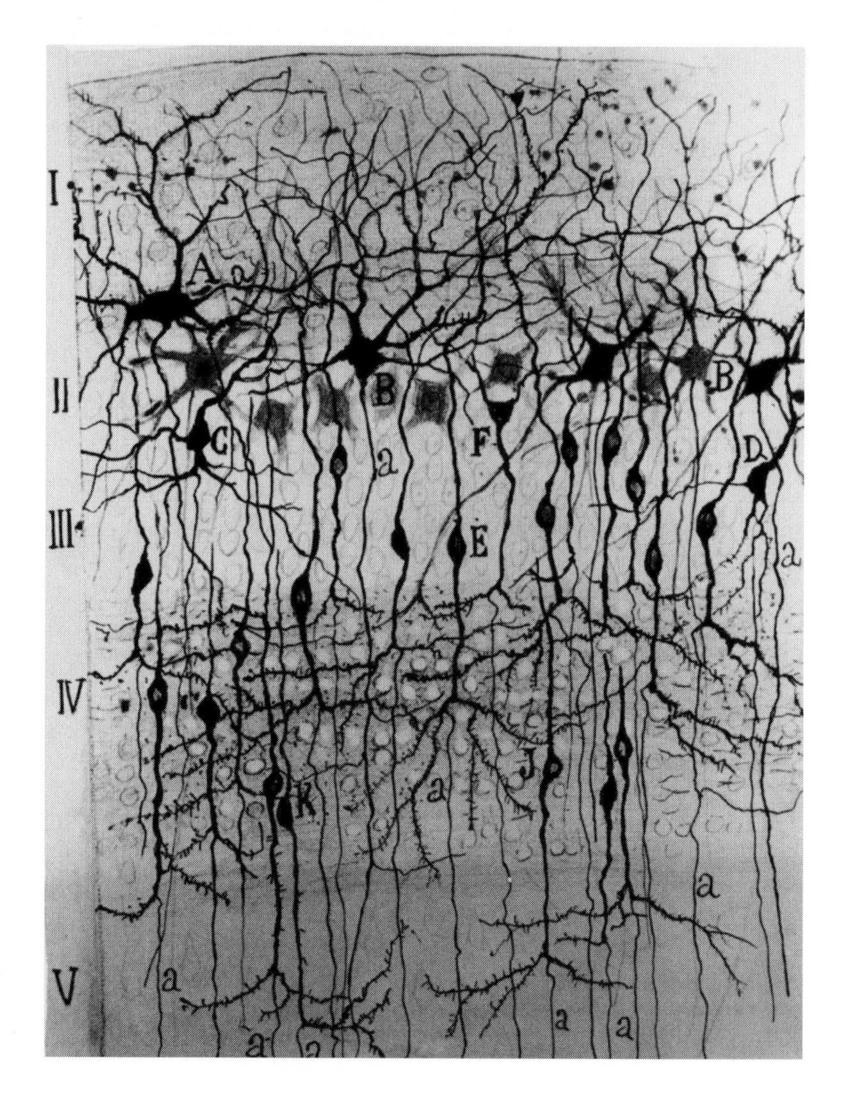

Figure 4.1
This reproduction of one of Santiago Ramón y Cajal's original anatomical drawings shows the neurons in the cortical area designated 29c by Korbinian Brodmann. The lower case letters **a** indicate the axons of the various neurons, amongst which are one with a horizontal axon (**A**), large stellate neurons (**B**), small pyramidal neurons (**C** and **D**), a superficial-layer fusiform neuron (**E**), and deep-layer fusiform neurons (**J** and **K**). In contrast to most other types of cell, neurons do not divide after birth, so the nervous system's complement of these cells must serve a lifetime.

to distinguish it from other types of cell.[4] We must now focus our attention on the neuron's remarkable ability to conduct signals.

Origins of agitation

Emil Du Bois-Reymond has been called the father of electrophysiology.[5] He was active in the latter part of the nineteenth century, and he set himself the ambitious task of explaining the phenomenon of animal electricity, discovered by Galvani.[6] Things had not been standing still since Galvani's time, of course, and Hermann von Helmholtz had even succeeded in demonstrating that the velocity of a nerve signal is not comparable to that of light, as had been surmised by Isaac Newton, but rather a far more modest thirty metres per second, or thereabouts. He established this by measuring the difference between the response delays for electric shocks given to the hand and the foot, the disparity arising because signals from the foot must travel a greater distance before reaching the brain.[7] Du Bois-Reymond hypothesized that a nerve process is constantly in a ready-to-fire state, rather like an indefinitely

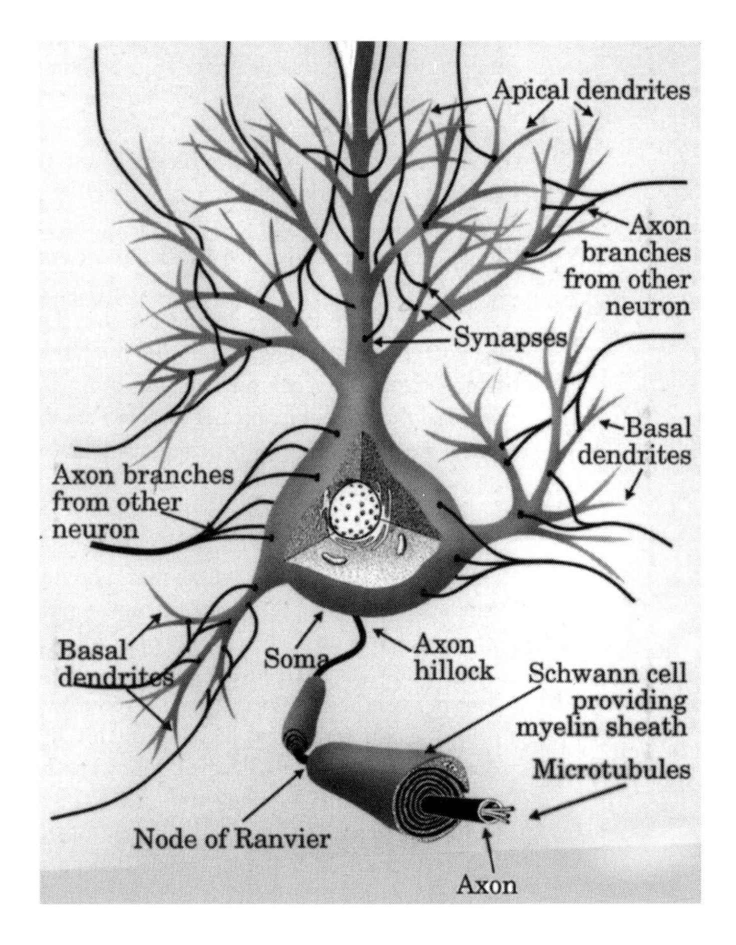

Figure 4.2
Information flow in the nervous system is mediated by neurons, which are special because of their property of excitability. Their membranous outer surfaces have elongated extensions called processes, and these act as conduits for electrochemical signals. The two types of process are the dendrites, which conduct signals toward the centrally located soma, and the axon, which conducts signals away from the latter if the sum of its inputs has exceeded a certain threshold value. Axons may be several centimetres long, and the nerve fibres that mediate passage of information to and from the brain, and also between other parts of the nervous system, are composed of bundles of these processes. Within the brain, the axon divides into several parts, the subsidiary members being known as collaterals, and both the main axon and the collaterals show considerable branching at their extremities. Axonic conduction is particularly rapid when that process is surrounded by a myelin sheath, as shown here. The dendrites (and sometimes also the soma, or even the initial stretch of axon – known as the axon hillock) receive signals via the synaptic contacts established with them by the extremities of other axons. These synapses are made either on the shank of the dendrite, as in the cases shown here, or on spines, which are small elevations on the surface of the latter.

reusable fuse, and he believed that it is maintained in this state by a voltage difference between its inner and outer surfaces. This has come to be known as the *resting potential* (i.e. resting voltage), and for the typical neuron it lies around 0.1 volt, the inside being electrically negative with respect to the outside. Du Bois-Reymond postulated that the resting potential is caused by the polarizing effect of nearby molecules. His own electrical measurements on the thigh of a frog suggested that a nerve signal is actually a *wave* of 'negative variation', rather than a continuous electrical current. As we are about to see, this idea was sound enough, but he was mistaken about the sign of the wave. It is positive rather than negative; the wave is one of *depolarization*.

Julius Bernstein suggested that the origin of Du Bois-Reymond's resting potential was to be found in the presence of *ions* (i.e. atoms that have an electrical charge by dint of having lost or gained one or more electrons) in the aqueous media inside and outside the typical neuron. It was already known that these media are rich in *sodium*, *potassium* and *chloride* ions; that is to say that they are essentially salt solutions. Bernstein also suggested that the passage of the nerve signal is somehow mediated by a *change in the permeability* of a process's bounding membrane. His ideas were enlarged upon by Ernest Overton, who argued that the resting potential arises from the imbalance implicit in the fact that there are more sodium ions outside the neuron than inside, whereas for the potassium ions the reverse is true. Excitation of one region of the membrane, Overton conjectured, causes a local change of permeability of the membrane to the ions, and their sudden passage is tantamount to an electrical short circuit. This, in turn, causes excitation of the adjacent area, and so on, and the upshot is a *self-propagating wave*, just as Du Bois-Reymond had predicted. The ability of a neuron's membrane to conduct electrical signals is now referred to as *excitability*.

It was to be many decades before this could be verified, and there were other issues already clamouring for elucidation. It had been irrefutably demonstrated that a neuron signals to other neurons by sending out electrochemical pulses, but it was not known how the transmitted messages were coded. Was there something analogous to the Morse code, for example, with the pulses being either long or short, or was the cryptographic system even more complex? As it turned out, the code was remarkably simple. Keith Lucas and Edgar Adrian demonstrated that nerve pulses are all of a standard duration, and that they all have the same amplitude.[8] There were subtleties, however. If the strength of the (depolarizing) stimulation is insufficient, no pulses are emitted. It is only when the depolarization exceeds a certain *threshold* that pulses are dispatched from the neuron's somatic region, and thereafter travel along the axon at the velocity first measured by von Helmholtz.

Lucas and Adrian had thus established that the neuron functions in an *all-or-nothing* fashion; there is nothing analogous to the Morse code's dots and dashes.[9] If the stimulus continues to be applied for more than a hundredth of a second, and most stimuli last considerably longer than that, more than one impulse is emitted. There is in fact a sort of code, because *the greater the magnitude of the stimulus, the higher is the frequency with which the impulses are*

the dynamics of the atomic motions in the molecules responsible
ulse impose an upper limit on this frequency: it cannot exceed
npulses per second. The threshold for pulse emission, for the typ-
lies around 0.05 volt in the positive (i.e. less negative) direction.
that nothing will happen until the voltage across the neuron's
rane, in the somatic region, is increased from roughly −0.10 volt
05 volt. This depolarization of approximately 50 millivolts is a
er to bear in mind.

actual mechanism underlying the generation of the nerve
ction potential as it is now more commonly called, was established
gkin and Andrew Huxley, through observations on a particularly
xon in the nervous system of the squid *Loligo forbesi*. As was dis-
hn (J. Z.) Young, this axon is wide enough to be easily pierced by
ctrode, designed to allow determination of voltages in salt solu-
its that permitted measurements of events with
th of a second, Hodgkin and Huxley were able to
rmeability when known voltages were applied.[10]
ct a scenario that nicely dovetailed into much of
ticularly that of Du Bois-Reymond, Overton and
at the impulse is initiated by a sudden *influx* of
terminated, about a thousandth of a second later,
f potassium ions. We thus see that it is not strictly
potential an electrical signal; it should rather be
cal signal.

uxley's guess as to what mediates these ionic fluxes
ted by detailed biochemical investigations. They
molecules in a neuron's bounding membrane that
ning through them, holes that change their shape
at is present across the membrane. Their idea was
mally not be wide enough to permit passage of ions
mbrane to the other, but that for suitable voltage
uddenly open wider, letting a few ions slip through.
ions per unit area of membrane would be sufficient
n potential, and Hodgkin and Huxley were able to
ons that produced an impulse velocity in good agree-
ntally observed value. Corroborating evidence for
lso emerged; the molecules of certain snake venoms
e been investigated by X-ray diffraction, and these
ss protruding regions that would fit snugly into the
by blocking them and paralysing the victim's nervous
ng development in this field, produced by the work
Sakmann, and known as *patch clamping*, enables the
y monitor the opening and closing of individual

ently been found that the mode of signal conduction
uite different to what occurs in an axon. The signals

are graduated according to the size of the stimulus, in contrast to the axon's all-or-nothing response, and they travel far more slowly; the typical transit time from the extremity of a dendrite to the somatic region lies around five to ten milliseconds, whereas the axon would conduct its signal over the same distance in about ten microseconds. And whereas the action potential that travels along an axon loses none of its potency as it moves away from the somatic region, the signal in a dendrite gradually dies away with increasing distance of travel; this means that only sufficiently intense dendritic signals will survive the inward journey toward the soma.

The situation in the typical neuron can be summarized as follows. At any instant, its tens of thousands of dendrites are receiving signals from surrounding neurons, and these signals are giving rise to slowly moving electrochemical waves that suffer diminution as they progress towards the neuron's soma. When they reach the soma, their effects are somehow added (by a mechanism which is still only imperfectly understood), and if the sum exceeds the threshold value, action potentials are emitted, without any subsequent attenuation, out along the axon. The typical dendritic signal is only about half a millivolt, so a couple of hundred such signals must impinge upon the soma at about the same time, if the threshold is to be exceeded. The frequency with which the action potentials are emitted from the soma depends upon the amount by which the threshold has been exceeded (see figure 4.3).

When the action potentials reach branching points in the axon, they give rise to onward-travelling action potentials, each of which is just as strong as the original action potential. Each action potential involves the passage through the membrane of just a few sodium and potassium ions per unit area of membrane, so with the large concentrations of the latter which are present in and around a neuron there is no risk of serious depletion even if the axon is used repeatedly for many seconds at its maximum frequency. Prolonged use would gradually lead to a decreased signalling ability, however, if it were not for the replenishing efforts of another species of protein molecule which returns sodium ions to the outside of a neuron, and potassium ions to the inside. Such proteins, the so-called *active ion transporters* (or *active pumps*), discovered by Jens Skou[12] in 1956, are obviously extremely important to the nervous system.

Influential connections

The job of the neuron is to send signals to other parts of the organism. The targets of these signals fall into three broad classes. The axons of motor neurons convey messages to muscle fibres. The axons of other neurons inject signalling molecules into *capillaries*, through which they are transported to other locations that can lie several centimetres from the signalling neuron. The sending neurons are in this case referred to as *neuroendocrine cells*. Finally, neurons can send messages to one another. We must now turn to these important message-passing processes, and start by taking a closer look at an axon's terminal region. Although only a single axon emerges from the somatic region of a neuron, this process invariably exhibits much forking farther along its length. To start with, there are the major divisions which give rise to the *axon collaterals*. These are usually not particularly numerous; they might number some dozens. Nearer the extremities of the axon and the extremities of its collaterals, a much more extensive division occurs, and this produces the *axon branches*.

The latter are essentially just as numerous as the dendrites, an approximate equality which is rather suggestive, for if the goal is to match axon terminus with dendrite terminus, there is no numerical barrier to this happening. Although the situation is not that precise, axons do make contacts with dendrites. But these contacts are not usually end-to-end affairs; an axon terminus might interact with a dendrite anywhere along the latter's length. The contact is not intimate right down to the molecular level, however. The

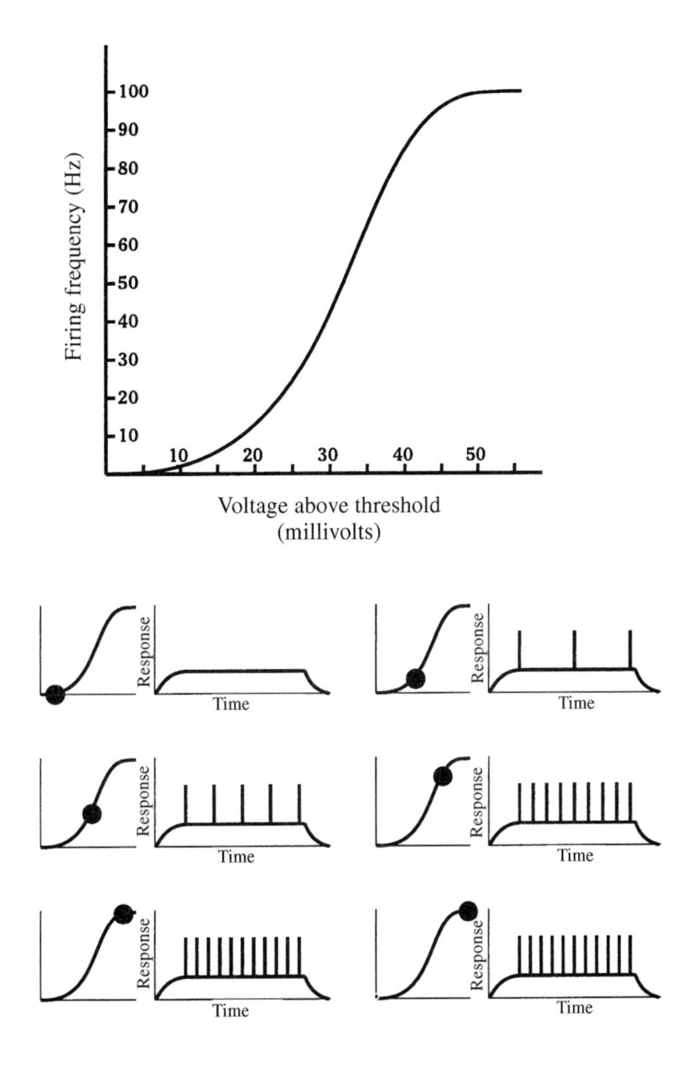

Figure 4.3
A neuron emits action potentials out along its axon only if the depolarization of its bounding membrane, in the vicinity of the axon hillock, exceeds the threshold value. When this is the case, the pulses are dispatched at a frequency that depends upon the amount by which the threshold has been exceeded. This frequency does not increase indefinitely, however, because a limit is imposed by the speed with which the atoms in the protein ion channel molecules can relax back to their original (pre-pulse) positions. This fact implies that the voltage–frequency characteristic of the neuron has a sigmoidal shape, as shown in the upper figure. Typical emissions from a neuron are shown on the right of the lower figure, the operating voltage in each case being indicated by the small black circle's position on the sigmoidal curve shown at the left. The sigmoidal curve is often assumed to represent the input–output characteristic of each unit in an artificial neural network. In both real and artificial systems, the elements currently exerting the greatest influence will be those in which the summed input exceeds the threshold by the greatest amount, the advantage being particularly large if the emission frequency is at the saturation value.

ending of an axon merely approaches to within about twenty-millionths of a millimetre, which is roughly forty times the diameter of a typical small molecule. As mentioned earlier, it was Charles Sherrington who coined the name *synapse* to describe this type of junction, and his picturesque description of neural networks provided the present book with its title.[13] He noted that the existence of synapses effectively joins the individual neurons into a vast mesh.

The question of what occurs at synapses had actually arisen before Sherrington put these near contacts on a firm footing. In 1877, Emil Du Bois-Reymond speculated about the possibility of chemical secretion from the region of an axon branch that interacts with a muscle fibre. Such a junction is a special form of synapse, and Du Bois-Reymond was the first to suggest chemical transmission of nerve signals. The first experimental endorsement of this idea was indirect, and it was unwitting in that the scientist involved drew erroneous conclusions. It came about through observations on the action of a certain poison, called *curare*, a substance used by South American Indians on the tips of their arrows. Its effects on the nervous system were investigated by Claude Bernard.[14] He injected some of the poison under the skin on a frog's back and verified its paralysing action. Bernard had made a dissection so as to expose the muscle of the hind limb and its associated nerves, and he found that attempts to make the muscle contract by direct stimulation were successful, whereas no reaction could be elicited if it was the nerve that was activated. He deduced, fallaciously as it turned out, that the curare was destroying the actual nerve.

Thomas Elliot supplied the correct answer, about fifty years later, his work being carried out on cats. He discovered that injection of *adrenalin* causes the bladder to relax, a function that was already known to be under the control of the *hypogastric nerves* in the natural situation. Elliot rightly concluded that adrenalin must be the substance responsible for the chemical transfer of the nerve's message, and he also suggested that the location of its release must be the very extremity of the nerve. His deduction was soon endorsed by Otto Loewi, whose flash of insight came in a dream.[15] This time, it was innervation of the frog heart that provided the inspiration. The *vagus nerves* were known to slow down the heartbeat by liberating a substance related to the chemical compound *muscarine*.

Loewi's idea was to arrange for some of this substance from the heart of one frog to flow into the heart of a second; he then demonstrated that stimulation of the vagus nerve of the first frog slows its heart down, whereafter it slows down the heart of the second frog, but only after a delay which corresponds to the substance's transit time in the interconnecting tube. Loewi subsequently went on to isolate the chemical messenger whose molecules make the short voyage across what has come to be known as the *synaptic cleft*, that is the small gap between axon and dendrite (see figure 4.4). The chemical is *acetylcholine*, and Loewi had succeeded in identifying the first member of what subsequently transpired to be a family of signalling substances; they are now collectively referred to as *neurotransmitters*. At the present time, several dozens of these chemicals have been isolated and characterized, and the list is probably not complete.

We now have a quite detailed picture of *synaptic transmission*, even though certain facts about the actual mechanism remain to be clarified. The neurotransmitter molecules are not merely suspended in the intracellular fluid (the cytoplasm), in the region of the axonal extremity. They are packed in small round envelopes bounded by the same type of fatty membrane that forms the skin of the neuron itself. These packets of neurotransmitter are called *vesicles*, and Bernard Katz and Ricardo Miledi showed that they all

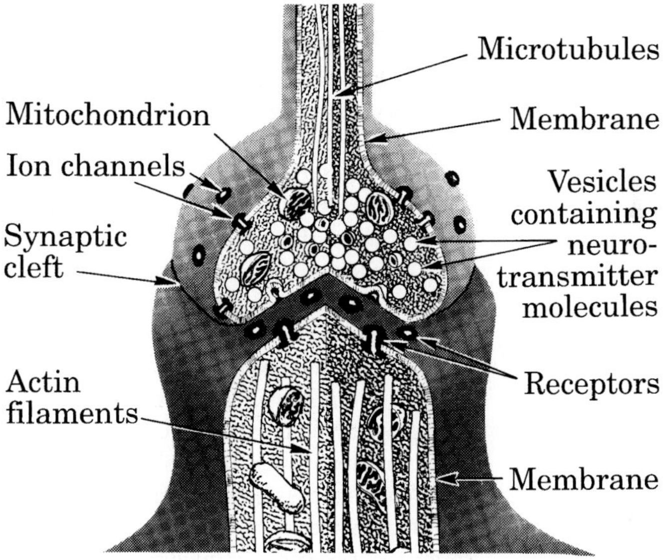

Figure 4.4
Neurons send information to other parts of the organism in three different ways. They can release molecules into capillaries, whereafter the message is broadcast by circulation and diffusion; they can dictate the movement of muscles via the synaptic contacts established with the membranes of the latter; or they can signal directly to other neurons by other types of synapse, as illustrated here. When the terminal region of an axon branch receives an action potential from the soma, neurotransmitter molecules are released from the presynaptic membrane. They diffuse across the synaptic cleft, dock with receptor molecules in the postsynaptic membrane of another neuron's dendrite, and generate an onward-travelling signal. The neurotransmitters are stored in membrane-bounded units known as vesicles, a few (and often only one or even none) of which fuse with the presynaptic membrane upon arrival of the signal. The various proteins destined for service in the outer membranes as excitability mediators or receptors are produced in the rough endoplasmic reticulum, and then transported through the (presynaptic) axon or (postsynaptic) dendrite. The neurotransmitters are manufactured by other proteins in the neuron's cytoplasm. In a human, the nervous system uses more than a quarter of all the energy consumed as food, the energy required by a cell being produced in the mitochondria. The synapse shown here has been established with a spine, which is a knob-like elevation on a dendrite, but synapses can also be made directly on the dendrite's shank, when no spines are present. Spines appear to limit the diffusion of ions in the postsynaptic space.

contain approximately the same number of molecules. When the nerve ending is in its quiescent state, the vesicles are somehow kept from coalescing with the axon's *presynaptic membrane*, but when an action potential arrives at that region, some of the vesicles (usually just one, and sometimes none) fuse with the membrane and thereby discharge their payload of neurotransmitter molecules into the synaptic cleft.[16] The liberated molecules then diffuse across the gap, dock with the (protein) *receptor molecules* that are located in the *postsynaptic* membrane of the next cell's dendrite, and induce a depolarization. (The actual biochemical details of this latter process are still under investigation, and they will be discussed later.) Because this is happening in a dendrite, and not an axon, the result is a gradual drift in voltage rather than an all-or-nothing action potential.

We now lack just one more piece of the puzzle. It has been implicit, thus far, that the signal passed from the axon terminus to the following dendrite, via the synapse, represents a positive stimulus. It is true that there is the complicating factor of the trans-membrane voltage being negative (inside, with respect to the outside), but this merely means that a positive influence will manifest itself as a decrease in that negative voltage, that is to say as a depolarization; the net effect will still be to encourage the passing on of a signal. The additional point that must now be taken into account is that some synapses mediate the passage of what could be called a negative signal. They involve types of neurotransmitter and types of receptor which collectively cause a *hyperpolarization* of the dendritic membrane (*gamma-aminobutyric acid*, or GABA, is prominent amongst such neurotransmitters), the effect on the postsynaptic membrane being to make the voltage *more* (rather than less) negative. A synapse that mediates such a negative message is said to be *inhibitory*, while one that passes on a positive influence is called an *excitatory* synapse. (*Glutamate* is the most common excitatory neurotransmitter in the cerebral cortex.) The existence of inhibitory synapses was established by Henry Dale, and the discovery's importance could hardly be exaggerated.[17] It was believed until recently that all the synaptic junctions entered into by a given *presynaptic* neuron are of one type: all excitatory or all inhibitory (see figure 4.5). There have now been indications that this might not always be the case, however.

Variations on the theme

The picture painted by the previous two sections is accurate, but things have been deliberately simplified, for the sake of clarity. Before moving on to contemplate the fascinating phenomena that arise in networks of interacting neurons, we ought to consider some of the variety present in the actual situation. It is this which enables the system to function with a subtlety that would not be possible if all neurons were as rudimentary as I have been making them out to be.

Let us start by returning to the question of myelination. The axons of some neurons are surrounded by a membranous coating known as *myelin*. This is in the form of a coil, rather like that of a watch spring except that it is

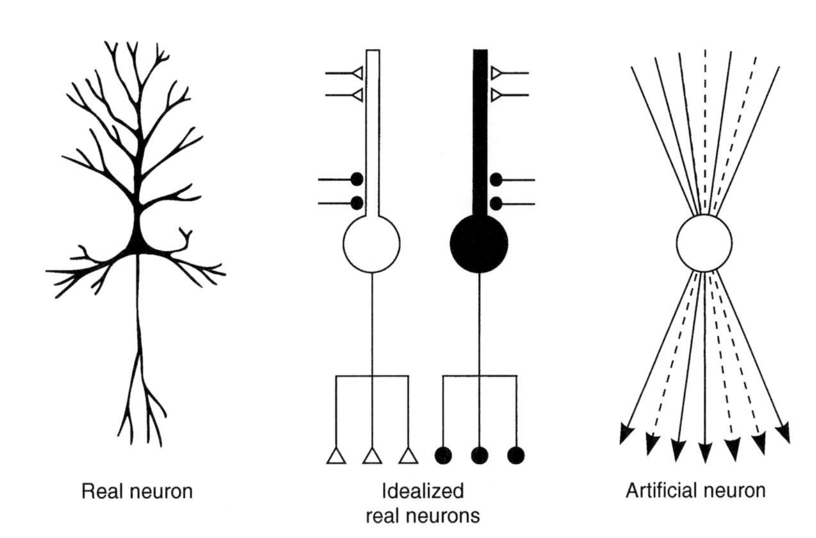

Figure 4.5
The symbols shown here permit ready discrimination between the various versions of neurons, despite their crudity. This convention will be adhered to throughout the book, with additional special cases being described as they arise. The neurons coloured white are excitatory, while the black ones are inhibitory. This distinction is not made in the case of an artificial neuron, the excitatory (full lines) and inhibitory (dotted lines) effects being embodied at the synaptic level.

greatly extended in the direction at right angles to the plane of spiral. The membrane is produced by *glial* cells, which actually come in three different varieties: *oligodendrocytes*, *Schwann cells* and *astrocytes*.[b] The membranes of the former two types of cell have a capacity for stretching themselves out and encircling, python fashion, nearby neurons. The oligodendrocytes do this to white-matter neurons in the central nervous system, and the Schwann cells achieve the same thing for neurons in the peripheral nervous system. The processes of the star-shaped astrocytes establish contacts with both capillaries and neurons, and they appear to play a dual nutritive and ion-controlling role.

The myelin hinders the free flow of ions in the vicinity of a neuron's extracellular space, and it thus acts as a sort of insulation. Because there is a limit to the length of axon that a single glial cell can myelinate, there are gaps in the myelination along the length of the axon, and these are known as *nodes of Ranvier*. It is only at these nodes, therefore, that the transmembrane flow of ions can occur which supports the action potential, and this causes the latter to jump from node to node. This *saltatory conduction* along an axon is faster than for the non-myelinated case, and diseases such as multiple sclerosis are devastating because of the inappropriate demyelination that accompanies them.[c] Glial cells serve in other respects. They provide a sort of scaffolding which guides developing neurons, and when development is complete they

b / There is actually a fourth class of glial cells known as *microglia*. These appear to be responsible for scavenging debris of expired neurons.
c / The word *saltatory* comes from the Latin word *saltare*, which means to leap. It has nothing to do with the sodium and chloride ions in the aqueous environment of a neuron.

remain in place to provide support. This support is both (passively) mechanical and (actively) biochemical, the latter being a sort of housekeeping that maintains the (aqueous) concentrations of the various molecules at their appropriate levels.

An important complication to our simple picture of a neuron is that there are channels other than those that provide trans-membrane passages for (positively charged) sodium and potassium ions. It had long been known that there are channels for (negatively charged) chloride ions, and more recently it has transpired that there are also channels which permit passage of calcium ions. These bear a double positive charge, and they are implicated in a number of processes occurring at synapses. Amongst these are the mechanisms of neurotransmitter release at the presynaptic membrane and the synaptic changes accompanying the storage of memories. There are actually several different types of potassium channel, one of which is influenced by the local concentration of calcium ions. The channel in this *calcium-dependent potassium channel*, as it is logically called, remains open for a rather longer time after membrane depolarization than is the case for the commonest type of potassium channel. All this makes for a rich subtlety of function at the synaptic level.

The positive influence of calcium ions on the efficacy of synaptic transmission is an important example of what is referred to as *potentiation*. The enhancement lasts for at least several minutes, and it can persist for up to an hour or so. It is due to the build-up of calcium concentration in the postsynaptic region. At certain types of synapse, the effect can even last up to several days, and it is then known as *long-term potentiation* (*LTP*). Both forms of potentiation can be provoked by what is known as *tetanic stimulation*.[d] We have already noted that neurons can be driven to emit trains of action potentials, at frequencies of up to roughly 500 per second, and in some extreme cases up to twice that value. When these impinge upon the presynaptic membrane, they lead to the release of unusually large quantities of neurotransmitter, and the postsynaptic effects build up and lead to the potentiation.

Our picture of a neuron is too simple in another important respect because it ignores the variety of shapes and special properties seen amongst real nerve cells. Nowhere is this more apparent than in the various types of *receptor cell* which serve the peripheral nervous system. But this is not surprising, when one considers how varied are the stimuli to which they have to respond. The *photoreceptors* located in the retina, which serve vision, must react to incident electromagnetic radiation. They are of two types: the highly sensitive *rods*, which at not present in the centrally located fovea, and the temporally discriminative and wavelength-sensitive *cones*, which are particularly dense in the fovea. Neither of these emits action potentials, the responses being of the graded type that we earlier identified with dendrites. Then there are the different types of *mechanoreceptor* that serve audition, touch and balance. Their membranes are directly sensitive to mechanical stimulation, which trans-

d / A burst of action potentials in the presynaptic neuron is referred to as a *tetanus*. This should not be confused with *tetanus toxin*.

plants itself to the particular types of ion channel found in these cells. Relatives of these imbue *thermoreceptor* cells with sensitivity to temperature changes, and yet others, called *nociceptors*, with the capability of transmitting the signals involved in the sensation of pain. Finally, there are the *chemoreceptors* which serve taste and smell. Their membranes contain protein molecules which react to the docking of the smaller molecules that mediate gustation or olfaction. The reaction involves release of what are called *second messengers*, which are passed on to ion channels that are thereby opened. Each type of receptor has a size and shape compatible with the geometry of the respective sensory system.

Because there will not be room to deal with them further in this book, we ought to note a few more brief facts about the receptors that serve taste and smell. The receptor cells which serve the latter are located in the *olfactory epithelium*. This lies in the dorsal posterior recess of the nasal cavity. These receptors send nerve fibres to the *olfactory bulb*, which can very loosely be equated with the thalamic nuclei that serve other senses (or, more strictly, certain *pre*-thalamic nuclei). The olfactory bulb is located immediately dorsal to the olfactory epithelium, and it is thus sandwiched between the latter and the frontal lobe, close to the uppermost regions of the nostrils. Information from the olfactory bulb ultimately reaches the *amygdaloid complex* and the *entorhinal cortex*. The taste receptors are located in epithelial cells that are clustered in sensory organs known as taste buds. Buds for the four basic tastes, *sweet*, *salty*, *sour* and *bitter* occupy different positions on the tongue, from the tip to the root, in that order. Axons from the buds project to the thalamus, in a more direct manner than is seen in the visual, auditory and tactile senses. The ultimate goal of gustatory information is two *taste areas* of the cortex, located in the *post-central gyrus*.

Another important variant in neural structure is seen in the dendrites of some cells, which are said to be spiny. This appellation refers to the small protrusions with which the dendrites of these neurons are dotted. They resemble the thorns on a rose bush, and when they are present they are the favoured sites for synaptic contacts. The *dendritic spine* appears to owe its presence to the tubular proteins that lie (with the appropriate orientation) just under the cellular membrane. The discovery of these structures initially posed neuroscience with a problem, because it was difficult to see what advantage they would give a neuron. Indeed, their rather small diameter would seem to put the neuron at a disadvantage because of the implied increase in electrical resistance. It has recently been demonstrated that the secret of the spine lies in the relative isolation that it accords the synapse (or synapses) established with it; the changes in potentiation resulting from a synaptic event stay localized to the vicinity of the synapse.[18]

Neuronal dictatorship

The field of neuroscience passed a major milepost in 1943. In that year, Warren McCulloch and Walter Pitts published the first detailed analysis of the properties of a network consisting of neuron-like units.[19] These were imagined to function in an on–off (i.e. *binary*) fashion, each neuron's activity being controlled by a threshold. From what we have already learned, this was a rather crude assumption because real neurons function in a more graduated manner. But it would have complicated matters considerably if McCulloch and Pitts had allowed their units to respond with a *perhaps*, rather than just a *yes* or a *no*.

The two pioneers were able to demonstrate that such a network, despite the rather modest demands being made of its constituent elements, is capable in principle of matching any stimulus to any response. It thus functions as a general computation device, and no task is beyond the reach of its talents. But they found that the circuits required to solve some problems can be quite complicated, and the question arose as to whether such idealized neural networks bear any resemblance to what is present in the real brain. A McCulloch–Pitts circuit might function as a general computer, but this in itself does not prove that the brain works in the same way. Indeed, it does not provide any evidence that the brain computes, at least not in the conventional use of that term. Descartes was responsible for the oft-quoted *cogito ergo sum*; had he been alive today, it is not obvious that he would have been disposed to add *computo ergo sum*.

The work of McCulloch and Pitts nevertheless deserves contemplation. We may start by considering what happens at a single *binary neuron*, and take this to be fed by just two input synapses from other cells. There will be no loss of generality if we assume that the threshold for this neuron is unity, and in the same vein we may let the two inputs, A and B say, adopt only the values unity or zero. Now if A is unity while B is zero, the total input to the neuron will be unity, and because this equals the threshold value, the neuron will become active, emitting nerve impulses out along its axon. If A is zero and B is unity, the same thing will occur. And if both A and B are active, the neuron will again be active. The neuron is thus carrying out, on its two inputs, the *logical operation* known as *INCLUSIVE–OR*; the neuron becomes active if A or B or both A and B are active. If, instead of the value unity, the threshold of the neuron is set at two, activity will *not* be generated if *only* A *or* B is active. This higher threshold requires that *both* A *and* B be active. The neuron will then be performing a different logical operation on its inputs, namely the *AND operation*.

With relatively few constraints on our binary neuron, with respect to its threshold and inputs, we already have two different logical operations at our disposal. Adding something we encountered in the previous section, namely that inputs can also be inhibitory, further operations can be achieved. For the sake of argument, let us assume that input A is excitatory while B is inhibitory, and let the threshold revert back to unity. If only A is active, the neuron becomes active, whereas activity in only B will this time not provoke activity in the neuron because of B's inhibitory character. If both A and B are active, their influences will mutually cancel, and the neuron will be inactive because the net input will be zero, which is below the threshold value.

It is not difficult to see that even a modestly sized circuit, made up of these types of *logic unit*, could perform quite sophisticated operations on its input. And McCulloch and Pitts added another feature which made their circuits even more powerful, namely inputs possessing what could be called the power of *veto*. There is now ample anatomical and physiological evidence for such inputs. They are established by synaptic contacts that occur either on the thickest part of the dendritic tree, just where this joins with the neuron's soma, or on the initial stretch of the axon, which is known as the *axon hillock*. One can imagine other subsidiary logical operations being carried out in the farther reaches of a neuron's dendritic tree, only to find themselves overridden by such a veto.

McCulloch and Pitts looked upon their work as having established 'a logical calculus of the ideas immanent in nervous activity', and at the time their paper appeared, it seemed that they had divined the gist of brain function. Their efforts were extended by Eduardo Caianiello, who worked out several fundamental properties of large networks composed of what are now referred to as *McCulloch–Pitts* (i.e. binary) *units*.[20] Only several years later did it transpire that this approach had serious drawbacks. The circuits required to perform some manipulations were very complicated, and it also emerged that the binary neuron represented a rather suspect simplification. As we have already seen, the output of a real neuron is not a single signal but rather a train of impulses whose frequency is related to the size of the overall input. Even a single neuron is thus able to function in a more subtle manner than was envisaged by McCulloch and Pitts.

It later transpired that there is a relatively simple logical operation that is beyond the discriminatory powers of a modelled neuron, namely the *EXCLUSIVE–OR operation* (often abbreviated to XOR). We will be taking a closer look at this surprising Achilles heel for the McCulloch–Pitts approach in chapter 6. The work that McCulloch and Pitts carried out in 1943 was thus a landmark rather than a signpost. It stimulated research into the way a network of interacting neurons might function, but it did not serve as a reliable guide because it suffered from severe limitations. In their subsequent studies, McCulloch and Pitts turned away from networks of logic elements, advocating instead mechanisms that bear more resemblance to *analog computing*. Because that mode of computation has much in common with the type of *parallel distributed processing* (or PDP) that some now believe to underlie the mechanisms of the real brain, the work of these two pioneers must nevertheless be regarded as the starting point of the modern era. This much will be manifestly clear as our story continues to unfold.

Spreading the responsibility

Perhaps the most serious criticism that can be levelled against the original McCulloch–Pitts analysis is that it pictured critical decisions as being made at the neuronal level (see figure 4.6). It is true that decisions of a kind are made by individual neurons. But the McCulloch–Pitts model went much farther than this; it allowed decisions influencing the *overall network function* to be made at the cellular level. Each step in a chain of logic usually exercises peremptory control over the behaviour of the entire sequence, and McCulloch and Pitts allowed individual neurons to wield such power. Viewed in this light, the work of Karl Lashley around 1950 is seen to provide a welcome counterweight.[21] He found evidence for the *distribution* of control, and his picture of the brain hinted at what might be called a democracy rather than a dictatorship. The question remained, however, as to which of these opposing concepts was correct, and it is for this reason that we should consider the evidence.

Lashley's favoured technique was *ablation*, which is the gradual cutting away of tissue. If individual judgments made at the neuronal level exercise decisive control, rather than mere influence, such tissue attrition should remove entire links in the brain's logical machinery. This is not what Lashley found. On the contrary, when he lesioned portions of rat cortex, he could pinpoint no faculty, however insignificant, that had been completely lost. Instead, he observed a gradual diminution over a broad range of functions, and he concluded that there is no evidence of localization in the brain. The antithesis of this attitude, in the case of vision, is epitomized by what Horace Barlow referred to as the *grandmother cell*, which emits its tell-tale impulses when, and only when, the retinas are presented with her image.[22] Lashley's experimental results were strongly indicative of distribution and this harmo-

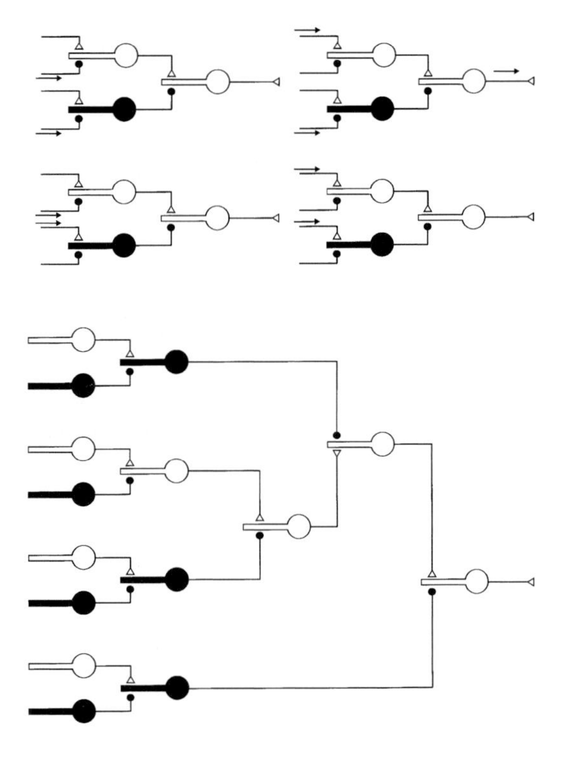

Figure 4.6
The operations that can be handled by logic elements of the McCulloch–Pitts type are illustrated by the upper four diagrams in this highly schematic figure. In keeping with the system used in this book, the neurons shown in white are excitatory while those shown in black are inhibitors. The arrows indicated where there is currently activity. If excitatory input or output is regarded as a YES statement, while inhibitory activity is taken to imply a NO, the circuit can be looked upon as carrying out logical evaluations. It is apparent that the output neuron in this three-cell circuit becomes active only through the combination of YES at the excitatory input neuron and NO at the inhibitory input neuron. The single diagram in the lower part of the figure indicates the sort of complexity that might be required to keep track of the multiple inputs in a more involved logical task. For yet more intricate assignments, the circuit might even have to incorporate feedback connections. Despite their obvious interest, circuits of the McCulloch–Pitts type could not underlie brain function because they would be too vulnerable to the loss of individual neurons.

nizes with the idea of parallel processing that is now so much in vogue. The implied *robustness* has proved to be a most useful feature of the computational strategy based on such processing.

But the mere emergence of this desirable attribute does not constitute verification, and we now know that Lashley's views were too extreme. There is plenty of evidence for relatively well-defined localization of function, as we have already discovered. We see it in the way the somatosensory cortex comprises a map of the body's external surfaces, and it is there in the motor cortex as well. There is further evidence for localization in the very existence of the various visual areas, the auditory areas, and so on. But this identification of faculty with cortical region admittedly refers to a scale that is larger than that of single neurons, so the Lashley view can still be salvaged. If many cells handle essentially the same type of information, in a given specialized region, a compromise is possible which incorporates aspects of both localization and distribution; it is possible to marry the best of both concepts. The question then remains as to how many neurons might be active in parallel, and Lashley was in no doubt that the figure was 'literally millions'. Let us take a closer look at his experimental findings.

Lashley's experimental yardstick was what he referred to as the *habit*, by which he meant either one particular facet of, or sometimes the totality of, an experimental animal's behaviour. As he removed progressively more of the cerebral cortex, he watched for loss of habit. Some of his results now seem surprising. In an attempt to determine the smallest amount of visual cortex that is capable of sustaining vision-related habits, for example, he found that discrimination of optical figures could still be learned when only one-sixtieth of the visual cortex remained. Lashley naturally took this as an endorsement of his non-localization thesis. But some of his other experiments were less clear cut. He trained rats in a maze and then checked to see if they retained their ability to navigate such a labyrinth if a portion of their cortex was destroyed. He observed little impairment of performance for 5–10 per cent ablation, but when about 50 per cent of cortical tissue has been removed, the maze-running skill is lost completely. Moreover, relearning often required much more practice than did the original learning. Lashley was able to quantify this observation. He found that he could draw a smooth curve through the experimental points, revealing the systematic increase of relearning time as progressively more cortical material was removed; the curve had what is referred to as an exponential form, the rate of increase itself steadily increasing.

Another of Lashley's results was also quite surprising. He found that if a *blind* rat is trained to run a maze, and its primary visual area is then removed, it experiences great difficulty in relearning, even though it obviously used no visual clues during the initial learning phase. His explanation of this curious fact was that the rat forms *concepts* of spatial relationships in visual terms, these concepts being integrated in the visual cortex. More generally, Lashley found that lesions to a single associative area of the cortex do not result in loss of habit, so long as the corresponding primary sensory area remains intact.

Lashley used the term *engram* for the physical changes that accompany the storage of a particular memory, and his lifelong hunt for the elusive engram was akin to a personal crusade for neurophysiology's holy grail. Ultimately accepting the fact that this search had drawn a blank, he settled for the (not inconsiderable) achievement of being able to state what the engram was not. *It is not possible*, he wrote, *to demonstrate the isolated localization of a memory trace anywhere within the nervous system. Limited regions may be essential for learning or retention of a particular activity, but within such regions the parts are functionally equivalent. The engram is represented throughout the region.*

Here was the strongest statement yet of the parallel distributed processing idea. And Lashley went on to state something else that was equally profound. *The so-called associative areas*, he continued, *are not storehouses for specific memories. They seem to be concerned with modes of organization and with general facilitation or maintenance of the level of vigilance.* He regarded this as further evidence of extreme non-localization, and he argued that there can be no group of cells that is reserved as the repository of special memories. The same neurons which retain the memory traces of one experience must also participate in countless other activities, and recall, in the Lashley scenario, involves the synergic action or some sort of resonance among a very large number of neurons. In retrospect, one sees that he laid the foundations of the presently accepted view of memory. What was missing in Lashley's time was the idea or concept that would enable us to pull all the threads together into a coherent picture. The following sections will chronicle various aspects of this quest, and reveal how far Lashley's crusade has progressed.

Pieces of neural circuitry

Until now, we have been discussing neuronal processes in general terms, without reference to the spatial arrangement of their connections. The point has arrived at which we should consider some of the ways in which groups of neurons can interact. And this is an appropriate place to introduce some idealized symbols which will be employed throughout the balance of this book. It will prove desireable, right from the outset, to differentiate between real and artificial networks. The former distinguish themselves by the obvious need to comply with biological constraints, which we will contemplate shortly, while the latter are governed primarily by expedience. The symbols shown in figure 4.5 are sufficiently different to preclude confusion.

Even the diagram of what has been labelled a *real neuron* is rather simplified, of course, because the numbers and lengths (relative to the soma) of the various processes have been greatly underestimated. But the diagram does at least look biological. This is clearly not the case with what have been designated *idealized real neurons*. In each of these (white representing an excitatory type and black an inhibitor), the thousands of dendrites have been stylized as a single thick shaft extending from the soma, while the axon and its branches, correctly indicated as being much thinner, can be seen on the other side. Even these meager details have disappeared in the case of the *artifi-*

cial neuron; it is merely denoted by a circle that is both the goal and the source of connecting lines, which may be regarded as axon–synapse–dendrite composites.

The synapses associated with the idealized real neurons clearly require some comment. The open triangles represent excitatory synapses, while the black circles signify inhibition. Inhibitory synapses do tend to be established in regions of the dendritic tree *proximal* (i.e. close) to the soma while excitation is more commonly *distal* (i.e. remotely positioned).[e] Analogously, the excitatory or inhibitory nature of an input to, or output from, the artificial neuron is indicated by a full or dotted line, respectively. And it is here that we begin to encounter an important fact about the artificial neuron: it is permitted to have properties not commonly seen in the real variety. A real neuron can, and invariably does, receive both excitatory and inhibitory inputs, but its output must be exclusively one or the other, not both. (This is *not* the case in certain neurons of the retina, however.) We thus see that it is in respect of its output that the artificial neuron shown above contravenes biological principles.

The first piece of neural circuitry that we should consider is actually implicit in the above diagram, because it reveals the neuron as a device for reacting to correlations in its inputs. Up to this point, such reaction has been regarded as being mediated by the summation process occurring at the soma, but nothing has been said about the question of time. We have already seen that it can take as much as ten milliseconds for a signal to travel inward from the distal region of a dendrite, so it becomes relevant to ask when the various signals were injected at the respective synapses. One could imagine the threshold of a receiving neuron being exceeded only if the majority of incoming excitatory signals arrived in unison. This issue does not arise in the most common type of artificial neural network because this has a layered structure, all neurons in a given layer being assumed to receive their incoming signals simultaneously.

It is a moot point whether such a layered structure would also represent a reasonable approximation to the arrangement in a real neural network. The striate area of the visual cortex gets its name from the markedly stratified distribution of some of its neurons, and there is clear evidence of a layered organization throughout the cortex. But the layers do not resemble the geometrical planes of the typical artificial network; there is much mutual penetration. Even more significant is the fact that several different types of neuron are always present in neural tissue, irrespective of its location in the brain. These provisos apart, it is still worth noting that the two arrangements shown in figure 4.7 are equivalent; in both cases, one group of neurons can be apprised of correlations in the outputs of another group.

e / In some cases, the inhibitory synapse is made *after* the soma, at the very beginning of the axon. This location is known as the *axon hillock*, and the inhibitory synapse apparently exercises veto control over the neuron's output.

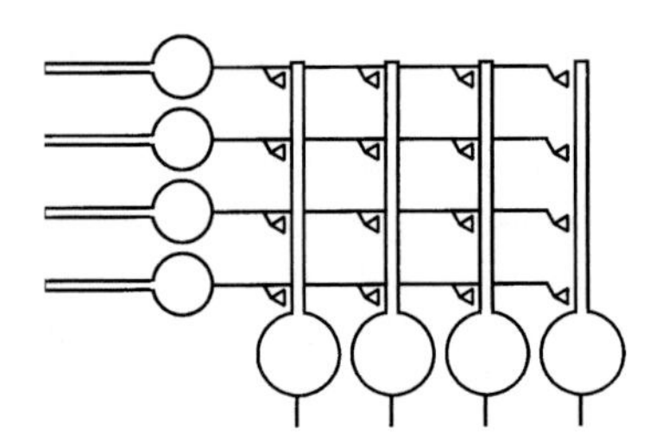
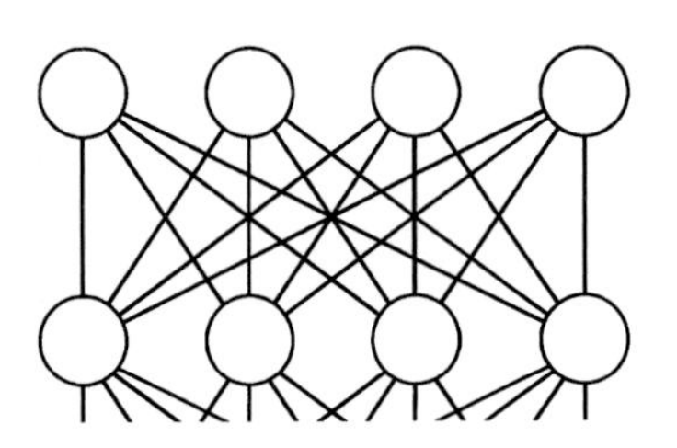

Figure 4.7
Although these two diagrams give a good impression of the neural arrangements in real and artificial networks, they are unable to do justice to the large number of synaptic connections normally encountered in the two types of circuit.

One should not be sceptical about the geometric regularity of diagrams that purport to represent real neural circuits. Despite their neatness, they may still faithfully represent the actual arrangement of connections, provided there is an underlying topological equivalence to the real thing. The situation is not unlike that which used to be encountered in electronics, before the advent of printed circuits. In the days of the vacuum tube, the inside of a wireless set, for example, bore scant resemblance to the tidy drawings found in the user's manual. The connecting wires were naturally permitted to take short cuts, and there was hardly a right angle to be seen.[f]

When drawing diagrams of neural circuits, of both the real and artificial varieties, the most difficult thing to convey is the actual number of connections involved. For many applications of the artificial type, the required number of units is several hundred, and this implies thousands of synaptic connections. The situation is far worse for real networks, and one has to bear in mind that the corresponding circuit diagrams are always gross simplifications. Diagrams that show a dozen or so synaptic contacts will frequently be trying to represent circuits which really involve tens of thousands of such junctions. The situation is further exacerbated by the fact that the actual degree of connectivity in real networks is usually quite low. It is virtually impossible to convey the correct impression with a diagram when the synapses are *sparse* yet numerous. How can one realistically depict a situation in which one per cent of a hundred thousand axon branches make synaptic contacts with a group of neurons?

With such limitations in mind, let us contemplate some important configurations of real neurons.[g] A slight modification of one of the above circuits gives the circuit shown in figure 4.8, in which the size of the triangular symbol indicates the relative efficiency (often called the *strength*, or *weight*) of a synapse. If equivalent input signals travel along the various axons in this figure, it will be those impinging on the strongest connections that will have the

f / This reminds me of the time I saw a boy suddenly clutch his mother's arm, *en route* from Holborn to Covent Garden on the London Underground's Piccadilly Line. He had been inspecting the idealized diagram of routes displayed on the carriage wall, and had remarked that we were approaching a sharp bend.

g / Artificial neural networks will make a comeback in chapter 6.

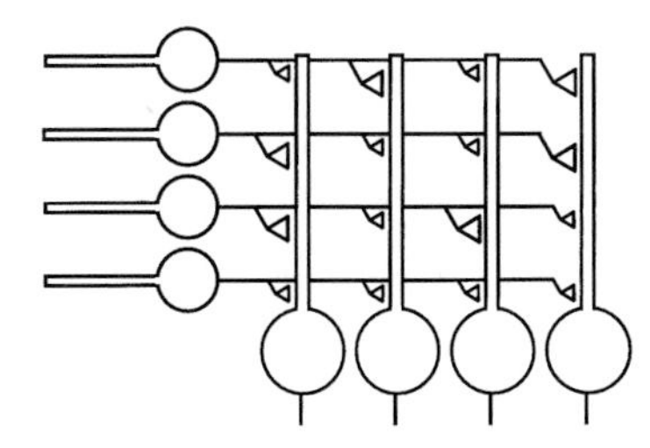

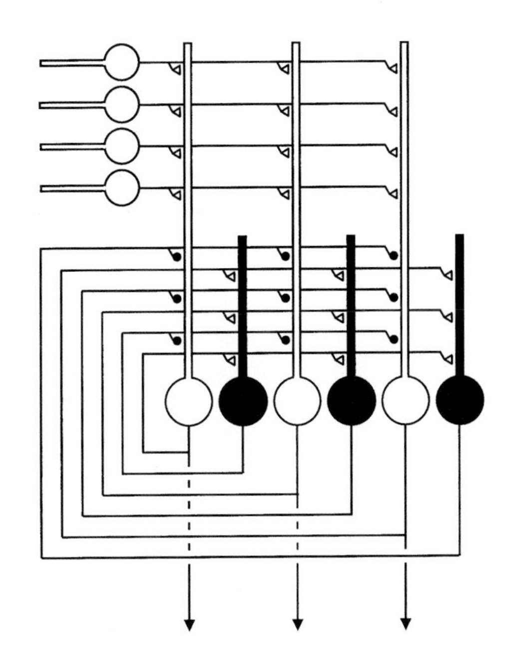

Figure 4.8
Lack of uniformity with respect to synaptic strength is a key feature of the typical neural network; it provides such a circuit with the ability to store memories.

Figure 4.9 (right)
If it were not for the ubiquitous presence of inhibitory neurons, the activity level in the typical neural network would rapidly reach its saturation value.

greatest influence. Since the inputs will usually *not* be equivalent (and some may well be zero), the situation can be quite complicated. Indeed, it nearly always is. The question of how different synapses could have unequal strengths must be deferred to the next chapter, but we may note here that this is the way in which *memories* are stored.

Another circuit worth our attention is shown in figure 4.9. It incorporates inhibitory neurons, which are seen to be activated by their excitatory counterparts.[h] And these inhibitors, having thus been triggered, return inhibition to those same excitatory neurons, thereby dampening down the activity in the latter. This is a very common situation, and it would be difficult to exaggerate its importance. This is the way in which the activity level of the brain is kept under control, and *epileptic seizure* is an example of what can happen when such control malfunctions. The diagram gives a reliable impression, despite its highly schematic nature, because the inhibitory neurons usually *are* dispersed amongst the excitatory variety. It is for this reason that they are commonly referred to as *inhibitory interneurons*. A noteworthy consequence of their presence is the introduction of another time factor, because the sending of excitation to these neurons and the return of inhibition from them is obviously not something that occurs instantaneously; such a round trip probably takes upwards of ten milliseconds to complete.

The inhibition that we have just discussed is an example of what is

h / The apparently longer dendrites on the lower set of excitatory neurons have been drawn that way merely to simplify the layout of the diagram. No biological significance of dendritic length is implied.

known (quite logically) as *feedback inhibition*. Another variation on the theme is seen in *feed-forward inhibition*. In this case, the inhibitory neurons are activated by other more remotely located excitatory neurons. The circuit shown in figure 4.10 displays the sort of neuronal arrangement that is involved, and one sees that this alternative could give inhibition with a diminished time delay; the inhibition no longer comes from an out-and-back route. There is an even more extreme variant of feed-forward inhibition, in which the delay in the inhibitory route can become zero, or even negative (that is to say that the inhibition then comes before the excitation). It is made possible by the independent activation of the excitatory and inhibitory routes. If the latter is activated first, it can set up a sort of cloud of inhibition through which the subsequent excitation has to penetrate. An example of such circuitry is shown in figure 4.11.

The inhibitory path is activated at A and the excitatory at B, and there is nothing to prevent A coming before B. An additional remark is required

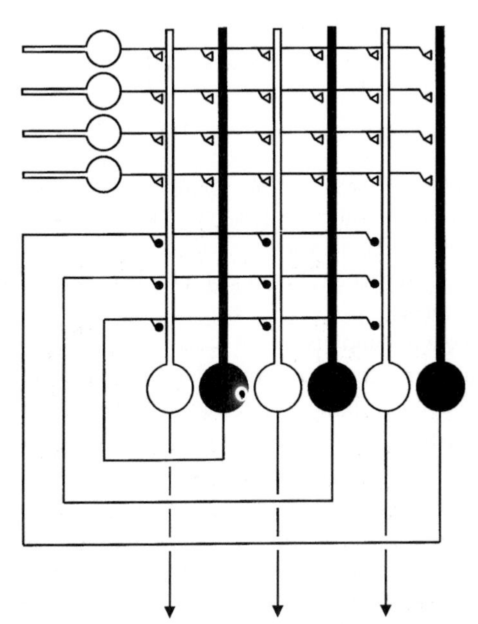

Figure 4.10
Feed-forward inhibition produces its counterbalancing effect more rapidly than is the case in the feedback variety.

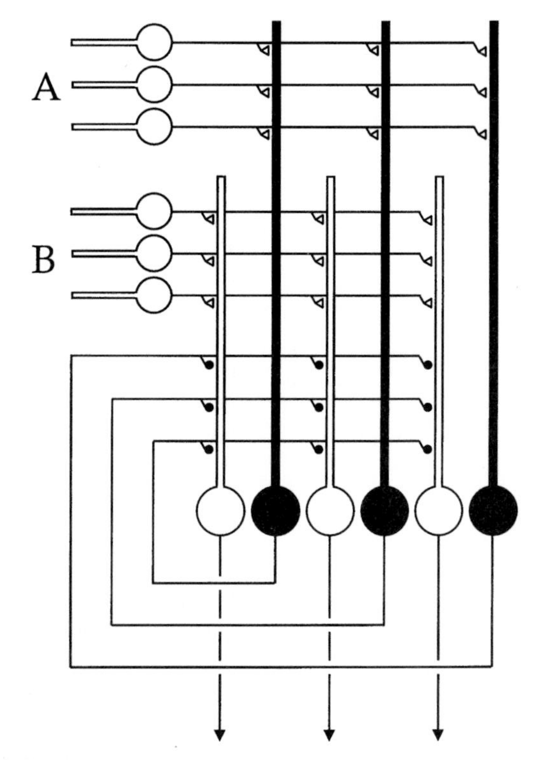

Figure 4.11
The special type of feed-forward inhibition shown here probably plays a key role in the winner-take-all mechanism. This type of early arriving inhibition is able to set up a sort of opposing cloud, through which only the strongest excitatory signals can penetrate.

concerning my use of that word *cloud*. If the activation at A is sufficiently large, the inhibitory neurons will each fire off a volley of action potentials rather than the odd one or two. And because the emissions of the various inhibitory neurons will not be in step with one another, the upshot will be inhibition that can last for a few tens of milliseconds.

The result of this, for the excitatory route, can be what is known as a *winner-take-all* situation. Imagine that the signals impinging on the excitatory neurons do not all arrive at the same time. If and when the activating effects of the early arrivers exceed the thresholds of the excitatory neurons, despite the opposing influence of the inhibitory cloud, those excitatory neurons will begin to dispatch signals to more distant neurons, via their axons.[23] The situation could very loosely be compared with trying to shine a beam of light through a cloud of fog; one succeeds only if the beam is sufficiently powerful. (If there is also feedback inhibition, which is not included in the above diagram, the excitatory neurons will simultaneously activate the inhibitory interneurons, via their axon collaterals.) Those excitatory neurons that do not reach the threshold, under these circumstances, will have lost the race. Whence the idea of winner-take-all, although *winners-take-all* would be a more appropriate term since more than one neuron will usually survive the cut. (A similar situation will occur when the various incoming signals arrive at about the same time, but with different strengths. It must be remembered, however, that the strength of a signal manifests itself through the *frequency* of the action potentials.) It seems likely that the various stages of sensory processing would all exploit the winners-take-all strategy, in order to reduce the amount of information that has to be handled, and possibly stored in memory.

There are actually two different forms of feed-forward inhibition. In the type illustrated in the above diagram, the sole purpose of the A route is to set up that inhibitory cloud. The alternative can arise when two excitatory groups of inputs are in mutual competition. The change required in the above diagram would then be that the A route merely inhibits the B route *en passant*, before proceeding onward with its main business. Axon collaterals from group A would activate the inhibitory interneurons *en route*, in a feed-forward fashion as shown, and the latter would inhibit the efforts of group B, either directly or by inhibiting *relay neurons* that lie between the B group and their ultimate goal. It is possible that the neurons in group A would go out of their way to achieve such pre-empting inhibition, projecting axon collaterals over a considerable distance to reach the appropriate location. One can think of the underlying neuronal arrangement as a spoiler circuit in that the activity of one route spoils the chances of the other path.

Examples of both feed-forward and feedback inhibition are seen in the circuits that control muscles. The skeletal muscles are arranged in opposing pairs, known as extensors and flexors, and the muscle of each pair has its own activating route. Working backward, the muscle is activated by a motor neuron, and the motor neuron is activated by an innervating afferent neuron. Simultaneous activation of both the extensor muscle and the flexor muscle is prevented by feed-forward inhibition fed from the innervating afferent neuron of one route to the motor neuron of the other. Self-regulation of either route is provided by feedback inhibition, in which the motor neuron is inhibited by (inhibitory) neurons that have been activated by the motor neuron's own collaterals.

As I shall be discussing in chapter 9, feed-forward inhibition of the spoiler type could play a role in the process of attention. It is possible that a hierarchy of spoiler mechanisms could also produce a macroscopic counterpart of those discredited McCulloch–Pitts networks, with the single neurons replaced by functional groups. All the neurons within one of the groups would be doing the same type of thing, even though they were handling different informational elements, such as the various parts of a visual scene. The neurons of another group would similarly be serving a different attribute of the input. In the event that the things being dealt with by the two groups were in conflict, the feed-forward inhibition would give one precedence over the other. In chapter 9, I will be arguing that this could occur when the visual system accords priority to moving objects, over stationary items. It does not appear too far-fetched to suggest that this could be an important organizing principle amongst the various areas of the cortex that serve a given sensory modality.

We are not done with our review of useful neural circuits. Most of those that remain to be discussed involve memory, however, so we will wait to deal with them until the next chapter. There are just two other bits of circuitry that should be mentioned here. One that is both common and very important, is the inhibition of inhibitory neurons. This is usually referred to as *disinhibition*, and it is a prominent feature both of the cerebellum and of those structures at the top of the brainstem which control the level of arousal. A piece of neural circuitry involving disinhibition is shown in figure 4.12, this diagram suffering from the deficiency common to all these illustrations, namely that it shows only a minute fraction of what would be present in reality. Since the net effect of inhibiting an inhibitor would be to raise the level of activation in the recipient neuron, one might wonder why the brain should prefer the more circuitous course; why not achieve the same result by direct excitation? The answer appears to lie in the greater degree of control that can be accomplished via the former path.

As a final point, we ought to take a more detailed look at something mentioned briefly earlier in this chapter, namely the *veto* function. It is one of the rare examples of a single neuron being able to control, all by itself, the output of another neuron. It accomplishes this through a synapse that is rather special, because of its position. It lies at the initial segment of the axon, that is to say on the axon hillock, rather than on a dendrite. The situation is thus as seen in the simplified diagram of figure 4.13.

Even if the net input to the receiving neuron exceeds the threshold value, the synapse on the hillock can exercise a veto over the neuron's output. The importance of this mechanism lies in its high specificity, compared with the rather indiscriminate nature of the other examples of inhibition that we have been contemplating.

This brief survey has touched on the most essential of the rudimentary types of neural circuit. I have been calling them circuits rather than networks to distinguish between groups of neurons that fleetingly carry out a contributory function and those responsible for overall processes. In particular, memory has played no part in what we have been discussing. We will turn to networks that do capture memories in the next chapter, after considering some of the complexities that this section has had to ignore. As a final point, I would like to note that everything described in this review of circuits is going to be used later in this book. But now that we have taken a look at the way these circuits function, later diagrams can leave out the explicit details. When different brain structures variously excite and inhibit one another, and when the situation is illustrated with a diagram, it will often suffice to indicate the influences by single broad arrows, appropriately coloured black or white.

The complexity of reality

The rudimentary neural circuits that we have just considered will serve us well in the chapters to come. They are implicit in much of the literature on the subject, and the underlying concepts are staple ingredients of numerous theories of brain function. It must be borne in mind, however, that the picture given above sacrifices detail for the sake of clarity. One could hope that these simple

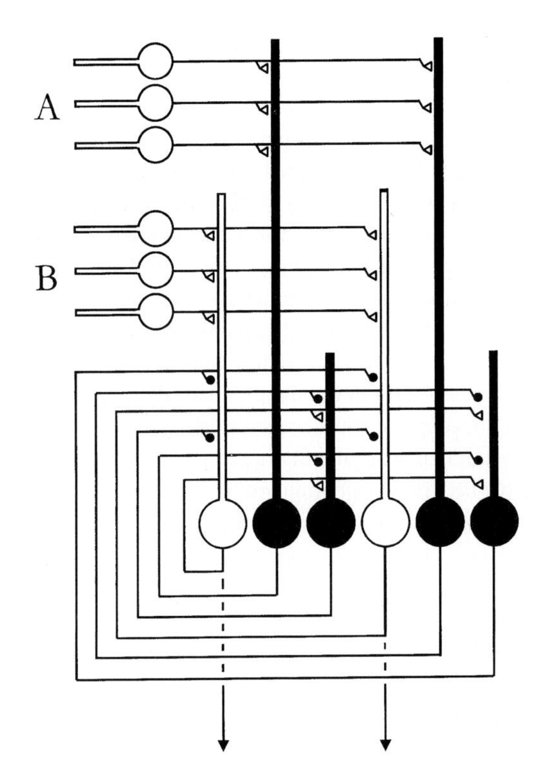

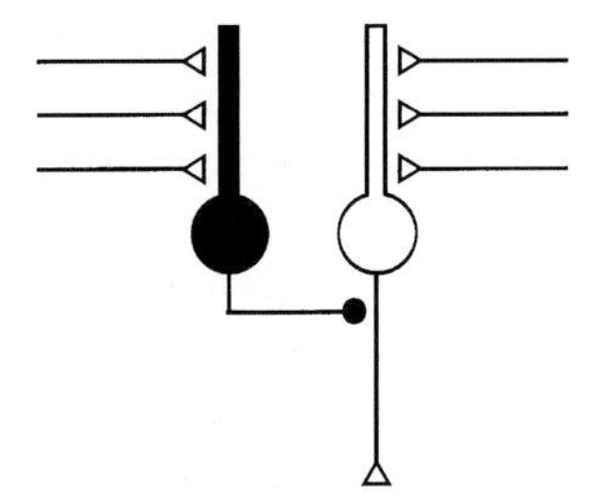

Figure 4.13
The most extreme form of inhibition is seen in the veto circuit, in which the output of an excitatory neuron is inhibited at the very point at which action potentials would otherwise be dispatched along the axon. In an important variant of this arrangement, seen in the cerebellum, the neuron being vetoed is itself inhibitory.

Figure 4.12 (left)
In disinhibition, inhibitory neurons are themselves partly or fully inhibited, and this enables the overall excitatory level to rise. The secondary inhibitors could thus be said to function as gain controllers.

circuits might nevertheless provide the basis for a brain theory that captured the bare essentials, and that incorporation of more detail would merely transform a utility model into a de luxe version. It is not clear whether such optimism is justified, however. Let us consider some of the things that have been ignored.[24]

Even our treatment of the individual neuron has been rather simplified. We cannot be sure that the signals arriving along the various dendrites really are summed arithmetically. A more complex process of consolidation may be at work. If this is the case, it would endorse the view of those who claim that the neuron itself functions like a sophisticated computer; it might be more of a black box than we have been giving it credit for. And there are other complications. For a start, some neurons have no well-defined axons. They merely have processes that appear to both receive and send signals, examples being the *horizontal cells* and the *amacrine cells* found in the retina. The processes of these cells do not transmit action potentials, and the same is true of two other prominent cell types in the retina: the *receptors* and the *bipolars*.[i] In this respect, they resemble the local interneurons that we considered earlier.

Then again, not all synapses are of the axon-to-dendrite type; some axons synapse directly onto the somatic regions of other neurons, and some dendrites synapse onto other dendrites. Another complication is that the signals transmitted out along an axon are sometimes of the graduated type more commonly observed in dendrites; these signals do not have the usual all-or-nothing character. Conversely, there is some evidence for limited action potential activity in some specialized dendrites. Interestingly, however, these complications are not particularly common in the cerebral cortex, and ignoring them might not be as perilous as one might fear.

Another source of complexity is the multiplicity of neurotransmitter types; over a score of these have now been discovered.[25] It is usually assumed that a neurotransmitter molecule's influence is always the same, irrespective of where it acts, but this will not be the case if the various receiving neurons possess different types of receptor. Thus although glutamate and aspartate usually excite, as does acetylcholine, there are situations in which it can inhibit. Gamma-amino butyric acid (GABA) and glycine are two of the most common inhibitors. These signal-mediating molecules can be divided into *peptide*[j] and *non-peptide groups*, and no axon in the mammalian brain has been found to release two different non-peptide neurotransmitters.

It appears that the peptide neurotransmitters act in a manner that is quite different from that seen with the non-peptide group. They merely *influence* synapses rather than mediate synaptic action directly. And this influence

i / The structure of the retina will be considered in more detail in chapter 7.

j / The peptide is familiar as a product of the digestive breakdown of proteins. It consists of amino acids, which are the fundamental building blocks of all proteins. The necessary background on proteins is given in chapter 5.

lasts for a longer time than is the case for normal synaptic transmission; it can persist for several seconds. Then again, a peptide neurotransmitter can exert control over neurons that lie well beyond the immediate vicinity of the point of release. These peptide molecules are transported by *diffusion*, as are their counterparts that make the short voyage across synaptic clefts, but the distances are much longer, and the journey consequently takes much more time. Such is the case when the neurotransmitters are released by the neuroendocrine cells, which as we have seen discharge their signal molecules directly into capillaries. Finally, and most complexly, some neurons have been found to release more than one type of peptide neurotransmitter.

These are some of the complications that must be kept in mind when one attempts to evaluate the credibility of any model of brain function. In particular, they must be given due consideration when contemplating the various mechanisms by which memories are stored. It is to such mechanisms that we must now turn, and the entire following chapter is devoted to them. We will learn that memory processes are numerous, and that they operate over a wide spectrum of characteristic times. By including such processes in our scenario for the brain, we will be giving it the opportunity to function as something more sophisticated than the black boxes that we encountered in chapter 2. And we are going to discover that the potential increase in sophistication is unexpectedly large; by the end of the next chapter, we might feel inclined to scribble **Pandora's** on our box's lid.

5 Lasting impressions

**If any one faculty of our nature may
be called more wonderful than the rest,
I do think it is memory. There seems
something more speakingly incomprehensible
in the powers, the failures, the inequalities
of memory, than in any other of our
intelligences. The memory is sometimes so
retentive, so serviceable, so obedient –
at others, so bewildered and so weak –
and at others again, so tyrannical, so
beyond our control! – We are to be sure
a miracle in every way – but our powers
of recollecting and forgetting do seem
peculiarly past finding out.**
Jane Austen (*Mansfield Park*)

Indoctrinating organisms

Neurons, broadly speaking, come in two varieties. Some excite and some inhibit, their influence on other neurons being via synapses in each case. In the previous chapter, we learned of ways in which individual neurons can be consolidated into networks, and we considered some of the basic properties of such assemblies. The fact that there is a choice of type for each synapse gives ample scope for complexity. There is still something missing, however; there is no obvious way in which a network could store memories. In principle, one could measure the electrochemical response at any position, due to the imposition of a given set of electrochemical stimuli, but this response would never vary. For the same set of stimuli, applied to the same places, identical responses would be observed. Memory requires something else in addition. It must involve some sort of change in the system. The response to a given stimulus must somehow depend upon whether the system has been exposed to that stimulus previously.

In 1949, Donald Hebb published a book entitled *The Organization of Behavior*. It has become one of the pillars of neuroscience, because of the seminal ideas it contained.[1] One of these was a surprisingly simple prescription for the changes that provide a neural network with the ability to store memories. Hebb suggested that the transmission efficiency of a synapse, what is now simply referred to as the synapse's *strength* (or sometimes, its *weight*), is modifiable by use. The exact wording of the relevant passage in his volume was as follows

When an axon of cell A is near enough to excite a cell B and repeatedly or persistently takes part in firing it, some growth process or metabolic change takes place in one or both cells such that A's efficiency, as one of the cells firing B, is increased. The most obvious and I believe much the most probable suggestion concerning the way in which one cell could become more capable of firing another is that synaptic knobs develop and increase the area of contact between the afferent axon and efferent soma.

Hebb thus suggested that the synaptic contact between two cells would be strengthened if the one was persistently contributing to the other's emission of nerve impulses. The strengthening of a synapse was perceived as resulting from nearly simultaneous activity in the presynaptic and postsynaptic neurons. Implicit in his conjecture was the further suggestion that if the firing of one cell was only rarely (or even never) accompanied by the almost simultaneous firing of another cell with which it made a synaptic contact, then that contact would be weakened. This type of synaptic modification has come to be known as Hebbian learning, and it has been invoked in countless theories of brain function.

The idea is impressively straightforward. Just as water running down the side of a mountain will gradually cut a channel, which subsequently becomes the preferred route, so will experience tend to increase the transfer efficiency of certain synapses, and weaken others, thereby creating favoured pathways for later nerve signals to follow. And just as the drying up of a river during a drought does not change its course, so too will the adopted neural pathway survive the periods when that part of the brain is inactive.

Some care is required with the river analogy, however, for it is valid only in so far as the distribution of the incident rain is reasonably uniform. And by the same token, the above discussion of Hebbian learning assumed a standard injection of nerve impulses, the sort of thing that might arise from a fixed sensory input, for example. It was the obvious need to be able to account for a wide variety of different sensory inputs that prompted Hebb to put forward another hypothesis in his book, one that is every bit as important as the learning mechanism that now bears his name. He suggested that for a given set of sensory stimuli only a very small fraction of all possible neural pathways would be in use, and he further proposed that these active routes would lie adjacent to one another, or very nearly so, because of their mutual electrochemical influences. Such a tight group of interrelated neurons, firing in or close to unison, has come to be known as a *neuronal assembly*, and the existence of these clusters relieves us of the problem of keeping track of what is going on in the bulk of the brain's roughly hundred thousand million nerve cells; at any one time, Hebb was postulating, the great majority of them would simply be silent.

Hebb's hypothesis of modifiable synaptic strengths is so attractive that it seems almost obviously correct. At one fell swoop, he had presented neuroscience with a beautifully simple mechanism that was able to explain the main features of memory. The way in which new memories are laid down in the neural network, and the manner in which they can be recalled, were suddenly revealed to be transparently simple, and it required little of the imagination to see that his conjectured mechanism could easily be extended so as to permit individual memories to be superimposed upon one another.

A number of years were destined to pass, however, before Hebb's brilliant insight was to receive experimental support. And in the meantime, other proposals for synaptic modification were made which received endorsement relatively quickly. Before going into the details of these, it would be a good idea to consider synaptic change in a broader context. In our earlier discussion of chemical black boxes, we saw that quite complex circumstances prevail if a set of chemical reactions involves several compounds, and we also learned that further complications arise if semipermeable membranes are available for compartmentalizing the interactions. And in the preceding chapter, we noted that a neuron's bounding envelope is such a semipermeable membrane, the passage of such ions as sodium, potassium and calcium being mediated by channel-forming proteins. It also transpired that the terminal region of an axon displays voltage-controlled permeability that is timed to within about a millisecond. This mechanism provides the basis for cascades of chemical reactions that are remarkable for the precision and timing of their regulation. They are strongly dependent on other membrane-spanning proteins, some of which are also channel formers while others involve more subtle transfer of chemical information. Given the possibilities inherent in a system with several different interacting species and a variety of mediating proteins, one should not be surprised if mechanisms other than the one proposed by Hebb also turned out to be useful intermediaries of memory. This is indeed the case, and evidence of their existence was accumulating even before the validity of the Hebb idea had been established.

Prominent amongst these alternatives was one put forward by Eric Kandel and Ladislav Tauc in 1963. It postulated that the strength of a synapse could be increased without the need for the postsynaptic neuron to be active, provided that a third neuron is able to influence the presynaptic member of the pair.[2] This is an example of a possibility that was raised during our earlier consideration of chemical black boxes, namely modification of a modifier. Evidence for the correctness of this suggestion was supplied by observations on a relatively lowly creature whose taxonomic name is *Aplysia californica* (see figure 5.1). This is a rather large aquatic snail with a nervous system that comprises only about 20,000 neurons. It has acquired the nickname sea hare not because of its speed across the seabed but because of the large ear-like structures that stick out from its head. *Aplysia* has been the subject of prolonged study by Kandel and Tauc, together with their colleagues, Thomas Abrams, John Byrne, Thomas Carew and Edgar Walters.[3] Their work required the preliminary demonstration that this creature is capable of displaying the basic forms of classical conditioning that had been established through Ivan Pavlov's work on dogs, half a century earlier.[4]

It is not *Aplysia's* large ear-like protuberances that are of interest in this respect; four of its other externally visible attributes are far more important, namely the tail, the siphon, the gill, and a centrally located structure called the mantle, which is bordered by the gill. The gill is naturally a vital organ in that any threat to its integrity will imperil the creature's oxygen supply. Both the siphon and the mantle serve as detectors of impending threat, and any disturbance to them promptly induces the gill to contract. The former two structures are both served by sets of sensory neurons, each of these groups forming synapses with the motor neurons that are responsible for gill withdrawal. The two sets also make synaptic contacts with assemblies of interneurons.

To appreciate what was accomplished by Kandel and his colleagues, we ought to briefly recall what Pavlov's research had revealed. Using a dog's tendency to salivate at the prospect of food, he first identified what is known as an *unconditioned response*. He defined this as a reaction in which response and stimulus are physiologically related. Such properties of food as dryness and texture are good examples of stimuli which fulfill this criterion. An unconditioned response is innate and highly effective. In a *conditioned response*, on the other hand, the stimulus[a] has no obvious physiological role, a good example of this category being colour. Pavlov discovered that conditioning is made possible by exposing a dog to a situation that pairs a conditioned stimulus with an unconditioned stimulus, the strongest effect being observed if the former precedes the latter by about half a second. He also demonstrated that the dog's capacity for salivation diminishes if it is repeatedly shown food to which it is denied access. The dog's salivation response is said to have become *inhibited* under such circumstances.

Let us now return to the work of Kandel and his colleagues. If the siphon (or the mantle), instead of being strongly provoked, is subjected to the gentle flow of water from a fine jet, the gill-withdrawal response gradually gets weaker. This is known as *habituation*. They then went on to demonstrate that *Aplysia's* repertoire of responses also includes the opposite of habituation, namely *sensitization*. When the tail is given a sharp pinch, the gill withdraws, this reflex being of the unconditioned type. If gentle (to-be-conditioned) stimulation of the siphon is paired with tail pinching, while stimulation of the mantle is carried out in an isolated (unpaired) manner, each process being repeated five times (say), subsequent stimulation to the siphon is found to produce a larger response than that observed for stimulation of the mantle. Conversely, if mantle provocation is repeatedly paired with tail shock, this becomes the conditioned response, and it is the siphon which elicits the smaller reaction. Interestingly, the greatest effect for either of these two alternative conditioning processes was observed when the conditioned stimulus was applied about half a second before the tail shock. Appreciable departure from this 'time window' resulted in a markedly diminished result, just as Pavlov had found in his experiments on dogs.

This set of observations was interesting in its own right, but the real value of the research lay in what it was able to reveal about the underlying

a / Kandel and Robert Hawkins have noted that a conditioned stimulus should more correctly be referred to as a *to-be-conditioned* stimulus.

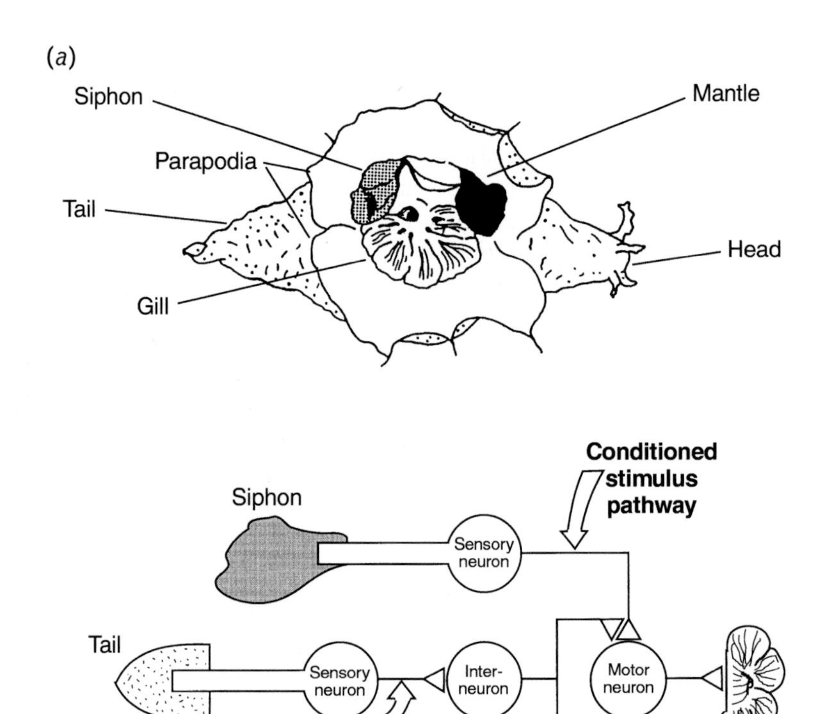

Figure 5.1
(*a*) The aquatic mollusc *Aplysia* (above) has a nervous system comprising only about 20,000 relatively large neurons, but it displays some of the types of conditioning seen in more sophisticated creatures, and is thus a convenient subject for investigations of memory mechanisms. In the highly schematic diagram below, groups of parallel neurons are represented by a single neuron. If either the siphon (or the mantle) is touched, the gill withdraws immediately, but prolonged gentle stimulation of either area, as with a small jet of water, causes habituation, and touching the siphon (or mantle) then elicits no response. The change is due to a gradual diminution of efficiency in the synapse between the siphon's (or mantle's) sensory neurons and the gill's (withdrawal reflex) motor neurons. The siphon's (or mantle's) normal influence on the gill can be restored by pinching the creature's tail. This activates the modulatory interneurons, which have synapses that are capable of sensitizing the siphon–gill (or mantle–gill) synapses. (*b*) (opposite page) The latter process is mediated by the release of serotonin into the cytoplasm of the presynaptic region of the sensory neurons' axonal extremity. This activates adenylyl cyclase, to which calmodulin becomes bound, and the result is an increase in concentration of the second-messenger molecule cyclic AMP. Finally, the cyclic AMP activates protein kinase, and this enhances neurotransmitter release into the synaptic cleft. These and other biochemical steps were revealed by the work of Eric Kandel and his colleagues, who have been major contributors to such investigations (see also figure 4.4).

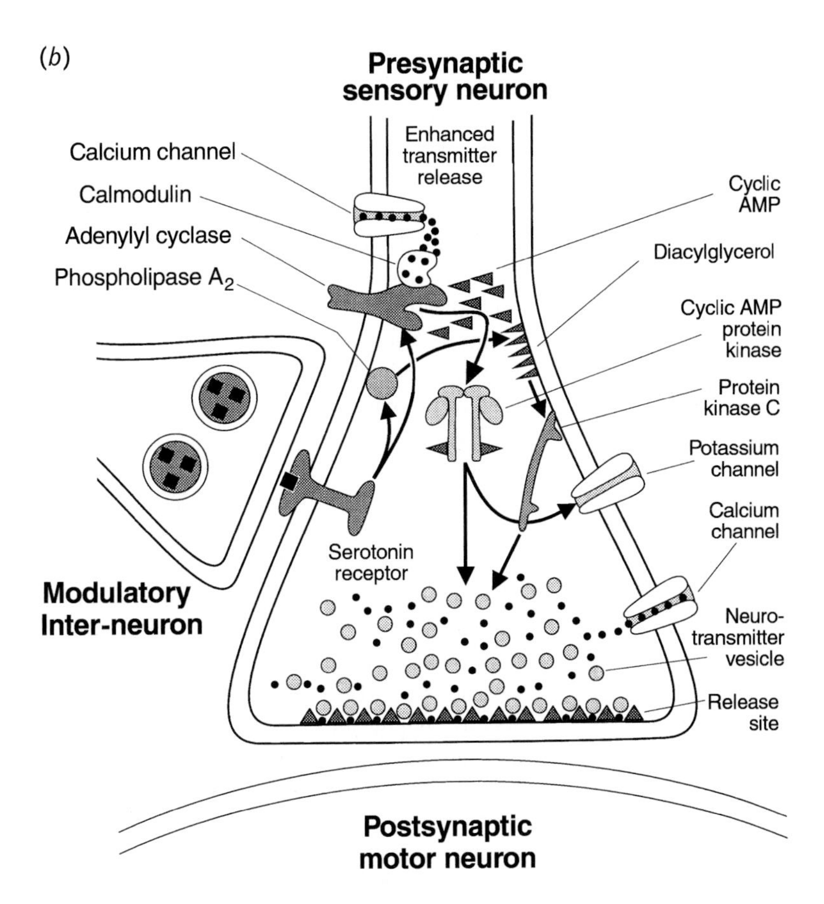

biochemical changes at the synaptic level. The neural circuitry that imbues the sea hare with its various types of response is quite simple, as nervous systems go. It is not the simplest imaginable, however, because a single neuron with its dendrites embedded in the siphon (or mantle) and its axon signalling to muscles that activate the gill would, in principle at least, be sufficient to facilitate the gill-withdrawal reflex. But a circuit that primitive would not permit the discriminating response that had been revealed in the creature.

In the somewhat more complicated arrangement actually found in *Aplysia*, the neuron that has its dendrites embedded in the siphon (or mantle) sends its axon not directly to the gill muscles, but rather to a synaptic contact with another neuron which does indeed have its axon in contact with the gill region. As we have already seen, the former neuron is known as the sensory neuron, while the latter is called the motor neuron. (In practice, the route is not served by single neurons of these two types; that would leave the nervous pathway much too vulnerable. There are actually several pairs of sensory and motor neurons, and they act in parallel. For the sake of simplicity, however, the discussion will invoke only one neuron of each type.)

It is this slightly more complicated circuitry which makes the various types of conditioning possible in *Aplysia*, and Kandel and his colleagues were able to establish that biochemical changes occur at the various synapses. When the gill-withdrawal reflex becomes habituated, for example, this is due to a change in the synapse between the (siphon or mantle) sensory and motor neurons. It had already been shown that, when a nerve signal reaches that region, calcium ions enter the immediate presynaptic region through its bounding membrane. Habituation, it was discovered, develops because there is a gradual diminution in the membrane's ability to pass these calcium ions.

The opposite happens in the case of sensitization, namely an enhancement in the membrane's capacity for allowing the passage of calcium. It is the presence of the interneurons which makes this possible. The dendrites of another sensory neuron are located in *Aplysia*'s tail, while its axon makes a synaptic contact with such an interneuron, which in this case is referred to as a *facilitating interneuron* (or *modulatory interneuron*). The latter makes a synaptic junction with that part of the siphon (or mantle) sensory neuron's axon that is immediately adjacent to the latter's synapse onto the gill-withdrawal motor neuron. This therefore involves a synapse onto a synapse; it permits the *modification of modification* that we encountered in the discussion of chemical black boxes. Provoking the tail causes sensitization because a signal is transmitted from that region's sensory neuron to the facilitating interneuron, whereupon the latter releases the neurotransmitter serotonin. This influences the synapse between the siphon (or mantle) sensory neuron and the gill-withdrawal motor neuron, increasing its ability to take in calcium ions. The several other biochemical changes that accompany this conditioning are reasonably complicated. We will have to go into their details, however, because these show how the synaptic modification mechanism acquires its time window. Ever since the pioneering work of Pavlov, it has been clear that such temporal discrimination is of the very essence.

The best way of appreciating what Kandel and his colleagues have achieved at the biochemical level is to work backwards from what is referred to as *protein phosphorylation*, and in order to understand that fundamental process, we will have to start with a brief review of the nature of proteins themselves. These molecules consist of one or more chains of units called amino acids, strung together like beads on a necklace, and the molecule's three-dimensional configuration is determined by the actual sequence of those elementary building blocks. Given that there are twenty different (common) types of amino acid, the possible variety in proteins is enormous. The (structural) proteins of interest here comprise several hundred amino acids, and each position along the chain (or chains) is a possible site of error, or *mutation*. Mutations can take the form of substitutions (of the correct amino acid by an incorrect one), deletions or insertions. Mutations change the shape of the molecule, although some have far greater influence than others.

Many proteins have been selected by evolution (which itself proceeds via mutation) because of the way in which they mediate chemical events, and we have already seen that the membrane-bound channel-formers are put to good use in neurons. An equally important class, in the neural context, are those that transmit signals across membranes by dint of their changes of shape when other, usually smaller, molecules become attached to them. In the native state, these protein activators are specific chemicals. But it is usually the case that a protein can also have its shape modified by certain other molecules which bear a structural resemblance to what we can think of as the proper activators.[b] These other activators fall into two broad classes: there are those that cause shape changes compatible with the protein's useful function, these being called *agonists*, while others known as *antagonists* oppose that function. A certain type of agonist, interacting with a certain type of neurotransmitter receptor molecule, will play a major role later in our story.

In the 1950s, Edwin Krebs and Edmond Fischer discovered that a key process in a wide range of metabolic processes is the attachment or detachment of a phosphate group (which consists of phosphorus and oxygen atoms) to a protein molecule. The importance of such molecular events lies in the associated change in the protein molecule's shape. This process is known as *reversible phosphorylation*, and in the present context the significant version of it is the addition of phosphate groups to (protein) potassium channels. The agency that carries out this attachment is an enzyme known as protein kinase.[c] Working backward through the sequence of reactions, the activation of protein kinase is performed by cyclic AMP (cyclic adenosine monophosphate), while cyclic AMP is produced from ATP (adenosine triphosphate) with the help of another enzyme called adenylyl cyclase. It is the latter, finally, which is the target of the serotonin that is liberated at the synapse between *Aplysia*'s modulatory neuron and siphon (or mantle) sensory neuron.

All these chemical events having now been adequately chronicled, we can return to the vital issue of timing. Phosphorylation of a potassium channel reduces the stream of potassium ions through that molecule, and this prolongs the action potential. It is the arrival of such an action potential that permits the passage of calcium ions into the presynaptic membrane, so a prolonged potential means a greater influx of those ions. This has two consequences. First, it enhances liberation of neurotransmitter into the synaptic cleft. Second, calcium ions bind to the molecules of another protein called calmodulin, and the calcium–calmodulin complex happens to enhance serotonin's influence on adenylyl cyclase. Taken together, these mechanisms are the reason why a properly timed signal in the siphon (or mantle) sensory neuron will be able to cooperate with the signal from the facilitating interneuron. The work of Kandel and Abrams on this chain of events can be rationalized if one makes the final, and certainly reasonable, assumption that it takes about half a second for the elevated level of calcium concentration to take effect. Here, after many decades, is a molecular-level explanation of classical conditioning.

It was perhaps a good thing that experimental endorsement of Hebb's proposal was not forthcoming immediately after its publication in

b / It is this vulnerability to molecular mimicry that underlies the side effects associated with many pharmaceuticals.

c / The word protein here refers to the enzyme's target molecule, but the enzyme itself is also a protein, of course.

1949. Had this been the case, the motivation for finding other mechanisms such as the one just described may have been lessened, and this would certainly have been detrimental to neuroscience. In noting that synaptic modification can occur in ways that do not require almost simultaneous activation of the presynaptic and postsynaptic neurons, which is what Hebb favoured, we become alive to the wider possibilities provided by chains of membrane-mediated chemical reactions. We come to appreciate, in fact, that the Hebb and Kandel–Tauc schemes may not have exhausted the possibilities. Meanwhile, it is satisfying to record that Hebb's mechanism finally received verification in 1986, by Holger Wigström, Bengt Gustafsson and their colleagues,[5] and independently by Stephen Kelso, Alan Ganong and Thomas Brown, both observations being made in the hippocampus.[6] And in 1987, Alain Artola and Wolf Singer observed a variant of the effect in the visual cortices of rats.[7] Before we can properly appreciate these later advances, however, it will be necessary to consider the background for what is known as long-term potentiation.

Freezing the moment

Long-term potentiation (LTP) is the name that Timothy Bliss and Terje Lømo gave to a mechanism they discovered in the hippocampus, in 1973. They found that the strengths of certain synapses in that part of the forebrain can be increased by a high-frequency series of action potentials, the enhancement persisting over a period of many days.[8] A remarkable feature of LTP is its pathway specificity; the synaptic strengthening is observed only along the active route. This is reminiscent of that second conjecture published by Hebb, in which he emphasized the importance of sparsely occurring neural assemblies. An explanation of how LTP could depend on correlated activity in the presynaptic and postsynaptic neurons will again require our patience with a piece of recondite chemistry. It also brings us to an exciting discovery.

The most common excitatory neurotransmitter in the brain is L-glutamate, which is the laevo form[d] of the amino acid *glutamate*. In the early 1960s, during investigations by Jeff Watkins of various agonists of the natural glutamate receptor, it was discovered that the compound *N-methyl-D-aspartate* (NMDA) can give a particularly strong response; stronger, indeed, than L-glutamate itself.[9] It later transpired that there are several different forms of glutamate receptor, and they were named after those agonists that elicit the strongest response. Other prominent examples are the *quisqualate* type and *kainate* type receptors. It has now become common to classify these large membrane-bound molecules as being NMDA receptors or non-NMDA receptors, but one must bear in mind that it is nevertheless a glutamate receptor that is being invoked.

d / The *laevo* (L) form of a molecule rotates the plane of vibration of polarized light toward the left, looking in the direction of the oncoming beam, while the *dextro* (D) form rotates that plane toward the right. The letter N is used to denote the *normal* form of a molecule.

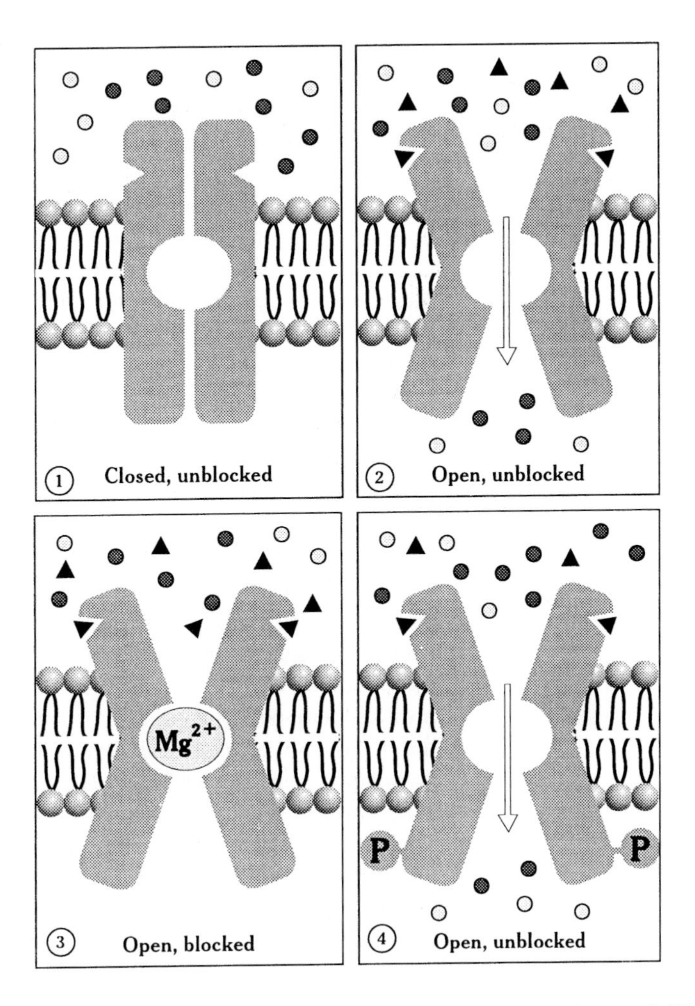

Figure 5.2
The *N*-methyl-D-aspartate (NMDA) receptor (shown straddling the lipid-bilayer membrane) belongs to the family of glutamate receptors, but it differs from the other members of that group in having unusual cooperative properties. When the channel through this membrane-bound protein molecule is in its open state, it permits passage to both sodium (shown as small light circles) and calcium ions (dark circles), but this requires more than the customary docking of a glutamate molecule (black triangles) with a site on its extra-membrane surface. The additional factor arises from the fact that the channel is normally blocked by a magnesium ion. The blockage can be relieved by depolarization of the membrane. This gives the receptor its cooperativity in that both factors, glutamate docking and depolarization, are required for the inflow of ions. When this dual condition is fulfilled, the postsynaptic influence lasts for about a fifth of a second. As indicated in this modified version of diagrams originally due to Charles Stevens (*Current Biology*, 1992, **2**, 497–9), phosphorylation of the receptor can act as a stand-in for depolarization. The individual parts of the diagram indicate (1) the channel unblocked by membrane depolarization, but closed in the absence of glutamate; (2) the channel in the open state induced by glutamate, and unblocked by depolarization, and therefore permitting the passage of sodium and calcium ions; (3) the channel in the open state induced by glutamate, but blocked by a magnesium ion because of the lack of depolarization; and (4) the channel open because of the presence of glutamate, and unblocked despite the lack of depolarization because the receptor is phosphorylized.

The NMDA-type glutamate receptor gained even more prominence when it was discovered that it possesses a remarkable capacity for capturing correlations between signals. It has a calcium channel that is normally blocked by a single magnesium ion, so the docking of a glutamate molecule with the receptor does not produce the usual influx of ions, as it does in the non-NMDA receptors (see figure 5.2). The magnesium block can be removed, however, if the membrane (in which the receptor is located) is depolarized. Operation of an NMDA *receptor* thus requires simultaneous depolarization and the arrival of glutamate molecules, and the resultant influx of calcium ions is reflected in a change in the voltage across the postsynaptic membrane which reaches a maximum within about twenty milliseconds, and which lasts for about two hundred milliseconds (that is, about a fifth of a second). This latter number is rather intriguing, because elementary cognitive processes are usually reckoned to require about the same amount of time. The necessary depolarization can result from the action of non-NMDA receptors. It is the NMDA receptor's dependence on two simultaneous and different types of event which enables it to capture correlations.

We can now return to LTP, because as Gary Lynch and his colleagues have shown, the passage of calcium ions through the unblocked NMDA channel is a prerequisite for this component of memory.[10] There is a fascinating complication, however, because it transpires that the process ultimately involves both postsynaptic and presynaptic mechanisms. Calcium influx through the NMDA channel is admittedly sufficient for establishing LTP, but presynaptic events are needed if LTP is to be maintained. This is the first evidence we have encountered of a process that runs in the 'wrong' direction, but such retrograde action has now been well documented by, amongst others, John Bekkers and Charles Stevens, Roberto Malinow and Richard Tsien, and by Timothy Bliss himself.[11] The hunt for the molecule that carries the retrograde message finally led to it being identified as *nitric oxide* (NO), the contributing sleuths including Georg Böhme, Ottavio Arancio and Thomas O'Dell, Paul Chapman, and Daniel Madison and Erin Schuman.[12]

The capstone to this very important story was provided by the efforts of Eric Kandel and his colleagues, and by Scott Small and Min Zhuo.[13] They found that the backwardly travelling NO produces LTP only if it arrives at the presynaptic membrane when that structure is in the act of receiving action potentials. In effect, this adds a further correlation-capturing ability to synapses that possess NMDA-type receptors in their postsynaptic membranes. And it is not difficult to see why this latter component of the synaptic scenario should be useful to the nervous system. For when a particular postsynaptic membrane has been provoked into activity, the NO that it returns to the presynaptic membrane will reward only those routes which were contributing to that postsynaptic state. This selective facilitation guarantees that it will be only the active routes that are favoured, which is in keeping with Hebb's concept of the neuronal assembly.

Recent studies of rats have strengthened the belief that LTP's role in the hippocampus of this animal is to serve spatial memory. Richard Morris had devised a more discriminating version of the traditional labyrinth, known as the *water maze*.[14] It consists of a circular tank with water too deep for the rat to touch bottom, but equipped with a submerged platform that the rat can rest on. If the water is rendered opaque, the rat will gain this refuge only if it can remember the platform's location. Normal rats can achieve this with little difficulty. Through the combined efforts of Alcino Silva, Susumu Tonegawa, Yanyan Wang, Richard Paylor and Jeanne Wehner, together with Charles Stevens, whose other contribution was mentioned earlier, it was shown that rats whose calcium–calmodulin dependent protein kinase bears a certain mutation encounter serious trouble with the water maze.[15]

We have now considered synaptic function in considerable detail. This has been necessitated by the extremely important part that synaptic transmission plays in the passage of signals in the nervous system. It must be emphasized that not all such transmission events lead to changes in synaptic strength. But when such plasticity is present, correlations in neural activity are taken note of by the system, which thereby learns. We have seen that these correlations can be exclusively presynaptic, as in the case of the adenylyl-cyclase-mediated classical conditioning in *Aplysia*, or presynaptic–postsynaptic, as in the NMDA-mediated Hebbian learning in the rat. And we have seen that the temporal windows of these two fundamental mechanisms are about half a second in the first case and something below about a fifth of a second in the second. The latter is so short that one could say, with only slight exaggeration, that the NMDA-type glutamate receptor enables the brain to capture the moment.

In the longer term

Up to this point, our discussion of synaptic plasticity has concentrated on events that occur within a fraction of a second. Memories would be of little use if they did not last for considerably longer times, of course, so we have to ask whether these modifications could persist for periods much in excess of the time it takes to establish them. Given that diffusion plays a key part in the mechanisms examined in the preceding section, and noting that diffusive processes over distances typical of the synaptic level occupy a small fraction of a second, we might have little confidence in the durability of these changes. Fortunately for the organism, it has a whole battery of other biochemical processes that can be exploited to transfer fleetingly captured correlations into its permanent repertoire of responses. These include the redistribution of proteins, both within the synaptic membranes and in the regions adjacent to them, the restructuring of the synapses themselves, the sprouting of new synaptic contacts, and even the activation of certain genes, this latter process leading to the production of other protein molecules.

These biochemical consequences of a brief learning event take place at various rates, and thus manifest themselves in structures that appear on different time scales. Within seconds of the primary synaptic incident, there is an increase in glucose metabolism. Subsequent to this, and during the following minutes, one observes a cascade of processes, including activation of a gene known as *C-fos*, phosphorylation of some of the proteins mentioned earlier, an increase in receptor binding, and an increased emission of groups of action potentials on the part of the neurons implicated in the relevant neuronal assemblies. On the scale of hours, these events give rise to changes in connectivity between the individual neurons, this being the result of a number of processes. The latter include synthesis of proteins, amongst which are *tubulin*, a structural molecule that helps to determine the shapes of dendrites and their spines, and membrane-bound *glycoproteins*. The latter are proteins that possess chain-like sugar segments which protrude into the extracellular fluid. They are known to mediate cell–cell recognition, not just for neurons but for many different types of cell. Ultimately, over a span of days, one sees changes in the physical appearance of the synapses, including a crowding together of the receptors, an increase in the number of spines, and an increase in the spine head diameter. These developments are not independent of one another. Jean-Pierre Changeux and Antoine Danchin showed that when new synaptic connections are being established, they can be stabilized by activation of the postsynaptic neuron.[16] Indeed, in neural populations which are still sprouting new synapses, those of the latter which do not get such stimulation simply degenerate.

We ought to pause at this point and consider some experiments on chick memory carried out by Steven Rose and his colleagues, because they do not obviously harmonize with what has been described.[17] The training paradigm was a simple one. A chick will peck at a small bright bead, but if this has been dipped in methylanthranilate, a bitter substance, it will rapidly develop an aversion to the bead, and remember its undesirability for many days. Rose and his team had first established which regions of the chick brain are involved in remembering these events, by exploiting the 2-deoxyglucose technique that we encountered in chapter 3. The relevant parts turned out to be the intermediate medial hyperstriatum ventrale (IMHV), the lobus parolfactorius (LPO) and the paleostriatum aumentatum (PA). They then investigated the influence on memory of lesioning the IMHV. If the right-hand lobe of this structure was damaged and the chick was trained the next day on the bitter bead, testing an hour later showed no impairment of its ability to recall the nasty experience. When the left-hand lobe was destroyed, however, the training and testing procedure revealed that the chick was now suffering from amnesia. Clearly, the chick's left-hand IMHV is necessary for learning.

Rose and his associates then reversed the schedule, training the chick for bitter-bead aversion an hour before the lesioning. Testing the chick a day later, they were surprised to find that it could remember the bead's negative associations. This showed that the left-hand IMHV is not necessary for recall. They then directed their attention to the LPO, first making the lesion prior to training the chick on the bitter bead. Testing three hours later exposed no amnesia. Switching the protocol, as before, they discovered that training before lesioning the LPO, and testing a day later, produced clear amnesia. For this other brain component, therefore, the opposite is true: the LPO is not necessary for learning, but it is necessary for recall.

It is not particularly unusual to find such division of labour in an animal brain, with one region responsible for learning and another for recall. Later in this chapter, we will discuss a similar division in humans. The continuation of these chick studies did reveal something quite surprising, however. Rose and his colleagues sought to shed light on what is known as *delayed sickness aversion*. If animals are presented with a novel food, or familiar food under novel circumstances, and later given a lithium chloride injection to induce sickness, they mentally link the sickness with the food even though the latter was consumed some considerable time before the nausea. Such association of events separated by tens of minutes is difficult to reconcile with the mechanisms described in the previous sections. And it is noteworthy that chicks, which show this behaviour, are intermediate to the lowly *Aplysia* and the obviously more advanced rat.

Before considering the final experiment carried out by Rose and his group, we should note that glycoprotein synthesis is metabolically blocked by the compound 2-deoxygalactose. It was not difficult to demonstrate that this substance also inhibits the formation of memories. Chicks pecked at a neutral bead, that is to say one which had not been dipped in the bitter substance. Half an hour later, they were given lithium chloride injections and became sick. When they were later confronted with the neutral bead, they avoided it. If the sequence of events was repeated, but with administration of 2-deoxygalactose at the time of pecking, the chicks did not associate the sickness with the neutral bead. A final version of the experiment involved pairing the 2-deoxygalactose with the lithium chloride injection rather than with the initial bead pecking, and in that case the subsequent aversive reaction was reinstated.

In view of what was stated earlier regarding the narrowness of the time windows involved in learning, these results are quite surprising. They indicate that a chick can retain the impression of a neutral stimulus for at least half an hour, and then associate it with sickness. Moreover, they showed that this memory-in-waiting requires synthesis of glycoproteins, which are involved in the relatively slow process of synaptic restructuring. Rose and his colleagues had exposed the presence of a mechanism that fits neither the half-second presynaptic nor the few-millisecond presynaptic–postsynaptic mould, and they stressed that there are memory mechanisms which lie beyond the simple scenario that was sketched in the previous sections. These results are still rather new, and their ramifications continue to be explored.[18] It is too early to attempt to place them in their proper position, in what may well be an even larger spectrum of learning mechanisms. It is not because our picture of learning processes at the molecular level is complete, therefore, that I now turn to a more macroscopic manifestation of memory.

In chapter 3, we learned that the somatosensory cortex comprises a map of the body's external surface. It is rather distorted, admittedly, but it is a map nevertheless. Adjacent parts of the body, we learned, appear in adjacent parts of the map. The question arises as to whether the features of this map are permanent, or whether they too can be changed by some or other learning process. An experiment carried out by Michael Merzenich has provided the answer, and it is an emphatic *yes*.[19] He trained a monkey to touch a rotating disc with only its three middle fingers. Examination of the animal's somatosensory area after several thousand repetitions of this task revealed that the amount of cortex serving the three digits had expanded, at the expense of the regions serving those that had been idle. This result indicates that neuronal plasticity prevails at a level much higher than that of the individual synapse.

It is not clear whether it also shows that passive neurons are in danger of extinction – a *use it or lose it* principle at the cellular level. The Merzenich result could be just as readily explained if inactive axonal branches passively await recruitment. Indeed, this view has received endorsement from a recent investigation of a cat's visual system, carried out by Charles Gilbert and Torsten Wiesel.[20] They had previously determined the *receptive fields*[e] of several neurons in a patch of the primary visual cortex of their experimental animal, and they then used a narrow laser beam to create a small patch of damage on the relevant retina, at a location corresponding to the centrally located members amongst those neurons. The receptive fields of those same neurons were subsequently examined again. The cortical neurons corresponding to the damaged area of the retina had gone silent.

This was only to be expected, of course. What surprised Gilbert and Wiesel, on the other hand, was that the receptive fields of the neurons surrounding the now-silent area had expanded, so as to compensate for the loss. Even more impressive was the fact that this visual counterpart of Merzenich's tactile result was observed to develop within hours of the retinal lesion. The most reasonable rationalization of these findings sees many of a neuron's synaptic connections with its neighbours as being silent, but ready to perform as soon as they are called upon. It is also interesting to note that the deliberately damaged area of the retina was effectively a second optic disc (or blind spot); that compensatory expansion of the receptive fields of the relevant cortical neurons is probably paralleled in the natural optic disc.

Keeping things tidy

The mammalian brain accomplishes a bewildering number of different functions with consummate ease, and one could be forgiven for doubting whether the system described in the preceding sections and chapters would be capable of such variety. It has been emphasized that the number of neurons in the brain is legion, but if they are all doing much the same sort of thing, sophistication would not necessarily follow. There are sources of diversity which we have only briefly touched upon, however. This is an appropriate point at

e / A cortical neuron's receptive field is the extent of sensory space to which it responds. For a neuron in the primary visual area, this field is a (roughly conical) solid angle subtended at the pupil, its cross-sectional angle being a few degrees.

which to take a closer look at them, and see how they mediate the nuanced behaviour that is the higher animal's preserve.

To begin with, there are the myriad spiky protrusions which dot the dendrites of some types of excitatory neuron. I identified them as dendritic spines, but said nothing about their function. When they are present, they are the preferred targets for the terminal regions of the relevant axon branches, which establish synapses with them. It has always been assumed that these structures must serve a useful purpose, but there has been a suspicion that the system might be paying the price of a slightly diminished performance because of the additional electrical resistance implied by the small diameter of the typical spine; such a constriction might be expected to attenuate the signals passing through it. A recent analysis carried out by Donald Perkel and David Perkel has shown that this might not be the case, however. They calculated the spine resistance and found that it could not impose a serious hindrance to signal transmission.[21] The question remains as to what spines are for. There is a growing feeling that their role might indeed be one of restriction, but in the chemical rather than the electrical sense. The point is that molecules involved in the postsynaptic processes that implement synaptic modification would not be able to perform their various functions if they were free to diffuse away into the main body of cytoplasm. If that happened, the changes justified by a local correlative event would become globally influential, and this might not be in the interest of the system; it would not seem to serve discrimination. A spine's function could be to allow a specific synapse to keep to itself the changes that it has mediated.

It would actually be easy to underestimate the advantages which could accrue from such confinement. Even in a system that had recourse to only a single type of neurotransmitter, it would of course give an important crispness to the patterns of synaptic modification that underlie memory. But the mechanism would really come into its own if the dendrites of a given neuron were capable of receiving messages transferred by several different neurotransmitters. There is now ample evidence that this is the situation that normally prevails. Although a particular neuron may be restricted to emitting only a single species of such molecules, its dendrites will usually possess receptors for several different species. It will thus be able to simultaneously accept messages not only from many different neurons of a given class, but also from neurons belonging to several different classes. In such a heterogeneous situation, the localization provided by a spine will clearly be to the system's benefit.

To understand how Nature has managed to imbue the nervous systems of higher animals with such an arsenal of different types of neurotransmitter, we need look no farther than something that has already been mentioned, namely mutation. We have encountered two different classes of glutamate receptor, for example, and described them as being of the NMDA and non-NMDA varieties. In fact, about a dozen variants of the glutamate receptor have now been identified, and the list may not yet be complete. Why so many? Although definitive statements on this question would be premature, it seems likely that mutations are constantly occurring in the genome, and these are naturally passed on to an individual's progeny. In many cases, such modifications are fairly neutral, and they merely add to the proliferation, but every now and then a mutant form of a receptor will be highly beneficial to the organism, and it will give its possessor the sort of competitive edge that provides the basis for Darwin's evolutionary drive. I have not arbitrarily chosen the NMDA-type glutamate receptor as a case in point; there are those who believe that the emergence of this mutation played a critical role in the appearance of consciousness itself. If that was so, the mutation was naturally a signal event, but we will have to suspend judgement on that proposition until we discuss the ins and outs of consciousness. There are several other things to be considered before we can move on to that central issue, and top priority must now be given to a closer look at neural networks.

Caught in the net

Although neuronal assemblies were alluded to in the preceding sections, our discussion has been primarily focussed on events occurring at the level of the single synapse. The point has come at which we ought to step back and take a broader view of synaptic plasticity. We need to know how the myriad changes at the local level conspire to modify the behaviour of the brain as a whole. In turning to that important issue, there is a fact that we ought to bear in mind, namely the distinction between *recognition* and *recall*. As anyone who has played Kim's Game will know, if one is shown a collection of objects, or a list of words, and subsequently asked to name them, it is difficult to recall more than about a dozen of the items. This is in sharp contrast with the fact that our powers of recognition are quite extraordinary.

In 1973, Lionel Standing reported the results of an investigation in which subjects were briefly shown a series of pictures, and later asked to identify them during exposure to a longer sequence which also included pictures they had not seen before.[22] He discovered that most subjects managed remarkably high scores in such recognition tests, even when those tests spanned many hundreds of pictures, and he found that the typical performance remained impressive even when the testing was carried out several days after the initial exposure. The reason why recognition is so much easier than recall is that the former can draw upon the brain's ability to associate. Let us take a look at how it might accomplish this useful feat, and start by drawing an important distinction between the brain and a computer.

The individual instructions contained in a computer program are carried out sequentially, but this does not mean that they are necessarily executed in the order in which they appear in the program listing. Control can be, and frequently is, transferred to a part of the program that lies remote from the current operational instruction. In order for this to be accomplishable, however, it is necessary for that distant instruction to be identifiable, and this is done by the use of a label or address. Likewise, the items contained in the computer's current memory can be accessed only if they are similarly ticketed. It is normal for such an address to consist of a string of numerical digits, just as part of a postal address are the digits that indicate the street number. A street number tells one nothing about what goes on inside the building, and the numbers on the envelope of a letter reveal nothing about the letter's contents.

Such a system of labels works very well in traditional computation, and there is no motivation for changing it. In the context of the brain, however, it is clear that such arbitrary tags would give rise to difficulties, because the cortex would have to find a way of unambiguously relating the label to the operational instruction, or to the item in memory. There is plenty of evidence that this is not the way real memory works. We have all had the experience of trying to recall something which, until it is finally remembered, seems to be *on the tip of the tongue*. It is instructive to consider the steps of such protracted recollection. To begin with, it is especially important to note that one always starts with a cue that is related to the item being sought. It is difficult to think of a situation in which this would not be the case, for what then would be the motivation for racking one's memory in the first place?

The desired item often springs to mind almost at once, of course, but when it does not, there is usually a period during which related things are recalled. Sometimes these contribute to the final recollection, by filling in further contextual details, but they might also take the form of wrong items that are dredged up from memory because of a superficial similarity to the thing being looked for. Even the recollection of these wrong items is enlightening, however, because they invariably bear a resemblance to the correct answer. The act of incorrectly remembering has a strong contextual component too, therefore.

These diverse attributes of recollecting point to the conclusion that we recall by association. This can be put more strongly, because it is difficult to think of a case in which what is being remembered is not simply an association. And this suggests that memory is used only for the storage of associations. This would make our memory system quite different from that employed by a traditional computer, because in the case of the cortex, with its stored associations, items are always pigeon-holed at least in pairs, and sometimes even in triplets or higher multiplicities. There is always a convenient contextual tag, and we could say that items in memory are recoverable through prompting with their constituent parts. This strategy of recollection is embodied in the term given to this type of memory: it is said to be *content addressable*. The difference between remembering through labelling and remembering through association is nicely illustrated by comparing the recovery of one's hat from a cloakroom and from a lost-property office; the former involves submission of a ticket, while the latter requires description.

The real brain's memory is, then, associative and content addressable. This sets a goal for theories of brain function. Any model which fails to emulate such behaviour should be summarily discarded. The converse is not necessarily true, however. Mere compliance with these two requirements is not in itself a guarantee that the corresponding theory faithfully reflects the functioning of the brain. This is an important caveat, and one that has occasionally been ignored. There have been model networks which attracted attention through their powers of association and recall, even though they violated certain biological principles. There is not room to discuss them here. We should rather consider examples which were admittedly rudimentary, but nevertheless biologically plausible and enlightening.

The first of these was conceived by Karl Steinbuch,[23] in 1961, and he called it *Die Lernmatrix* – the learning matrix. Although he took his inspiration from biology, Steinbuch was just as interested in the hardware implementation of his network, which was essentially a system of crossing conductors with adaptive junctions between them. The neural equivalent would be a series of axons crossing a series of dendrites, the two sets interacting via modifiable synapses. Steinbuch's network therefore had the form shown in figure 5.3, the symbols being those that were introduced in the preceding chapter. The small triangles have been given different sizes, also as before, in order to indicate that the corresponding synapses have different strengths (although for ease of drawing, only two different sizes have been used – to denote weak and strong). We note that all these synapses are excitatory, so they must emanate from excitatory neurons, which are off the diagram, to the left. The dendrites are similarly parts of excitatory neurons, and the entire system is thus without inhibition. I have used Steinbuch's original nomenclature, with the input and output axons labelled E and B, respectively. These letters stand for the German words *eigenschaften* (attributes, characteristics or properties) and *bedeutungen* (meanings, responses or outcomes), and he saw his network as implementing classical conditioning.

The learning mechanism was quite straightforward. An input signal was presented to the E axons, its form being characterized by the pattern of activity in those conduits, and the B axons were simultaneously forced to have the desired output pattern. The synaptic strengths were then adjusted according to the rule that they were to be increased if both presynaptic and postsy-

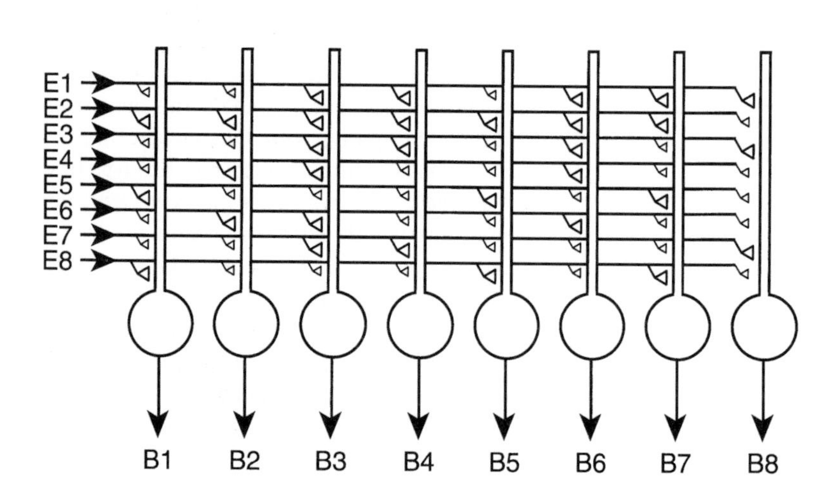

Figure 5.3
A simple associative network of the type described by Karl Steinbuch in 1961.

naptic neurons were active, and decreased if the postsynaptic neuron was inactive when the presynaptic neuron was active. No change was made if the presynaptic neuron was inactive. The amount of the adjustment was not critical, because the network's ultimate behaviour was subsequently controlled by adjusting the thresholds of the output neurons. This first association having been impressed upon the network, the set of E axons were presented with a new input signal, which was forced to associate with a new output signal, the synaptic strengths being re-adjusted according to the previously used prescription. The procedure was repeated for as many times as there were input–output associations to be captured by the network.

We note that this network, despite its simplicity, stores memories in the manner we would expect, in view of what has been discussed previously. And it is particularly noteworthy that the memories are superimposed upon one another, a given synapse participating in the storage of several individual memories. Anyone confronted with this network for the first time could be forgiven for believing that the memories would have become inextricably mixed up by this learning procedure, but this is not the case. Recall is achieved by simply presenting a chosen member of the set of input patterns to the E axons, this time without simultaneously forcing an output pattern on the B set. The network obliges by nevertheless reconstructing the desired output pattern at the B axons. The system naturally has its limitations. Although we need not go into the details here, it is clear that the number of patterns that can be stored in the network is dictated by the number of E and B axons. The larger these numbers, the greater is the network's capacity.

Because Steinbuch was primarily concerned with hardware and problems of technical interest, his original network was designed with electrical components, and it possessed a symmetry that would not be present in the brain. His connections were variable resistances, and these could naturally function in both directions. A small part of his configuration is illustrated in figure 5.4, in which the little arrows indicate the possibility of adjustment. When the full network was properly trained, and a learned pattern was presented to the E lines, the correct response appeared at the B lines, just as it would for the biological equivalent. Unlike in the latter case, however, the B

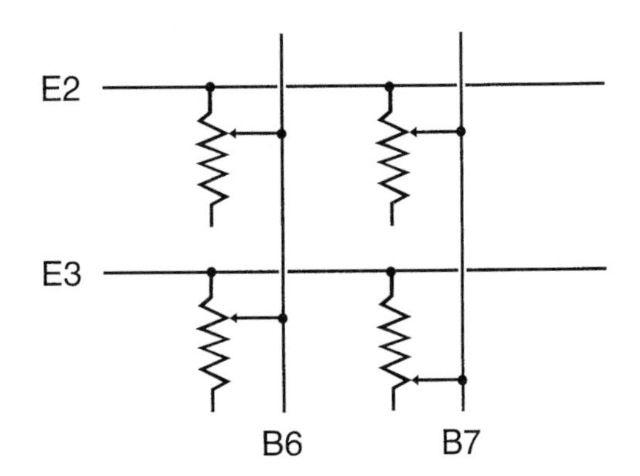

Figure 5.4
Karl Steinbuch's electrical version of an associative network involved variable resistors, which were equivalent to the modifiable synapses of a neural network.

Figure 5.5
The associative neural network discussed by Willshaw, Buneman and Longuet-Higgins.

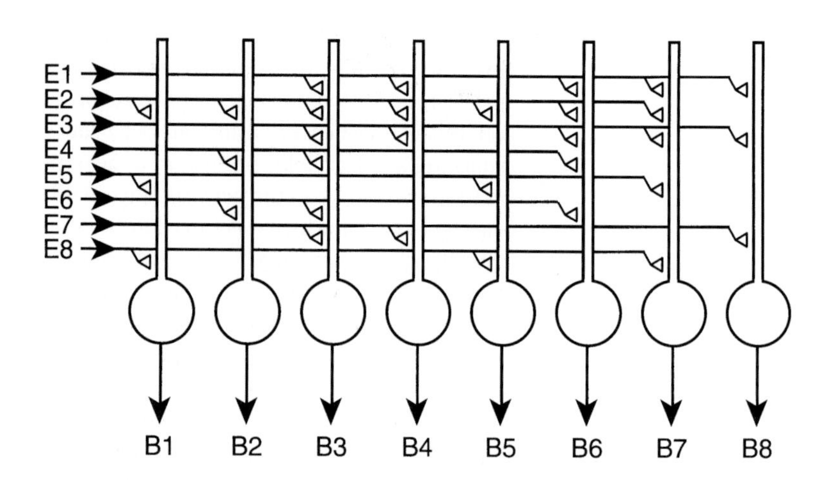

lines could also be used for input, and the correctly associated response would then be observed at the E lines. The lack of such reciprocity in the case of the biological net stems from the fact that a synapse functions in only one direction. Amongst the practical applications that Steinbuch perceived for his network, mention should be made of automatic pattern recognition, speech recognition, machine translation, information storage and retrieval, and chemical analysis by means of spectrograms, the characteristics of which would be recognized by his approach.

I have not yet described the way in which the network actually achieves its aims. The mechanism is slightly more transparent in a variant conceived by David Willshaw, Peter Buneman and Christopher Longuet-Higgins.[24] Their learning rule was even simpler than the one used by Steinbuch. The only event capable of modifying a synapse was simultaneous activity in the presynaptic and postsynaptic neurons, and this led to an increase in synaptic strength by a standard amount. The latter was also the maximum value the strength could adopt, so a given synapse reached its saturation point the moment it first mediated transmission between simultaneously active neurons. And until that happened, the synaptic strength was zero. Figure 5.5 shows the actual example published by Willshaw and his colleagues.[f] All the synapses in this picture have the same strength, and synapses with zero strength have simply been left out. The network has been trained on four sets of paired stimulus and response. In the first of these, lines E1, E3 and E7 were paired with B3, B4 and B8, while the remaining pairings were E2 E5 E8 with B1 B5 B7; E2 E4 E6 with B2 B3 B6; and finally E1 E2 E3 with B4 B6 B7. After the learning phase, all these pairings are captured in the array of synaptic strengths, and they can be retrieved individually and almost without

f / To facilitate comparison, I chose the equivalent synaptic pattern to illustrate the Steinbuch network. The appropriate thresholds would produce identical behaviour in the two circuits.

mutual interference. The word *almost*, here, refers to the fact that there is a single retrieval error. As the reader can readily verify, if the individual signal strengths and synaptic strengths are all taken to be 1, and the thresholds are all assumed to be 3, presentation of the *three*-neuron pattern E1 E2 E3 results in the *four*-neuron response B3 B4 B6 B7. This is the only error, but it does indicate that the network is approaching its maximum capacity, and any attempt to teach it further input–output matchings would lead to a higher rate of mistakes upon retrieval. Indeed, when all possible synaptic connections have been made, the network is maximally ambiguous.

There is something rather peculiar about the way in which these two networks operate. Nothing has been said about how the B patterns are to be established. We know that neurons become active if they have sufficient input, of course, and this is what happens when an E pattern is used as the stimulus, after the training is completed. During the learning phase, however, by what means is a desired B pattern achieved? This would seem to require a second, independent, input. If this too came via several synapses, however, it is not obvious that there would be the required degree of control. The necessary accuracy would be realized if the learning state of each B neuron was dictated by a single master synapse, but such an arrangement is rather remote from biological reality.

This problem with the origin of the B neurons' imprinted states exposes a danger that is always lurking in discussions of the brain's networks: there is a temptation to make these too clever. The difficulty is a fundamental one, for it is all too easy to slip into the trap of requiring that certain neurons possess wisdom. No one nowadays believes what was once regarded as self-evident, namely that there is an inner being in mission control, keeping track of the incoming stimuli and responding accordingly. Such a *homunculus*, as it was called, has long since been exorcised by neuroscience. But its ghost has remained, in the guise of the neural arbiter that knows all the answers. This problem is epitomized by the perceptron, which we will be meeting in the next chapter.

It would be premature to give up on the simple associative network, however, for we have the alternative of equipping each dendrite with an additional *set* of input synapses, rather than merely with a extra one. The entire network would then have the appearance indicated in figure 5.6, with no geometrical significance being attached to the fact that one set of axons enters from the right and one from the left; this has been done simply for clarity. The letter E used to label the incoming axons in Steinbuch's network has been replaced by the letters CS, which denote the stimulus to be classically conditioned. The new axons have been labelled US, indicating unconditioned stimulus, and the R at the output axons stands for response. These changes have been made to emphasize the connection with the biological systems described in the preceding sections of this chapter. In either form, this type of network associates two different stimuli, and it is accordingly said to be *heteroassociative*.

It is important to note that things do not really become more complicated when single B inputs are replaced by multiple US inputs to each of the

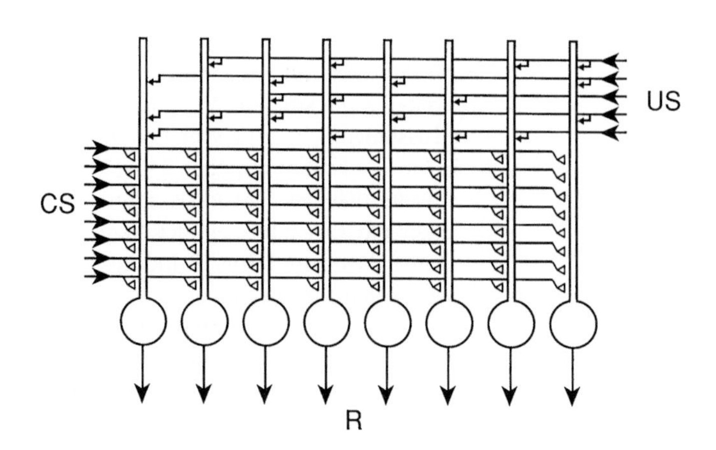

Figure 5.6
An associative neural network in which a group of similar neurons receive inputs from two different sources, one corresponding to an unconditioned stimulus (US) and the other a (to be) conditioned stimulus (CS).

output neurons. If the associative network is well removed from sensory input, its inputs will inevitably be composites that simultaneously embody many of the sensory input's attributes.

The idea behind adding the US axons is to let them dictate the response, and their synapses with the dendrites of the output neurons are assumed to be non-modifiable. Another prerequisite for the appropriate functioning of this augmented network is that the CS axons are not permitted to influence the state of the output neurons during the learning phase. If these dual restrictions did not apply, the situation would become quite confusing, with different combinations of CS and US signals giving the same response. A far more favourable situation would prevail if NMDA-type glutamate receptors were present in the synapses that the CS axons make with the dendrites. The network would then function in the desired manner, provided that the US signals arrived at just the right time to provoke the depolarization needed by the CS axons' neurotransmitters. Subsequent arrival of the CS signals would then launch glutamate molecules at the receptors when the latter were in the appropriate (magnesium unblocked) state. The remaining details would be as before, with the CS signals being able to elicit the appropriate response, unaided, after completion of the learning phase.

There is something quite important that we should note about the networks discussed so far. In every case, the output neurons have merely functioned as passive mediators. The behaviour of the network has always depended upon the nature and timing of the incoming signals. The output neurons have thus been acting in a manner analogous to that of a piece of film, which has no control over the image that will ultimately be imprinted upon it. As we have seen, this does not prevent the network from having the power of association, but the association has been of a rather primitive and limited type. It does not possess something which we have already argued must be present in the brain, namely *content-addressability*. The essence of the latter lies in the way in which fragments of a memory cooperate in recalling the missing pieces, so we see that this is something that requires collective behaviour. And collective behaviour in a neural network demands that the neurons send signals to one another. In a word, it implies *feedback*.

Teuvo Kohonen has analysed networks that incorporate feedback, and he has shown that even relatively simple forms can be content addressable, or *autoassociative*. And although the networks had simple structures, they were nevertheless biologically plausible.[25] The feature that is new, compared with what has been discussed earlier in this section, is a branching of each axon, and we know that this occurs in reality; it is what produces axon collaterals. In the case of the first of Kohonen's networks, the collaterals are of a special type, because they bend back and make synaptic contacts with the dendrites of their own neurons, and with those of neighbouring neurons. In such a case, they are referred to as *recurrent collaterals*. A network that incorporates this additional attribute is illustrated in figure 5.7.

I have deliberately returned to the simple type of one-synapse input, for each neuron, to avoid obscuring the way in which the network goes about its business. Equally sized triangles have been used to denote that the synapses initially all have the same strength. This would be the situation before the network has been trained. Let us now follow the training process,

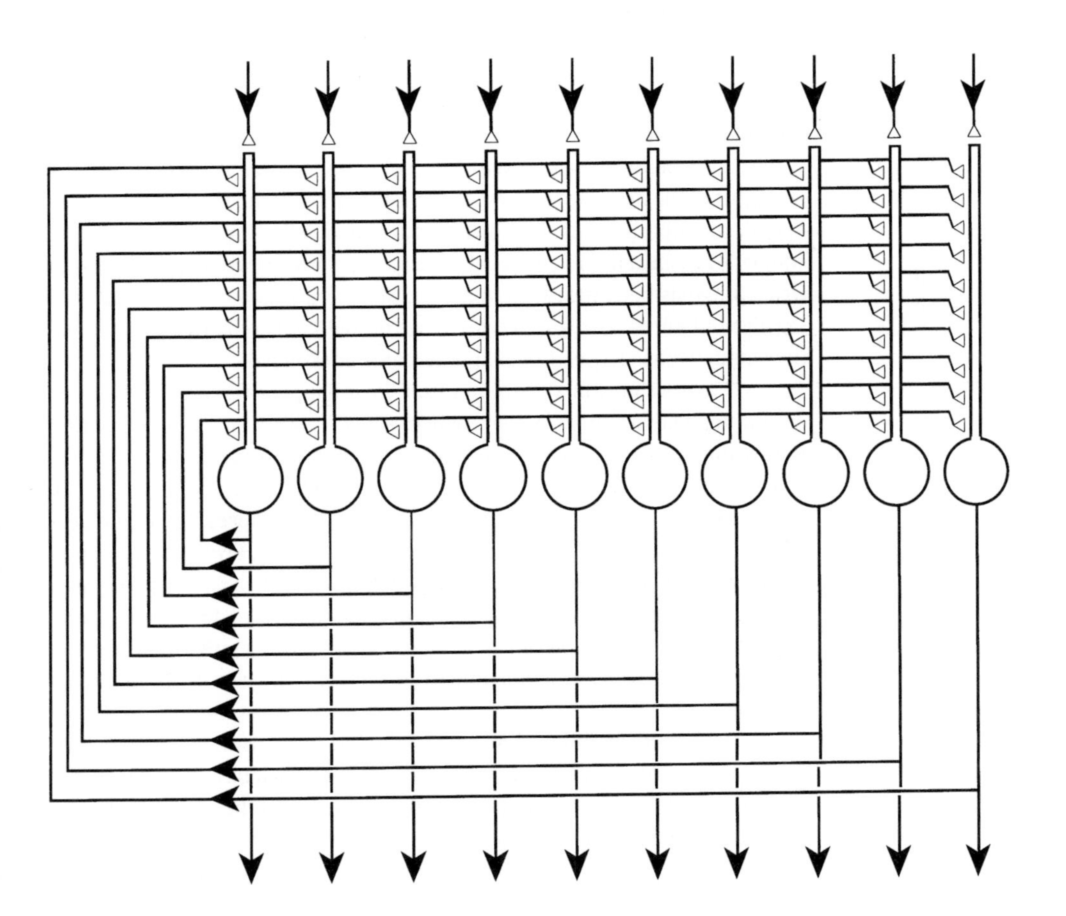

Figure 5.7
An autoassociative neural network. The key feature of the network is the presence of recurrent collaterals.

in more detail than previously. We need some nomenclature to characterize the initial input signals to the various dendrites, which enter at the top via the master synapses. It is convenient to use the digit 1 to indicate activity, and 0 for quiescence.

Let us take the input pattern to be 1010101010, for the ten neurons, reading from the left. These are the activity levels initially injected at the ten master synapses, and the same activity pattern must appear at the ten axons after sufficient time has elapsed. In reality, the time delay would be in the vicinity of five to ten milliseconds. A replica of the input pattern is then transmitted back to the dendrites, via the recurrent collaterals, and the individual signals give rise to neurotransmitter molecules that impinge upon dendritic membranes which have been depolarized within the last five to ten milliseconds. These statements naturally apply only to those parts of the circuit that have actually carried activity. If the neurotransmitters are glutamate molecules, and if the glutamate receptors are of the NMDA type, conditions will be favourable for synaptic modification. Although certain details of the underlying processes are still lacking, let us assume that this mechanism can lead to either strengthening or weakening of the individual synapses. New signals will now have been injected into the dendrites, this time by a different route, so the neurons will be in a position to emit further action potentials, if their thresholds have been exceeded. In practice, however, the signals rapidly lose their power, and the modified circuit falls into inactivity. Its synaptic pattern will now be as shown in figure 5.8.

As before, I have used different sizes of triangle to indicate the strengths of the various synapses, and I would recommend the reader to check that all the modifications have been understood. Where an active input was coupled with an active response, the synapse has been strengthened, whereas weakening has been the result in every case of a mismatch (active input with inactive output; inactive input with active output). Where both the input and the output were inactive, the synapse remains unchanged.

We are now in a position to see what happens when the network is confronted with an input pattern which *merely resembles* the one used during the learning phase. Let us assume that this test pattern is 1010101001. It differs from the original only in the final two digits, which have been interchanged. When the test pattern is presented to the ten master synapses, five of the neurons will be activated, while the others remain quiescent. Five of the recurrent collaterals will be activated, therefore, and they will be in a position to impart new activation to five of the neurons, namely the first, third, fifth, seventh and tenth. We can assume that this new activity arrives sufficiently after the original activity that the latter is not able to contribute to the states of the various neurons.

The activity induced in each of the five neurons in question will be the sum of five contributions, and each of these will itself be the product of the signal strength in the axon collateral and the strength of the synapse. The signal strengths will all be equal, of course, because the signals in the initial pattern were all equal. The synapses, on the other hand, are not all equal. This was the whole point of the training; the network's memory lies in the *non-uniformity* of the synapses. The final thing to be decided is the magnitude of the threshold. Because we have not been putting numerical values to the synapses, the threshold can be regarded as being adjustable, and inspection of figure 5.8 shows that the new output of the network could easily be 1010101010. The ninth neuron will have become activated despite its not receiving any input from the initial pattern. The reason for this is not difficult to find. It arises from the fact that this neuron is nevertheless being fed by four active collaterals, through four strong synapses. That the tenth neuron, conversely, is switched off is attributable to the fact that four of its activity contributions are arriving through weak synapses.

Kohonen's network is thus able to autoassociate, that is to say it will respond with a previously stored pattern even if it is presented with only a fraction of the latter. Just how large that fraction has to be will depend upon the degree of connectivity in the network, but we need not go into such details here. Let us instead end this section by turning briefly to another important issue addressed in Kohonen's investigations. Things to be remembered often take the form of a series of events, rather than a single item. Obvious examples are the succession of tones that make up a melody, and the concatenation of phonemes in speech. In these cases, recall has a temporal aspect, and the question naturally arises as to how the brain achieves this.

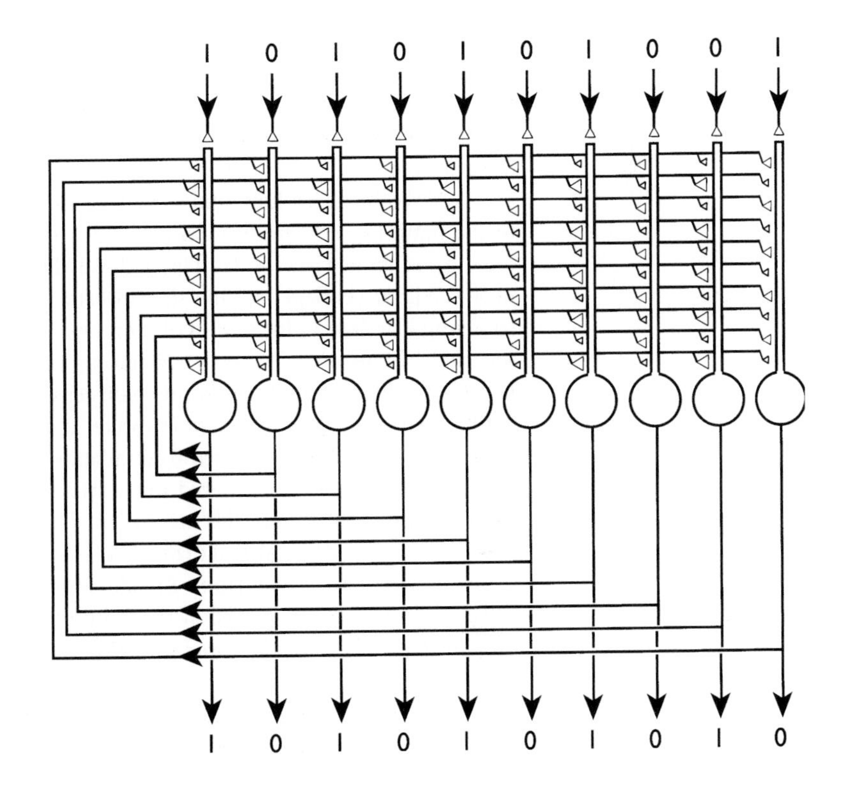

Figure 5.8
An autoassociative neural network in which learning has taken place. The size of each (excitatory) synapse denotes its strength. The numerals indicate activity (1) and quiescence (0).

Kohonen pointed out that networks which incorporate feedback are ideally suited for the storage of *temporal sequences*, and he noted that the time scale over which this storage can be effective is determined by the magnitude of the delay involved in the feedback mechanism. The crux of the matter is that the signals generated by the neurons at a particular time, and fed back by the recurrent collaterals, must act upon the dendrites of some of the neurons in the network while other neurons in the same assembly are in the act of receiving subsequent input signals. If both these sub-groups of neurons have recurrent collaterals, their signals will interact, and the outcome will be a sort of autoassociation in the time domain.

From temporary to permanent

The case for memory being stored in the pattern of synaptic strengths is, as we have seen, a strong one. But this fact alone tells us nothing about the actual location of memories in the brain. There are upwards of a hundred million million synapses. Are they all modifiable by experience, and are memories thus distributed throughout the entire structure? Firm answers to these questions are not yet available, and the situation is complicated by the fact that there appear to be different forms of memory. The distinction between short-term and long-term memory is reasonably familiar, but we are about to discover that this vital faculty comes in a variety of other, more esoteric, forms. There is growing evidence that these are mediated by different regions of the brain, and progress is also being made concerning the way in which they influence one another. Memory processes involving the temporal lobes have been studied with particular vigour in recent years, and a brief description of what has been achieved will show how this type of physiological cartography is carried out.

We have seen how studies of patients with lesions to different parts of the brain led to identification of various faculties with specific cortical regions, prominent examples being those that serve vision, hearing and touch. Wilder Penfield was the first investigator to accomplish the corresponding localization for a form of memory.[26] During the 1940s, he probed the exposed cortices of patients undergoing corrective surgery for epilepsy. Because the brain itself possesses no pain sensors, Penfield was able to carry out his tests while the patients were awake, and they were thus able to report their reactions to the mild electrical stimulation that he applied to various cortical regions. Penfield discovered that excitation of either of the temporal lobes elicited recall of episodes in a patient's past, as if a tape recording was being replayed, and he referred to these areas as the *experiential centres*.

About a decade later, Brenda Milner[27] obtained more specific evidence of the key role played by the temporal lobes, through observations on a male patient in his late twenties, H.M.,[g] who had been suffering from

g / It is customary to refer to patients only by their initials, so as to preserve their anonymity.

crippling seizures in that part of the cortex. H.M.'s condition was considerably ameliorated by removal of the medial portions of his temporal lobes, but it soon became apparent that he was now the victim of a serious memory defect. His ability to recall events that occurred prior to the operation was clearly unimpaired, but Milner discovered that he had lost the ability to store new long-term memories. His general mental faculties had not been diminished, and he was still able to form short-term memories. What H.M. now lacked was the mechanism by which the record of recent events is transferred into long-term memory. His social interaction with the doctors and nurses was quite normal, for example, but he could not remember them from one day to another.

Milner's further investigations of H.M., together with those carried out by Elizabeth Warrington and others undertaken by Lawrence Weiskrantz, all on patients with similar lesions, added a subtlety to the original findings.[28] These invalids *are* still able to acquire long-term memory, but only of the somewhat restricted type that we encountered earlier in this chapter when we were considering the classical conditioning of *Aplysia*, that is to say sensitization and habituation; they can still learn new motor skills. These studies had revealed that there are at least two different forms of long-term memory. The type exemplified by H.M's deficit requires the participation of consciousness, and it had now been shown that there is a second form that is unconscious and altogether more automatic. Thanks to subsequent work by Neal Cohen and Larry Squire, and also by Daniel Schacter, the distinction was sharpened and the two memory mechanisms were given the labels *declarative* (or *explicit*), for the conscious form, and *procedural* (or *implicit*) for the type that does not require conscious participation.[29]

Procedural memories are acquired slowly, through repetition, and are often produced through the application of sequences of stimuli. They influence the type of performance that is often referred to as being mindless: routine sets of motions in which conscious involvement can actually be a hindrance. Declarative memory is quite different. It is formed rapidly, frequently by just a single exposure to a situation, and it requires the participation of consciousness. It invariably involves simultaneous stimuli, which have to be evaluated rapidly and put into the context of what has been experienced previously. This clearly draws on the ability to associate. From what has been described, regarding the observations on H.M. and similar patients, declarative memory is predicated on an intact medial region of the temporal lobe, whereas this is not the case for the procedural form. The memory mechanism studied in *Aplysia* by Kandel and his colleagues was procedural.

More recently, it has proved possible to home in on the actual structure within the temporal lobe that appears to be critical in the special form of amnesia displayed by patients like H.M. Studies of humans, monkeys and rats, carried out by Mortimer Mishkin, and also by Squire, together with David Amaral and Stuart Zola-Morgan, showed that the region particularly susceptible in this respect is the *hippocampus* (see figure 5.9). Moreover, because the impairment involves only the transfer of items from short-term to long-term

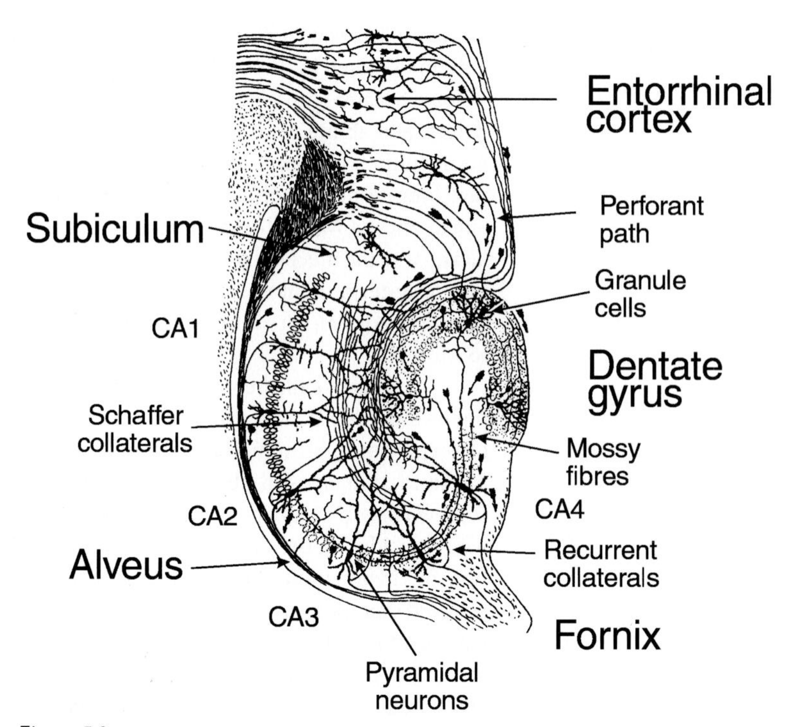

Figure 5.9
The hippocampus is essential for the storage of new declarative memories, whereas items already present in long-term memories can be recalled even when the hippocampus has been damaged or surgically removed. A declarative memory is formed rapidly, frequently by just a single exposure to a situation, and it requires the participation of consciousness. Procedural memory develops slowly, through repetition, and it does not require consciousness; it is the type of memory observed in *Aplysia.* The picture of the hippocampus on this page was produced by Santiago Ramón y Cajal, almost a hundred years ago, but explanatory legends have been added. The highly schematic diagram on the opposite page shows the positions of the main groups of neurons, only three cells being shown at each location, for simplicity. In each case, the hippocampus is depicted in cross section; its long axis lies at right angles to the plane of the paper. It is noteworthy that the connections between the different areas form a closed circuit, which runs from the entorhinal cortex, to the dentate gyrus, to the CA3 area, to the CA1 area, and finally back to the entorhinal cortex via the subiculum. But the circuit is complicated because most of the areas send onward projections to more than one of the following areas. (Connections from CA3 on one side of the brain to CA1 on the other side, via the fimbria fornicus, are via the commisural fibres.) Declarative memory invariably involves simultaneous stimuli, which have to be evaluated rapidly and put into the context of what has been experienced previously. This clearly draws on the ability to associate. The hippocampus appears to be able to influence many of the cerebral cortex's association areas, including the visual, auditory, parietal and frontal association cortices, with all of which it shares reciprocal projections.

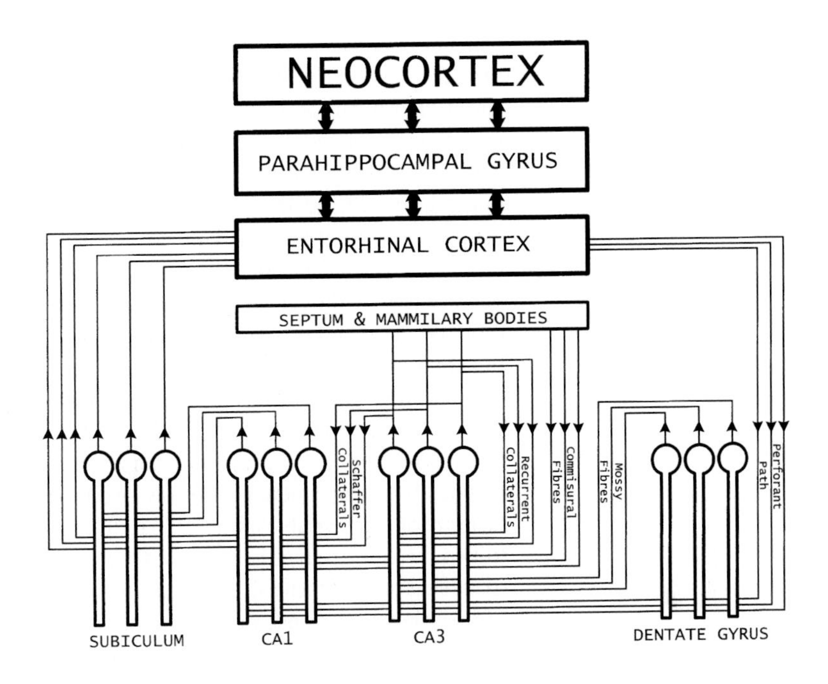

memory, the task of the hippocampus appears to consist of processing experiences and of subsequently exporting a suitable record of them to other parts of the cortex.

An investigation made by Squire and Amaral, in collaboration with Arthur Shimamura, was even more specific; it pinpointed the source of the trouble to what is known as the CA1 region of the hippocampus.[30] A patient R.B. had become amnesic (at the age of 52) in 1978, as a result of ischemia[h] that arose as a complication during open-heart surgery. When R.B. died five years later, of congestive heart failure, the researchers received the consent and encouragement of his family to minutely examine the brain, just a few hours after death. Their main finding was that both CA1 regions (i.e. those lying on either side of the brain) were totally devoid of neurons, the depletion extending along the entire rostral–caudal axis of the hippocampus in each case.

To put these observations in proper perspective, we ought to pause here and consider what makes the hippocampus such a special structure. It is sometimes depicted as if its two parts had a worm-like shape, but this is an artefact of the understandable desire to differentiate it from the rest of the cortex. In fact, the hippocampus lies along the edge of a centrally located region known as the entorhinal cortex, and the latter exchanges reciprocal nerve-fibre connections with many other cortical areas, including the orbitofrontal cortex, the perirrhinal cortex and the parahippocampal cortex, all of these

h / Ischemia is a shortage of blood in a region of the body.

being polysensory associative regions. Its edge location makes a special structural demand on the hippocampus: its neurons must either represent a dead end or they must return signals back to the cortex. It is naturally the second option which is adopted, and the return route actually prescribes a loop, the signals travelling from the entorhinal cortex through the dentate gyrus, onward, in turn, to the CA3 and CA1 regions, and ultimately back to the entorhinal cortex via what is known as the subicular complex.[i]

We need to consider just one more level of anatomical detail to appreciate that the job of transferring short-term memories to somewhere more permanent has been entrusted to a structure of substantial sophistication. Study of the neural connections in the hippocampus proper (also known as Ammon's horn, from the Latin *cornu Ammon* or ram's horn, which it resembles) dates from the pioneering efforts of Santiago Ramón y Cajal, at the end of the nineteenth century,[31] and the present nomenclature is that recommended by a subsequent prominent investigator, Rafael Lorente de Nó. Two facts are of considerable interest. First, the axons of the CA3 neurons give rise to two different sets of collaterals. One set, known as the *Schaffer collaterals*, make synaptic contacts with the dendrites of the CA1 neurons, while members of the other set synapse onto the dendrites of the CA3 neurons themselves, that is to say that they are *recurrent collaterals*.[j] As Edmund Rolls[32] has emphasized, the presence of the latter arrangement means that the CA3 neurons can function as an autoassociative neural network, of the type due to Kohonen which we considered earlier in this chapter.

The second thing of special interest has been discussed by Thomas Brown and Anthony Zador, who drew particularly on the prior work of Sam Deadwyler, Robert Hampson, Thomas Foster and Gregory Marlow.[33] It is that the pattern of connections which link the various points on the route from the entorhinal cortex, through the hippocampus, and back to the entorhinal cortex do not merely join adjacent regions. Far from it; several of the regions project not only to the next region but also to one or two regions after that. The entorhinal cortex, for example, projects to the dentate gyrus, CA3 and CA1, while CA3 projects to CA1, the subiculum and the entorhinal cortex. Should we be surprised by such complexity? Probably not, when we consider what an important function the hippocampus fulfills.

In attempting to fathom the workings of this fascinating piece of neural circuitry, we will benefit from two other pieces of information. The first concerns the distribution within the hippocampus of NMDA-type glutamate receptors. They are profusely present in the CA1 region, while lesser amounts are seen in CA3. The pathway between the dentate gyrus and CA3, on

i / The CA2 and CA4 regions are now regarded as having no separate significance.

j / Some diagrams of the hippocampus, such as the one shown in figure 5.9, do not distinguish between these two types of axonal division, the recurrent collaterals being drawn as if their continuations simply become Schaffer collaterals. The distinction may be insignificant.

the other hand, appears to be almost free of these receptors. The second suggestive fact is that rhythmic behaviour can easily be detected in the hippocampus. Its frequency lies in what is known as the *theta band*, which spans the range from 5 to 12 cycles per second. The amplitude of this oscillatory activity is highest in the CA1 region, but it is also clearly detectable in the dentate gyrus. Indeed, it is present in the entorhinal cortex too.

The dual appearance of NMDA-type receptors and theta activity in the CA1 region is probably not fortuitous. We earlier noted that the ion channel through the NMDA-type receptor remains open for almost a fifth of a second, when a proper correlation has triggered it, and this duration is comparable to the time interval between the theta pulses. The latter are believed to be generated in a region outside the hippocampus (possibly in a region called the septum, which need not concern us here), but there appears to have been an evolutionary advantage in the NMDA-type receptors being present in the CA1 region, in order to exploit this appropriately paced rhythm. The upshot could be the capturing of a further correlation. If this is the case, the NMDA-type receptor would then be effectively relating three different events.

Just what these might be brings us back to that unusual multiple connectivity. I believe that it mediates something very profound, something which I tend to think of as a sort of *temporal integration*. Consider, for example, the two different routes from the entorhinal cortex to the CA3 region. One passes through the dentate gyrus whereas the other is direct. Because the passage through the dentate gyrus will consume a certain amount of time (as it would through any group of neurons, indeed), this means that there will be two sets of signals arriving *simultaneously* at the CA3 region which left the entorhinal cortex at *different* times. The temporal difference will be due almost entirely to the time it takes signals to pass along the dendrites of the dentate gyrus neurons, and it will be roughly 10 milliseconds. A similar temporal integration will occur at the CA1 region, and indeed at all other stations on the looped route through the hippocampus.

What could be the purpose of a mechanism which seems to militate in this manner against the concept of good timekeeping? I believe that it could be the brain's way of ensuring that only signals persisting for several tens of milliseconds can exert an influence. We have already seen that NMDA-type receptors can capture correlations. The temporal integration might be a way of stringing such correlations together into a continuous succession; it could provide a sort of glue in the time domain. Perhaps this is what is required to ensure that the temporal sequence in an episodic memory survives the transfer from short-term to long-term memory. When we hum the succession of notes that constitutes a familiar melody, we may be taking advantage of a sophisticated mechanism that involves NMDA-type receptors, the theta rhythm, and the peculiar wiring of the hippocampal circuit. We cannot rule out that the mechanism is essentially the relatively simple one proposed by Kohonen for autoassociation in the time domain, which we considered earlier in this chapter, but I suspect that something more subtle, and more flexible, is going on.

There is a final point that ought to be made, before we take a temporary leave of the hippocampus. Despite its obvious importance, this brain structure cannot be the seat of consciousness, because patients such as H.M. and R.B. still possessed that faculty despite their debilitating lesions. In fact, their cases demonstrate that the hippocampus does not even play an essential role in consciousness. Where, then, does consciousness reside? Recent investigations of another region, which lies rostral to the hippocampus, indicate that it might be a good candidate. It is the frontal lobe. Let us continue by considering the evidence.

At the blackboard

The key step in mental arithmetic is retention of the result of previous operations while a new one is being carried out. When adding a string of numbers, for example, one has to remember the current sum while adding to it the next number in the sequence. The key step in the use of verbal language is retention of what has just been spoken while suitable new words can be selected and arranged for one's continuation. These are just two examples of the effortless procedure that is going on during much of our waking time: mental retention of items while new ones are either received through our senses or retrieved from memory. The mechanism that enables us to carry out this obviously important process is now referred to as *working memory*. It also involves erasure, in that items need not be held on to beyond their immediate usefulness. In mental addition, the current sum ceases to be of interest when a new number has been added to it, and we have no need to remember every word we use during conversation. It is for this reason that Marcel Just and Patricia Carpenter made an appropriate choice when they referred to working memory as *the blackboard of the mind*.[34]

Evidence that has been accumulating in recent years suggests that working memory requires the participation of the *frontal lobes*. The main source of this information has been observations on people with injury to that part of the cortex. Phineas Gage, whose sad acquaintance we made in chapter 3, was perhaps the most famous sufferer of general damage to the frontal region, and we noted that it totally changed his personality. But personality is a rather diffuse attribute; we must consider the impact of well-defined lesions on mental processes that are easier to characterize.

One of the techniques most frequently used in the study of working memory is known as the *delayed-response test*. The subjects have often been monkeys, because they appear to possess working memories that are reasonably similar to our own; they suffer similar deficits when they have sustained damage to their frontal lobes. The essence of the delayed-response test is that it gauges the animal's behaviour in a situation requiring reaction to something experienced in the recent past. In a typical experiment of this kind, a monkey is first trained to give priority to a centrally located spot on a fluorescent screen, and to fix its gaze on this irrespective of what other objects are displayed elsewhere. The monkey then fixates on the spot while a single target object, a cross for example, is briefly flashed in one of the corners. After a further time lapse of several seconds, the central spot is extinguished and the investigator checks to see if the monkey then shifts its gaze to where the target had appeared. It will do so only if it has been able to retain an internal representation of the target's location.

The part of the frontal lobe known as the prefrontal cortex clearly plays a part in this ability because monkeys with lesions in that area of the brain cannot remember where the target appeared. As Patricia Goldman-Rakic[35] has put it, *out of sight* for these animals means *out of mind*. The ability is also lacking in infant children and in monkeys below the age of two months, and it is satisfying to note that synaptic development in the monkey's prefrontal cortex is most rapid between the second and fourth months of life. Joaquin Fuster, and also Kisou Kubota and Hiriaki Niki have investigated activity in the prefrontal area at the neuronal level, during delayed-response testing. They found that different neurons respond at different phases; some become active as soon as the target is presented, while others are active during the few-second delay period, and those of a third group remain dormant until the monkey initiates its eye-shifting response.

Goldman-Rakic, Shintaro Funahashi and Charles Bruce, have shown that each neuron in the second of these three groups has what they call a *memory field*: it reacts during the delay period only if the target happened to fall within a certain visual area relative to the centre of gaze.[36] Collectively, the neurons in that second group look after the spatial aspect of working memory. These neurons lie in the vicinity of a region known as the *principal sulcus*. It was subsequently demonstrated, by Fraser Wilson and James Skelly, that there is another area somewhat below the principal sulcus in which the neurons respond to the attributes of objects rather than to their spatial positions.[37] Examples of such features are shape and colour. For either set of neurons, it is the case that initiation of activity follows automatically when the target has been briefly exposed, the only exception being in the event of a sufficiently strong distraction.

The question naturally arises as to what these neurons are accomplishing with their activity, as they bridge the interval between target stimulus and motor response. The answer appears to lie in the interactions between the principal sulcal and other brain regions. It had been established anatomically that the principal sulcus is reciprocally connected to several other areas, prominent amongst which are the major sensory regions, the premotor cortex, and the hippocampus. Moreover, Goldman-Rakic and Harriet Friedman have used the 2-deoxyglucose technique to demonstrate simultaneous activity in the principal sulcus and the hippocampus during a delayed-response test, with its attendant demands on the ability to rapidly update information. Just as importantly, they showed that the neuronal activity level in those same regions is much lower when the monkey is merely being required to associate.

One of the sensory regions discovered to be implicated in this shared activity is the *parietal lobe*. This is significant because that area is known to make a decisive contribution to the general sense of awareness. A person who has sustained damage to the parietal cortex in one of the cerebral hemispheres ignores the opposite side of the body; half the hair goes uncombed and, in the case of a male, half the beard goes unshaved. The involvement of the hippocampus is equally intriguing, given the structure and possible function of that structure, as was described earlier in this chapter. Goldman-Rakic and her colleagues believe that the hippocampus consolidates new associations, whereas the prefrontal cortex somehow retrieves the products of previous associations from elsewhere in the cortex. What is still lacking, however, is a detailed picture of how the activities in the various areas, and especially the interactions between them, conspire to produce the faculty that working memory provides. The fact that the mechanism involves both the brief retention of things present and the rapid retrieval of things past is suggestive of something fundamental. It would seem to touch on the fascinating issue of how one *knows*.

It appears to be possible that *everything* required to characterize observable features of the external world is served by one or another region of the prefrontal cortex. We have already seen that there are regions which variously handle position, shape and colour. Robert Knight has located another which looks after auditory stimuli.[38] And he has investigated the electroencephalographic (EEG) responses of patients with prefrontal lobe lesions to what is sometimes referred to as *the oddball paradigm*. The subject is exposed to a regular series of sounds, such as a succession of clicks, and every so often one of these is substituted by something unexpected. The latter might be a click of a different tone, or it might even be the omission of a click altogether. When normal people are subjected to such a disturbed series, their EEG traces show a characteristic surge in amplitude at around 300 milliseconds. The corresponding traces for patients with prefrontal damage have a much smaller amplitude around the 300 millisecond mark, and in some cases this part of the trace is missing altogether.[k]

Structural damage to a cortical area naturally represents a rather gross disruption of the system. More subtle disturbances occur when the trouble lies at the biochemical level. Goldman-Rakic and her colleagues have established that the prefrontal cortices of monkeys are rich in dopamine, which is a prominent member of the catecholamine family of neurotransmitters. It has transpired that upsetting the distribution of dopamine in the prefrontal cortex impairs working memory, in much the same way as is observed with physical lesions. In aging monkeys, such an imbalance occurs naturally, and poor performance in delayed-response tests is one of the consequences of advancing years. If these senior animals are given dopamine injections, their abilities are restored to the level seen in young monkeys.

It is interesting to note that dopamine imbalance had previously been linked to schizophrenia, and it has been established that malfunctioning

k / A strongly truncated amplitude around the 300 millisecond mark is also observed in many autistic patients. This may indicate that autistic people suffer from a deficit in their working memories.

of the prefrontal cortex is the cause of that affliction. It is especially noteworthy that the typical characteristics of schizophrenia, such as thought disorders, inappropriate or truncated emotional responses, lack of initiative, reduced attention span, and an inability to plan ahead, bear a striking resemblance to the abnormalities seen in patients with damage to the prefrontal region. Schizoid patients can manage routine procedures, and their reflexes are intact, but they perform poorly in situations requiring verbal information or symbol manipulation. It was consequently not surprising when Sohee Park and Philip Holzman demonstrated that they also do rather badly on delayed-response tests.[39]

Schizophrenic patients have been subjected to a particularly revealing type of test, namely a measurement of their eye-tracking ability. It has been discovered that they are quite bad at anticipating the projected trajectories of moving objects. This too is rather suggestive in that the eye-movement neuronal groups involved in such predictive tracking are located in the posterior region of the prefrontal cortex. Indeed, Charles Bruce and Martha MacAvoy have shown that the same deficit is present in monkeys with damage to that cortical area.[40] In view of the fact that these various pieces of evidence are in such harmony, Patricia Goldman-Rakic has suggested that the time is ripe for a change of view regarding the schizophrenic syndrome.[41] She prefers to see the impairment as a failure of the link between representational knowledge and behaviour. The neural connections in the prefrontal region are continually updating our inner models of reality, because of the constantly changing circumstances in which we find ourselves. Those connections serve short-term memory, and thereby guide behaviour. If they fail in this task, perception becomes fragmented and a continuous train of events is seen as a series of snapshots. This produces the familiar symptoms of schizophrenia because behaviour is then dominated by the impulses of the moment, rather than by the evened-out influences of past and current information.

Let us close this chapter by taking a step or two backwards, to get a broader view of what has been our central theme, namely memory. We have been confronted with a series of antithetic pairs. First there was short-term memory, as opposed to long-term memory. Then there was declarative (explicit) memory, as distinct from procedural (implicit) memory. And we have just been considering working memory, which presumably has its counterpart in what could be called storage memory. Larry Squire has further discussed semantic memory, which contrasts with episodic memory, and there are also active memory and passive memory.[42] It is still too early to say whether all these types really are different, and we do not know whether there are still other forms waiting to be discovered. It will be exciting when a proper synthesis becomes possible.

And what about consciousness? We have taken a close look at two important structures, as well as their interactions with other regions of the brain, but it could not be claimed that we have been able to identify consciousness with either of them. For despite the severity of their situations, H.M. and R.B. still possessed consciousness although they lacked the hippocampus.

This conclusion seems inescapable, given that these two patients were able to converse with the people who psychologically examined them. Similarly, the prefrontal cortex cannot be the repository of consciousness because Phineas Gage, and all those similarly afflicted, have retained this faculty even though they had lost cortical tissue in that area. And schizophrenics, with their equally serious biochemical deficits in that region, are nevertheless conscious when they are not asleep. If we are searching for a Holy Grail of consciousness in some specific location in the brain, therefore, we have so far drawn a blank. It is true that we have yet to take a detailed look at the other cortical areas, but one could be forgiven for gaining the impression that consciousness cannot be pinpointed in that manner. Perhaps it is more spread out, with no single brain region carrying the sole responsibility for this supreme attribute. It is to such a distributed type of processing, therefore, that we must turn in the chapter that follows. Before doing that, however, we should take a brief look at an aspect of organization in neural networks. It is related to something that has been much in focus of late, namely complexity. In the context of what we are discussing here, it touches on the vital question of how order is salvaged from potential chaos.

Getting organized

We have been considering the mechanisms whereby memories can be stored in the brain's neural networks. We have also touched on the question of how the different types of memory task are distributed amongst the cortical and sub-cortical regions. The discussion was couched only in the most general terms, however, nothing being said about how the various items stored in a particular form of memory are actually shared between the myriad synapses. This omission has not been an oversight, for the fact is that this degree of detail is still inaccessible to neuroscience; Lashley's engram continues to be elusive. There have nevertheless been several attractive suggestions as to how efficient pigeon-holing could be achieved, and one proposed by Teuvo Kohonen has enjoyed considerable support.[43] His analysis is fairly mathematical, as are many analyses of cortical function, but I will try to put across the underlying idea by starting with a simple analogy.

There is a conundrum that runs as follows. If you encircle a number of cigarettes with the clenched thumb and fingers of one hand and tightly squeeze them, the cigarettes all lying parallel and roughly at right angles to the fingers, what pattern will they form? The answer is that they will end up as a close-packed array of hexagons, not unlike a honeycomb. Such a pattern is dictated by simple geometry, and by the fact that the cigarettes cannot interpenetrate. The latter fact, in turn, is ultimately related to the action of one of Nature's four fundamental forces, namely the electromagnetic force, which governs the interactions between atoms. At the microscopic level indeed, if we replace the cigarettes by atoms, similar considerations dictate that the atoms in crystals are arranged in patterns just as regular as that of the honeycomb. But things are more complicated at that level, because the symmetries and strengths of the forces vary from one element to another; crystals of different chemical compositions will often display different symmetries.

If one deforms a crystal by an amount not exceeding what is known as the elastic limit, the atoms will spring back to their appointed places when the deforming stress is removed, and the crystal will thereby regain its original shape. One could say that the atoms remember where they ought to be, this memory clearly residing in the forces through which they interact with one another. There is no limit on the time during which the forces are operative, so the memory of shape can be regarding as being essentially permanent.[l]

The neurons in the cortex are not free to move around, but they do have interactions which, once established, are also reasonably permanent. These interactions are of course those mediated by the synapses, and they can be regarded as storing knowledge of arrangements reasonably analogous to those stored by their interatomic counterparts.[m] Kohonen realized that the arrangements achieved by synapses can apply to anything storable by memory. One can look upon his idea as replacement of geometrical shape by a more generalized form, and such replacement is often referred to by the term *mapping*. We are of course familiar with one usage of this concept, namely the representation of the curved surface of our spherical Earth by the flat type of picture that one sees in an atlas. We also readily accept the distortions familiar in such mapping. In the simplest projection, Greenland looks huge, and the poles are transformed from points to lines that are just as long as the equator. This is usually looked upon as a small price to pay for the convenience of a flat representation, the compensating advantage being the preservation of continuity; places adjacent to each other on the Earth's surface are also adjacent to each other on the map, which thereby becomes a reliable aid to navigation.

The conceptual leap achieved by Kohonen was the realization that a neural network can be used for such mapping even when the thing being mapped has no geometrical significance (see figure 5.10). The one requirement is that the elements of this thing have mutual and unequivocal relationships that are just as fundamental as those that applied to points on the Earth in the above example. The essence of Kohonen's type of mapping is that the relationships between those elements are reflected by the synaptic relationships between the neurons which were representing them. And one could say that his task lay in finding a mapping which minimized the distortion; he strove for a situation in which elements of the thing being mapped were as close as possible to their conceptual neighbours.

l / I am ignoring the phenomenon of creep, in order to keep the argument simple.

m / The conditions for storage are far more favourable for neurons than they are for atoms, because synapses can be established with other neurons that are quite distant. Atoms 'forget' their original positions if they are displaced through distances greater than the nearest-neighbour distance. Crystals have elastic limits; neural networks do not.

Questions naturally arise as to what produces the actual arrangement of elemental representations in the cerebral cortex, assuming that a mapping really exists of course, and also as to how it can be probed. In the case of our Earth, the counterparts of the former were the work of continental drift and erosion, while the latter was a job for the cartographer. The arrangement of atoms in a crystal is established during solidification from the melt (or vapour) or deposition from solution, and determination of the resulting

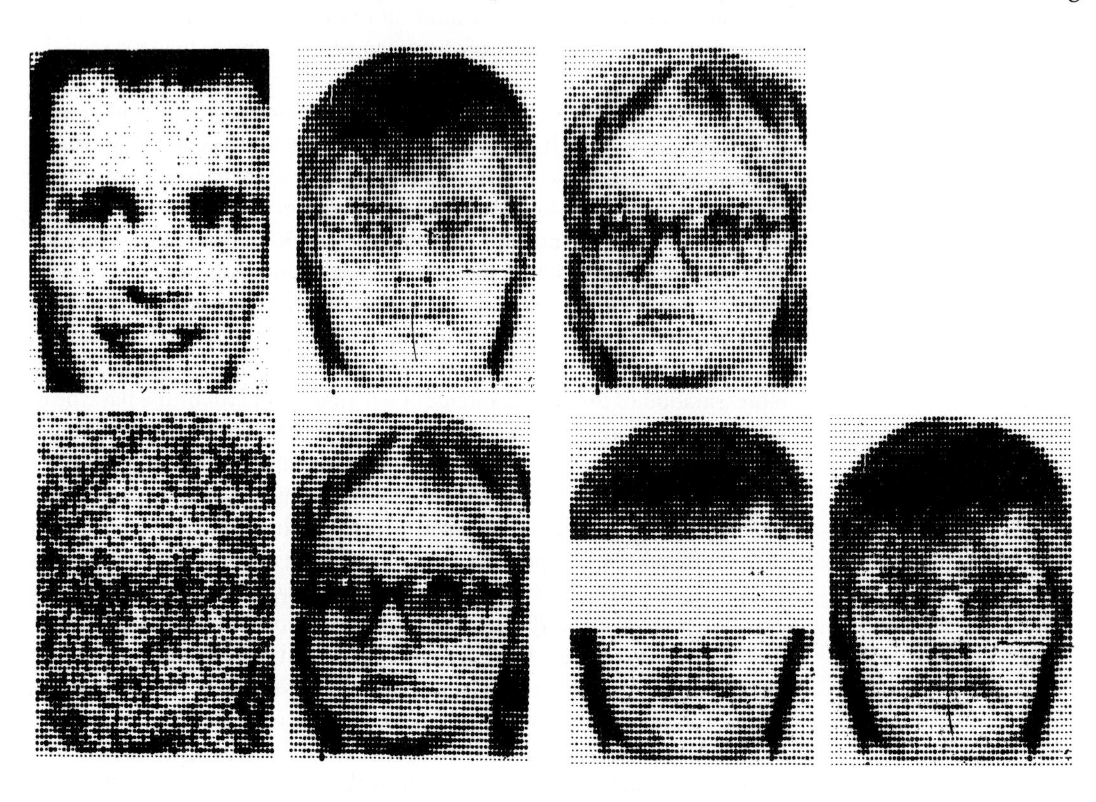

Figure 5.10
Teuvo Kohonen was one of the first to investigate both autoassociation and self-organization in neural networks. In the former process, the individual neurons are reciprocally connected to one another via synapses whose strengths are determined by the memories stored in the network. In the simplest approach, each memory is a pattern of ones and zeros, superimposed on the array of neurons. This defines the incremental changes in the synaptic strengths, synapses joining two active neurons being strengthened and those linking an active neuron with an inactive neuron (or vice-versa) being weakened. Subsequent memories are stored in the network by the same procedure, the new incremental changes simply being added to the current values of synaptic strength. The individual memories are thereby superimposed on one another. The instantaneous situation in the network can be depicted by representing the ones by spots and the zeros by blanks, as in the above example. If the network is exposed to a degraded version of one of the stored patterns (lower set, left end), it is able to recover the original (upper set, right end, and lower set, second from left) provided the degrading was not too severe, but only if the network has not been overloaded with patterns. Similarly, the trained network can recover a full image (upper set, middle, and lower set, right end) if it is presented with only a part of the image (lower set, second from right).

pattern is achieved by X-ray diffraction. In the case of the brain, the pattern is formed by neuronal activity and synaptic modification, and the nature of the map can be explored by physiological techniques. Prominent amongst the latter are the direct measurement of activity using electrodes to monitor changes in voltage of individual neurons. This method was touched upon in chapter 3, and it was the one used to demonstrate that the primary visual cortex indeed maps the composite field visible to the retinas.

In taking a closer look at the formative mechanism, we will again be helped by the atomic analogy. The force between two atoms is zero if the distance between them happens to be what is known as the equilibrium value. If the atoms get closer than this, they repel each other; they are pushed apart toward the equilibrium separation. If they stray farther apart than the equilibrium distance, the force becomes attractive and they are pulled back toward that equilibrium separation. Amplitude of atomic motion increases with rising temperature, and this is why the restraining influence of the interatomic forces can be overcome at sufficiently high temperature, as is the case in melting. Similarly, when a solid has been formed, a controlled heat treatment can nudge malplaced atoms into their appropriate (equilibrium) positions. The metallurgist calls this annealing.

The interactions between neurons likewise fall into two categories. There are excitations and there are inhibitions. But if we loosely equate excitation with attraction and inhibition with repulsion we find that the spatial arrangement of these contributions is just the opposite to that seen in the atomic domain; neurons tend to excite other neurons lying close to them and inhibit those that are more distant.[n] This difference is without significance, however, because the neurons themselves do not move. As I have already emphasized, it is the *connectivity* which provides the basis for the development of the pattern, that is to say of the mapping. Indeed, other spatial distributions of the excitatory and inhibitory components could also serve as the basis of mapping, one with a different spatial pattern admittedly, but a mapping nevertheless.

Kohonen simulated the development of various maps by an approach that was quite analogous to annealing. He replaced the atomic realm's thermal agitation by activation of the participating neurons, and *the statistical correlations between the firing patterns* of the latter was the agency by which the connectivity gradually developed (see figure 5.11). Examples of the maps he produced in this fashion include an arrangement for the phonemes which underlie speech. We do not know whether or not this bears any resemblance to the actual situation. Electrode probing of the type used to reveal the visual map in monkeys has to be ruled out on ethical grounds; one does not stick electrodes into the human brain, of course. If current progress with the

n / If one plots the spatial deployment of synaptic influence, interpreting excitation as positive and inhibition negative, the resultant graph looks like a *Mexican Hat*, which is the term commonly applied to this distribution.

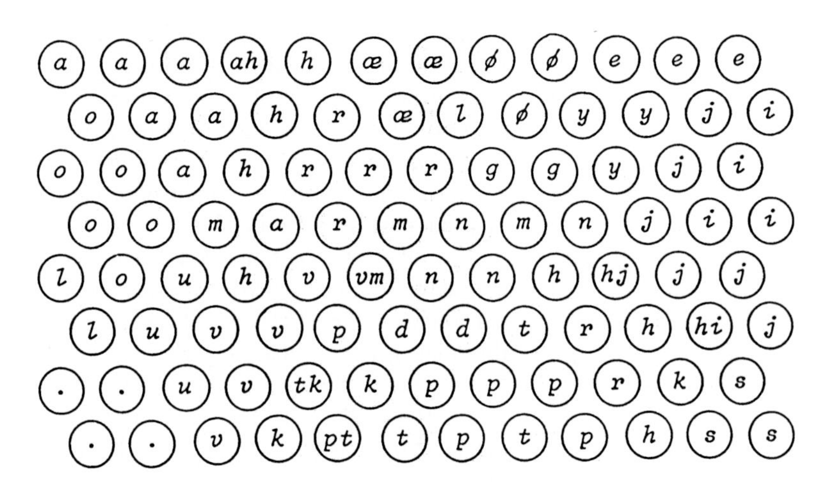

Figure 5.11
In the more complicated case in which the individual items of a memory are not all the same, reliable retrieval of a desired item can be achieved if the entire set of items has been unambiguously associated with an array of neurons. This is accomplished by arbitrarily assigning one item to each neuron and then gradually changing the synaptic strengths while the neurons are activated in patterns that reflect the statistical relationships between the items. This process is known as self-organization, and in the example shown here, a Kohonen network has produced an arrangement of phonemes which reflects their mutual relationships (Kohonen, *Proceedings of the IEEE*, 1990, **78**, 1464–79).

MRI technique is maintained, this may ultimately uncover a phoneme map. It would be difficult to exaggerate the significance of such an event.

There is one final point that should be made in connection with mapping. It concerns the question of dimension. The mapping of the Earth's surface onto the flat page in one's atlas is relatively straightforward because they are both two-dimensional. Difficulties arise when the mapping involves a change in the dimensionality, but they can be overcome. A classic example is the transmission of two-dimensional pictures by the one-dimensional signals of television broadcasts. The trick used in this case is to split the picture up into a series of lines, which are scanned raster-fashion and transmitted in sequence. As a matter of fact, the sequence does not even have to reflect the geometrical arrangement of the picture elements; any sequence will do as long as there is no ambiguity regarding which picture element is being transmitted at which instant. This is the basis of one form of coding, which can be used to prevent television viewing by those who have not paid the appropriate licence.

This brings us to the heart of the dimension issue. It shows that two things having different dimensionalities can always be made compatible provided that a means can be found of relating their elements in an unambiguous manner. In the cortex, as Kohonen realized, this is achieved by an appropriately trained set of synapses. It is this which effectively releases the brain from the apparent bounds imposed by the two-dimensionality of the cortical surface;[o] there is almost no limit to the number of conceptual dimensions that

o / Although the cortex has thickness, there is good reason to regard this as providing sophistication of signal-handling ability, rather than an additional geometrical dimension. The processes occurring at different cortical depths are not equivalent, as will be discussed in chapter 7.

the cortex can manage.[p] It also means, however, that mere inspection of the way elements have been allocated different places on the cortical surface does not give one the whole picture, because information that is potentially just as important resides in the pattern of synaptic strengths.

There is a final point that can be made, since this book also addresses the question of computer architecture. The *Connection Machine*, fabricated by the aptly named Thinking Machines Corporation, exploits these same principles. Its individual units are wired up in a configuration known as a *hypercube*, which is to say a cube in more than three dimensions. This does not confront the designer with a difficult challenge, because one can readily achieve with electrical connections just what the cortex achieves with its synapses. The computer is housed in a cubic box, about a metre along each edge, and its components are laid out in a normal cubic arrangement, that is to say in a simple three-dimensional configuration. One could be forgiven, therefore, for wondering where the *hyper* comes in. The answer lies in the pattern of the connections, and these would have to be carefully examined if one wished to discover the machine's operational dimensionality.

p / In practice, the limit is set by the number of synaptic contacts established by each neuron, but this is many thousand; the scope of the cortex is indeed very large.

6 From percept to concept

In a comparative study of human performance and machine performance, it must be realized that the human being does some things much better than the machine and some things worse. The human system is not as precise nor as quick as a computing machine. On the other hand, the computing machine tends to go to pieces unless all details of its programming are strictly determined. The human being has a great capacity for achieving results while working with imperfect programming. We can do a tremendous amount with vague ideas, but to most existing machines vague ideas are of absolutely no use.
Norbert Wiener (*Nerve, Brain, and Memory Models*[1])

Computers and brains

Brains are often likened to computers and computers are often likened to brains. The front cover of the 7 February 1983 edition of the magazine *Newsweek* bore the title of one of its articles, 'How the Brain Works', and as a subtitle it used 'The Human Computer'. References to electronic brains are an equally familiar feature in the popular press. Are these valid comparisons? There is a certain resemblance, admittedly, since both brains and computers perform the functions of information processing, storage and retrieval. A closer look at their details suggests that the similarity is only a superficial one, however. Let us consider the facts.

Operations in an electronic *digital computer* are carried out in the *binary system*. The state of each of its switching elements, at any instant, is described by the simple choice *on* or *off*. These are represented numerically by the binary digits 1 and 0, and their logical equivalents are *yes* and *no*. A neuron is capable of a continuous set of different responses, as we saw in chapter 4, but because the key to a neuron's participation in physiological function appears to lie in the influence it can exert, it might be the case that the choice essentially boils down to *very active* or *not very active*. It is probably not in this respect, therefore, that computers and brains are so different. The disparity first becomes appreciable when one considers the relevant numbers.

Today's typical large computer comprises about a hundred million transistors in its printed circuits, whereas the neurons in the human brain outnumber this figure by a factor of about a thousand. There are many more neurons in a single brain than there are people on Earth. Moreover, the typical transistor in a computer has connections to only three or four other transistors, and this contrasts sharply with the roughly ten thousand (and in some cases one hundred thousand) synaptic interactions between a single neuron and its brethren. Taking these two factors together, we see that the brain's neurons outstrip a computer's transistors to the tune of about a million times as many inter-unit connections. The degree of interaction in the brain is truly enormous. The advantage lies on the computer's side, however, when it comes to switching rate. A transistor can be flipped from one state to another within about a nanosecond (i.e. a thousand millionth of a second), while it takes about a millisecond to achieve the same thing in a neuron. In this case, therefore, the difference is again a factor of about a million, but it lies on the other side of the ledger; what the computer loses on the roundabout, it makes up for on the swings.

A more significant difference between the computer and the brain concerns the manner in which the individual processing steps are carried out. Despite the recent development of massively parallel processing in computers, these machines nevertheless suffer from the fact that many of the computational steps are performed sequentially. The move toward parallelism merely means that the multiple central processing units (CPUs) in these machines simultaneously handle equivalent calculations in a serial fashion, fetching, processing and returning data to the memory banks. What used to be referred to as the von Neumann bottleneck[a] has only been replaced by multiple bottlenecks.[2] The neural circuits in the brain carry out their tasks in a fundamentally different manner. Numerous quite varied steps are carried out simultaneously, in a manner more reminiscent of that employed by an *analog computer*. It will be worth our while to briefly consider such machines.

The mechanical version of the analog computer was invented by Vannevar Bush in the 1930s, when it was known as the differential analyser. Analog computers perform arithmetic manipulations upon numbers that are represented by some physical quantity, and in the mechanical variant these were the dimensions of the angular rotations of shafts and gear wheels. In the now more common electrical type, the operational variables are usually represented by currents and voltages. The essence of analog computation is input data which varies with time, this being immediately subjected to the relevant processes of addition, subtraction, multiplication, division, integration and function generation.

The rate at which these operations are carried out is essentially dictated by the speed of the electrical processes that simulate them, amongst which the charging of condensers is probably the slowest. The computations can nevertheless be carried out very quickly, and the output usually takes the

a / This takes its name from John von Neumann, who first drew attention to its existence. Although the electronic digital computer was invented by John Vincent Atanasoff, in the late 1930s, the present form of stored-program machine is generally credited to von Neumann.

form of a deflection indicated on a dial, a graph plotted by a pen, or a trace plotted on a cathode ray tube. At a typical moment during an analog computation, the great majority of the machine's connecting wires are carrying current, and the various dials and graphs give a remarkably realistic impression of the physical attributes of the system being investigated. In a study of the atmosphere, for example, simultaneous traces of the temperature, wind strength, humidity and barometric pressure give one a vivid feel for the forces that mould the weather. The drawback of the analog computer lay in its lack of precision. Unless one goes to prohibitive expense, three significant figures is about all that this type of machine can manage, whereas sixteen-digit calculations are routine for the digital variety.

It seems doubtful whether synapses and dendritic currents are particularly finely tuned either, and in this respect too, the brain resembles an analog computer. Its individual processes are probably carried out in a manner that demands no great precision, the advantages coming from the fact that large numbers of them are executed in unison. At any instant, the somatic regions of millions of neurons are simultaneously receiving signals through their dendrites, and a fraction of them are dispatching signals out along their axons, as a consequence, toward the dendrites of other neurons. It is this massive parallelism which enables the brain to 'compute' so rapidly. And with so many routes in operation between any two neurons, the individual inaccuracies are not significant because they tend to cancel each other out. It is this multiplicity of routes, moreover, which gives the brain its *robustness* against synaptic loss, and indeed against the demise of whole neurons. As a result, brain function is said to *degrade gracefully*. In contrast, the program serially executed on a digital machine aborts the moment one of its steps fails.

The brain has to carry out its functions rapidly, of course, because its possessor would otherwise be at a serious disadvantage. Speed of reaction was vital to our ancestors, both in the catching of prey and in the avoidance of predators. In our technological society, with its use of all manner of fast machines, the premium on fast thinking is, if anything, even greater. When we have the time to use our logic, on the other hand, our mental processes are sequential, and it is not surprising that we have designed our digital computers to process information in the same fashion. The steps in our logic do indeed resemble the decision units invoked by McCulloch and Pitts, which were discussed in chapter 4, but their error was to try to implement such steps at the level of the individual neuron. As a result, McCulloch–Pitts circuits lacked the above robustness. The question remains as to what can be used in their place, but unfortunately the answer is still pending. We will be closer to understanding the neuronal basis of our logic when we have a better grasp of the machinery that underlies our thought processes.

Meanwhile, there have been commendable attempts to simulate logic on the digital computer, this being one aspect of the broader field of artificial intelligence (AI). The goal of AI is to understand human intelligence, and also to emulate it. Amongst the routes to these overall aims are the manipulation of symbols in the description of objects, events and processes, in the drawing of inferences, and ultimately in the solving of problems. The most difficult aspect of AI is probably the study of mental faculties through the use of computational models, not the least because a complete description of such faculties would, perforce, have to account for the processes of vision and speech recognition, as well as for the structure of language. One product of AI is the *expert system*, which uses symbolic knowledge and inference in attempts to reach irrefutable conclusions. A very simple form of this general approach is seen in the parlour game *Twenty Questions*, which proceeds by the sequential elimination of alternatives by a series of binary choices, each question being answerable only by a *yes* or a *no*. The hierarchical logic in such an approach has what is referred to as a *tree structure*.[3]

We have already noted that the structure of real neural circuits suggests that they do not function as such trees. The degree of connectivity displayed by the typical neuron is far greater than would be practical for such logical hierarchies. This chapter is going to be concerned with an alternative approach that usually goes under the *neural network* banner, an alternative label being *connectionism*.[4] Before getting down to that, however, I ought to add a few comments about a term which will be cropping up from time to time. It is *computer simulation* (which Teuvo Kohonen has also referred to as emulation). Reference to this burgeoning scientific technique has already been implicit in the discussion of the two types of electronic computer. It is important to draw a distinction between simulation and animation, since both are commonly achieved by computer. Animation is merely representation, whereas simulation is essentially a form of remote experimentation, carried out by manipulation of numbers which describe the properties of interest of a group of elements. The manipulations follow duly established physical laws. This is the origin of their realism, and of the fact that simulation now enjoys widespread support. Simulation typically follows the dynamical behaviour of a system, and in the present context it is used to study the functioning of neural networks, of both the real and artificial types.[5,6,7,8]

Given the wide availability of digital computers, and their advanced degree of development, it is not surprising that they are the preferred vehicle for simulations. They are now used extensively in simulations of the dynamics of large arrays of neurons, which are linked up in a manner that faithfully reflects the arrangement in one or another region of the brain. It is worth emphasizing that simulation of part of the brain by computer does *not* imply that one regards the brain as a computer. After all, when one simulates the Earth's atmosphere in an attempt to understand the weather, it is clear that one is not equating the atmosphere with a computer. If misunderstandings do arise when a similar thing is done with a brain model, it is because of the colloquial linking of the two things that we remarked upon earlier. I mention this by way of introducing an aspect of the story which really does have potential for confusion. *Artificial* neural networks are sometimes referred to as *neurocomputers*, and it is important to differentiate between them and the real variety. In this book, I have been at pains to emphasize this distinction whenever the risk of confusion threatened. There is just one final point to be made, before we start our discussion of networks. One does not have to build a neurocomputer in order to study its mode of operation. One can *simulate* it. And the most convenient tool for such simulation, as I have been saying, is a digital computer. So a quite familiar feature of current investigations in this subject is the *simulation by digital computer of the behaviour of what is essentially an analog computer*. With all misunderstanding thereby precluded, we can now get down to business.

The perceptron

We have reached the point at which we should consider the nature of neural networks, and examine the way in which they function. When considering such things, it is important to discriminate between modelled brain circuits and what could be called brain-inspired strategies. But it has to be admitted that the distinction is easier to make in retrospect than it is when a promising network bursts upon the scientific scene. Networks which were originally conceived as facsimiles of those in the real brain have frequently been found to be lacking, despite the ingenuity embodied in them. There is no frowning on such failed attempts. On the contrary, everything that is at least consistent within its own constraints is potentially useful, and the documented evaluation of each network's performance is simply added to the ever-growing store of information in this area.

Those who once saw their networks as holding the key to the mind need feel no embarrassment at finding them reappraised and shifted into a different category. The switch from one pigeon-hole to another does not imply relegation. A neural network originally perceived as representing a region of the brain, may subsequently prove to be more relevant as a computational strategy. I am now going to turn to just such a network, known as a *perceptron*, a variant of which is currently the most used of all artificial neural networks. It has a quite illustrious pedigree, and I will give a brief sketch of its origins and, later, of some of its ramifications. My main task will be to give an account of the way in which it works, however, and I will start with some preliminary remarks.

The job of any neural network, real or artificial, is to convert an input to an appropriate output. In the visual system, for example, the pattern of optical densities falling on the retina is transformed into a pattern of nerve signals that permits the storage of an association between that image and other sensory inputs; a face seen with a name heard, for example. A network is thus an input–output machine, the sort of thing that we have been referring to as a black box. One of the most celebrated examples of such a device was the machine built by Alan Turing and his collaborators for code-breaking during the Second World War. Turing's machine translated a coded message into one that was intelligible by carrying out a series of operations on the input. In devices of this general type, the operations can be as simple as those used in parlour games, such as omitting every other letter from a passage of text, while difficult codes might even employ strings of randomly generated numbers.

The input–output machine of interest here is one that builds upon, and extends beyond, the binary mode of functioning displayed by idealized neurons. As we saw earlier, Warren McCulloch and Walter Pitts demonstrated that an assembly of neurons operating in the on–off fashion dictated by their thresholds is capable, in principle, of matching any response to any stimulus. In practice some tasks would require McCulloch–Pitts circuits of daunting complexity, and it is not surprising that attempts were made to produce strategies that offered the advantage of conceptual transparency. This was the case with the perceptron, in which decision-making units are arranged in two layers. The adjacent layers are connected via synapses, in the same way that real neurons are connected. Marvin Minsky actually built a hardware version of such a network in 1954. Its 400 vacuum tubes performed as idealized neurons, the thresholds being dictated by the electrical biases imposed on these components. The thresholds determined which operations were carried out on the input, both input and output being patterns of binary digits, and the most difficult aspect of Minsky's work lay in adjusting these thresholds.

The term perceptron was coined by Frank Rosenblatt in 1958, and his goal was certainly an understanding of real neural networks, rather than construction of a brain-like machine. This is made clear by the sub-title of his major publication on the subject: 'A probabilistic model for information storage and organization in the brain'.[9] Rosenblatt posed three fundamental questions, and then set about trying to answer them through investigations of the behaviour of his perceptron. The questions were: *how is information about the physical world sensed, or detected, by the biological system? in what form is information stored, or remembered? how does information contained in storage, or in memory, influence recognition and behaviour?* And Rosenblatt's aspirations were equally well-defined, for he ended his article with a sentence that read: *by the study of systems such as the perceptron, it is hoped that those fundamental laws of organization which are common to all information handling systems, machines and men included, may eventually be understood.*

A schematic diagram of a perceptron has a simplicity that belies the subtlety of its operation. And the standard manner of drawing its arrangement of neurons and synapses might even seem rather restrictive, because the neurons merely appear to be strung out along lines. An example is shown in figure 6.1.

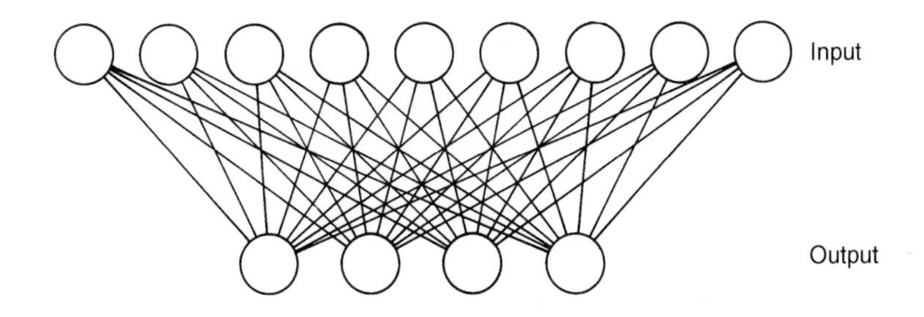

Figure 6.1
The perceptron comprises two layers of neurons, and one layer of synaptic connections, via which the activity pattern in the input neurons dictates the output response.

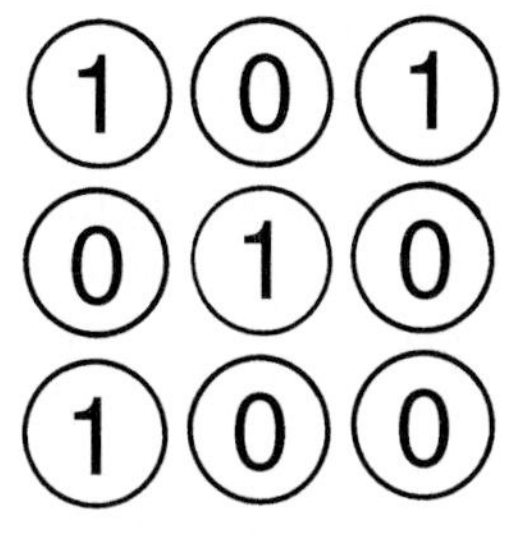

Figure 6.2
Provided the assignments are not ambiguous, a one-dimensional layer of neurons can represent an arbitrarily complicated spatial arrangement of cognitive elements. In the case shown here, the nine input neurons of figure 6.1 are representing a three-by-three array.

It must be borne in mind, however, that the neuronal arrangement is depicted in this way only because of the limitations imposed by the (two-dimensional) printed page. The nine neurons deployed on a line in the above diagram could just as correctly have been drawn as a three-by-three array, as indicated in figure 6.2. And if the two alternative values of the binary digit (1 or 0) are interpreted as black or white, such an array takes on the form of a picture, a rudimentary one admittedly, but a picture nevertheless, as can be seen in figure 6.3. Indeed, such a picture is no more primitive than that used in many of today's digital displays, which actually require the black–white decision to be made in only seven different locations. Examples of the numerals displayed in that system are shown in figure 6.4, and they are unambiguous despite their crudity.

Every neuron in the input layer of the above perceptron is connected to every neuron in the output layer. In Rosenblatt's original work the connections were initially assigned strengths that were chosen at random, and they were subsequently adjusted during a period of 'training', in which matching pairs of input and output patterns were imposed on the perceptron. (Using a term that we encountered in the previous chapter, such matched pairs may be also referred to as mappings.) Synapses that were contributing to a correct pattern match were 'rewarded', that is to say strengthened, while those which were in discord with that goal were 'punished', that is to say weakened.

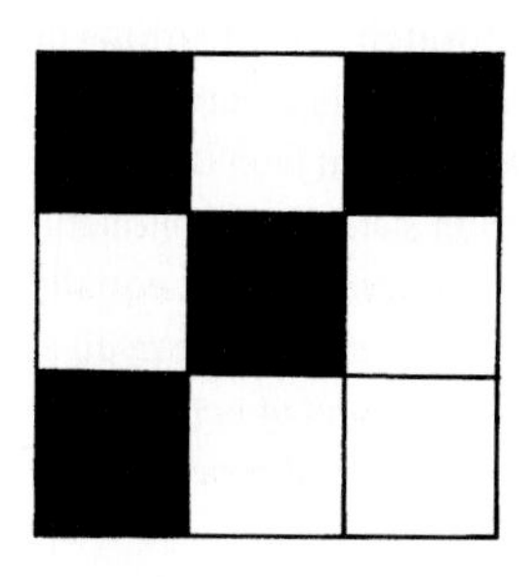

Figure 6.3
If activity (designated by 1 in figure 6.2) and lack of activity (designated by 0) are interpreted by different colours, such as black and white, the overall activity pattern can be made to represent a picture. The example shown here is very simple, but the same principle could be used to code more complicated pictures such as those reproduced in newspapers.

Rosenblatt demonstrated that a perceptron, with its two layers of neurons, has many merits. When fully trained, it can properly recognize a number of different inputs, recognition meaning that the device gives the response that it was trained to give.[b] Even more impressively, when the perceptron is presented with an input pattern *which it has not seen before*, this will be associated with the appropriate learned memory, or class of memories. The memory of a perceptron is *distributed*, which means that any association will generally make use of a large proportion of the possible synaptic paths between a given input neuron and a given output neuron, and removal of a fraction of these paths does not have an appreciably detrimental effect on the perceptron's powers of discrimination and recall.

Adjustment of the hoard of synaptic interconnections between the two neuronal layers was a difficult and tedious undertaking, however, and in 1969 Minsky and his colleague Seymour Papert declared that the perceptron concept is intrinsically flawed.[10] They showed that some input–output matchings simply lie beyond the network's discriminative powers. This was a most serious shortcoming, because it indicated that not all logical operations could be handled. And Minsky and Papert went on to show that if one tries to improve matters by making the perceptron more complicated, it loses its capacity for learning from its mistakes. This is because there is no longer an unambiguous relationship between the mismatch of patterns, at a specific

b / The actual number of pattern matches that can be stored in a perceptron's memory is, not surprisingly, related to the number of interconnections it comprises.

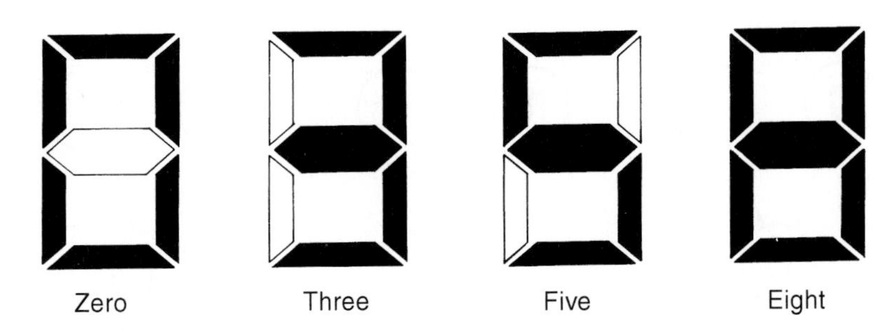

Figure 6.4
The cognitive elements represented by the neurons in a perceptron do not all have to have the same shape. The elements employed in the numerals shown here come in two different shapes. In more advanced versions of the perceptron, the individual elements can even represent different concepts.

stage of the training period, and the direction in which the various synapses are to be adjusted. Although our aim is merely to see how a perceptron works, it will be worth our while to first consider just one of the things that caused Minsky and Papert so much trouble. We will return to other shortcomings of the perceptron later on.

A surprising headache

Minsky and Papert discovered certain deficiencies in a perceptron, with its two neuronal layers interconnected with a single layer of synapses. Given the impressive capabilities that had nevertheless been demonstrated in this computational scheme, one could be forgiven for guessing that its inadequacies would be subtle and of only minor consequence. It thus comes as a surprise to find that such a perceptron fails at a task that seems rather easy. One of the strategy's Achilles heels turned out to be something known as the *exclusive–or problem* (XOR), and the perceptron that suffers from this particular imperfection could hardly be simpler, for it involves just three binary neurons, two in the input layer and one in the output layer. The arrangement is thus as indicated in figure 6.5, and the weakness shows up when we try to adjust the connections of this rudimentary network so as to permit certain input–output mappings. There are two input neurons, and because each of them can adopt either the ON or the OFF mode, which are conveniently designated by the binary digits 1 or 0, there are a total of four possible input combinations. These are: 0 and 0; 1 and 0; 0 and 1; and 1 and 1. With just a single neuron in the output layer, there are only two possibilities: 1 and 0. There are thus eight possible input–output matchings, but only four of them can be learned by a given network; we cannot ask a network to capriciously respond sometimes with a 1 and other times with a 0 at the output layer, when its two-neuron input layer is exposed to the *same* combination 1 and 0, for example.

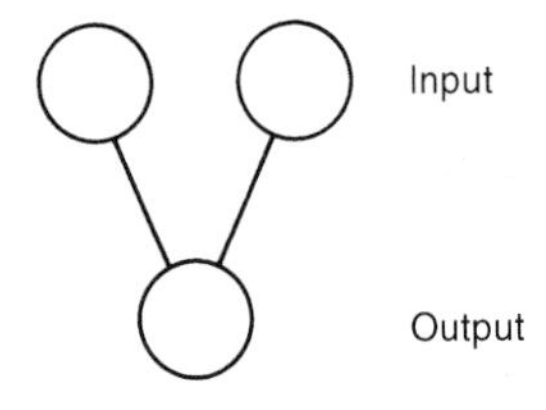

Figure 6.5
In this very simple perceptron, the single output neuron captures correlations in the activities of the two input neurons. The transformation of the input into the output, carried out by a perceptron, is referred to as a mapping.

Let us chose as the desired matchings four of the eight possibilities, therefore, and suppose these to be as shown in figure 6.6. This set of mappings is said to produce a network that functions as an *exclusive–or filter*; a 1 is produced at the output neuron if there is a 1 at *either* the left-hand *or* the right-hand neuron in the input layer, but not if *both* these neurons are active.

The trouble with the above set of four matchings is that *they cannot all*

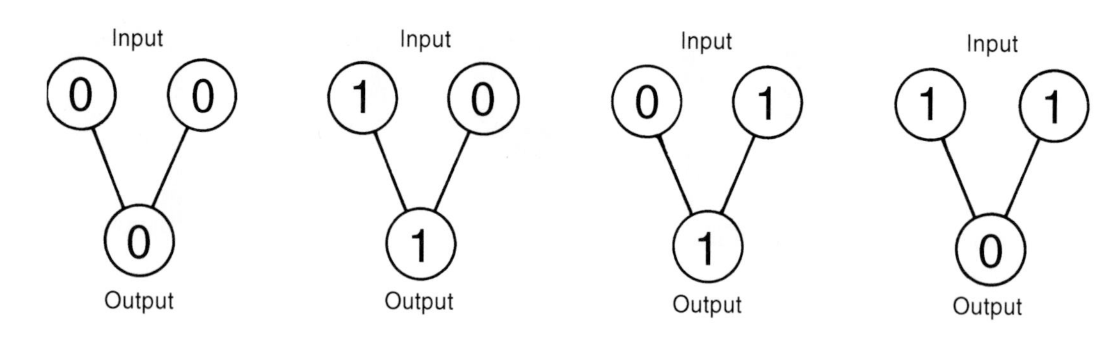

Figure 6.6
The four mappings depicted here collectively constitute what is known as the exclusive–or function. The output neuron responds to activity in either one of the input neurons, but not to simultaneous activity in both of them.

be simultaneously realized for any combination of excitatory and inhibitory neurons functioning in the normal manner. A neuron sums its inputs, and it becomes active or inactive depending upon whether that sum is greater or less than the threshold value. This process, occurring at the single output neuron, will determine whether we are to denote that cell's activity with a 1 or a 0, and we will find below that at least one of the above situations is inconsistent with the others. At least one of the mappings is thus beyond the network's capabilities. We should note, moreover, that there is nothing artificial about the mapping that is causing this trouble. If we imagine, as we did earlier, that the active and inactive neurons merely denote dark and light regions in a graphical representation, the exclusive–or problem is seen to frustrate us in the innocent quest of mapping one complete picture into another; it will cause some parts of the picture to be mapped wrongly, spots appearing in the output where there should be no spots, and spots missing where there should be spots. The problem will thus be both common and serious.

Let us now analyse the situation fully, to convince ourselves that the problem really is unavoidable. We have already seen that the neuronal activities can be conveniently represented by the binary digits 1 and 0, and it is obviously appropriate to designate excitation and inhibition by + and –, respectively. We also note that the nature of the output neuron, excitatory or inhibitory, is irrelevant because this merely determines what would be sent onward from that cell, not the state of its activity. There are sixteen different possible combinations, only four of which can apply to any given pairing of neuronal types, excitatory and inhibitory. For the sake of completeness, let us contemplate all sixteen possibilities. These are depicted in figure 6.7.

The four possible situations that can occur for a given combination of synapses appear on a single line in this diagram, and the effect of the threshold will be to make the output neuron inactive if the relevant sum is zero or negative.[c] We are now in a position to assert that it is quite impossible for this rudimentary neural network to function in a manner that would embody the exclusive–or proposition; there is no combination of excitatory and inhibitory input cells that would permit the output neuron to be active if just

c / In fact, the threshold could be raised to just below 1 without changing this result.

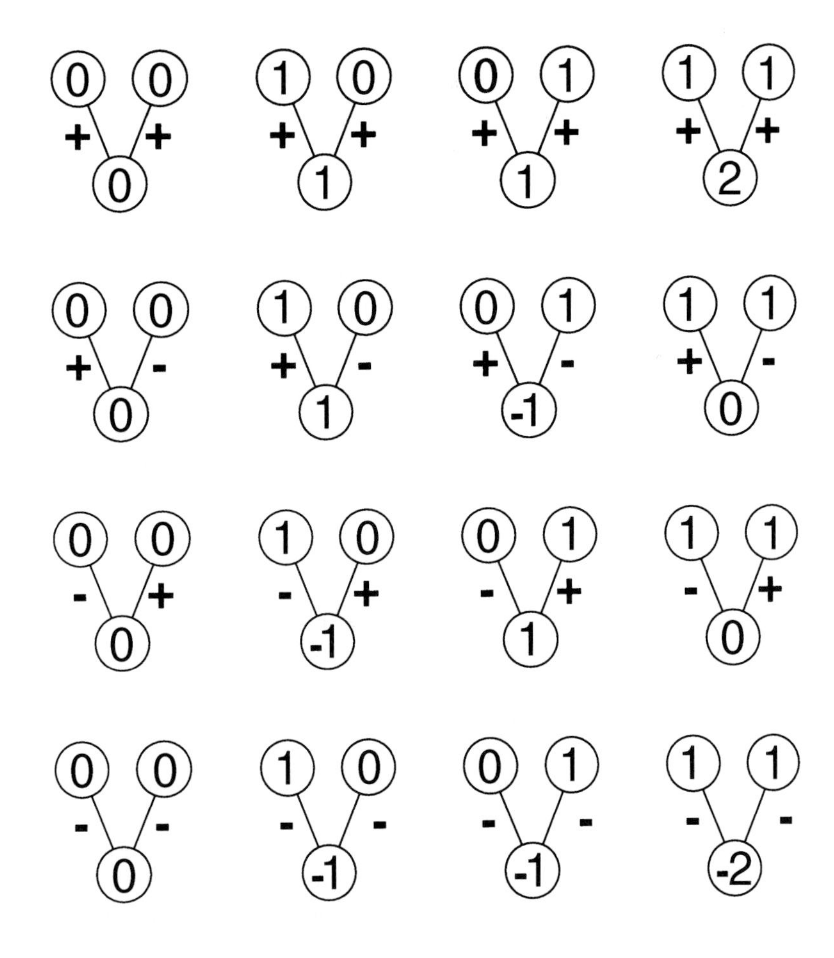

Figure 6.7
The four different combinations of excitation (+) and inhibition (–) possible in a two-synapse perceptron give rise to four candidate scenarios for implementing the exclusive–or function. As can be seen from this diagram, in which each scenario is allocated its own row, none of them captures all four mappings.

one of those cells was active, but not if both of them were simultaneously active. And it is important to emphasize that changing the value of the threshold would not change this situation.[d] It could certainly alter the pattern of 1s and 0s appearing at the output neuron, but it would never allow us to salvage the exclusive–or function.

Mathematically inclined readers will no doubt have noted the similarity between the small calculations implicit in the above examples and what are known as simultaneous equations. The activity level of the output neuron is obtained by summing two terms, each of which is the product of an input activity and a synaptic strength. One is reminded of those problems in elementary algebra, in which the variables are given the time-honoured labels *X* and *Y*. We also recall that those same two letters figured prominently in our introduction to graphs, and the impossibility of obtaining the exclusive–or function with the primitive network discussed above can indeed

d / The reader might like to work out what the situation would be if, for example, the threshold was changed to –0.5.

be demonstrated graphically. As the number of input neurons increases, we would have to recruit more and more such labels, and the corresponding multi-dimensional graphs would be difficult to visualize, let alone draw. It is partly because of this complexity that the inadequacies of the two-layer perceptron remained concealed for so long.

Working through this little computational exercise has been useful in at least two ways. It has enabled us to see for ourselves that a simple neural network in which the neurons are arranged in two layers is incapable of providing a quite undemanding function. But the non-trivial amount of analysis required to establish this fact, for what is surely a minuscule network, has also given us a feeling for the magnitude of the analytical task that the pioneers had set themselves when they first set out to probe the perceptron's properties. Small wonder, therefore, that the limitations did not reveal themselves at the outset. And it is gratifying that their commendable efforts have now been rewarded. Let us take a brief look at how this salutary vindication was achieved.

Concealed wisdom

In 1957, a mathematician named Andrei Kolmogorov conceived a theorem that proved to be of major importance to neural networks. It emerged in the field of topology, and Kolmogorov himself had possibly never even heard of neural networks. Nevertheless, it had a direct bearing on the deficiencies of those having only two layers of neurons. The theorem demonstrated that any input can be matched to any output if the network comprises at least *three* layers of neuronal units. Unfortunately, the Kolmogorov work gives no prescription as to how the synaptic strengths are to be adjusted, to produce the desired input–output mapping, but its insistence on a third layer of units does provide a strong hint. The exclusive–or problem is a case in point.

The minimal change that one could make to the three-neuron perceptron analysed in the previous section would be the addition of a fourth neuron lying between the input and the output layers, and we will now see that this indeed provides that primitive network with the desired power of discrim-

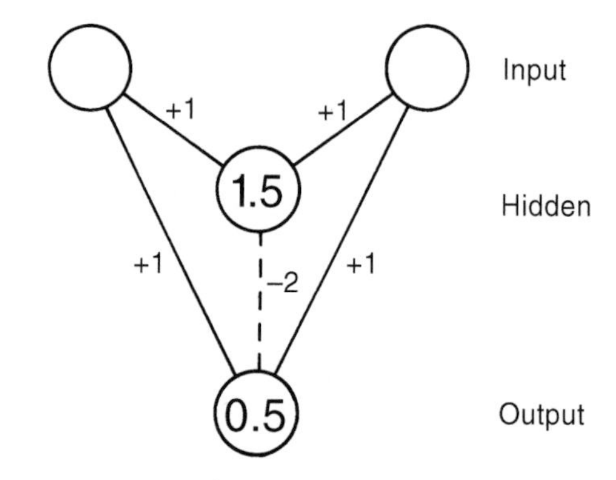

Figure 6.8
The perceptron shown in figure 6.7 acquires an exclusive–or capability if one extra neuron is added to the circuit, at a layer lying intermediate to the input and output layers. This diagram shows the position of the one-neuron hidden layer, and also the required synaptic strengths and threshold levels. The reader can verify the exclusive–or functioning of this perceptron by calculating the output response (1 or 0) for each of the possible input combinations 00, 10, 01, and 11.

ination. Tiers of neurons lying between the input and output levels have acquired the designation *hidden layers*. Let us consider the situation in which there is a hidden layer consisting of a single computing unit. The modified cellular arrangement of our little network thus becomes the one shown in figure 6.8, and we see that it includes a feature that is certainly not general in networks of this class, namely direct connections between the input and output layers. That such connections occur in this case simply reflects the fact that we are trying to correct the defects discussed in the previous section, and this is achieved by adding that extra neuron rather than by making more gross alterations to the circuitry.

The above diagram includes more details than were given in those that appeared in the previous sections. The actual synaptic strengths are specified, and an indication is given as to whether these are to be excitatory (full lines) or inhibitory (dashed line). Finally, the numerical value of the threshold is given for each neuron, where this information is relevant.[e] These details are sufficient for calculating the network's response to any input pattern of excitation, and it is a straightforward matter to demonstrate that the desired input–output matchings, namely (0,0) → 0, (1,0) → 1, (0,1) → 1, and (1,1) → 0, are obtained. The network is thus now successfully carrying out the exclusive–or operation that was beyond the capacities of the one comprising only two layers of neurons.

It is not difficult to see that the single hidden neuron is making the decisive difference. Because of its artfully adjusted threshold, this unit is responding only when both the input neurons are active, and its curative influence derives from the fact that it then passes double-strength inhibition down to the single output neuron. This inhibition, in turn, is sufficient to offset the excitations that travel along the above-mentioned direct routes between the input and the output, and the net effect is the successful switching off at the output neuron when both input neurons are active.

Unfortunately, as mentioned earlier, the theorem proved by Kolmogorov does not provide a recipe for adjusting the strengths of the synapses, so this would still be a challenge when the network is larger than the one considered here. Not surprisingly, this issue has been the subject of much speculation, and we are about to encounter the clever solution that emerged a few years after Kolmogorov's contribution.

Progress through regress – back propagation

We have seen that addition of a neuronal layer between the input and output layers obviates certain difficulties. These arise because a two-layer network is incapable of capturing all possible features of the input. And those difficulties

e / In the present case, this threshold has actually been written into the space normally reserved for the indication of neural activity; in later diagrams, a scheme will be used which permits both the bias and the activity to be indicated simultaneously.

were illustrated by carrying through the analysis of a very simple network comprising just two input neurons and a single output neuron. We then saw how a single neuron inserted at the mezzanine level, known as the hidden layer, suffices to give that network exclusive–or discrimination. The resulting circuit had the complication that its input layer made connections to both the hidden layer and the output layer. There is nothing to suggest that this architecture is anatomically indefensible, and the networks present in the brain even possess a feature not present in this simple example,[f] namely massive *feedback*. But such multiplicity does make for complexity, especially if the various layers all consist of many neurons. It is thus usually the case that computational networks have synaptic connections only between adjacent layers. Indeed, the majority of the networks that are now the workhorses of all manner of cognitive tasks invariably employ just two layers of synaptic connections, one lying between the input and hidden neuronal layers, and the other lying between the hidden layer and the output.

Following the early work on perceptrons, the aim was to find the elusive prescription for determining the synaptic strengths that would optimize a network's performance in handling a given class of input–output mappings. As is often the case in the scientific endeavour, the answer was hit upon by several individuals independently and almost simultaneously. The groundwork was carried out by Bernard Widrow and Marcian Hoff, in 1960.[11] The gist of their approach was to let an imagined teacher inspect, at a given moment during the adjustments, what the network was producing at its output layer, in response to a given input. In general, if the synaptic strengths still departed from the ideal, this output would not be that demanded by a perfect input–output mapping. Widrow and Hoff looked upon the difference as an *error*, whose magnitude was readily calculable. By linking this error to the amount by which the synaptic strengths were altered, they were able to home in on an adjustment that corresponded to zero error. It is worth stressing that the Widrow–Hoff corrections *continue to be made* even though the network might already be properly classifying every input in its learning set; the process stops only when the net error has fallen to zero. It can be shown that a similar thing happens in human learning. Psychophysical measurements have demonstrated that a subject's reaction time continues to decrease even when the responses to test tasks are already the appropriate ones.

The strategy now favoured for adjusting the synaptic strengths is known as *back propagation*, and it is a descendant of the Widrow–Hoff method. It had already been described by Paul Werbos, in his doctoral thesis in 1974, and it resurfaced in the hands of several others about twelve years later.[12] David Parker[13] published a description of the method in 1985, and the following year saw further articles by Yann Le Cun,[14] and by the group of David Rumelhart, Geoffrey Hinton and Ronald Williams.[15] One begins the process by imposing a suitably chosen pattern on the network's input layer of neu-

f / What I am designating a simple network is often said to be of the *feed-forward* type.

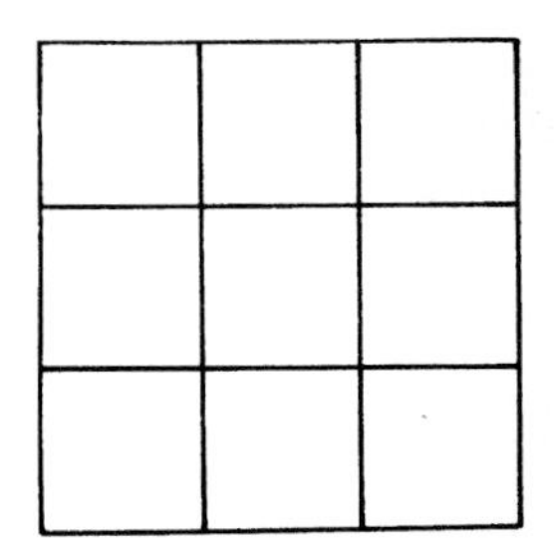

Figure 6.9
The three-by-three grid of pixels used to test discrimination between the letters T and C.

rons, the synapses having been initially set at small strengths randomly distributed either side of the zero level (that is to say, with both positive and negative initial values). The resultant effects on the neurons in (first) the hidden layer and (then) the output layer are calculated, by the usual expedient of summing the input to each neuron and then checking to see if it exceeds the (signal-transmitting) threshold. This part of the process is usually referred to as the *forward pass*.

The resultant output is then compared with what would have been obtained had the input–output mapping already been perfect, and the deviations from the desired values are then used to specify the amount by which each synaptic strength should be altered. This is only part of the story, however, because it would not be right to simply adjust all the synaptic contacts feeding into a given cell by the same amount. On the contrary, *the size of the alteration is made proportional to the contribution that each synapse is making to the receiving neuron's activity level*. This strategy ensures that the blame for an error is apportioned according to culpability. These error-related adjustments are made in the reverse direction, starting at the output layer and working backwards towards the input layer, whence the term *back propagation*. The entire process of forward pass, followed by calculation of error, followed by *backward pass*, is repeated again and again until the network is able to correctly respond with all the mappings used in the training set. In practice, one monitors the gradual evolution of the various parameters, and this can give an idea of the way in which the network acquires its expertise.

An example will serve to show what happens to a network as it is trained, and how this training gives the network its cognitive ability. To begin with, let us consider the relatively simple task of discriminating between the letters T and C, drawn on a primitive grid of nine squares (or pixels), arranged in a three-by-three array, as shown in figure 6.9. We have already seen that such an array can be represented as a single line of neuronal units. The numbering on the two versions given in figure 6.10 makes this equivalence quite clear. In an attempt to make things difficult for the network, we can draw each of the two target letters by blackening the same number of pixels, namely five. The two letters thus give rise to the cases indicated in figure 6.11.

As with all implementations of the technique, we have to decide what data to include in the training set, and we will later also need meaningful data to act as test cases. A little reflection reveals that such data are at hand,

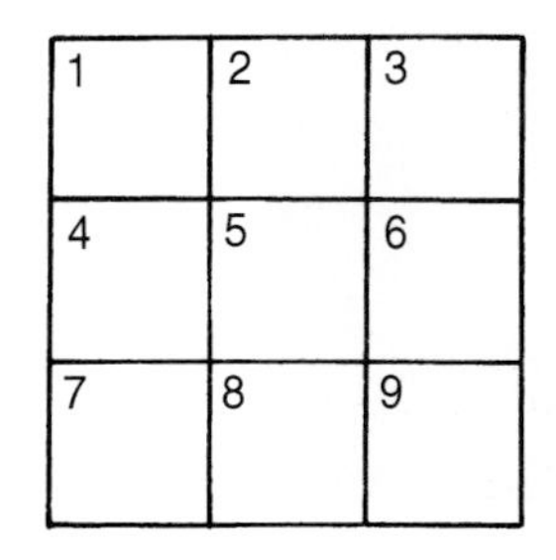

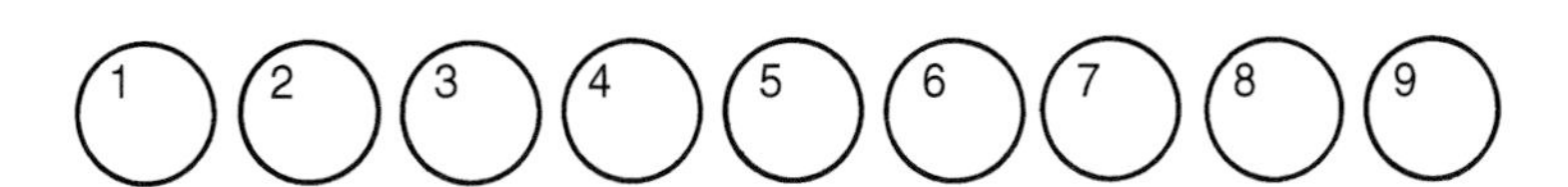

Figure 6.10
The assignment of pixels to the various input neurons of the perceptron is as indicated here.

Figure 6.11
One of the four possible orientations of the letter T that can be accommodated in the three-by-three grid, together with one of the eight possibilities for the letter C.

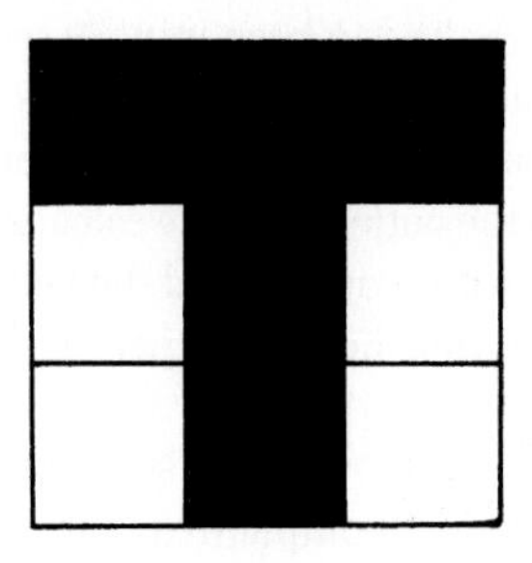

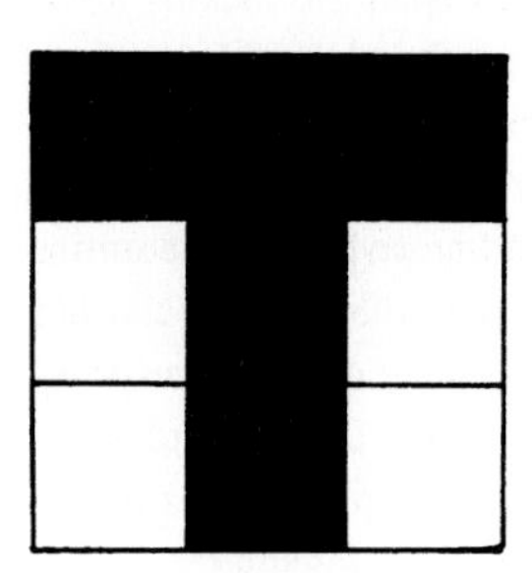

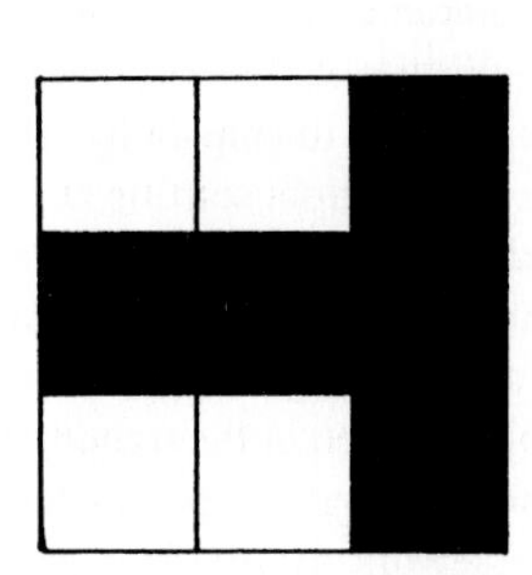

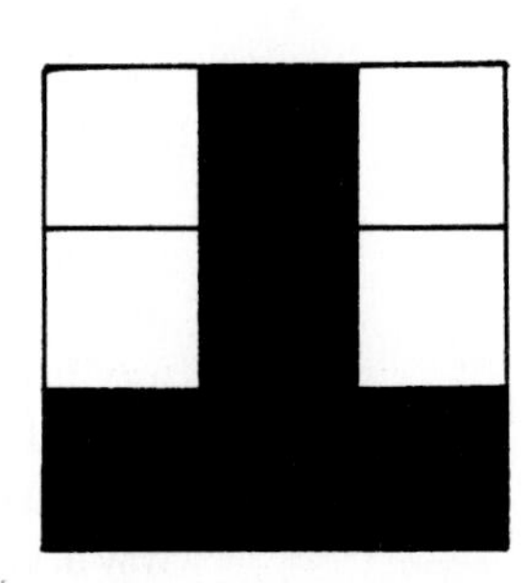

Figure 6.12
Three of the four possible orientations of the letter T in the three-by-three grid can be used as the training set, while the fourth possibility can be retained as the subsequent test pattern.

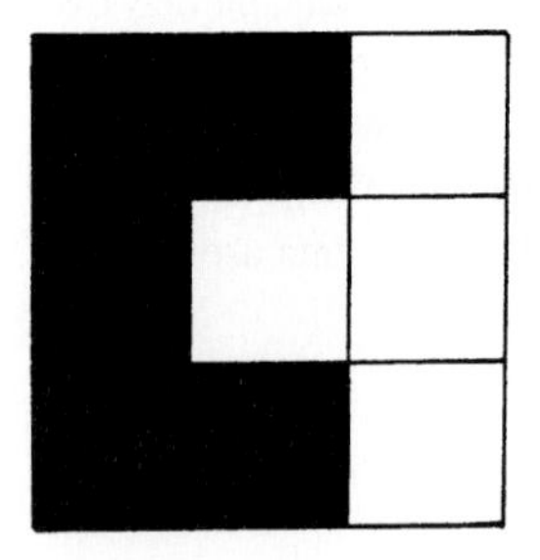

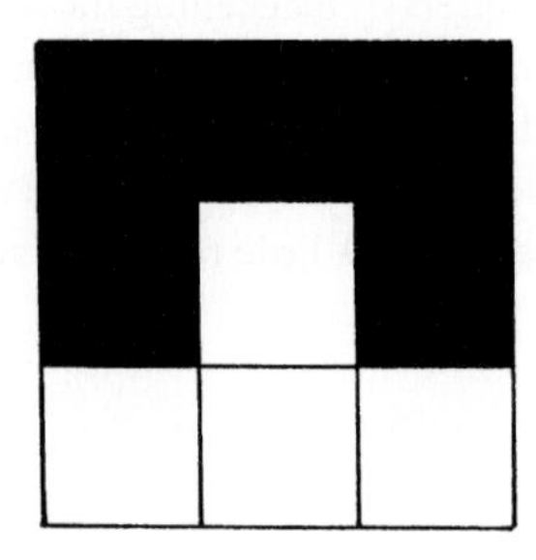

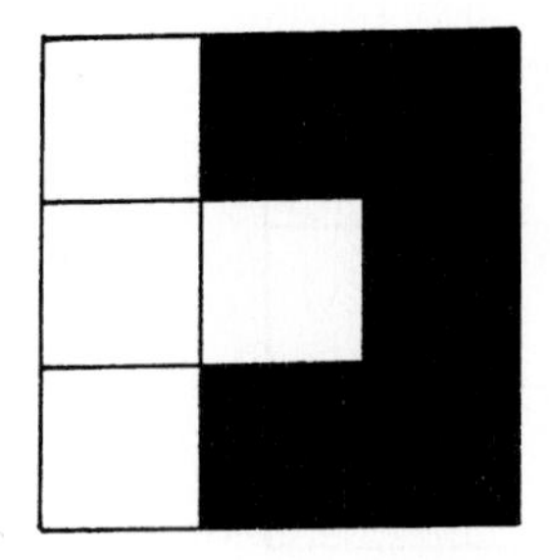

Figure 6.13
Three of the eight possible orientations of the letter C in the three-by-three grid can be used as the easy training set, while the fourth easy possibility can be retained as the subsequent test pattern. The difficult Cs all run through the central pixel, and they are thus not used in this particular version of the T–C discrimination problem.

because each of our represented letters can be rotated by various multiples of ninety degrees so as to present a different pattern of 1s and 0s to the layer of input neurons. For the Ts, we could use the three cases shown in figure 6.12 for the training set while for the Cs, we could similarly use the three possibilities shown in figure 6.13.

There is of course one other way of placing a T in the three-by-three array, but we must save this for the test. In the case of the Cs, the total number of possibilities actually jumps to eight, but we will initially confine our attention to those four Cs that occupy only the outer-lying squares in the array. As with the Ts, we use only three of the four Cs for training. We will subsequently put the (hopefully) trained network through its paces by exposing its input layer to the remaining T and the remaining C, to see if it can correctly classify them.

The procedure regarding what is to be exposed to the input layer, during the various stages of the enterprise, is now clear, but there are other decisions still to be made. One of these concerns the output layer. How many neuronal units should we have there? Given that one neuron is capable of indicating two alternatives, such a single unit would be adequate, in principle at least. Those two choices are usually thought of in terms of *yes* and *no*, so we could arrange matters such that a YES indicated a T while a NO meant a C (i.e. not a T). Although this would be a workable plan, it is usually the case that a network's discriminative powers are increased if one arranges for what is called *algebraic separability*.[g] For our purposes here, we can interpret this term as meaning that any given unit in the output of the network should only ever be associated with a single quantity; its YES or NO settings should refer only to one thing. We therefore choose to have *two* units in the output layer: one to handle the T, and one to look after the C.

We still have one more decision to make, and this concerns the hidden layer. How many idealized neurons should it contain? This is a much more difficult issue, because it is not amenable to the type of logical reasoning that serves as a reliable guide for the architectures of the input and output layers. And it is true to say that our understanding of the hidden layer is still far from complete. There are certain guidelines, however, which stem from considering what might happen in limiting cases. Let us take a brief look at those extremes. It is reasonably obvious that using a hidden layer comprising just one neuron will expose us to the risk of overburdening that single unit. It is true that the example of the XOR network discussed earlier did involve such a minimal hidden layer, but it must be borne in mind that that particular network was rather special; there were additional synapses which connected the input layer directly to the output layer, and the single hidden unit was

g / The illustration given earlier for the type of digital display commonly embodied by liquid crystal devices was deliberately chosen because it does not possess such separability; this is immediately apparent from the fact that ten different digits can be displayed despite the fact that there are only seven pixels.

therefore playing only a partial (and corrective) role. A network in which the input is connected only to the hidden layer would be a barren affair if that hidden layer contained only a single neuron.

The other extreme, in which the number of neurons in the hidden layer is very much larger than the numbers in the input and output layers, would also be inappropriate, but for rather different reasons. It would be rather expensive on computer running time, of course, but quite apart from that, it would degenerate to what is often referred to as a *look-up table*. The point is that with so many units available in the hidden layer, it would not be necessary for them to *cooperate and capture the potentially important correlations*, as did the hidden layer's neuron in the XOR network. Far from making things easier, therefore, an excessive number of units in the hidden layer will actually diminish the network's powers of discrimination. The ideal number must therefore lie somewhere in between; it must not be too small, and it must certainly not be too large. But just what it should be is not altogether clear. There will be more to discuss about the hidden layer later, but for now let us decide on a number of units which is intermediate between the nine in the input layer and the two in the output layer. For reasons that I will explain shortly, let us choose to have three.

The architecture of our network is now fixed, and it is to have nine input units, three hidden units, and two output units. It will thus have the appearance indicated in figure 6.14, where all the synaptic connections have been drawn in. In the trained network, those synapses will not all have the same strength, of course, and indeed some of them will even be negative. It is useful to have these strengths indicated on the diagram, so this has been done (in figure 6.15) by letting the thickness of each connecting line be proportional to the connection strength, and by indicating its sign, negative or positive, through the use of dashed or full lines, as earlier.

The standard approach is to let the synapses have initial strengths that are small and roughly equal. But not precisely equal, because the network would then have difficulty in deciding which units are to look after which

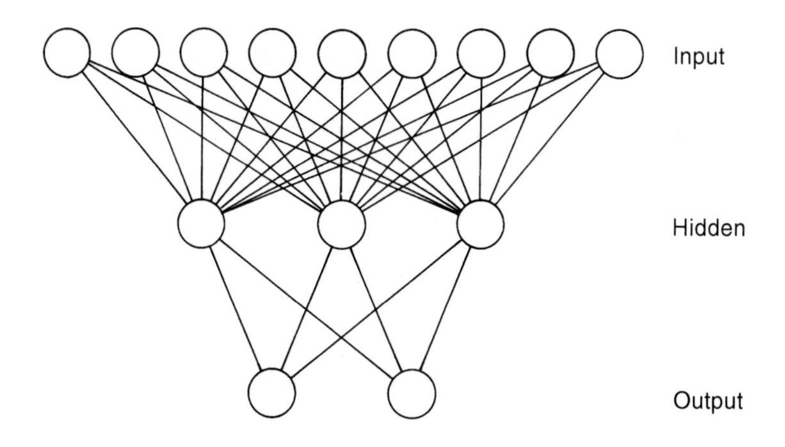

Figure 6.14
The architecture of the 9–3–2 network used in the easy version of the T–C discrimination problem. No training has taken place, so the lines merely indicate the positions of the synapses, not their strengths.

items or concepts during the training phase; it is advisable to start with what is referred to as *broken symmetry*, to give the slight lack of uniformity its proper name. Moreover, since some of the synapses will finish up with negative strengths, it is also customary to let approximately half of them actually start out with such values.

Although this is too complicated a subject to be discussed at length here, it transpires that networks function even more efficiently if each neuronal unit is allocated an extra variable parameter, which acts as a sort of bias. This additional factor can be thought of as representing the threshold of the neuron, indeed, but in this case a summed input that lies below the threshold does not produce a zero output. Although this was not discussed in chapter 4, when we considered the neuron, such behaviour is biologically plausible; a neuron can be spontaneously active in the absence of recent synaptic input, the probability of such a random event depending upon the proximity of the neuron's voltage to the threshold level. The other important feature of the neural network, in its modern version, is that the output of each neuron is graduated, that is to say that it is not merely an all-or-nothing response; as the summed input increases, so does the output. But the rise in the latter is gradual, and it smoothly increases toward a saturation, or limiting, value. This too is in accord with biological reality, and it was described in chapter 4 with the help of an appropriate diagram (see figure 4.3).

We have now considered the basic aspects of the network, and we are ready to observe it in action, during its training phase. This proceeds as follows. The first of the T patterns is presented to the input layer, the representation being given by the numbering system specified above. It is 1 1 1 0 1 0 0 1 0 and it comprises five 1s, as settled on earlier. The five active neurons pass signals down to each of the three neurons in the hidden layer, these signals being modified by the relevant synaptic strengths, some of which have been arbitrarily given negative values. The inputs to each of the hidden neurons are then summed, and these sums are modified by the corresponding bias values. (At this first step, however, there will not actually be any modification, because it is customary to start out with all the biases set to zero.) The modified sums are now used as the inputs to what is called a *squashing function*. This too might sound complicated, but the squashing function merely ensures that the output does have the graduated and saturation qualities referred to earlier; its form was indicated in figure 4.3, when we were discussing the output signals from a typical neuron. The suitably squashed outputs of the hidden neurons are now used as the signals to be passed on to the output neurons, and again these signals are modified by the strengths of the corresponding synaptic connections.

Signals thus ultimately reach the output neurons, where they are again subjected to the three processes of summation, modification by the relevant bias, and squashing, so as to produce the final output. It is especially important to note that the effect of squashing is to give an output that cannot exceed a certain value, which can be scaled to unity for convenience. Such a final output can thus be directly compared with the target output that we have specified, namely 1 0 because we are dealing with a T. Had this first input corresponded to a C, the desired output would have been 0 1.

We have reached the most vital part of the process, for this first presentation, because it is now that the learning process begins. Unless by pure chance the various synaptic strengths were given just the right values, which is extremely unlikely, the signals emerging at the output layer will not match those of the above target values. Those emerging signals might, for example, have the values 0.43 and 0.71, in which case they would differ considerably from the desired 1.00 and 0.00. Back propagation is effected by taking the two errors, namely 0.57 (=1.00 – 0.43) and –0.71 (=0.00 – 0.71), multiplying them by what is referred to as the *learning rate*,[h] and using the result to change the strengths of the synaptic connections that feed into the two output neurons. When the synaptic strengths have thus been changed (and also the biases, by a similar corrective procedure), the entire process is repeated for the synapses between the hidden-layer neurons and those in the input layer (and again also for the biases of the hidden-layer neurons).

All of this constitutes just one back-propagation step. Because these small corrections will usually not be sufficient to give the correct answer if the same input pattern is again presented to the input layer, one could use that pattern again and repeat the back-propagation procedure. One could indeed go on presenting the same pattern until the network had learned to give the correct response at its output layer, but this turns out to be an unwise approach to the training task. It proves to be far more efficient in the long run to switch immediately to a different pair of input and output patterns, and let the network learn a little of that mapping too. In the case that we have been considering here, the next pattern corresponds to the first of the Cs, and the activities at the input layer will be 1 1 0 1 0 0 1 1 0 which of course comprises five 1s, as before. The above procedure is again followed, and the final squashed output signals will not, in general, be equal to the desired pair of values 0 1. The subsequent back propagation again makes its small corrections to all the synaptic strengths and biases, and the network thereby learns a little more of its task.

We continue with the same routine, presenting the remaining four patterns in the training set, namely 0 0 1 1 1 1 0 0 1; 1 1 1 1 0 1 0 0 0; 0 1 0 0 1 0 1 1 1 and 0 1 1 0 0 1 0 1 1, and these too will not in general produce the desired output pairings of 1 0; 0 1; 1 0 and 0 1 respectively. Back propagation thus makes further small corrections to all the variable parameters, and the network continues on its inexorable path toward the fully trained state.

When all six of the training input patterns (and their corresponding six training target patterns) have been presented to the input layer (and the output layer), the entire process is repeated, over and over again, until all the errors have been reduced to essentially zero. The network will then be in the

h / The learning rate is a measure of the amount by which the synapses are adjusted for each back-propagation step. It cannot be made arbitrarily large because of the risk of overshooting. It is usually optimized at about 0.25 per step.

fully trained state, and presentation of any of the training patterns at the input layer will immediately result in the correct target pattern appearing at the output layer. In our example here, the synaptic strengths, and signs, as well as the biases, turn out to have the values indicated in figure 6.15.

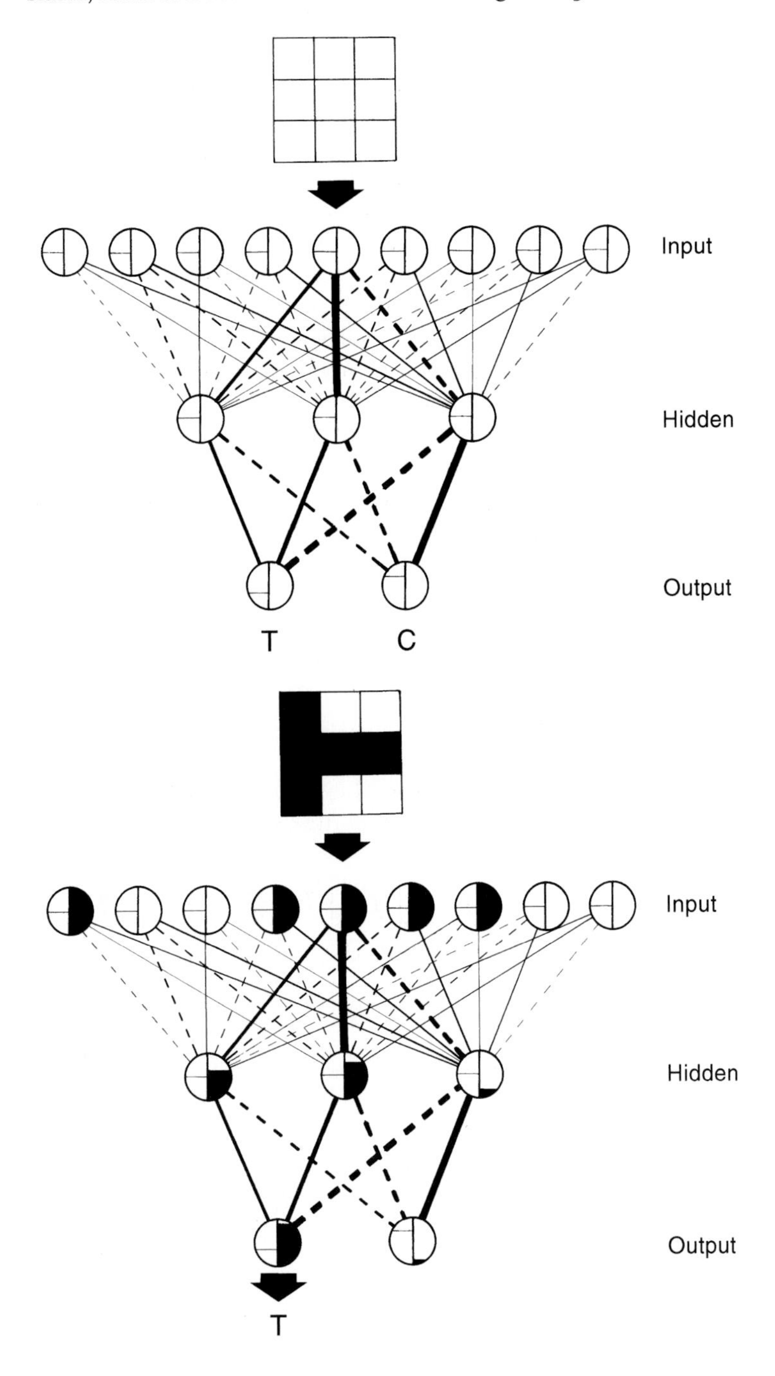

Figure 6.15
The 9–3–2 network after completion of the training on three of the four possible T orientations and three of the four possible easy C orientations. The thicknesses of the lines indicate the strengths of the excitatory (full) and inhibitory (dotted) synapses. The horizontal lines in the circles indicate the bias (threshold) levels in the corresponding neurons.

Figure 6.16
The response of the trained 9–3–2 network upon presentation of the remaining T orientation at the input layer. The blackened portions in the right-hand halves of the circles indicate the current activity levels in the corresponding neurons, and it can be seen that the network is correctly classifying the input pattern.

As promised earlier, the strengths are now indicated by line thickness, while excitatory and inhibitory connections are indicated by full and dashed lines respectively. The (positive or negative) biases are also shown, on the left side of each symbolic neuron, by the vertical positions of the short horizontally drawn lines, relative to the centrally placed zero point. (The right side of each symbolic neuron has been reserved for an indication of that unit's present activity level, and we will return to that aspect of the story later.)

We have now reached the exciting moment at which the network is to be shown something that it has never seen before! As agreed upon earlier, this is to be either the remaining T pattern, or the remaining C pattern. Let us choose the T, namely 1 0 0 1 1 1 1 0 0, for which the target output is 1 0. The presentation results in the situation captured in figure 6.16, the right-hand side of each symbolic neuron indicating its present activity level. The latter can be readily compared with the corresponding bias.

The activity levels emerging at the two output neurons can easily be read off, and they prove to be 0.91 and 0.08. These are certainly quite close enough to 1.00 and 0.00 for us to draw the conclusion that a T has indeed been used as the input. Similarly, presentation of the remaining C pattern 0 0 0 1 0 1 1 1 1 results in the values 0.01 and 0.97 appearing at the two neurons of the output layer. These are as close to 0.00 and 1.00 as we need them to be, in order to correctly conclude that a C has been presented to the network. Figure 6.17 shows that situation.

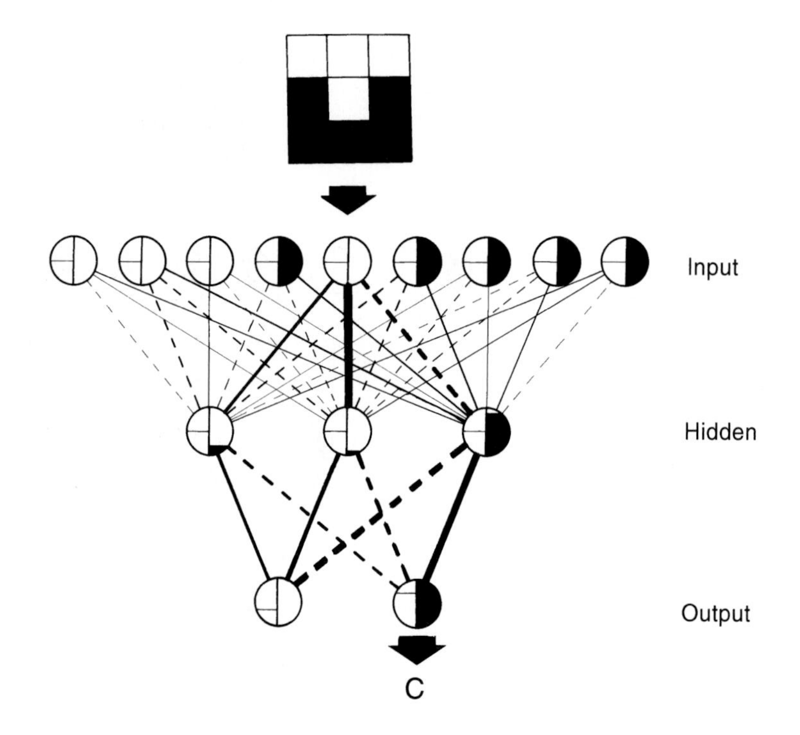

Figure 6.17
The response of the trained 9–3–2 network upon presentation of the remaining easy C orientation at the input layer. It is apparent that the network is again correctly classifying the input pattern.

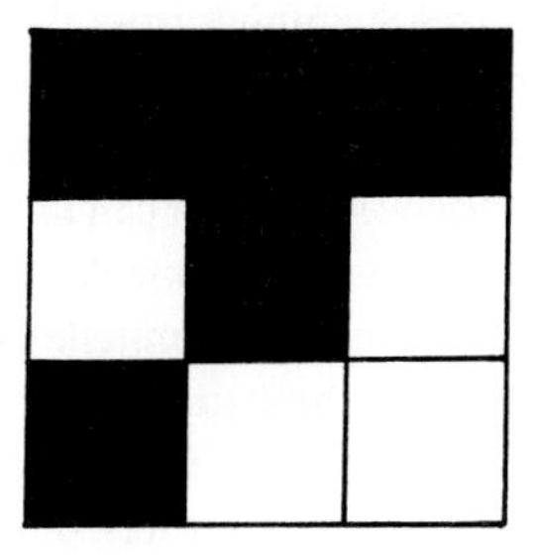

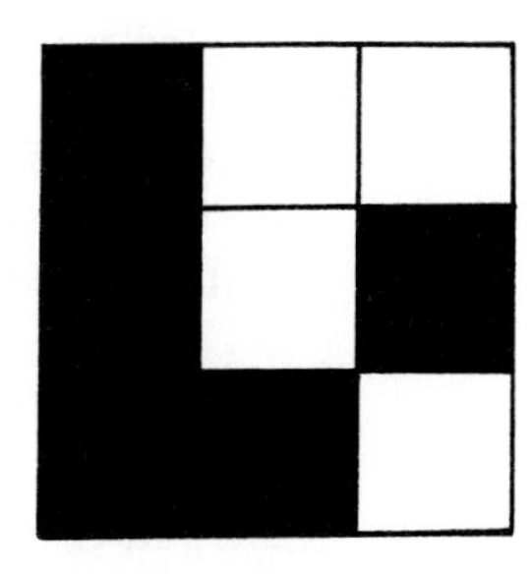

Figure 6.18
Examples of incorrectly drawn patterns which the trained 9–3–2 network classifies as a T (left) and a C (right), respectively. Either pattern can be transformed into a perfect T or a perfect C by movement of the appropriate single black square, so the discrimination is unsound. Inspection of the synaptic strengths in figure 6.15 reveals the underlying cause: the central input pixel is carrying out the discrimination virtually single-handedly.

It might seem that this exercise is rather trivial, because we seem to be asking the network questions to which we already know the answers. This is not the point, however, for properly trained networks are found to have this ability to classify even when there are a large number of possible patterns in the training set. And a trained network can display generalization in that it will successfully classify an item even if this deviates somewhat from the type of input used in the training phase. In the above network, for example, an incorrectly drawn T or C will still be recognized, provided that the discrepancy in its written form is not too large. Examples of incorrectly drawn letters which are still recognized as being sufficiently close to the real thing are shown in figure 6.18, the correct answers being T and C, respectively. Investigation of just how much latitude can be tolerated is an interesting pursuit in its own right, but there is not room enough here to go into such issues. There are more pressing matters, and the fact is that there are serious things that this first brush with the T–C problem has been ignoring. These inadequacies will be exposed through contemplation of the hidden layer, to which we must now turn.

Hidden layerology

Having just guided the reader through a considerable amount of detail, I now have a confession to make. The network we have been discussing is something of a fraud. It cannot be denied that it manages to discriminate between T and C, but it does this in a manner that could hardly be called clever; it simply looks at the central pixel in our three-by-three array and concludes that the input is a T if that pixel is occupied! That this is the case can readily be seen by inspecting the synapses of the trained network, as shown in the above diagrams. The connections stemming from the central neuronal unit of the input layer are much stronger than those from any of the other units; they play the clearly dominant role.[i] During the back-propagation training process, our network has obviously discovered that it can take a short cut; it has 'realized' that it can afford to ignore what is presented to all but one of the pixels.

My choice of a hidden layer comprising just three neurons was based on a guess. The three-by-three input grid comprises three different classes of pixel: corner pixels, of which there are four, (middle of) edge pixels, of which there are also four, and finally central pixels, of which there is only one. I was counting on the possibility that the gradual training of the network would result in one of the hidden-layer neurons becoming a sentinel for the central pixel, while the other two took charge of the corner pixels and the edge pixels, respectively. This is just what happened, but it was a surprise to find the network virtually ignoring the corners and edges altogether.

For the chosen architecture, therefore, our network is not being confronted with a sufficiently tough challenge. We need to give it something more difficult to work on, and a more exacting task is already at hand in the form of those four remaining Cs. In view of what has transpired, we may now call them the *difficult* Cs. The three of them that we could use for the training set are shown in figure 6.19, and we see that they too involve that central pixel. Having already done the groundwork, we can jump directly to the performance of the network when it is fully trained on three of the four Ts and three of the four difficult Cs. Confronted with the remaining C, it responded in the correct manner. When it was shown the remaining T, however, it erroneously perceived that letter too as a C. This breakdown is apparent in figure 6.20.

The network had in fact been trained for a much greater number of passes than in the previous case, to give it the best possible chance of success, but this was all to no avail. Examination of the details of this new diagram is nevertheless rewarding because we see that the strong synapses (actually axon–synapse–dendrites, of course) are now more spread out in the network. The central input neuron is no longer dominating the network's behaviour. One can also see why the network is failing to make the desired discrimination: there is almost no difference between the activity levels of the three hidden-layer neurons.

i / In chapter 4, the point was made that a neuron's influence on the other neurons it synapses with will be greater the higher the frequency of its output pulses. The variability of this frequency is captured in the use of the squashing function, but the influence on another neuron is also determined by the strength of the intervening synapse. If the product of the two factors is small, the transmitting neuron could be said to whisper *attend-to-me*; if the product has a medium value, it requests more forcefully *Attend-To-Me Attend-To-Me*; when the product is very large, the neuron demands *ATTEND-TO-ME ATTEND-TO-ME ATTEND-TO-ME*.

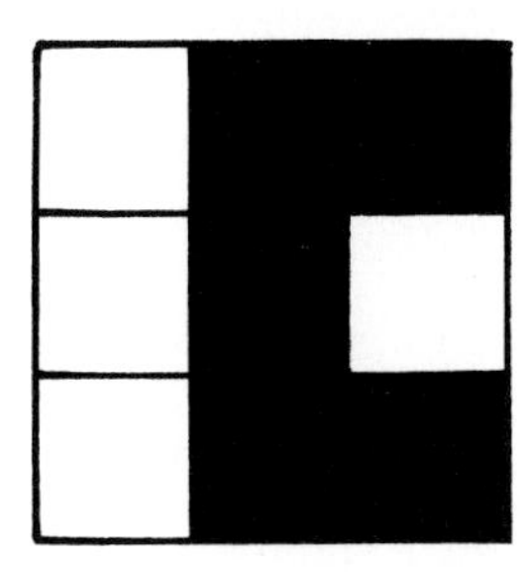
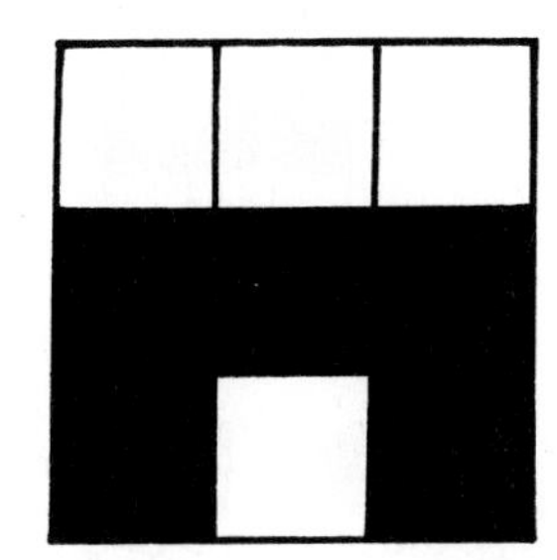
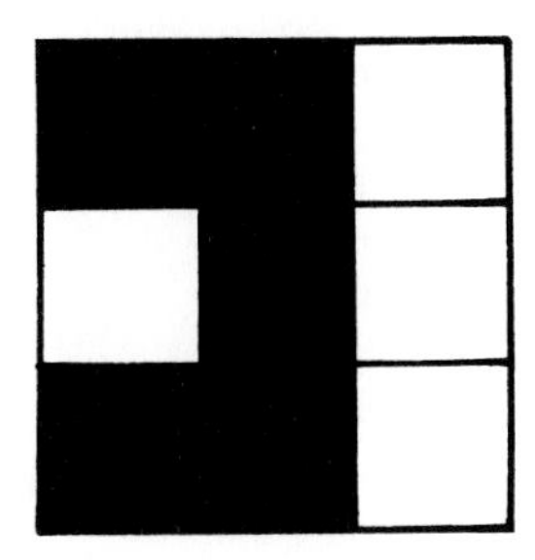

Figure 6.19
Three of the four possible orientations of the difficult Cs, in the three-by-three array of pixels. They are designated difficult because, in common with all the Ts, they involve activation of the central pixel. This makes discrimination far more difficult.

Since the numbers of neurons in the input and output layers are fixed by the nature of the task, the only factor that remains negotiable is the size of the hidden layer. To avoid our description becoming too wordy, we can introduce a shorthand designation for a network's architecture. We can refer to the above network as being of the 9–3–2 type, the numbers indicating the sizes of the input, hidden and output layers, respectively. In an effort to manage the T–C problem when the Cs are of the difficult variety, let us enquire whether a 9–5–2 network would do the trick. The answer is that it can; it correctly identifies both the remaining T and the remaining C. Figure 6.21 shows it performing in the desired manner in the latter case.

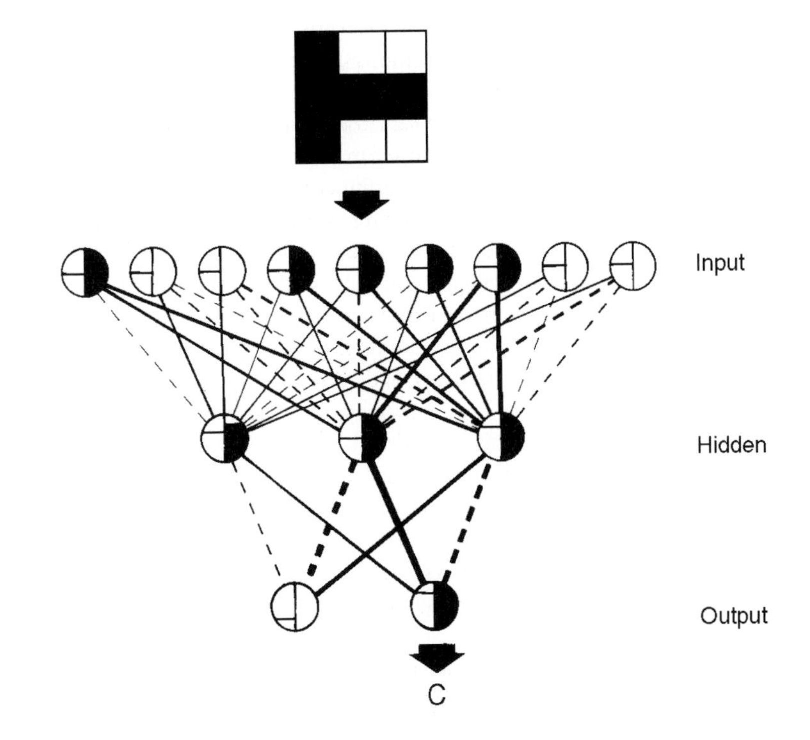

Figure 6.20
The 9–3–2 network, fully trained on three of the T orientations and three of the difficult C orientations, failing to classify the remaining T. As can be seen from the more even distribution of synaptic strengths, compared with the situation shown in figure 6.16, the central pixel is no longer playing the dominant role; on the contrary, its influence is relatively weak.

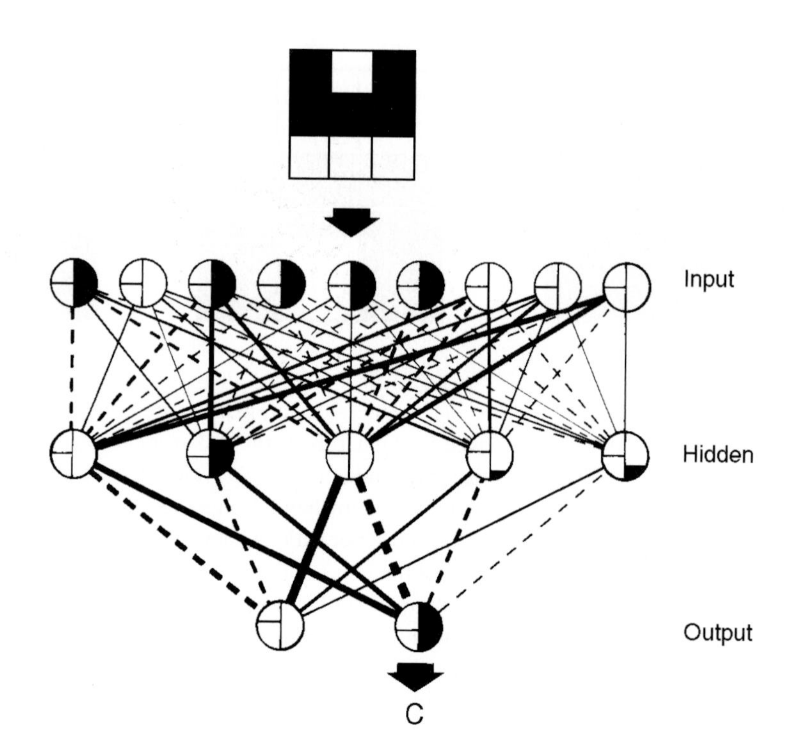

Figure 6.21
A 9–5–2 network, fully trained on three T orientations and three difficult C orientations, is able to correctly classify the remaining difficult C, as shown here, as well as the remaining T. Its success, compared with the performance of the 9–3–2 network, is directly attributable to the presence of the two additional neurons; they increase the ability of the hidden layer to capture significant correlations.

The details of this 9–5–2 network are really quite revealing. As was the case with the unsuccessful 9–3–2 network, the stronger synapses are distributed amongst many of the neurons, the notable exception being the centrally located neuron of the input layer (which of course represents that centrally located pixel in the three-by-three picture). The synaptic connections emanating from that neuron have now become very weak. It has virtually no influence on the cognitive process. During the training phase, the network has latched on to the fact that the central pixel is always active, irrespective of whether the input layer is being exposed to a T or a C, and the corresponding synapses could be said to have atrophied as a consequence. It is also worth noting that the distribution of activity amongst the hidden-layer neurons is now again uneven, as it was when the 9–3–2 network was succeeding with its trivial task of distinguishing between the Ts and the easy Cs.

Increasing the size of the hidden layer proved to be efficacious in this example. The question naturally arises as to whether we can make things even better by making the number of hidden-layer neurons still larger. This is something on which it is difficult to generalize, because it appears that each case must be treated on its merits. The job of the hidden layer is to capture correlations and to act upon them, as we have seen, but the amount of correlation present in the input will depend upon what that input is.

Let us see what happens to our difficult version of the T–C problem when we further increase the size of the hidden layer, beyond those five neurons that produced success. I am going to suggest that we contemplate the discriminative abilities of a 9–126–2 network! Now why on earth should I choose such a peculiar number as 126? What I have in mind is a little combinatorial calculation. It asks how many different ways there are of selecting 5 items from a total of 9 (that is, how many different ways there are of colouring five of the nine pixels black). The answer is 126. I intend to allocate one hidden-layer neuron to each of these combinations.[j] If a 9–126–2 network is confronted with the difficult variant of the T–C problem, it fails. The reason is not difficult to find. The hidden layer is now large enough to permit each combination of black pixels to be treated as a separate case; it can afford to allocate one neuron to each of the 126 possible arrangements, and the test cases of the final T and the final C will correspond to hidden-layer neurons whose synapses have been starved of attention. The network has become oblivious of correlations, and it has therefore lost its ability to generalize. Using the term introduced earlier, we must conclude that its hidden layer has become a mere *look-up table.*

The large size of this network makes it impossible to display in the same manner as those shown earlier, but figure 6.22 captures the spirit of the thing even though the central 120 neurons in the 126-neuron hidden layer have had to be replaced by dots, and all the synapses have been omitted. It shows the network failing to identify the test C, as predicted.

The lesson to be learned from this exercise is therefore that the hidden layer must be neither too small nor too large, but beyond that it is difficult to say just what its size should be. Indeed, as I noted earlier, the optimal number of neurons in the hidden layer depends upon the type and strength of the correlations in the input, and because this is not always the same, each case must judged separately. There are cases, indeed, in which the degree of correlation in the input is such that a hidden layer is not even necessary. Should we conclude that the network handles the T–C problem in a totally satisfactory manner, therefore, provided that we use the 9–5–2 version? The answer is *up to a point.* There is still a quite severe limitation to our network's performance, because it is not robust against a reversal of the colours.[k] In the region where I live, those traffic signs which indicate place names come in three varieties: local places are shown in red on a white background, while places indicated on the two different types of motorway sign are in white on green and white on blue. One is so use to such schemes that one hardly notices the reversal between dark colour on a light background and light colour on a dark background. But such reversal leaves the otherwise so successful 9–5–2 network stumped. We see the figures shown in figure 6.23 as showing a white T and a white C on a black background. The network trained

j / In general, the number of ways in which *m* items can be selected from a total of *n* items is given by $n!/m!(n-m)!$ the exclamation sign indicating *factorial.* The factorial of 5 is 5×4×3×2×1. Five items can thus be selected from a total of nine in 9!/5!(9-5)! different ways. This gives us the number 126.

k / I am regarding black and white as colours, in this context.

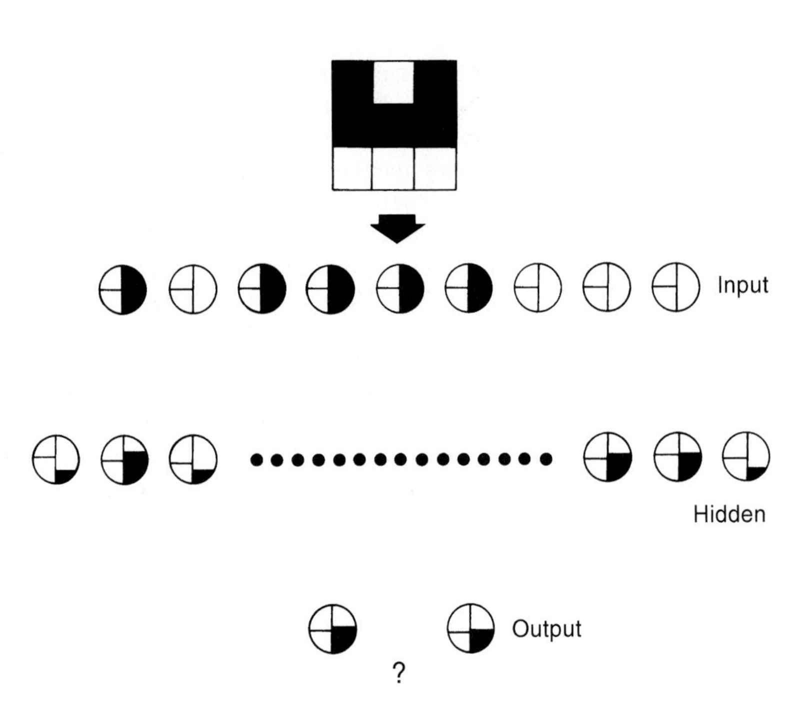

Figure 6.22
A 9–126–2 network, fully trained on three of the T orientations and three of the difficult C orientations, is unable to classify either the remaining T or the remaining C. Its hidden layer has become a look-up table, and it has thus lost its ability to detect correlations in the input patterns. Most of the hidden-layer neurons and all of the synapses have been omitted for the sake of clarity.

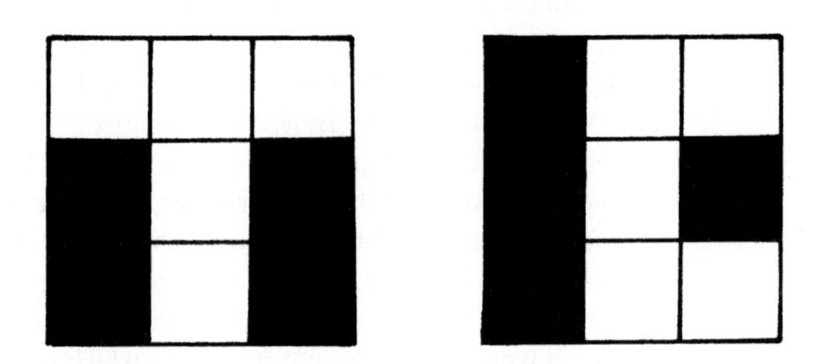

Figure 6.23
The simplest interpretations of these two patterns is that they display a white T and a white C, respectively, on a black background. The fully trained 9–5–2 network illustrated in figure 6.21 classifies the white T as black C, and the white C as a black T, because the centrally located pixel exercises only minor influence in that network.

on the black Ts and the black Cs on white backgrounds fails to appreciate the change. It perceives the white T as a black (difficult) C, and the white C as a black T. The reason for its error is not difficult to find. It stems from the fact that the 9–5–2 network has learned to downgrade the central pixel, as we discussed earlier, because it was always active, irrespective of whether the input layer was being shown a T or a difficult C. Small wonder, therefore, that it reacts as if the missing portions of the black C and the black T were actually present. Once again, we find the network failing in a relatively simple challenge. In terms of the enlightenment it affords, however, this failure is a salutary one because it provides a timely reminder of how much is entailed when the visual system inspects apparently simple pictures like the ones that we have just considered. Determining what to regard as object, and what to relegate to background is in reality a difficult challenge.

Although there are aspects of the T–C problem which are certainly not trivial, it is nevertheless a simple challenge compared with many that neural networks are now being confronted with. As the task for which a network has been trained becomes even more complex, it is increasingly difficult to make clear identifications of which jobs specific neurons are undertaking. This, of course, underlines the usefulness of these computational strategies, for if the distribution of responsibilities amongst the hidden-layer neurons was always as transparent as it is in the above cases, there would probably have been an easier way of going about things in the first place. This, of course, is the essence of what is now referred to as *parallel distributed processing* (PDP), and we must bear in mind that the usefulness of a trained network is not dependent upon our knowing how it achieves its remarkable results. As we shall see in the other examples we will be moving on to, this is even more the case as the task becomes more complicated.

Percept and concept

The three-by-three array used as a basis for the Ts and Cs in the previous section is certainly a modest little affair compared with the visual fields that our retinas have to deal with. But it was also being handled by a minute number of input neurons compared to what we have in our retinas. The question arises as to whether the allocation of one neuron per pixel in the above network is generous or parsimonious. How would one decide this issue? A logical first step could be to determine the *graininess* of the most detailed scenes we are able to examine, and then to compare this with the density of *receptor cells* in that part of the retina which we use for such observation. The situation is rather more complicated, however, because we see not with our receptors but with the entire visual machinery. We might be able to arrive at an answer by considering the early part of the visual system in a little more detail.

The light entering the eye through first the cornea and then the lens is focussed on the retina, which consists of neurons of five different types (see figure 6.24). Only one type is light sensitive, namely the receptors, which come in two varieties: the *rods*, which detect brightness, and the *cones*, which are less sensitive but which code for colour as well. The outputs of these receptors are not fed directly to the brain. On the contrary, there is preprocessing of the visual signal before it reaches the cerebral cortex, and much of this actually occurs in the retina itself. Thus the receptors synapse onto the *bipolars*, and the bipolars synapse onto the *ganglion cells*. But these contacts are not merely made on a one-to-one basis. There is much evidence of both convergence and divergence, each receptor helping to stimulate many ganglions, and each ganglion being fed by many receptors.

Further complicating things, there are two other prominent types of retinal neuron: the *horizontal cells* and the *amacrine cells*. Unlike the neurons of the other three types, the processes of members of these latter two classes lie in the plane of the retina, rather than at right angles to it, the horizontal cells laterally spreading the signals excited in the receptors, and the amacrine cells

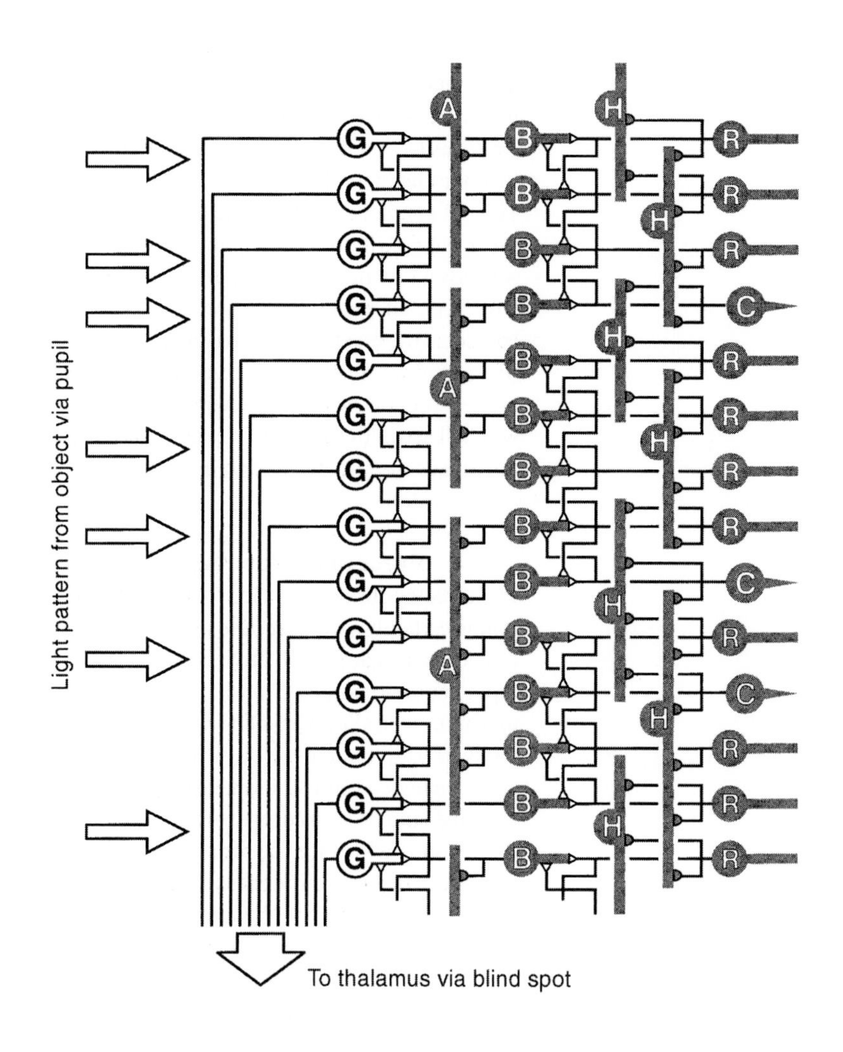

Figure 6.24
The mammalian retina has a complicated structure that comprises neurons of five different types, as indicated in this highly schematic diagram. Oddly enough, and contrasting with what happens in the eye of a frog, for example, the image impinging on the light-sensitive receptors has to pass cells of the other four types. The rod receptors (R), which are sensitive to brightness, and the cone receptors (C), which monitor contrast, motion, size and wavelength, feed signals to the bipolar cells (B), and the latter send signals to the ganglion output cells (G). Signals from each neuron are passed on to several neurons of the next type, so the system involves much convergence and divergence. Moreover, the neurons of the remaining two classes, the horizontal cells (H) and the amacrine cells (A), mediate the lateral spread of signals, helping to eliminate the effect of (optical) noise and enhance the contrast at discontinuities in illumination. Only the ganglion cells emit action potentials, the responses of all the other neurons being of the more graduated kind normally associated with dendrites. (All neurons except the ganglions are coloured grey to reflect this fact.) Horizontal and amacrine cells do not have axons. Many of the synapses in the retinal circuit permit two-way passage of signals, and those that do so have been drawn as semi-circles rather than triangles. The receptive field of a given ganglion cell is determined by the number of receptors which can influence its output.

doing the same job, farther along the path, for the bipolars and the ganglions. The neurons of these two classes thus function as coordinating agencies. The upshot of this arrangement is that each retinal output neuron, that is to say each ganglion cell, has a catchment area on the retinal surface which extends over several dozens of receptors. That retinal area, in turn, corresponds to a certain area in the field of view, and this is referred to as the ganglion cell's *receptive field*.[l,16] We could be surprised by so much structure in the retina itself, but the eye manages some impressive feats of discrimination and adaptation. Think, for example, of our ability to see in a remarkably broad range of illuminations; this is a consequence of the mediatory and self-adjusting role played by those horizontally disposed cells.

The axons of the ganglion cells are elongated in the general direction of the centre of the head, and they collectively form the *optic nerve* (and later on, the *optic tract*), but this does not connect directly with the cortex either. Instead, it passes to part of the centrally located structure known as the *thalamus*, the part in question being the *lateral geniculate nucleus* (LGN). It is from the LGN that the partly processed visual signals are sent to the early parts of the visual cortex, the LGN axons being collectively referred to as the *optic radiations*. Just as for the optic nerve, there is one of these for each eye, but we will ignore binocular effects in what follows. Neurons in the early areas of the visual cortex also have receptive fields, and these are larger than those of individual ganglion cells because there is a further spreading out of synaptic connections where the optic radiations reach the cortex. The LGN is actually a rather busy junction box because it also receives return signals (via what are known as *corticofugal fibres*) from the cerebral cortex, and indeed from several other brain structures.[m] The function of this neural circuitry is now beginning to be understood, but that is an issue which lies beyond our immediate needs here.

More to the point is the fact that Haldan Hartline,[17] Horace Barlow[18] and Stephen Kuffler,[19] and many who followed their pioneering lead, investigated the response characteristics of various cells in the visual system, and established that these neurons react optimally to uneven distributions of light and dark. If one inserts a measuring probe into the axon of a ganglion cell, one finds that the maximum response is observed when a certain area of the field of view is illuminated, and the measured signal is at its greatest when that illumination consists either of a bright central spot surrounded by a dark ring or, conversely, a dark central spot surrounded by a bright ring, depending upon which ganglion cell is actually being investigated. Both of these patterns are now usually referred to by the term *centre-surround fields*. Putting it simply, a ganglion cell responds either to a doughnut or to a doughnut hole.

l / Just as a neuron later in the visual pathway has a receptive field consisting of a number of neurons earlier in the system, so can a neuron in an early region be said to have a *projective field* consisting of a number of neurons in a later region.
m / Such *reverse projections* will figure prominently in the next chapter.

The typical centre-surround receptive field can handle far less detail than that encompassed in our T–C network's three-by-three array.

The details detected by individual receptive fields may be regarded as the elementary building blocks of which visual scenes are composed. Just how the visual system consolidates them into larger items is a subject that will have to await the discussion of such things in the next chapter, and also in chapter 9. Meanwhile, there is an important point to be made concerning the scale of receptive fields in the visual system. It is certainly not the case that a single one of these can distinguish between a written T and a written C. If that were the case, the eyes would be able to take in a much greater density of information than they can in practice. The nine-pixel pictures used above might seem to be simple things, but in reality the visual system has to use many receptive fields in order to scan such an array's contents.

Even though our T–C network's three-by-three array required much of our visual resources, it nevertheless had severe limitations. It permitted the letters very little room for manoeuvering. The letter T could only rotate, and although the Cs could also translate (i.e. move laterally), this translation was only through a distance equal to the width of a single pixel (when an easy C became a difficult C, or vice-versa). When we view real scenes, we have to be able to recognize things irrespective of their position in the visual field. The question thus arises as to whether a network could achieve a similar degree of flexibility. What happens if we increase the picture to an eight-by-eight array, say, and thereby give the Ts and Cs room to move about? The network would clearly need more neurons, to compensate for the increase in size, but would it still be able to distinguish between those two letters, given suitable back-propagation training?

This question was addressed by David Rumelhart, Geoffrey Hinton and Ronald Williams in 1986, and the answer was in the affirmative.[20] But it was a conditional affirmative, because good discrimination was achieved only if the neurons in the hidden layer were allocated receptive fields (see figure 6.25). In the examples discussed in the previous sections of this chapter, each neuron in the hidden layer received signals from all the neurons in the input layer. In the investigation carried out by Rumelhart and his colleagues, a given neuron in the hidden layer received signals from only a limited patch of the input layer. As I mentioned earlier, a full discussion of the visual system will have to wait until the next chapter, but it should be noted here that there is more to a receptive field than merely a projecting area. In general, some parts of the latter have an excitatory influence on the receiving neuron's activity level whereas other parts tend to inhibit that activity. In practice, this distribution of excitation and inhibition has circular symmetry in the early stages of the visual system, but other symmetries begin to impose themselves as one penetrates toward the higher reaches of the system.

Because they were attempting to emulate the performance of the overall system, Rumelhart and his colleagues were justified in incorporating these higher-order symmetries in their assortment of receptive fields. Some of these were more suited to detecting bits of Ts than bits of Cs, while for other

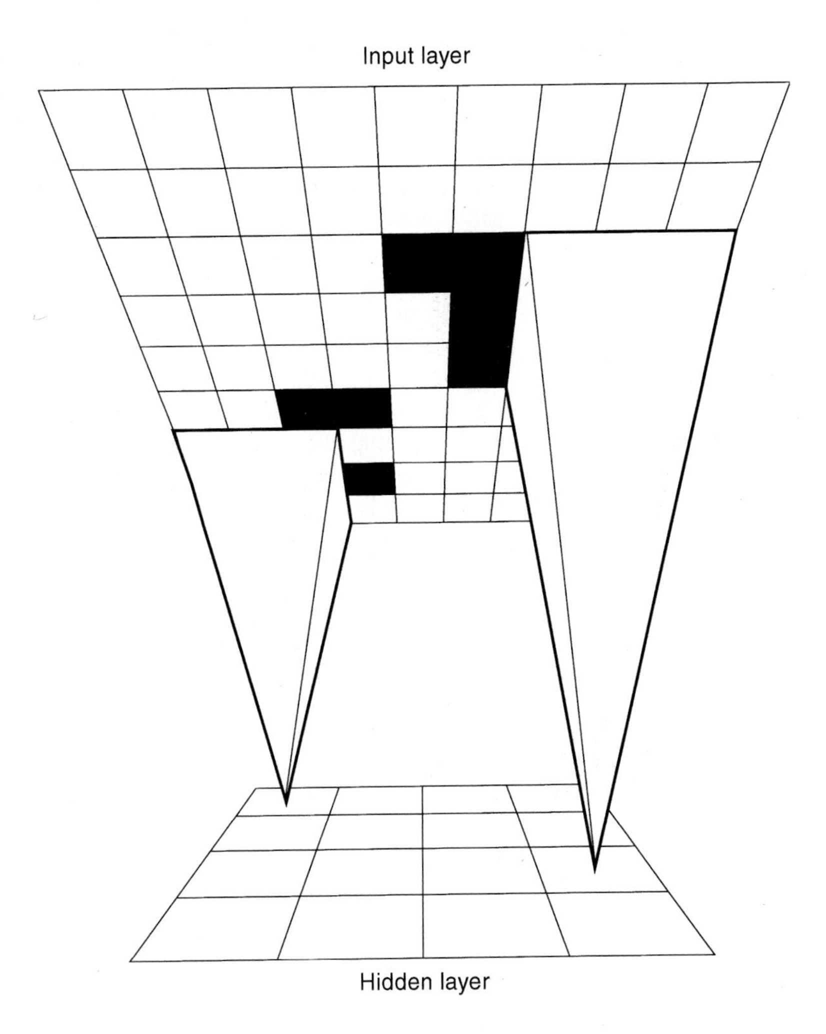

Figure 6.25
The problem of permitting a network to distinguish between the letters T and C, irrespective of their positions in an array larger than the minimal three-by-three grid, was solved by Rumelhart, Hinton and Williams through introduction of receptive fields. In this strategy, a given neuron in the hidden layer receives signals from only a limited area of the input layer, rather than from all that layer. In the illustration shown here, each receptive field is a three-by-three array, and adjacent fields overlap by just a single row of three pixels (or by a single pixel in the case of diagonally displaced fields). The hidden layer is shown as a plane, rather than as a line of neurons, but these arrangements are equivalent in network terms, as explained earlier. The two receptive fields shown here are covering two pixels of a letter C and one pixel of a letter T, respectively.

receptive fields it was the other way around. It was this that enabled them to achieve the desired discrimination, irrespective of where in the picture their Ts and Cs were placed; they succeeded in training a network that possessed what is referred to as *translation-invariance*. The smaller version of the problem, which was discussed in the preceding sections of this chapter had, of course, already achieved *rotation-invariance*.

Position and orientation, and for that matter proximity,[n] are of the essence in vision, and invariance to these parameters is clearly essential. Other sensory modalities have their own invariance requirements. We can recognize tunes, for example, irrespective of the key they are played in, and a surface that feels rough to the finger feels rough to the elbow as well. Our sensory

n / Proximity-invariance would be achieved by a network if it could distinguish a letter, say, irrespective of that letter's size.

systems generalize without us being aware of the fact. They effortlessly interpret the particular in terms of the general, thereby converting a *percept* to a *concept*. We have just seen that a simple neural network, in the hands of Rumelhart, Hinton and Williams, was doing the same thing. Is that comparison a valid one, however? Later in this chapter, we will have to attempt an evaluation of these and similar efforts, and ask how much relevance they have to the quest of understanding the brain. But before we embark on that, we ought to consider another achievement of such a network which caused a great deal of excitement.

Reader in English

In the opening chapter of this book, I referred to games of chess played against a silent computer. It was silent because it lacked a speech synthesizer, but even if it had been equipped with such a thing, its verbal reactions would probably have been drawn from a stereotyped set of responses. In 1986, Terrence Sejnowski and Charles Rosenberg trained a neural network to read English text.[21] The software that they produced to achieve this is now regarded as a classic example of a network's discriminative potential. They gave it the name NETtalk.

The network was trained by exposing it to a thousand of the most common English words, and to the phonemes of which these words are composed. The network was thus confronted with numerous mappings, and the interneuron connections were gradually adjusted until the fit was optimal, the adopted procedure being that of back propagation.

It would be pointless to present the input layer of such a network with a single letter, of course, because this is insufficient to define the letter's pronunciation. Take for example the sentence MARY HAD A LITTLE LAMB. The letter A appears four times, and it is given three different intonations. This is because pronunciation of a letter depends on the surrounding letters, and in NETtalk the context was defined by three letters either side of the centrally positioned letter. This defined what is referred to as the *input window*, and in the case of NETtalk it was thus seven letters wide. From what was stated in the preceding section, we could call this window a *receptive field*. There are 26 letters in the English alphabet, and there are also space and punctuation symbols to be allowed for. This brings the number of alternatives which can be presented at any input position to 29. Sejnowski and Rosenberg went for algebraic separability, as described earlier, and employed 29 neurons to code for these 29 alternatives. When any one of these was presented at a given position in the window, one neuron was activated while the remaining 28 remained inactive. The seven-position input layer therefore comprised a total of 203 ($=7 \times 29$) neurons. The various phonemes, stresses and syllable boundaries in the English language happens to be 26, and because the strategy of *algebraic separability* was also used for the single output position, this was the number of neurons required in that location. Sejnowski and Rosenberg used a hidden layer that comprised 80 neurons.

It is worth stressing that NETtalk differed from the Rumelhart–Hinton–Williams approach to the expanded T–C problem in one important respect. In that other work, many receptive fields were simultaneously being used to analyse for correlations in the input, and to feed the effects of those correlations into the neurons of the hidden layer. The receptive fields in that previous example were *static*; their positions remained fixed throughout. In NETtalk, there is only one receptive field (or window) and it *moves*; it systematically scans the text, as would the eye of a reader. If we use left and right parentheses to indicate the extremities of this receptive field, successive stages of the scanning of the above sentence may be indicated thus: MAR(Y HAD A) LITTLE LAMB; MARY(HAD A)LITTLE LAMB; MARY (HAD A L)ITTLE LAMB, MARY H(AD A LI)TTLE LAMB, and so on.

The 203–80–26 network employed in NETtalk acquired its expertise through the back-propagation method. The strings of words, and their corresponding phonemes, were stepped through the input and output layers during the training on the 1000-word set of mappings. I have not had the opportunity of inspecting the training set used by Sejnowski and Rosenberg, but I doubt whether it contained the words MARY or LAMB; I suspect that neither of them is amongst the one thousand most used English words. The word THE certainly is amongst the thousand, of course, and for the sake of argument let us assume that the same is true of the words CARROT, YOUR and CATCH. A typical situation, during the training period, would be as indicated in figure 6.26.

The network is being exposed to part of the phrase THE CARROT, and the letters and spaces currently in the window are (HE CARR). Appropriate pronunciation of the centrally positioned letter C is thus the target. The surrounding letters, three on either side, provide the context, and they show that the correct phoneme is /k/, which is voiced like the first letter in *ken*. At the next training step, the window will contain (E CARRO), and the central letter A will correspond to the phoneme /@/, which is voiced like the

Figure 6.26
A token representation of the 203–80–26 network NETtalk, trained by Terrence Sejnowski and Charles Rosenberg to map strings of letters onto the correct English phonemes. Twenty-nine neurons are required at each location in the seven-membered input window, rather than the four indicated. Similarly, there are twenty-six different phonemes for the network to choose between, rather than the three indicated. The network is shown correctly mapping the centrally located letter C onto the appropriate phoneme /*k*/.

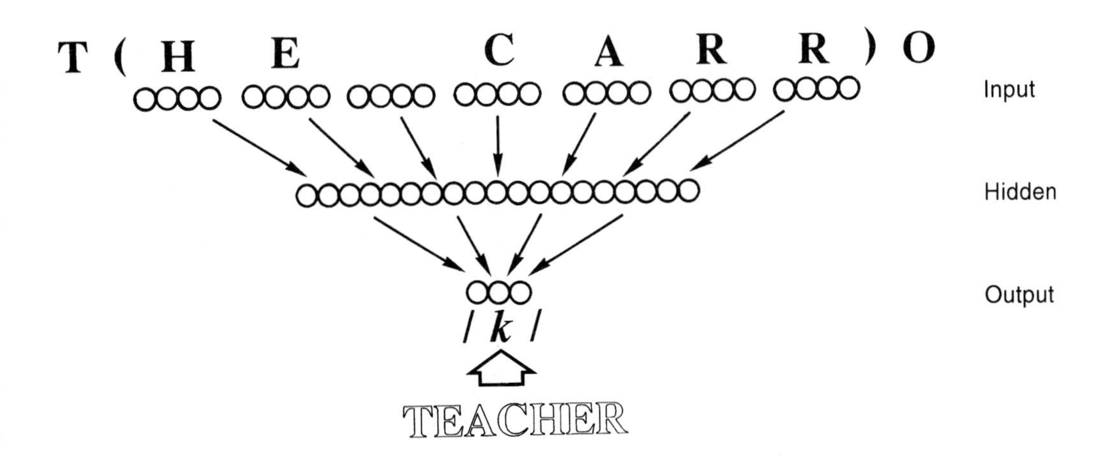

central letter in the word *bat*. If, after the training, the network is confronted with a window containing (UR CATC), during the reading of the phrase YOUR CATCH, for example, it will correctly respond with the phoneme /k/. It must be emphasized that the arrows in the above figure merely indicate direction, not the actual synaptic connections, of which the model has nearly twenty thousand. Likewise, there is not room to draw in all the neurons, and only a token number have been included at each position. The key to NETtalk's remarkable discriminatory powers lies in the hidden layer, of course, for it is here that the various correlations between features of the input are captured.

In NETtalk, the phoneme data that appear at the output layer are fed into a speech synthesizer, and listening to the ersatz human voice is an eerie experience.[o] During the early stages of training, the machine can manage nothing better than a continuous stream of gibberish, but its performance steadily improves towards the maximum of somewhat better than 90 per cent of all words pronounced correctly. In one demonstration, the pitch of the speech synthesizer was set so as to match the voice of a female child, and listening to it was quite moving. It was like observing the fumbling of a real child as it initially struggled to pronounce the various words properly, and then gradually became more confident, even with rather difficult words. The importance of NETtalk lay in its demonstration of the way in which a problem can be made suitable for handling by a neural network, the critical step in the procedure being *the choice of representation for the input and output data*.

One other point should be made concerning NETtalk, because it has a bearing on most implementations of the neural network approach, even though the degree varies from one example to another. This is that the size of the input window (receptive field) will occasionally prove to be too small. Such trouble could be obviated, of course, by the simple expedient of increasing the number of neurons in the input layer, but in practice it will not pay to thus cater for the once-in-a-blue-moon event. For many applications, it is sufficient to merely aim for a high rate of correct output predictions rather than to demand nothing less than perfection. In the case of NETtalk, for example, the word THE admits of two different pronunciations, as in the cases THE ONE and THE ONLY. When the letter E is centrally located in the seven-letter window employed by Sejnowski and Rosenberg, the windows will contain the stretches (THE ON) and (THE ON) for the two cases. The contents of the window in these examples are thus identical, so the desired differentiation in the intonations is beyond the capacity of the network. An increase from seven letters to nine letters in the window would cure the trouble in this instance. An even more difficult case occurs with the word READ. It too can be pronounced in two different ways, depending upon whether it refers to the past or present

o / It is a pity that letters of the ordinary alphabet are used to denote many of the phonemes; it would have been preferable to use other symbols. As it is, confusion is avoided by the use of diagonals as the initial and final symbols of the triplet employed for each phoneme; /k/, /a/, and so on.

tense, and the length of text required to remove the ambiguity can be quite long.

There is not room here to go into the many other interesting ramifications of NETtalk, so I will confine myself to just one more point. It was found that the different neurons in the hidden layer appear to have different jobs, as the network pronounces various phonemes. Sejnowski and Rosenberg made a thorough study of the hidden layer activities and they found that some of the neurons were mostly involved with vowels whereas others were clearly most active when combinations of consonants were presented to the central part of the input layer. Yet other hidden-layer neurons seemed to be called into action for specific combinations of both vowels and consonants.

A sheep in wolf's clothing

The three-layer neural network has become a highly valued cognitive tool, and its applications are widespread. To understand why this should be so, one has to interpret the word *cognitive* in the broadest sense, and be prepared to apply it to any issue which requires the perception of correlations. In the commercial sector, the perceptron has found employment in such disparate tasks as automatic recognition of hand-written characters and the assessment of credit and insurance risks. The former, which is of interest to postal services, is close to the problems discussed earlier in this chapter. In the cases of credit and insurance assessment, the input layer is fed data on the age, personal status, habits, economic situation, and various other details of the applicant, and the output layer of the trained network then indicates whether he or she is a good or bad risk. I will not attempt to produce an exhaustive list of other applications. Suffice it to say that the field is large. Anything requiring evaluation of data can in principle be handled by such a network, the main proviso being that suitable representations can be found for both the input and the output. It is worth stressing that a network is also useful when it is trained to produce *multiple answers*. The composite picture shown in figure 6.27 illustrates a typical example of this more exacting type of challenge. If the suitably structured network has previously been exposed to a sufficiently large set of drawn digits, and adequately trained on them, it will be able to reply with *one*, *two* and *four*.

The three-layer network nevertheless has severe technical limitations, one of which is often referred to as the *multiple minimum problem*. To appreciate the nature of this difficulty, one can imagine representing the current error at the network's output layer by the local height at certain coordinates of a contour map. In the geographical counterpart of such a map, the coordinates are simply those of latitude and longitude, while the height is usually measured with reference to sea level. In the version relevant to the neural network, the height is measured with reference to the zero output error of the perfectly trained network. The number of coordinates in the network is dictated by its size, and it will usually be relatively large. This makes the analogy to geographical landscapes a rather loose one, but it nevertheless provides

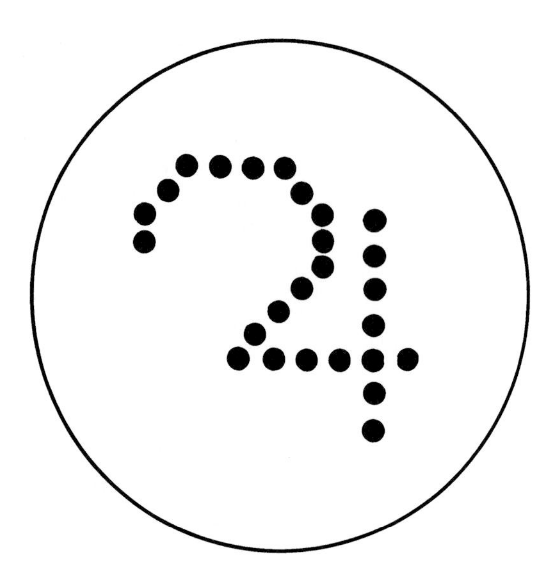

Figure 6.27
When this composite diagram is presented to the input layer of a suitably trained network possessing the appropriate architecture, it should be able to discern a 4, a 2 and a 1.

a useful visual impression of the essentials. Now when a network is trained by back propagation, the prescription for adjusting the synaptic strengths and somatic thresholds[p] is such that it is constantly searching for the direction along which the height decreases most rapidly; it is looking for the quickest route down the hill. As Marvin Minsky and Seymour Papert have emphasized,[22] such a *gradient descent* approach will always risk stranding the network in one of the landscape's local depressions. The height above sea level, as it were, may still be considerable, but the network is trapped because all directions lead uphill. As one might expect, there have been attempts to circumvent this problem, effectively by giving the network a pair of seven-league boots, but the boots have not been accompanied by information about the best direction in which to step out of the hole. Algorithms of this type can only resort to attempting numerous randomly directed steps, and for large networks this can be very costly in computational resources.

When we compare such a network with the real brain, other types of difficulty are seen to arise. We have already noted that equating the input layer with the retina, in the case of pictorial tasks, is altogether too facile. It is easy to underestimate the amount of visual machinery required to differentiate between those innocent-looking black and white squares used in the previous sections of this chapter. Then again, it is not easy to imagine the strengths of real synapses being amenable to the degree of fine adjustment required during back-propagation training of a network. Moreover, the training of a network usually requires thousands of forward-error-backward cycles of adjustment, and this is difficult to reconcile with the fact that humans can

p / Strictly speaking, the network uses scale-setting biases rather than thresholds, but we can safely ignore the difference in this argument.

learn things during a single brief exposure, if those things are sufficiently important. An even greater difficulty arises in connection with the output layer. One must ask *what agency in the brain is able to function as the network's teacher?* Some ingenious suggestions have been made as to how the errors that appear at the output layer, during both training and failed recognition, could be automatically assessed at the neuronal level, but none of them has won widespread support. One could say that the artificial neural network has a homunculus problem, if it is likened to the real brain. The basic trouble may well be that the neural networks in the real brain are so structured that there is no output layer of the type we have been considering here. There is much colloquial reference, nowadays, to the metaphorical *bottom line*, meaning the ultimate outcome of what is under consideration. Brain function often also leads to an outcome, but this does not have to imply that it appears at a bottom line of neurons.

Francis Crick[23] has drawn attention to other shortcomings of the network approach. The main thrust of his critical view was a philosophical one. He noted that those who write on the subject of neural networks often appear to be striving for powerful general principles for handling information, expressible in deep mathematical form, 'if only to give an air of intellectual respectability to an otherwise rather low-brow enterprise'. In contrast, he argued, evolution must perforce build on what has gone before. Given the Darwinian strictures, Nature must achieve its progress tinker-toy fashion, and any modification that better equips an organism for competition will be incorporated into the design irrespective of whether it is elegant or not. One could say that the teeth by which evolution inexorably ratchets itself upward are uneven, but they are serviceable for all that. Crick's final verdict on the matter is one that this book is attempting to comply with: if one wishes to find out how the real brain works, the best way is to look at its anatomy and try to understand how this determines its physiology.

Other profound notes of caution have been sounded by Jerry Fodor and Zenon Pylyshyn.[24] In essence, they demonstrated the limitations of a principle that underlies so much of network function, namely *association*. The idea of association is a venerable one. It may be found in Aristotle's writings on memory and reminiscence, around 400 BC, and we saw in the previous chapter that it was prominent in the work of Steinbuch, Kohonen and their successors about three decades ago.[25] The trouble is that association is useful only up to a certain level of complexity, whereafter it actually becomes a burden. The most accessible examples of such failure come from grammatical, as opposed to lexical, considerations, and they involve syntax and semantics.

Before considering what Fodor and Pylyshyn had to say on these matters, it would be just as well to remind ourselves of a few definitions. The word *lexical* is applied to the individual words of a language, and it makes no reference to their juxtaposition in sentences. When we use a dictionary, our input is a word while the output is a meaning, and it is not too far-fetched to imagine a network being able to deputize for such a compilation. The *grammar* of a language, on the other hand, deals with the relationships between patterns of words and meanings. It has two contributing divisions. *Syntax* deals with sentence construction, and its domain is the permitted arrangements of words, while *semantics* is concerned with the meanings associated with those arrangements.

The main thrust of the Fodor–Pylyshyn argument is that the connectionist approach is unable to handle the syntax and semantics of language, and the reason for this inadequacy lies in its inability to extract rules from examples.[q] The connectionist network, in its present stage of development at least, is not able to derive the general from the particular. This is a serious disadvantage, given that such rules are the very essence of syntax. Fodor and Pylyshyn cite some telling examples. If the sentence *John went to the store and Mary went to the store* is true, this logically entails the truth of the sentence *Mary went to the store*. It would not be difficult to teach a network such association between two sentences, but how would one go about teaching it to extract the underlying rule. Once we have learned this piece of syntax, we immediately see that the truth of the sentence *The moon is high and the night is beautiful* logically implies the truth of the sentence *The moon is high*. Getting a neural network to achieve such transfer of logic to a different set of words is still difficult in the present state of this particular art. For even with the ability to extract such rules, how is the network then going to tackle more difficult sentence constructions such as that in another of the examples cited by Fodor and Pylyshyn. The truth of the statement *The flag is red, white and blue* does not imply the truth of *The flag is blue*.

The failure of the traditional neural network to handle more than a small degree of novelty is seen to be serious even for phrases. One of Fodor's examples is the two-word case of *antimissile missile*. Once we understand that this refers to a missile which shoots down missiles, it demands little of us to appreciate that an *anti-antimissile missile* is a missile that shoots down missiles which shoot down missiles. And given our grasp of the underlying rule that affords such understanding, we have no difficulty seeing that an *antithief thief* is the type of person being invoked in the proverb *It takes a thief to catch a thief*. To use Fodor and Pylyshyn's own words, the ability to entertain a given thought implies the ability to entertain other thoughts with semantically related contents, and they therefore felt able to confidently claim that such arguments make a powerful case against mind/brain architecture being connectionist at the cognitive level.

Despite the apparent validity of these criticisms of connectionism, it cannot be denied that the brain's anatomy is strongly suggestive of parallel processing. It is equally clear that our minds function in a sequential manner. The steps by which human logic is exercised provide ample evidence that this

q / The word *connectionist* is now commonly applied to all neural networks that perform parallel distributed processing, and it thus embraces not only the three-neuronal-layer type but also other, possibly more sophisticated, cognitive strategies that employ the same general approach.

is the case. The two mechanisms are not incompatible. It is perfectly feasible that thought involves sequential events, each of which demands parallel distributed processing. Perhaps the best endorsement of this idea comes from the limitations of thought. Sentences such as *The house that Jack built* give us no trouble. But as Minsky and Papert have noted, we all fall down on expressions like *The cheese that the rat that the cat that the dog bit chased ate.* Finding out what goes with what in such cases can place severe demands on our ability to handle sequences (that is to say on our short-term memory), even though the individual items are no more complicated than those used in the shorter sentence.

Minsky and Papert produced a pair of apparently similar figures which nicely make the point in a visual manner. These have become classics in the field, and should be reproduced here. They are shown in figure 6.28, and apart from their different orientations, they look like a pair of twins. A suitably trained perceptron would surely arrive at just that conclusion. It is only when we carefully pan along the curves that we discover their profound difference. And it is this need to pan that reveals the sequential requirements of our minds. As Andy Clark recently put it, what counts is not the presence or absence of some static, text-like element but the ability of the system to use the information it possesses in a flexible and open-ended variety of ways.[26]

There is every reason to be optimistic that such flexibility will become common amongst future artificial neural networks. A good example of what can be achieved already is seen in the demonstration by Derrick Nguyen and Bernard Widrow that a modified network can be trained to back up an articulated vehicle.[27] This is relevant because it involved sequentially applied controls. In practical situations, the input often describes the current conditions while the output indicates the most desireable response. Anyone who has ever tried to reverse an automobile that is being used to tow a trailer will know that this is not an easy task. The normal reactions that come almost second nature from one's previous driving experience are simply not appropriate, let alone adequate, and one has to quickly learn a new set of responses. Even skilled truck drivers often have to make several attempts at positioning the trailer so that the final backing manoeuver can be smoothly executed, the successful last reverse being preceded by several forward and backward adjustments. Getting the procedure right without these preliminaries is even more difficult.

Nguyen and Widrow trained their network to perform such a truck-backing operation, and their simulated vehicle could be successfully reversed

Figure 6.28
This classic pair of figures, drawn by Marvin Minsky and Seymour Papert, exposes the inadequacy of the typical artificial neural network. Apart from their different orientations, the figures appear to be quite similar, and would be classified as being so by the typical network. In reality, the figures differ in a fundamental respect: that on the right comprises two topological components whereas that on the left has only one. The human cognitive system experiences a similar difficulty, and conscious attention has to be used in order to discern the difference. This effort involves short-term memory, in order to keep track of the lines and the spaces.

toward a chosen point on an imaginary ramp, despite the fact that it was fully articulated. The variable parameters that described the situation, at any instant, were the (fixed) position on the ramp up which the truck was to be backed, the position of the centre of the truck's rear end, the position of the swivel joint between the truck's cab and the trailer, and the angle between the long axis of the cab and the axis of the trailer. Figure 6.29 shows an idealized representation of such a truck, in side view, while the essence of the docking task is best appreciated by contemplating the situation in plan view.

The goal of the operation was to have the back of the truck lying parallel to the edge of the ramp, and simultaneously for the centre of that back end to be at the chosen point on the ramp. As we know from experience, any departure from this goal could have caused the truck to topple over as it was reversed up the ramp.

Nguyen and Widrow trained their network in two stages, this division being similar to what would happen in the training of a human driver. First, the relationship between various settings of the controls and the response of the vehicle had to be imprinted on the system. Thereafter, this relationship was incorporated in a network whose input at any instant was the present values of the above parameters while the desired output was new values that were on a feasible route to the goal. As with all network applications, the key lay in the choice of appropriate representations for the various parameters. A complete reversal of the truck was simulated in small steps. Nguyen and Widrow showed that backing up could be undertaken successfully even when the articulated vehicle started out in the most awkward of jackknifed configurations, as the sequence shown in figure 6.30 amply demonstrates.

The network used in this project comprised forty-five units in its hidden layer, and the question naturally arises as to how the expertise was apportioned between these pseudoneurons. In contrast to the examples given in earlier sections of this chapter, however, it was not possible to identify any aspect of the backing up with specific hidden-layer units. Even the smallest adjustments became the collective responsibility of many of these units. The other point to be made, by way of closing this section, is that the network architecture used by Nguyen and Widrow was quite reminiscent of that origi-

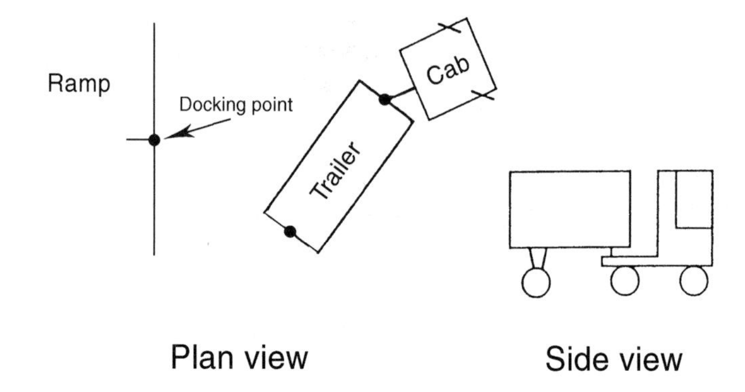

Figure 6.29
The two views of the articulated vehicle reveal the nature of the backing-up task tackled by Nguyen and Widrow's neural network.

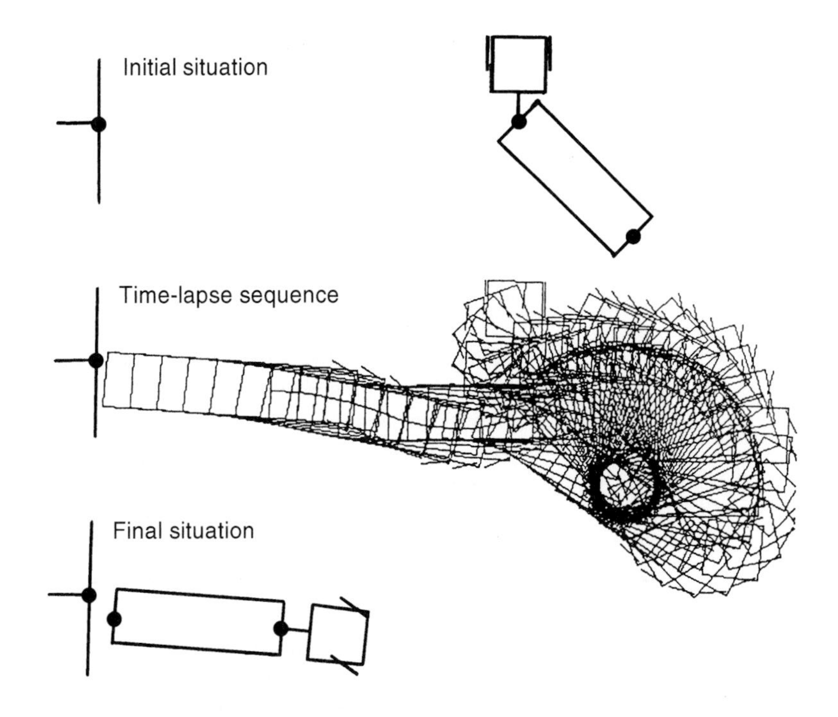

Figure 6.30
The sequence of positions shown in the central part of this diagram indicates the route which Nguyen and Widrow's articulated vehicle was guided through by their trained network. It can be seen that the backing-up was carried out almost to perfection.

nally put forward by Teuvo Kohonen to handle temporal sequences. Part of its output was fed into the network at the input layer. As I mentioned in the previous chapter, it is possible that something akin to this happens in the hippocampus.

Roots and routes[r]

Several sections of this chapter have been devoted to a reasonably detailed description of the current most common neural network strategy. In an attempt to provide a balanced view of the situation, I have also given an account of the inadequacies of the connectionist approach which a number of people have identified. It would be easy to gain the impression that these networks are like toys compared to the real brain. But one should bear in mind that the science of neural networks is still young. The early models and products of any human scientific endeavour invariably seem crude, sometimes even comical. The early versions of neural networks were indeed primitive, but so was knowledge of the brain at the time of their emergence. The field of neural networks continues to mature, and rudimentary ideas are giving way to more advanced concepts. A good case in point is that truck backer-upper developed by Nguyen and Widrow. This has a significance that reaches far

r / Readers less interested in the technical details of artificial neural network hardware may prefer to pass directly to chapter 7, at the first reading.

beyond the reversing of articulated vehicles. It was a demonstration that neural networks can be trained to have predictive capabilities in the mechanical domain, and such capabilities must be possessed by the movement-controlling systems of our own bodies. Such progress makes one optimistic that robots will become gradually better at emulating the capabilities of their human counterparts. In chapter 9, I will make the point that the power of prediction may have more general significance for brain function, and in chapter 10 I will be exploring some of the devastating consequences of its breakdown.

As knowledge of the real brain increases, then, so will the sophistication of neural network strategies. In order to get some idea of what might be accomplished in the future, therefore, it is instructive to contemplate what the interaction between brain science and network modelling produced in the past. There is not room here to do full justice to the subject, and I must limit myself to an admittedly personal view of some of the past mileposts. These have been briefly touched upon earlier in this chapter, but we ought to go into some of the technical details before returning to the real brain. Justification for doing so stems from our ultimate need to judge whether conscious computers are likely to emerge in the foreseeable future.

Frank Rosenblatt's speculations, before he set out to construct the first perceptron, made him realize that brain theorists had been making the cardinal error of attributing too much ability to the brain.[28] Given the respect usually accorded the brain's grey matter, this conclusion might seem odd. But Rosenblatt's scepticism was justified, because it was specifically addressed to the issue of logic. He argued that the brain is not able to directly mediate logical functions, because logic appeared only recently in mankind's culture. It is unlikely that this faculty became wired into the cerebral cortex when our species evolved from its predecessor, and then remained dormant for thousands of years. Rosenblatt perceived that the brain's forté is its remarkable power of association, and it was this that he set out to imitate in his apparatus.

His basic approach was quite straightforward. It attempted to mimic the gross architecture of part of the nervous system, namely that associated with vision. His scheme was rather idealized, however, and although the input layer was referred to as the retina, Rosenblatt was not claiming that his device really was a surrogate eye. His first network comprised three layers of units, in fact, the input 'retina' being a sort of generalized sensory surface; it could just as well have been taken to represent the sense of touch, or even hearing. The units in that first layer were referred to as *S-units*. A number of such units were connected to a second layer, comprising the *A-units*, the task of this being to capture associations between different parts of the input. Finally, the A-units were connected to a third layer of response units, the *R-units*, and the goal of the enterprise was to have given R-units activated for specific patterns of input to the S-units. In other words, Rosenblatt was attempting to build a machine with cognitive abilities. Because we have already considered such networks in considerable detail, we would obviously be inclined to identify the letters S, A and R with the input, hidden and output layers, respectively.

Although other strategies would probably have been workable, Rosenblatt decided at the outset that only one R-unit should be allowed to display activity at any given time. He therefore built in a set of inhibitory connections, via which the most active R-unit, at any moment, could suppress its nearest rivals. Such a tactic has subsequently become standard practice in this type of device, and it was referred to by its standard name, Winner-Take-All, in chapter 4. The actual implementation of this stratagem involved setting up *reciprocal connections* between the A-units and the R-units. Such connections were *not* a feature of the T–C network, of course; neither was the *lateral inhibition* in the output layer.

Rosenblatt's first working perceptron had an input layer consisting of 400 S-units, arranged on a 20 × 20 grid. There were 512 A-units, and just 8 R-units. This first machine was therefore a quite modest affair, which aimed at being able to discriminate between only eight different classes of input. It must be emphasized, however, that the latter were not merely eight different patterns, because the goal was to produce a machine that could generalize; it was hoped that the device would be able to categorize many other novel patterns presented to it, correctly linking them to those patterns in the training set to which they bore the greatest resemblance. This goal was, by and large, achieved.

The S-units were actually photocells, mounted in the picture plane of a camera to which the stimulus pictures were exposed.[s] Each S-unit was permitted to have connections with up to forty A-units (out of a possible 512, of course), the actually choice being made on the basis of a table of random numbers. Comparing this strategy with the one used by Rumelhart, Hinton and Williams for the T–C problem with translational invariance, one could say that the Rosenblatt approach employed *randomized receptive fields*. The connections between the A-units and the R-units naturally had to have variable strengths, otherwise the desired matching of stimulus and response would not have been attainable. This was accomplished by establishing the connections via motor-driven potentiometers.[t]

The training or learning procedure involved several stages, the general idea being to reinforce some connections at the expense of others, so as to build in a differentiated response. Following presentation of one of the predetermined input patterns to the array of S-units, and the striking up of activity in some of the A-units, some of the R-units would begin to show a response. As mentioned earlier, the R-unit displaying the most activity was allowed to

s / A photocell is a device for measuring light intensity; there is a miniature version of such an item in every camera that has a built-in exposure meter.

t / A potentiometer is a simple device for varying the output voltage, for a given input voltage, everyday examples being those that provide the volume control in a radio, gramophone, tape-recorder or television. Another example is seen in the type of wall switch that allows one to dim the illumination in a room.

win, and the activities of the A-units not connected to it were suppressed. An A-unit that was active when the stimulating pattern was presented automatically had its activity increased. Conversely, the activities of many A-units were decreased, because they were unfavorably disposed towards the presented pattern. Because the perceptron could thus be said to learn from its mistakes, it was displaying one aspect of intelligent behaviour.

Following completion of the learning phase, presentation of one of the training patterns would produce stronger activity in one of the A-units than previously, and this in turn would produce a greater response in one of the R-units. Similarly, presentation of one of the other training patterns would produce activity in a different R-unit. Most remarkably, however, presentation of a pattern that was not in the training set, but which bore a sufficient resemblance to one of the training patterns, also produced a strong response in the appropriate R-unit. A particular example of the later type of pattern would be a degraded version of one of the training patterns. This is the type of situation that would arise in practice if the input pattern was a picture that had acquired scratches because of mishandling. Such extraneous picture features are usually referred to as noise, and the perceptron functioned despite its presence.

As exciting as Rosenblatt's inaugural perceptron undoubtedly was, it did suffer from severe limitations. It was soon realized that certain learning tasks were beyond its capacity, even though these seemed to make quite modest demands of the device. Thus although the machine was organizing itself to recognize certain juxtapositions in the input patterns, during the training period, it was not clever at detecting the underlying logic in them. It was failing to form concepts, despite its dexterity at handling percepts. As Rosenblatt himself admitted, the perceptron was too literal and inflexible, and it was something of a dunce when it came to making abstractions. It was, to use Rosenblatt's own words, rather reminiscent of a patient with brain damage.

The appearance of the first perceptron sent a buzz through the scientific community, and when it was subsequently demonstrated, by theoretical analysis, that such a device would always converge on a stable set of values for its inter-unit connections, the way seemed clear for the production of truly intelligent machines. That some sceptics eyed these ominous developments as harbingers of Aldous Huxley's *Brave New World* could not dampen the buoyant optimism of the new brain technocrats. The artificial brain was apparently here to stay, and it seemed ripe for further development.

The speed with which the first perceptron undertook its learning tasks was not particularly impressive, however, and this was clearly a point that required attention if the technique was ever to become viable. A notable attempt to achieve just that was made at the end of the 1950s by Bernard Widrow and Marcian Hoff, and they drew their inspiration from the physiological properties of real neurons.[29] Indeed, they initially named their device to reflect this influence, the acronym being ADALINE, which stood for ADAptive LInear NEuron. Later, when the field of neural networks suffered some notable defeats, and fell from grace with those who allocated research grants, Widrow and Hoff recanted and changed the attribution to ADAptive LINear Element. This deft piece of literary legerdemain thus permitted them to retain the original acronym.

The appeal of ADALINE lay in its simplicity, which, in turn, derived from its closeness to biology. The basic concept, the adaptive neuron, functioned as a threshold logic unit, the connections to its input units being variable. Therein lay the device's ability to learn from experience. The idealized neuron summed the inputs that it received, the individual items in these inputs collectively representing patterns to be learned. In order to give the device an extra measure of adaptability, Widrow and Hoff included a bias factor, an embellishment that was later to become an almost standard feature of such learning machines.

Mimicking what happens in its biological counterpart, the idealized neuron emitted a signal if the sum of its inputs exceeded a certain threshold value, the output signal being set equal to unity for mathematical convenience.[u] If the sum of the inputs to a real neuron is less than the threshold value, it remains 'silent', but in the Widrow–Hoff scheme this situation resulted in the neuron emitting a minus unity. In this particular aspect, the choice was clearly motivated by mathematical expediency rather than a desire to faithfully reflect biological reality.

Physically, the ADALINE looked very much like many other pieces of electrical apparatus. There was a control panel with a four-by-four array of toggle switches, one for each member of the 16-element input pattern.[v] There was also a toggle switch via which the operator could tell the machine whether a particular input pattern was to correspond to a +1 or a −1. The other obvious feature of the control board was its battery of dials, these too being arranged in a four-by-four array, plus subsidiary dials related to the output. The system learned something from each pattern, and its experience was stored in the strengths of the connections between the inputs and the neuron.

The central idea of the Widrow–Hoff scheme was that the departure from ideal adjustment, at any stage of the learning process, could be looked upon as an error, and one which had as many components as there were input-to-neuron connections. Whence the need for the sixteen dials. This became a standard aspect of later machines, and we earlier saw that it is used in contemporary networks. To emphasize their aspirations of emulating the performance of a real brain, Widrow and Hoff used anthropomorphic terminology for the various functions of their device. They likened the collective efforts of the connections to a 'worker', while the agency that adjusted these connections was referred to as the 'boss'.

The learning procedure began with the presentation of the first pattern, which was of course reflected in the on–off arrangement of the four-by-four array of toggle switches. The output toggle switch was then set to

u / The output was thus *normalized*, to use the standard term for such rationalization.

v / Such picture elements would be called *pixels* in today's parlance.

indicate the desired output, the choice being essentially either *yes* or *no*. The error corresponding to the first pattern, for the given arbitrary settings of the connection strengths, could be read from the main dial, and the gain factors were then all changed by the same amount, so as to bring the error to zero. Because there were seventeen gain factors in all, sixteen for each member of the four-by-four array and one for the overall level, this step involved changing each of these by one seventeenth of the total error. The second pattern was then set up on the array of input toggle switches, and the adjustment process was repeated.

Presentation of the first pattern again, following this second set of adjustments, would not in general be expected to give a situation in which the error was zero. On the contrary, there would be a small residual error, but this could be brought to zero by repeating the error-annulment procedure. Further patterns were incorporated into the network in like manner, the small adjustments to the other patterns being undertaken in a systematic fashion. It was found that the network gradually homed in on a state in which all the errors were zero, and this of course corresponded to a situation in which all the patterns had been learned.

Widrow and Hoff tested their device on a series of patterns which were, perforce, simple; there is not much scope for imagination when it comes to colouring the various squares in a four-by-four array either black or white, even though it is larger than the three-by-three array we used earlier in connection with the T–C problem. In fact, they chose primitive representations of the letters T, G and F, the corresponding optimal settings for the output dial being +60, 0 and –60, respectively. It must be remembered that there are many ways in which each of these letters can be placed on such a chessboard grid, because both lateral shifts and rotations are possible. (To a certain limited extent, the letters can even be drawn with different sizes.) Widrow and Hoff found that the overall error in such a classification experiment fell to essentially zero by the time fifty pattern presentations had been made. The performance of the trained machine was quite impressive. It could correctly detect a letter which was written in a part of the array that had not been used during the training period, the requirement merely being that the same letter had been shown to other regions of the four-by-four grid.

Although the adjustments were made by a human operator, in the original trials of the device, the routine was a purely mechanical one which required no thought. It would thus have been possible to fully automate the corrections, and Widrow and Hoff proposed an electronic version in which the various gain values would be stored magnetically. They also foresaw the use of ferromagnetic films for this latter task, and they expressed the belief that microelectronic 'neurons' would ultimately be fabricated. Although this dream has essentially come to fruition, the form of today's counterparts of the ADALINE have rather different designs than the device produced by Widrow and Hoff.

When Widrow and Hoff dreamt of seeing their ADALINE reproduced as a microcircuit, a sort of Lilliputian piece of neural hardware, they could not have guessed the extent to which such miniaturization was going to be realized. Their work was carried out before the advent of large-scale integration of transistorized circuits, a development which has of course revolutionized computing in general. Although the general form of the printed circuit is reasonable familiar, its basis being the dual techniques of implanting atoms into the region adjacent to the surface of a solid and depositing thin films onto such a surface, the actual connection with a biological neural circuit is not at all obvious.

One might naively picture the deposited film to take the form a neuron, complete with dendrites and axon. That is not to say that the film's configuration would copy the organic shapes of biological neurons, of course, for it is clear that such features would preferably be idealized. It might be assumed, nevertheless, that the printed circuit of a neural network would look pretty much like the diagrams that have now become such a common feature of the neural network literature. This is not the way things are done in practice, because such a predetermined connectivity would rob the network of the very flexibility that is the neural network's greatest merit.

In order to understand how a neuron is simulated in this age of the printed circuit, we must return to the binary system, in which two alternatives such as black and white, or yes and no, are conveniently represented by the digits 1 and 0. An array of such bits (i.e. binary digits) is suitable for representing something to be processed by a neural system. This is reasonably obvious if the item is actually a picture, as we have already seen in this chapter, and inspection with a magnifying glass of a printed photograph reveals that it is indeed composed of discrete spots. Even non-pictorial items can be coded into a suitable array of bits, however. But this does not imply that the *binary system of numbering* is being used. In other words, a string of bits, when used to illustrate the functioning of a neural network, usually does not mean that the positions of the various digits has any numerical significance. With the printed circuit neural network we do encounter such a case, however, so it would be well to recapitulate the essentials of the binary number system, as it is exploited in the typical digital computer.

The central feature of a modern electronic digital computer is its facility for rapidly manipulating the physical representations of numbers, and in effect this amounts to the continuous and massive rearranging of bits. In order to do this, the physical locations of these bits must be identifiable, and this means that they must be labelled with addresses. A point of possible confusion arises from the fact that these addresses are also most conveniently written in binary form, so in what follows, special care should be made to distinguish between the two uses of the binary concept.

The speed of a modern computer derives in part from what is known as *random access memory*, usually referred to by its acronym RAM. The RAM can be thought of as a strategy for storing bits in, and retrieving them from, memory elements, such that all these elements are equally accessible. A similar tactic was employed in the earlier computers based on vacuum tubes; in the modern counterpart, the switching that permits the rapid access is managed by transistors. The size of a RAM element is defined by the number of bits that it can contain, typical examples being eight-bit and sixteen-bit elements, (and I use words here, rather than numerals, so as to minimize the risk of confusion). Let us consider the former case in detail. The number eight is the cube of the number two; it is the number two raised to the power of three. This means that there must be eight different ways of combining three bits. These eight ways, in increasing order of the corresponding number, are: 000, 001, 010, 011, 100, 101, 110, and finally 111. These are the addresses of the eight different 'boxes' comprised by an eight-bit RAM element.

Each of these eight boxes can hold either a 1 or a 0, and neither of these two digits normally has any relationship to the bits used in the eight addresses; it is simply convenient to use bits for both purposes. Turning to the miniaturized neural network, however, we see a case in which this sharp distinction is annulled. To understand why this should be advantageous, we first note that the bit offers a convenient way of denoting the state of a neuron, since a 1 can represent activity while a 0 will indicate lack of activity. And to obtain a full account of what is happening at a neuron, we need to know what both its inputs and its output are. The trick then is to let each of the three bits in the above address represent the inputs, while the content of the corresponding box represents the output. This would automatically imply that the eight-bit element discussed above would represent a neuron with just three inputs, a rather impoverished neuron. But although the connectivity would thus be much lower than in a real neural network, the system might nevertheless have useful capacities.

There is more to the story, but let us consider what has been achieved so far. If the address is 000, the above artifice would indicate that the neuron has no input, and in general there will only be as many inputs as there are non-zero digits in the address. Another direct consequence of what has been assumed thus far is that the larger the number of 1s in the address, the greater will be the net input to the neuron. Such a system still lacks sufficient flexibility, and it is not close enough to biological reality. The one additional step that we require is to bring synaptic effects into the scheme, and this can be achieved simply by multiplying the bits in the addresses by factors that reflect the state of the relevant synapses. We must bear in mind, moreover, that each of these can be either positive or negative, since some synapses are excitatory whereas others are inhibitory. Similarly, a real neuron will have a threshold, activity being generated only if the arithmetic sum of the inputs exceeds this value. To complete the picture, we need to note just one other point, namely that the eight boxes in our eight-bit RAM, with their associated addresses and contents, can be regarded as representing not eight different neurons but rather eight different sets of inputs and outputs of a single neuron.

In Widrow and Hoff's ADALINE, the modifiability of real synapses was reflected in the readily varied connections between the units of that primitive circuit. The question arises as to how such variability could be built into a printed circuit. The answer is that this is not the way things are usually done with such forms of neural networks. The point is that it is usually most practicable to fabricate a piece of neural hardware for a specific purpose, and the strengths of the various connections can be calculated during the design stage and then simply built into the circuitry. There are now many such dedicated pieces of neural hardware on the market, and the coming years promise to bring a veritable explosion of these devices, both with regard to numbers and to variety of application. The new era of neural computing is upon us.

The first hardware computer based on neural network principles to be fabricated, and the first to be marketed as a commercial product, was the machine developed in the 1960s by Igor Aleksander and his colleagues.[30] Its name, *WISARD*, captured in acronymic form both the surnames of the original research team and the goal of the apparatus, namely Wilkie, Stonham, and Aleksander's Recognition Device. During the thirty or so years of its existence, the machine has been steadily improved, and it remains today one of the most perfected of all neural network computers. Its broad appeal owes much to the graphic nature of its expertise, for its most prominent application has been to visual pattern processing; it can recognize the faces of people, and in so doing it appears to imitate what used to be regarded as a peculiarly human faculty.

In order to appreciate what WISARD was able to achieve, reference must be made to the pivotal studies of the visual system by David Hubel and Torsten Wiesel during the past three decades.[31] And to set that work in perspective, we should begin by recalling those studies by Hartline, Barlow and Kuffler, which revealed the centre-surround arrangement in the receptive fields of retinal ganglion cells. Hubel and Wiesel did for the cerebral cortex what those earlier investigators did for the ganglion cells. Sticking their measuring probes into the primary visual cortices of live, but anaesthetized, cats and monkeys, they were able to demonstrate the presence of neurons which also respond maximally to uneven distributions of illumination. But whereas the ganglion cells become most active when the relevant fragment of the visual image has circular symmetry, the neurons in the primary visual cortex are most sensitive to patterns of light and dark arranged along lines of well-defined direction. And just as importantly, Hubel and Wiesel discovered that some cortical neurons react only if such line patterns have specific orientations. Indeed, they even found that in some cases a measurable response could be elicited only if the pattern in question is moving in a particular direction. Amongst the favoured motifs are lines that separate light from dark, and bars of light on dark or dark on light.

This much is now very well established, thanks to the collective efforts of hundreds of other researchers who have followed Hubel and Wiesel's lead. What is not known, on the other hand, is how the other visual areas of the cortex handle the signals that are passed on to them from the primary visual area. One could imagine that more abstract information emerges in these regions, but just what form this might take remains unknown. If and when reliable light is shed on this issue, it will be a major event in the history of brain science.

Illustrations of all manner of things in our books, papers and magazines sometimes appear with full shading, but they still most commonly consist of mere lines. Think, for example, of how few strokes of the pen a good caricaturist requires to portray a familiar person. Our visual systems are obviously able to make do with such minimal detail, and the work of Hubel and Wiesel indicates, indeed, that the brain actually reduces incoming visual information to a skeleton of outlines. Subsequent studies by David Marr did much to confirm this conclusion. This gives a strong hint as to what a neural network should be aiming at, if it aspires to imitating part of the visual system; its primary goal must be to be able to recognize configurations of lines having various orientations and juxtapositions. This is precisely what the designers of the WISARD set out to do, and they can be said to have achieved a large measure of success. Let us consider some of the problems that were encountered *en route*, and the clever way in which Aleksander and his colleagues got around them.

Any picture can be regarded as an array of dots, and the positions of these with respect to one another is all that is required, in principle at least, to define the image. The positions of the dots can be conveniently described by reference to an imaginary background grid, and for the sake of illustration we will here consider a rather small example. Let us suppose that it consists of just sixty-four squares, deployed in an eight-by-eight arrangement. The grid will thus resemble a chess board, but one in which no distinction is made, initially, between the colours of the various squares; they can be thought of as all being white. Just as with the case of the T–C problem discussed earlier, we can draw pictures by colouring some of the squares black while the remainder are left white. A full description of the drawing requires specification of the colour which appears in each of the sixty-four squares, or pixels as we called them earlier. Because there are two choices for each pixel, the total number of different possible figures is equal to the number two raised to the sixty-fourth power. This is a prodigious number; it is about twenty million million million.

It would be quite out of the question to attempt to have a single neural network handle so much information in one fell swoop. It would also not be a particularly useful goal, because if a given picture were to be deemed recognized if, and only if, all sixty-four pixels were seen to have the correct colour, the associated artificial visual system would have no tolerance of error. An incorrect colouring in just one of the pixels would be sufficient for a drawing to be rejected, even though it would be clear to us that the resemblance to the original was quite striking. And even if we tried to allow for this by training the network to permit near misses, the number of such extra memory items would soon reach astronomic proportions.

To obviate such difficulties, it has become standard practice to subdivide one's grid so as to split it up into many smaller networks, each of which can easily be managed by the neural network approach. An eight-by-eight array of pixels can be divided up into sixteen subsidiary arrays, for example, each comprising just four pixels in a two-by-two array. There is a practical limit to such subdivision, however, because a subsidiary array that is too small will be dangerously susceptible to the very type of random error that the subdivision was designed to avoid. To see this, one has only to consider what happens when things are taken to the extreme, in which each subsidiary array consists of only one pixel; an error in this small array would change the colouring of all the pixels!

There is another problem that is rather more subtle, and this has to do with the sensitivity of the system. Let us consider a rather simple manifestation (see figure 6.31). The sixty-four pixels in our eight-by-eight array can be imagined as being numbered from the top left to the bottom right, such that the pixels on the top row have numbers that run from 1 to 8, those on the second row from 9 to 16, and so on until the bottom row, which consists of pixels 57–64. Now imagine that the neural network is being asked to recognize a vertical line drawn through pixels 4, 12, 20, and so on down to pixel 60. In this

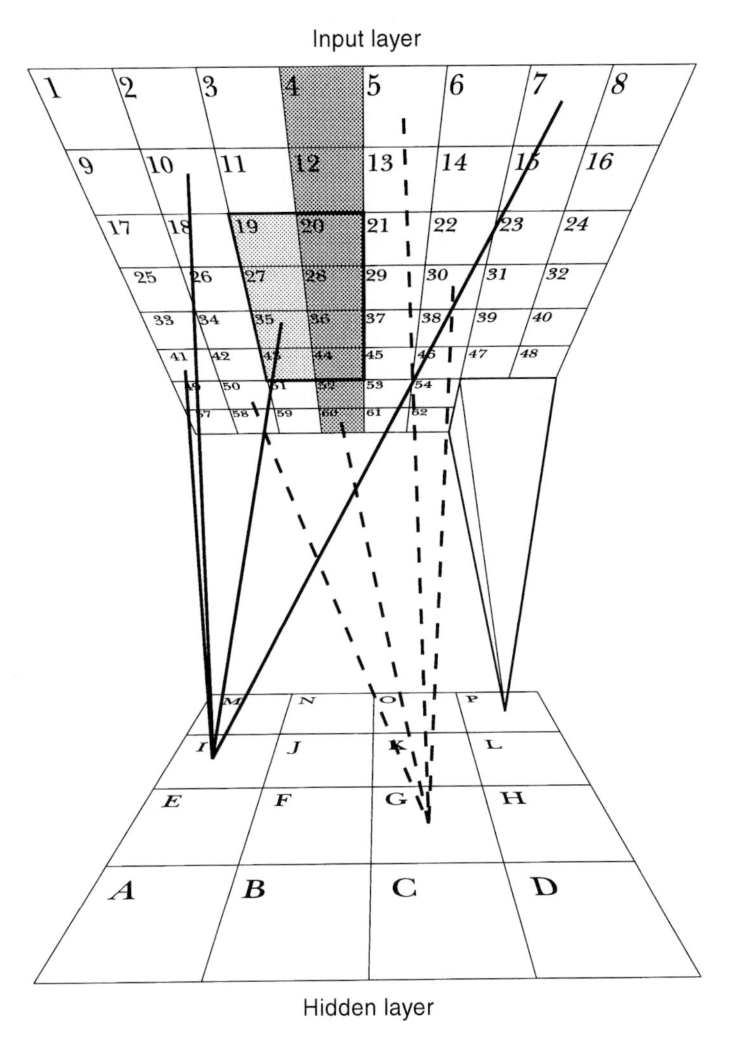

Figure 6.31
The receptive fields in the human visual system are arranged in an orderly fashion, such that adjacent positions in the field of view tend to activate adjacent neurons in the thalamus and in the primary visual cortex. Some artificial cognition systems, including WISARD, have exploited the fact that randomized receptive fields can provide more efficient feature detection. In the simple example shown here, three-quarters of the simple (non-overlapping) receptive fields consisting of two-by-two arrays would make no contribution to feature detection, these being those that feed hidden-layer neurons A, C, D, E, G, H, I, K, L, M, O and P. (Neuron P receives signals from neurons 55, 56, 63 and 64 in the input layer, as indicated.) If the fields are randomized, neurons I and G, for example, would play an active rather than a passive role in distinguishing between the dark shaded and light shaded figures. Because the real visual system consists of several areas, which signals reach in succession rather than simultaneously, and because there is a spreading out of connections between successive areas, it is possible that something like randomization of the effective receptive fields also occurs in the brain.

situation, if the subdivision of the entire array has been into sixteen two-by-two arrays, as suggested earlier, the line will pass through only four of them. All four pixels in each of the remaining twelve little arrays will be coloured white.

This might not seem to be a cause for concern, but what if the neural network was being asked to discriminate between the above eight-pixel vertical line, which runs the full height of the large array, and a thicker but shorter eight-pixel line, such as a two by four line occupying the pixels 19, 20, 27, 28, 35, 36, 43 and 44? Seen from the viewpoint of twelve of the sixteen little two-by-two arrays, the two motifs are identical, because in both cases all the pixels in those twelve arrays will be coloured white. Changes are observed only in the remaining four arrays, and this is a relatively small fraction. The neural network would have a hard time discriminating between the two pictures.

The designers of WISARD solved this problem in a most ingenious manner. When they linked up their little arrays to the underlying neural network, they did not follow the neat divisions that were described above. That is to say that they did not form the first mini-array from the pixels 1, 2, 9 and 10, the second from pixels 3, 4, 11 and 12, and so on. Instead, they deliberately mixed things up, randomly linking together pixels that lay some distance from one another in the large array. This trick decreases the chance that there will be mini-arrays in which all four pixels are always coloured the same, and which could thus be said to remain idle, as were twelve of the sixteen mini-arrays in the above example.

It remains to be said of WISARD that its pixels were exposed to suitable photosensors, and that they were linked up to RAMs in the manner discussed earlier in this section. The machine was able to learn numerous visual patterns, amongst the more useful of which were human faces. It does not demand much of the imagination to guess that such a face-remembering device would be of interest to the police; London's Scotland Yard was indeed one of the machine's first customers.

7 The grand design

As the brain is a machine, we must not flatter ourselves that we can discover the contrivance of it by any other means than are made use of for knowing other machines; and we have no way left but to take it to pieces and to consider what every part is capable of in a separated and in an united state.
Niels Steensen (*The Anatomy of the Brain*[1])

The burden of mandatory change

It is easy to get the impression that things which mediate sensory perception enter the head's inner reaches unchanged. Odour molecules pass farther up the nostrils than can be penetrated with one's little finger, and that same digit cannot touch the eardrum, which sound reaches with ease. If we are polite about our eating, the mouth closes around the materials that generate taste, without obviously modifying them as they disappear from sight. And anyone who has read elementary biology knows that the light rays which pass through the eye's pupil produce an image on the retina that is a faithful reproduction of the object being viewed. René Descartes demonstrated that this is so by cutting a window in the back of an ox eye, by covering it with paper, and by observing on the latter the inverted images of objects that the front of the eye was aimed at. It is only in the case of the fifth sense, touch, that the stimulus seems constrained to remain *outside* the body.

Touch is the odd one out in another way, however, because it is the only sense that involves bodily locations other than the head. Perhaps it was this fact which drove Descartes to conjure up a mechanism whereby tactile sensation could be passed up to the brain, essentially in its original form. He saw this sense as being served by a series of hydraulic conduits which ran from the body's extremities right up to the brain's ventricles. Pressure on the surface of the skin could thereby give rise to pressure in the brain.

What would be the point of letting these agencies reach the inner recesses of the brain, however, if there was nothing there to witness their arrival? At the time of Descartes, this was not seen as a difficulty. The *homunculus* (or little man) was always there in mission control, assiduously reading all the signals and passing the appropriate commands to the muscles. The trouble with this 'model' was that it did not explain how the brain of the homunculus functioned.

The truth is, of course, that these sense-mediating vehicles do *not* enter the body. The relative inaccessibility of the various receptors is merely a precaution that Nature has taken to keep them out of harm's way. Their fairly remote locations notwithstanding, none of them are *inside* the body, in the sense of being under the skin. The reason for this is that ordinary body tissue (as distinct from nervous tissue) would be a very poor conveyer of information. For example, if the molecules that mediate smell had to physically penetrate our outer covering, they would have to take their chances in competition with a host of other molecular species. And to get an idea of how effective vision would be if the incident light had to pass through flesh, one has only to try looking at a bright lamp through the loose skin that lies between thumb and forefinger; this material is weakly translucent, but it does not even permit one to discern outlines.

As we have already seen in earlier chapters, Nature's way of getting around this difficulty was to evolve the nervous system. The sensory agencies are stopped in their tracks, when they reach the appropriate receptors, and the information is transformed in nerve impulses. The molecules that provide us with smell, and those that give us taste, and the pressure that supplies us with touch, and the oscillations of air pressure that mediate hearing, and the rays of light that either emanate from or are reflected from the things we see, they all give rise to nerve impulses. And this confronts the body with a problem of considerable magnitude. Because just as diffusion would have brought our smell-mediating molecules into competition with other types of molecule in the body, with the threat of losing the information because of mutual mixing, so too is there the prospect of these different nervous signals getting mixed up. This is one aspect of the black box problem that we were contemplating in the second chapter: how must things be arranged inside the machine, to avoid such confusion? Viewed in this manner, the mandatory change from one signalling medium to another represents quite a burden.

The answer lies in the exquisite design of the nervous system, with its numerous different parts, its various neuronal types, its multiplicity of receptors, and its delicate balance between excitation and inhibition, which tends to keep signals from coalescing when it would be inappropriate for them to do so. Moreover, the successful development of the system involves beautifully orchestrated patterns of growth, selective elimination at certain stages, and the establishment of myriad synaptic connections. This is not a passive process, however. There has long been evidence that correct development of the nervous system is also dependent upon its being subjected to relevant sensory input. We will review the evidence for such mechanisms later in this chapter, in the specific context of vision.

Even without sensory input, the system is robust in its ability to determine what goes with what. In the 1940s, Roger Sperry cut through the optic nerves of an embryonic newt, rotated the eyes through 180 degrees, and replaced them in their sockets, holding them in place by suitable bindings until the connections between the severed endings had been re-established.[2] The rotation had naturally placed the wrong nerve endings opposite each other, and the important question was whether they would nevertheless join up despite this mutual alienation. Had they done this, the recovered newt would have behaved normally because the incorrect joining up of the nerve endings, *up* becoming attached to *down* and vice-versa, would have exactly compensated the rotation. This is not what happened. On the contrary, the newt lunged downward when prey flew overhead, and leaped upward when a potential meal passed under its nose. The rotation had therefore been alone in determining the net effect, and the nerve endings had somehow managed to find their correct partners.

The biological machinery that accomplishes this impressive piece of path finding still awaits final clarification, but an important contribution was made by Rita Levi-Montalcini when she discovered that the presence of a complex of proteins, collectively known as *nerve growth factor*, is vital to the correct routing of growing nerve processes.[3] Just what degree of accuracy this guidance achieves depends upon the species. Early in the present century, Richard Goldschmidt studied the nervous system of the intestinal parasite *Ascaris*, and found that the brains of these small worms always comprise 162 neurons, identically arranged. Then in the 1970s, Sydney Brenner and his colleagues established that the same is true of the 279 neurons in the brain of the nematode worm *Caenorhabditis elegans*.

Such precise circuitry is not seen in more advanced species, for the good reason that there simply is not enough genetic material in the chromosomes to differentially code for the connectivity of all the neurons. The standardization seen in the brain circuits of higher species is with respect to the *general pattern of connections*, rather than to that of the individual neurons. This fact has been well documented in recent decades by a variety of staining and tracing techniques, some of which make entire cells visible, while others are specific for particular types of process (or neurite). A variety of different microscopic techniques have been applied to this cartographic enterprise, including transmission electron microscopy, which is even able to reveal structural detail in the synaptic membranes.

And what have such studies disclosed? They have shown that the axonal collaterals and axonal branches of the typical neuron fan out so as to make connections with thousands of other neurons. This inevitably means that a given neuron will also receive signals from many other neurons. Taken at face value, that would seem to suggest that nerve signals nevertheless spread out and get mixed up! Does Nature thus thwart the intention of avoiding confusion in the brain's black box, or is there some way of salvaging the situation? In order to answer such a critical question, we will have to consider the actual wiring of our brains, and those of our close evolutionary kin. And we are not going to be able to avoid getting into considerable detail. The best place to start is the visual system, both because it claims the lion's share of the cortex (over half, in ourselves and our near relations) and because it is the best studied of all the brain's sub-structures.

Window on the mind

Anyone dissecting a human brain for the first time is invariably struck by its seeming uniformity, and this is why its anatomical divisions emerged much later than those of the rest of the body. Organs such as the heart, kidneys and lungs can be distinguished by the naked eye; the surgeon's job would be a lot harder if this were not the case. It is much more difficult to see the divisions between the various structures in the brain, particularly the sub-cortical components. It is true that finding one's way around the cortical surface is facilitated by the presence of the gyri and sulci, but below this one is confronted with the gelatinous mass which Alan Turing likened to a lump of porridge. The student of brain anatomy is often recommended to start with the visual system, and to trace the route that leads from the retina to the thalamus, and thence to the cortex itself. We could do no better, here at the outset, than to embark on a similar journey.

During my waking hours, I have the feeling that there are two distinct aspects of my person. There is my body, and there is a sort of inner presence, my *me-ness* it could be called, located a few centimetres behind my eyes. When my eyes are shut, there is a similar feeling that this inner presence is located between my ears. My working assumption is thus that my eyes are functioning as windows and my ears as funnels. Had I been born both blind and deaf, however, it is debatable whether the senses of taste and smell would have been sufficient to still make me identify with my head, given that they would have had to compete with the more dispersed sense of touch. It requires but little contemplation to appreciate that the putative eye windows must be more than mere panes of passive biological glass. If I enter a dimly lit room, my vision slowly accommodates to the altered level of illumination, without my conscious participation. If I suddenly look away from the typescript of a book, and try to focus on a distant object, there is again a brief period of adjustment in which my role is minimal. There are things going on in vision, therefore, that are both sophisticated and beyond our mental control. The study of these processes, and of the many others that contribute to vision, tells us much about how the brain's neural circuits function. The eye is not a window *for* the mind, it is a window *on* the mind. Let us see what it is trying to show us.

Starting at the retina, we note something that must have intrigued Descartes. If it was Nature's aim to supply the *homunculus* with a picture to look at, it perversely required him to do a headstand; the image is upside-down! Moving on to the receptors, which are merely neurons with an ability to convert light energy into electrochemical changes at the cellular membrane, it is instructive to consider the question of *acuity*. This can be measured with the aid of a *Landoldt C chart*, the pattern on which is a black ring with a thin sector missing, rather like a closed-up letter C. The chart can be rotated to different orientations, and the subject is asked to locate the gap for a number of trials, and at a series of distances. The subject's limit of acuity has been reached when the performance in detecting the position of the gap is no better than chance. Vision is said to be normal when the gap subtends an angle of 1 minute of arc (i.e. one sixtieth of a degree) at the pupil, and acuity is usually stated as a fraction which relates a standard testing distance of 6 metres (or 20 feet) to the distance at which a just discernable gap would subtend 1 minute. For the normal eye, that distance would be the standard 6 metres, and the subject's vision would be said to be 6/6 (or 20/20). A person with 6/3 (or 20/10) vision, for example, can resolve a gap at 6 metres which would subtend 1 minute at 3 metres, so the performance is better than normal.[a]

Let us keep that 1 minute of arc in mind and consider what it means in terms of the spacing between adjacent *receptor neurons* in the retina. They are of two types. The roughly 125 million *rods*, which detect brightness, are most dense (about 160,000 per square millimetre) at an angle of around 20 degrees away from the centre of gaze, and gradually diminish to almost none at the latter. The roughly 7 million *cones*, which additionally code for colour, are about five hundred times less sensitive than the rods, and they are most densely packed (about 140,000 per square millimetre) in the central region, which is known as the *fovea*. The fovea is about 1.5 millimetres in diameter, and it is the cones that will concern us because they are the ones which give us our acuity. From the figure for their density, it is easy to calculate that a 1 millimetre line in the fovea would span about 400 cones. Knowing the approximate diameter of the eye, we can also calculate that our angle of 1 minute of arc subtended at the pupil projects to a line about one-two-hundredth of a millimetre on the retina. This would span about two cones. The image on the retina of the just-resolved Landoldt C therefore has a gap which is equivalent to a mere couple of cones, and our inference must be that these receptors can carry out meaningful detection even when functioning individually. The system is remarkably efficient; it appears to get the most out of its neuronal components.

Things thus seem to be reasonably straightforward, but we are about to run into a complication. The receptor cells do not feed information directly to the cortex. Far from it. As we learned in the preceding chapter, the receptors synapse onto *bipolar cells*, and these in turn pass signals on to *ganglion cells*. But the junctions are not simply one-to-one affairs. There is a huge degree of convergence in that the roughly 130 million rods and cones feed their signals into just 1.2 million ganglion cells. At first glance, this looks as if the happy match between the 1 minute ideal visual acuity and the spacing between the receptor cells has been to no avail. But it should also be noted that each receptor feeds its signals into *several* ganglion cells. This is reminiscent of the PDP principle that was discussed in such detail in the previous chapter, and we will be considering that comparison more extensively later.

a / For comparison, the Moon subtends an angle of 30 minutes at the Earth's surface, so we can readily make out its details.

In passing, we ought to recall the other two types of neuron in our retinas, which both lie in the plane of the retina, and are thus oriented roughly at right angles to the receptors, bipolars and ganglions. It is the ability of these *horizontal cells* and *amacrine cells* to laterally spread the activity of the other three types which allows the system to adjust to widely differing levels of illumination. But we do not need to go into the details here.[4]

The axons of the ganglion cells collectively form the *optic nerve*, which carries the signals away from the retina toward two different sub-cortical structures. Before considering them, we should note that the human retina could be said to be back to front. The receptor cells are located nearest the back surface of the eyeball, away from the incident light, and the latter has to pass the cells of the other four types before reaching its goal. This presents the eye with a problem in that it has to find a way of letting the axons of the inward-facing ganglion cells exit through the retina, without causing too much disturbance. The solution is that they are immediately consolidated into the optic nerve, so that there has to be only a single exit point, which is offset from the fovea by about 17 degrees. This is the *optic disc*, and because it is devoid of both rods and cones it gives rise to the well-known *blind spot*. The fact that we are normally unaware of the latter, even when we have one eye closed and therefore lack the benefit of overlapping visual fields, is going to become a major issue when we consider consciousness in chapter 9. It is interesting to note that the retinas of some simpler creatures such as the squid and octopus have their retinas oriented the other way around, so they do not have blind spots.

The hundredfold convergence between the entry and exit sides of the retina, as expressed by the mismatch between the numbers of receptor cells and ganglion cells, is another manifestation of the profound change that occurs as signals travel inward from the body's outer surface. Because there is no internal agency to observe them, these signals do not need to have a simple relationship with what is observed in the outside world. The only requirement is that the signals do not get inextricably mixed up, so that the system loses track of what goes with what. When we look at an object, we merely sense it as having a certain shape and colour, and a specific location relative to our eyes. We see it *out there*. It gives rise to an image on the retina, admittedly, but there are no further images deeper in the system. The transformation is even more marked in the case of hearing. Sounds are represented in the air as pressure oscillations of various frequencies, but inside the head these different frequencies give rise to neural impulses at different parts of the auditory cortex. In this modality too, we sense something as being *out there*, and Eric Knudsen and Masakazu Konishi have established that there are *receptive fields for hearing* just as there are for vision.[5] Creatures so lowly that they neither see nor hear must be incapable of accurately locating objects at a distance. Some of them can no doubt detect things somewhat removed from themselves, because of the odour molecules arriving at their receptors, but this form of perception must be rather blunt.

If the visual system does its job by leaving observed things out there where they belong, what is being achieved by the inwardly travelling nerve impulses? The short answer is that they are enabling us *to detect what is where*, and the reliable fulfilment of that quest requires a machinery that is surprising in its complexity. The overall strategy is one of breaking things down to their component parts, and in the case of vision this process starts in the retina itself. The net effect of what the horizontal, bipolar and amacrine cells collectively accomplish is a ganglion response that is maximized for *non*-uniform illumination. This was briefly touched upon in the preceding chapter, but we now need a little more detail. Haldan Hartline, in 1938, was the first person to record the signals from single ganglion cells.[6] His studies were mostly carried out on frogs, but he also investigated the ganglion cells of the silt-seeking fish known as the mudpuppy, the turtle, alligator and shark. He found that the ganglion responses could be divided into three classes: *on-centre* type, *off-centre* type, and *on-off* type. These categorizations refer to the illumination in the cell's visual field required to give the maximum response, that field being the area within which any change in illumination gives a detectable change in the activity of the cell being recorded from. The off-centre type clearly needs some explanation; it is a ganglion which responds most strongly to a circular patch of light with a dark spot at its centre. Hartline found that both the on-centre and off-centre types give a *sustained response* while the light is on, whereas the on-off type gives a *transient response* at both the onset and the cessation of illumination.

In 1953, Horace Barlow[7] and Stephen Kuffler[8] independently reported that the receptive fields of the on-centre and off-centre ganglion cells in frogs and cats, respectively, can be divided into two concentric zones, one a centrally located circle and the other a surrounding annulus. This arrangement is now known by the adjective centre-surround. Barlow and Kuffler both found that illumination of the surround opposed the response elicited by illuminating the centre of an on-centre ganglion. The off-centre type of ganglion responds best to doughnut-shaped illumination, having the appropriate dimensions, while the on-centre type prefers doughnut holes. This implies that the response of either type of ganglion cell would be zero for uniform illumination, and this is found to be approximately so. Barlow also reported that a frog's on-off ganglions are very sensitive to moving stimuli, whereas they show virtually no response to stationary illumination. As he pointed out, this could be the neural basis for the fact that these animals react strongly to things which move but appear to be essentially blind to stationary objects. Barlow therefore suggested that these on-off ganglions can be regarded as *fly-detectors*.

The centre-surround arrangement is believed to be the underlying cause of a fascinating optical illusion discovered in 1869. The pattern, now called a *Hermann grid*, which provokes this illusion consists of a simple array of black squares, regularly arranged in a square lattice and separated by white lines.[9] Its overall arrangement is thus like a bird's eye view of the blocks and streets of a city (see figure 7.1). Inspecting this figure, one sees nebulous dark shadows at the corner intersections. These mysteriously evaporate when one

Figure 7.1
The Hermann grid produces a visual illusion that is related to fundamental processes early in the visual pathway. Nebulous shadows are seen at the intersections of the white strips, but they seem to evaporate if one stares directly at them. The retina's ganglion cells respond maximally either to a light central circle surrounded by a dark ring or vice-versa. The intersections in the grid's image on the retina come closer to the former of these optimal stimuli, and the slight departure from the ideal gives rise to the shadows. But it is still not known why they disappear when one looks directly at them.

looks directly at them, but they reappear when one's gaze is shifted. The explanation of this effect runs as follows. An on-centre ganglion responds maximally to its favoured doughnut hole because its central circle is being excited by illumination at the same time as its surround is being stimulated by the dark annulus. (This latter statement may sound peculiar, but we should bear in mind the unobtrusive role being played by the laterally deployed horizontal and amacrine cells.) At a grid intersection, this optimal illumination pattern comes close to being satisfied, but the four side-streets contribute four portions of inhibition and this is apparently sufficient to cause the evanescent shadow. In the street regions between the intersections, the optimal pattern comes closer to being satisfied, and there is now insufficient inhibition to cause the effect. If one inspects Hermann grids of different sizes, it is possible to find one which gives the strongest effect, and the dimensions of the pattern can then be used to determine the size of one's centre-surround receptive fields.

It is generally the case, as we have noted earlier, that the right hemisphere controls the left side of the body and vice-versa. Because there is inversion when light rays pass through the pupil, with left becoming right, and right becoming left, this makes things somewhat complicated. The left hand, for example, will normally be imaged on the right side of the retina, so the interests of coordination will be best served if signals from that part of the retina also pass to the right hemisphere. This is just what happens. The nerve fibres from the right sides of both retinas project to the right hemisphere, and those from the left sides project to the left. Because all the fibres from a given eye leave the eyeball in a single bundle, the optic nerve, this means that there must be a subsequent rearrangement. That is just what happens, the site of the sorting out being the *optic chiasm*. It is interesting to note, in passing, that the *suprachiasmatic nucleus*, which sits atop the chiasm, supplies the control for the body's daily rhythm, the circadian cycle. Because the chiasm is that place in the entire system which is most influenced by received light, because of the doubled number of fibres, we see that the suprachiasmatic nucleus could not have been better positioned.

The continuations of the visual nerve bundles, after the chiasm, are known as the *optic tracts*. Each of these divides into a major and a minor sub-bundle, the former passing to one of the two *lateral geniculate nuclei*[b] (LGN), which are located on either side of the thalamus, and the latter going to the *superior colliculus*, which is positioned near the top of the brainstem, at the rear and slightly above the cerebellum. Both of these structures are extremely important to the visual system, and because they are going to serve as prototypes for our more general discussion of brain function, we will have to consider them in some detail (see figure 7.2). Let us take them in turn. The thalamus acts as a dual interchange and relay station for several of the senses. (The parts of it which serve hearing are called the *medial geniculate nuclei*.) The dendrites of the neurons in each LGN receive axonal input from the fibres of the relevant optic tract, and their own axons project to the early visual areas of the cortex via the *optic radiations*. The thing that gives the LGN its special properties is that it receives input from three other sources. There are axons that enter the LGN *from* the cortex, and they are in fact more numerous than those that take the forward route. Then there are inhibitory axons from a structure known as the *nucleus reticularis thalami*, and finally there is other axonal input from something referred to as the *brainstem reticular formation*.[c] These latter two structures are supremely important in the context of this book, for they are believed to play decisive roles in the processes of attention and consciousness. We will be considering regulation by these structures later in this chapter.

Meanwhile, let us return to the main signalling route to the cortex. Just as is the case in the somatosensory and motor cortices, which were described in chapter 3, the primary visual cortex has a distorted map, which is this case 'depicts' the visual space in front of the eyes, and is thus related to retinal location. The fovea is by far the best represented area. In fact, despite the reduction factor of about a hundred referred to earlier, the receptor cells at the very centre of the fovea are represented on a one-for-one basis in the optic radiation. But we must remember that this is not the same as saying that each receptor is served only by its own radiation fibre, because that would ignore the dual factors of convergence and divergence, as well as the mixing influences of the horizontal and amacrine cells.

Given this mapping between the retina and the primary visual

b / – and to what are known as pretectal destinations.

c / The word reticular refers to the fishnet pattern seen on the surfaces of these structures.

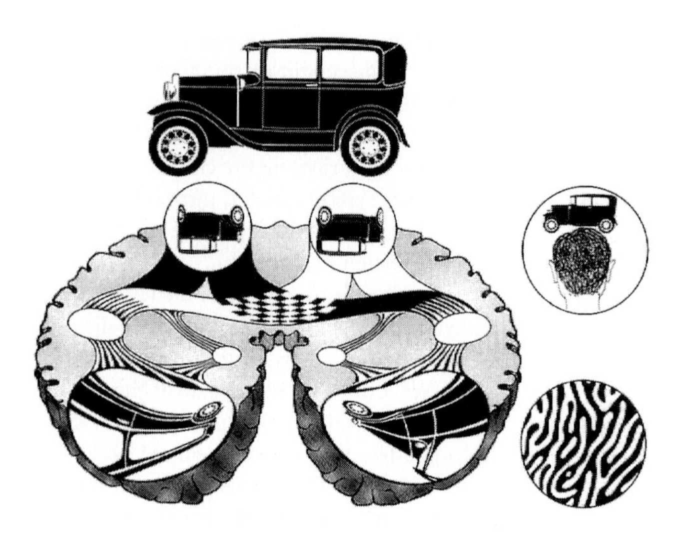

Figure 7.2
The brain's left side controls the body's right, and vice-versa. The visual system thus has a problem, because each eye lens produces an inverted image on the retina, as indicated in this *highly* schematic rear view (see upper inset) of a horizontal section through the brain. The solution lies in a division of each optic nerve, half the fibres passing to the early visual cortex at the rear of the same hemisphere while the remainder go to the corresponding area in the other hemisphere. They are not joined directly to those visual regions, however; the connections are via the lateral geniculate nuclei (the larger ovals in the figure – which in reality have more central locations). These also receive counter-running fibres from the early visual areas, and this permits the feedback control vital to the visual system's correct functioning. The smaller ovals represent the superior colliculi (also shown out of their proper positions, to simplify the picture), which give the system its gaze-holding ability when the head or an observed object are in motion. They also mediate one form of 'blindsight' when the primary route is damaged. Adjacent points in the primary visual cortex (and also in some of the other visual areas) receive signals from adjacent regions of the viewed object, so one can loosely refer to an 'image' at that location. This is highly distorted, however, and it is split down the middle. Moreover, the maximal response of cells in the primary visual area is to discontinuities of illumination, although no attempt has been made to indicate this in the figure. A conjectured internal observer would have an even harder job of interpreting what goes on in the higher visual areas, where the representation is even more abstract. Depth perception is produced by the left–right alternation of fibres feeding the primary visual area; the existence of this arrangement was established by David Hubel and Torsten Wiesel, who injected a radioactive substance into one eye, let it diffuse to the visual area, cut a slice from that region, and located the radioactivity-bearing cells by placing the slice against a photographic film (see lower inset, in which the left–right alternation is indicated in black and white).

cortex, mapping at the intermediate stage of the LGN is more or less unavoidable, and this is certainly the case. There is something else worth mentioning in connection with the LGN. It was discovered by Ernst Mach around 1860. A series of parallel stripes with systematically increasing depths of grey colour are painted on a flat surface with no intervening gaps. If this pattern is uniformly illuminated, one would expect each stripe to appear evenly coloured, but this is not what is observed. At each inter-stripe junction, a thin lighter line is observed on the darker side, and a thin darker line is observed at the lighter side, both of these lines running parallel to the junction. Mach's interpretation of this surprising illusion invoked laterally acting inhibition, of the type discussed in chapter 4 under the name feed-forward inhibition.[10] The fuller explanation now available involves something in addition, namely *spontaneously active cells*. Some neurons emit impulses even when they are receiving no input. The advantage they confer upon the system is that decreased input then gives a (negatively directed) response rather than no response at all. Near the boundary between two of the stripes, the discontinuity of shade causes an imbalance in the degree of lateral inhibition which extends across the dividing line. This results in a locally enhanced activity level on the dark side and a locally diminished level on the light side. These local effects give rise to the observed thin lines.

The *Mach band illusion* is not very important in the broader sweep of things, but it is instructive because it shows how easily the visual system can be fooled. It is also a useful, and reasonably direct, demonstration of the reality of inhibition. Let us now move on to something of very great importance indeed. We have seen that the early part of the visual system functions by breaking things down into their elementary components, achieving this by the simple expedient of subjecting them to centre-surround filtering. How do the later stages of the visual system use this partially processed information? A strong hint was supplied by the work of David Hubel and Torsten Wiesel, which was touched upon at the end of the previous chapter. They temporarily paralysed cats and monkeys, with their eyes open in a rigidly held gaze, and probed the activity of neurons in their primary visual cortices. These cortices are located at the rear of the head, and the cat's primary visual area is particularly accessible. They discovered that a given cell in this cortical area responds most vigorously to stimulation of a specific region of just one of the retinas, and by a line having a given orientation. When presented with a line deviating more than about 20 degrees from the ideal orientation, the response is markedly diminished.

Investigating the situation further, Hubel and Wiesel found that the primary visual cortex is divided up into bands about half a millimetre in width, lying parallel to one another. Within one of these bands, all the neurons respond to the same eye, while the adjacent bands to either side react to the opposite eye. Intriguingly, they also established that within one of these *ocular dominance bands* there is a systematic change, from place to place, of the orientation that gives a maximal response. This *orientation preference* varies most rapidly in the direction that lies roughly parallel to the boundaries between the ocular dominance bands (see figure 7.3). Hubel and Wiesel put forward a mechanism whereby the centre-surround receptive fields of several contiguous ganglion cells could be consolidated so as to produce a line-sensitive cell with a particular orientation preference, but their model still awaits definitive confirmation.

What *is* clear, on the other hand, is that this tendency to respond to lines is not something an animal is born with; there have been experiments which show that it develops during early life. One of these was carried out on newborn kittens by Richard Held and Alan Hein in 1963. The head movements of the animals were restricted by neck-yokes, and one of them was free to walk about whereas the other rode in a gondola.[11] The active kitten's grosser movements were imposed upon its passive partner by a rod and chain mechanism, and they were confined to a small room in which the walls were decorated with exclusively vertical stripes (see figure 7.4). Following a training period of several weeks, the two kittens were subjected to several tests, one of which is known as the *visual cliff trial*. This involves confronting the animal with a choice between shallow and deep drops when descending from a platform. The entire room in this second part of the investigation was again covered with vertical stripes, and an invisible glass floor prevented injury in those cases where the deep drop was chosen. The active kitten always chose the

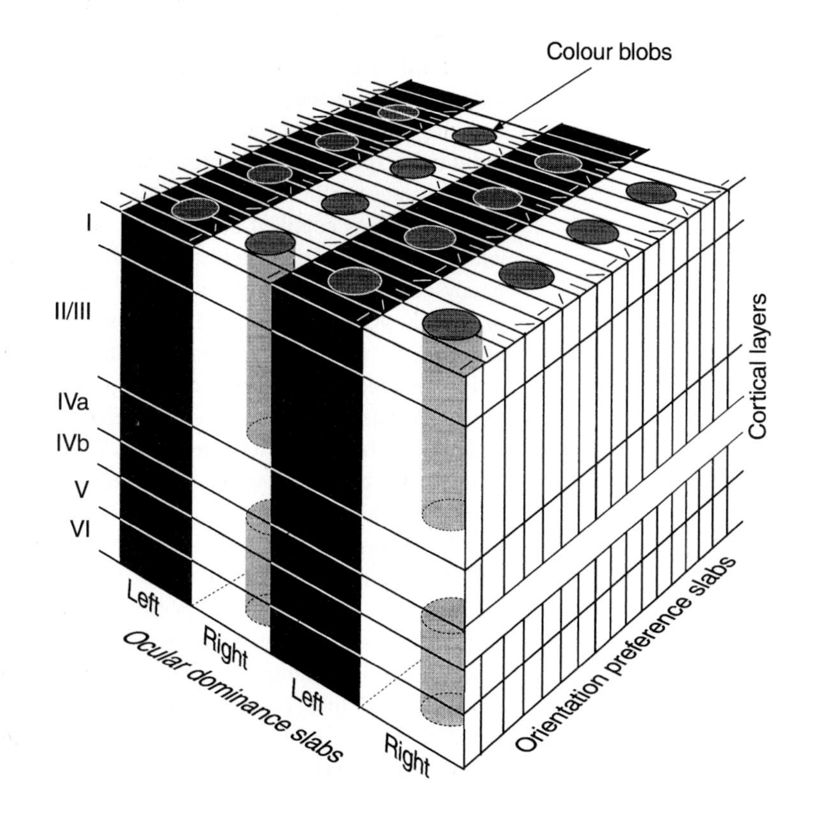

Figure 7.3
The pioneering efforts of David Hubel and Torsten Wiesel revealed the presence of ocular dominance slabs in the primary visual cortex, with a thickness of about half a millimetre. The direction of their left–right alternation lies roughly at right angles to the direction of varying orientation preference (indicated by the short lines). The latter variation is cyclically repeated, as indicated by the multiplicity of slabs in that direction. Subsequent studies by Hubel and Margaret Livingstone established the presence of additional cylindrical regions, within each slab, which exclusively serve colour vision. These heterogeneities are not visible in the highly schematic representation of the primary visual cortex shown in figure 7.2, and the graininess of the 'image' in that cortical area becomes apparent only at much higher magnification. The arrangement is different in the higher visual areas; in the second visual area, for example, the colour-mediating patches appear as lines rather than (approximate) circles in plan view. Also indicated in this picture is the division of the cortex into several layers, although no effort has been made to indicate what variations in neuronal structure are observed between the different strata.

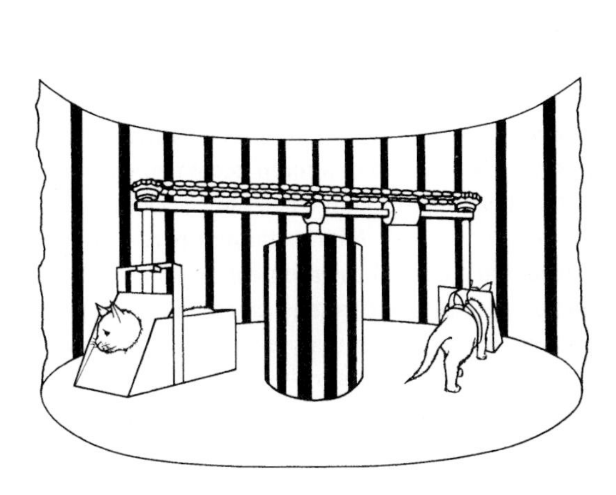

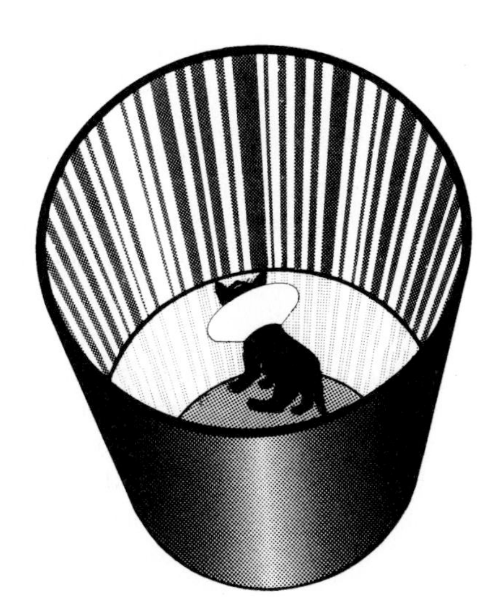

Figure 7.4
The coarser consequences of synaptic modification have been investigated with experiments on newborn kittens. In the work of Richard Held and Alan Hein (left), head movements were restricted by neck-yokes, and the animals were confined to a cylindrical room decorated with exclusively vertical stripes, the active subject being free to walk about whereas the passive member rode in a gondola. The active kitten's grosser movements were imposed upon its partner by a rod and chain mechanism; both therefore had the same visual experiences. The goal was to show that the acquisition of appropriate motor responses depends upon conditioning during early development, and the post-training tests included blink reflex to an approaching object, visually guided paw placement, and visual cliff trials. In the latter, the kitten was confronted with a choice between shallow and deep drops when descending from a platform, the entire environment again being covered with vertical stripes. (An invisible glass floor prevented injury during descents on the deep side.) The actively trained kitten always chose the shallow drop, whereas its motor-deprived sibling could not appreciate the danger on the deep side; the latter's brain presumably lacked synaptic connections that had recorded the visual–motor associations. In Colin Blakemore and Grahame Cooper's experiment (right), a newborn kitten was similarly confined to a vertically decorated environment, and a fitted collar even prevented it from seeing its own body. Within a few months, it had become quite clumsy, bumping into objects as it moved about. Direct electrical recording from neurons in its primary visual cortex revealed the reason: neurons sensitive to horizontal lines were sparse, whereas their vertically tuned counterparts were over-represented. This demonstrated, in addition, that nerve cells are able to change their preferred direction, through modification of the synapses via which they receive input signals.

shallow drop, whereas its passive colleague could not appreciate the danger on the deeper side. Having been deprived of motor experience, the latter kitten apparently lacked neurons that had recorded the visual-motor associations for vertical stripes.

An even more direct demonstration was reported seven years later by Colin Blakemore and Grahame Cooper.[12] They too worked with newborn kittens, each of which was similarly confined to a room decorated with vertical stripes, and they even fitted their animals with collars to prevent them from seeing their own bodies (see figure 7.4). After a training period of several weeks, the kittens were tested in a normal environment, and they had a tendency to bump into things. Electrical recordings taken from the primary visual areas of these horizontally deprived animals revealed that neurons sensitive to horizontal lines were sparse, whereas the population of vertically tuned neurons was far greater than normal. This investigation had verified that orientation preference develops with time, and it also strongly hinted that neurons are able to *change* their preferred orientation. If, following the period of unilateral indoctrination, a kitten is permitted access to a normal environment, it gradually acquires the ability to discern the orientations of which it had been oblivious.

The monkey cortex has a reasonably uniform thickness of about three millimetres, and it has long been known that it comprises six fairly well-defined layers. The convention is that these are denoted by Roman numerals, I to VI, numbered from the outer surface of the brain, and there has recently been a tendency to classify II and III as belonging to a single composite layer, designated II–III. Layer I is unusual in that it contains very few cell bodies (soma); its appears to be primarily the site of synaptic connections to neurons in lower levels whose dendrites extend into this layer. This is not to say that the neurons lie in planes, as if in a chest of drawers, but the orderliness is at least sufficient to permit ready discrimination between the various strata (see figure 7.5). Moreover, the different levels are populated by characteristic types of neurons. A great deal of anatomical work has gone into charting the connections between cells in different layers of a given cortical area, as well as the white-matter inter-area connections.[13] Then again, much is now known about the cortex's afferent fibres, which bring signals from sub-cortical regions of the brain. Finally, and most importantly, there is now an increasingly good picture of the connections through which the cortex sends signals to brain structures lying ventrally to it.

We will return to a detailed description of these connections shortly, because they have a very important bearing on the way in which the cortex functions. Meanwhile, let us close this coverage of the early visual system by considering two final types of fact. One emerged during the investigations carried out by Hubel and Wiesel. They found that the neurons in different cortical layers respond to different classes of oriented pattern. We have already remarked upon the cells which are best stimulated by lines. There are also those that prefer edges, and there are still others that appear to be able to discern corners. In another scheme, there are neurons for which the actual place-

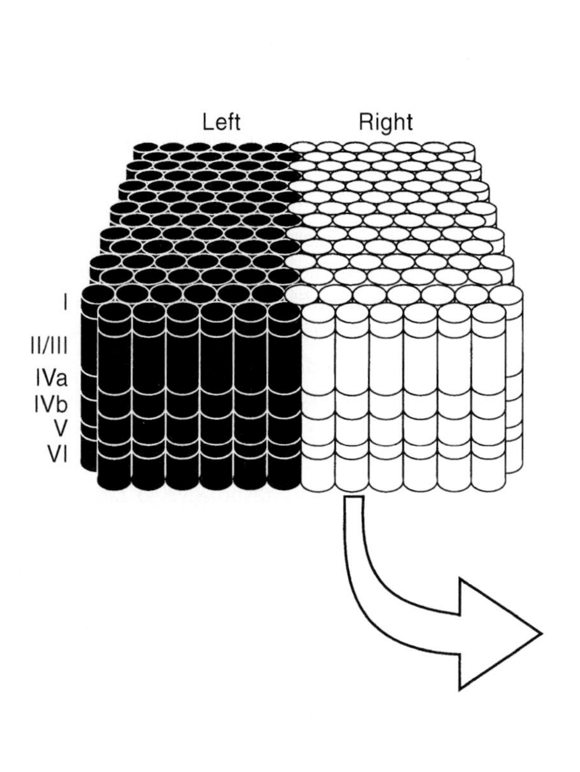

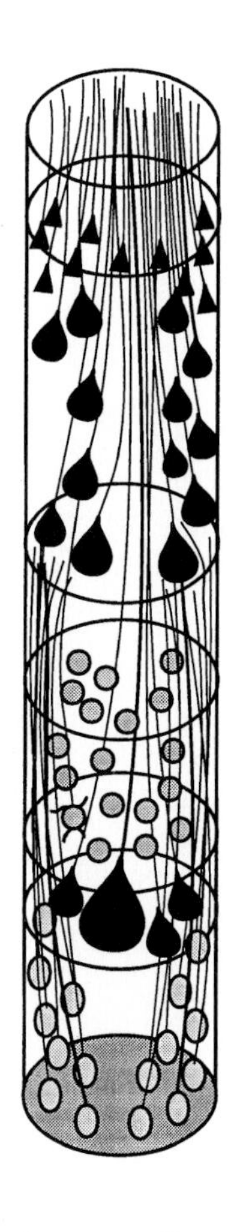

Figure 7.5
The cerebral cortex shows clear stratification, and in higher mammals six layers can be identified through their complements of neural type, density and distribution. Layer I contains few cell soma, and its main function appears to stem from the synaptic connections established there with the apical (apex-related) dendrites of pyramidal (pyramid-shaped) neurons whose soma lie in deeper layers. The traditional distinction between layers II and III is not as sharp as it used to be, and they are often looked upon as constituting a single composite layer, designated II–III. Layer IV shows clear subdivisions, the multiplicity of these tending to increase with increasing evolutionary rank. There are also divisions within the plane of the cortex, which is composed of mutually contiguous cylindrical regions, lying with their long axes at right angles to the cortical sheet. These so-called minicolumns are much narrower than the roughly 0.5 millimetre thick ocular dominance slabs (indicated here in black and white, as previously). In the cat, their diameters are about 0.05 millimetres, and they comprise about 200 neurons, distributed amongst the layers as indicated in this schematic diagram adapted from the work of Alan Peters and Engin Yilmaz (*Cerebral Cortex*, 1993, **3**, 49–68) and Peters and Bertram Payne(*Cerebral Cortex*, 1993, **3**, 69–78) (the large black cells are the pyramidal cells – this figure does not follow the usual convention of white for excitation and black for inhibition). The minicolumn appears to be the functional unit in the cortex, and the primary visual cortex of one hemisphere of the cat brain contains about 160,000 of them. In the monkey, there are about ten times as many minicolumns, and their diameters are only half the size of those in the cat.

ment of a line, say, in its receptive field is important, whereas there are others which are indifferent to position, as long as the orientation is correct. One complication to this picture is that the situations in cats and monkeys are not quite the same. Orientation preference and ocular dominance are seen in both creatures, but the placement within the layers of neurons showing the various propensities is not the same.

There is, then, this clear progression from the centre-surround responses of the ganglion cells, to the line preferences of what are known as *simple cortical cells*, and onward to the more complicated cases of neurons which prefer corners and the like. What is this chain leading up to? There seems to be a trend toward greater complexity as we penetrate deeper into the system, away from the retina. Does this indicate that there will be neurons in later areas that are able to detect quite complicated things such as tables and chairs? Horace Barlow, has suggested that this is the case and he has referred to the *grandmother neuron*, which is a cell that emits nerve pulses only when the retinas are exposed to one's grandmother.[14] There is evidence that neurons of this type are indeed present in the cortex, because reports have appeared in recent years on cells responsive to faces, hands and other such complex features.[15]

This cannot be the whole story, however, because any verbal declaration on our parts that grandmother had been recognized would have to involve association between visual appearance and the spoken word. From our discussion of the progression to higher visual areas, it seems logical to assume that the association area, or areas, lie still more remote from the retinal input, and that the areas responsible for articulation lie even beyond that. It is clear that more definite statements would require knowledge of the actual layout of the various cortical regions and, not the least, a reliable picture of the mutual connections. It is therefore exciting to note that such detail has been emerging during recent years. The time has come to consider the evidence.

To each his own?

Neuroscience has come a long way since the early map-making efforts of Brodmann and his colleagues. Although the division of the cortex into anatomically distinct areas was well known in their time, almost a century ago, no light had been shed on the manner in which the different areas interact. The situation has changed dramatically, and magnificently detailed charts of such connections are now available for both the cat and the monkey. It transpires that the typical cortical area does not share white-matter connections with all the other areas. Far from it; connections are usually made with just a few other areas. As we will be enlarging upon shortly, the arrangement can be regarded as being hierarchical. Knowledge of the corresponding situation in the human cortex is not as advanced, although there has recently been progress on this front too.[16] This multiplicity of areas raises the obvious question: what are they for? The answer appears to be emerging, as we will now see.

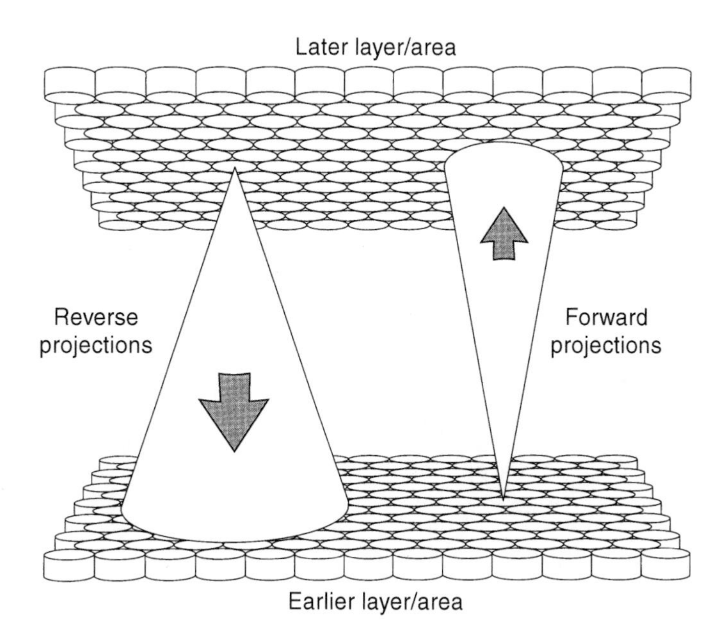

Figure 7.6
The neurons in a given minicolumn, and in a given cortical layer, have synaptic connections with the neurons in many different minicolumns in at least one other cortical layer, the latter being either in the same cortical area or in another area. These connections may be either excitatory or inhibitory. The spread of these inter-layer connections is described by what is referred to as the vergence angle. In this highly schematic diagram, cortical layers are shown as if they were isolated, rather than contiguous with other layers, for the sake of clarity. The diagram also indicates that inter-layer connections may be established in both the forward and reverse directions, and that the vergence angles need not be the same in the two directions. Semir Zeki and Stuart Shipp have shown that such imbalance of vergence angles often applies to the forward and reverse projections which link different cortical areas, the reverse projections actually showing the wider spread. Both Zeki (*A Vision of the Brain*, 1993, Blackwell Scientific Publications) and the author of the present book believe that this fact has a bearing on the question of perception.

One of the most interesting things to transpire in anatomical studies of the cerebral cortex has been that many of the inter-area connections run in both directions.[17] The latter characteristic is often reflected by the word *reciprocal*, although this does not imply that white-matter axons occur as counter-running pairs. The reciprocity referred to merely stems from the fact that the inter-area connections run in both directions (see figure 7.6). This has to be considered in conjunction with the spreading out that always occurs because of the multiplicity of contacts entered into by any neuron. A neuron in one cortical area, let us call it *alpha*, will be represented by numerous axon branches in one of alpha's target areas, *beta* for example. Likewise, and because of the reciprocity, a neuron in *beta* will exert its influence on many neurons in *alpha*, by dint of its many axon branches in that area. There is an important point about this reciprocity, however, and it has been stressed by Semir Zeki and Stuart Shipp.[18] The reciprocity is often rather lopsided. The axon branches of a neuron in *alpha* may be rather less spread out in *beta* than those of a neuron in *beta* are in *alpha*.

How are such inter-area connections charted? One technique that has been very popular in recent years exploits the ability of the substance horseradish peroxidase to diffuse over considerable distances within the processes of nerve cells. This substance is injected into cells in a cortical area of interest while the animal is still alive, and time is given for it to diffuse.[d] Thereafter, the animal is sacrificed and other cortical areas are examined to

d / This technique is obviously not used to study the human cortex.

(*a*)

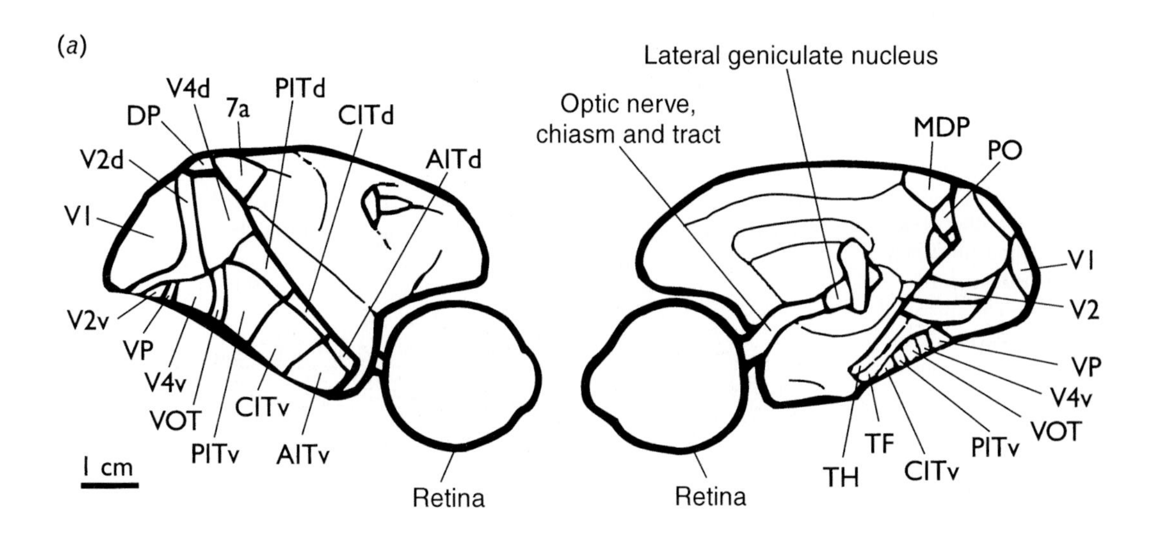

Figure 7.7
(*a*)–(*c*) The layout of the various cortical areas of the macaque monkey adapted from the work of Daniel Felleman and David Van Essen, whose papers on the subject have drawn on data published by a large number of other authors. (*a*) A lateral view of the right hemisphere seen from the right (left-hand diagram) and a midsagittal section of the right hemisphere seen from a vertically oriented median plane, looking toward the right (right-hand diagram). The early visual areas such as V1 and V2 are seen to be located in the occipital lobe, the letters d and v designating dorsal and ventral respectively. Prominent amongst the features shown in the temporal lobe are the anterior inferotemporal area (AIT), the central inferotemporal area (CIT), and the posterior inferotemporal area (PIT). The ventral occipitotemporal area (VOT) is located at the border between those two lobes. The medial dorsal parietal area (MDP) is a major feature of the parietal lobe, while the parieto-occipital area (PO) is another borderline region. (*b*) How one entire cerebral hemisphere appears when it is spread out on a table, with all the gyri flattened out. The retinae have been included, to show how they project to the visual cortex via the (appropriate) lateral geniculate nucleus, superior colliculus and pulvinar. The features seen in the lateral and sagittal views are easily identifiable, as are many other cortical areas. This spread-out view gives a good impression of the relative positions of the main entry regions for the various senses, as well as the position of the motor cortex. The frontal eye field (FEF) is of particular interest to the issue of working memory, because it appears to cooperate with several distantly positioned areas in providing the system with a sort of erasable sketch-pad or blackboard. The manner in which some of these areas are interconnected is illustrated in (*c*). The entry point for visual signals is shown at the bottom, while the hippocampus appears at the top. The central role played by the frontal eye field is reflected in its central location in this circuit diagram. Each line represents the millions of individual axons of a known inter-area (white-matter) route, no effort having been made to indicate vergence. Many of these routes comprise both forward and reverse projections. As can be seen from this final diagram of the series, the various cortical areas can be arranged in a hierarchy, running from the early visual areas to the hippocampus.

(*b*)

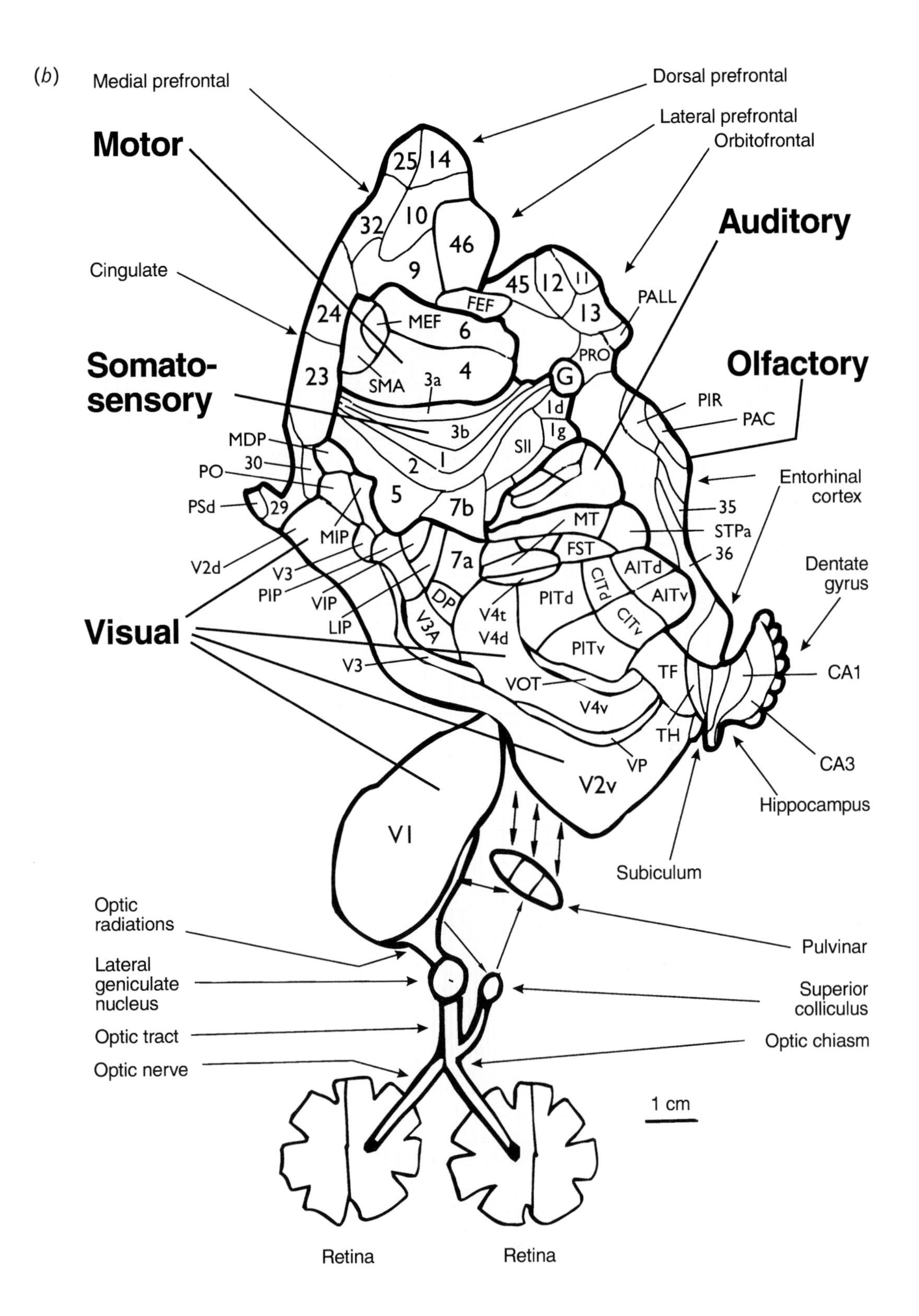
Medial prefrontal
Dorsal prefrontal
Lateral prefrontal
Orbitofrontal
Motor
Auditory
Cingulate
Somato-
sensory
Olfactory
25
14
32
10
46
9
45
12
11
13
24
FEF
MEF
6
PALL
PRO
23
SMA
3a
4
G
PIR
PAC
1d
1g
3b
SII
MDP
30
PO
PSd
29
1
2
5
7b
Entorhinal
cortex
35
STPa
36
MT
FST
MIP
V2d
V3
PIP
VIP
LIP
7a
DP
V3A
V4t
V4d
PITd
CITd
AITd
AITv
CITv
PITv
Dentate
gyrus
Visual
V3
VOT
V4v
TF
TH
CA1
CA3
VP
V2v
Hippocampus
V1
Subiculum
Optic
radiations
Pulvinar
Lateral
geniculate
nucleus
Superior
colliculus
Optic tract
Optic chiasm
Optic nerve
1 cm
Retina
Retina

(c)

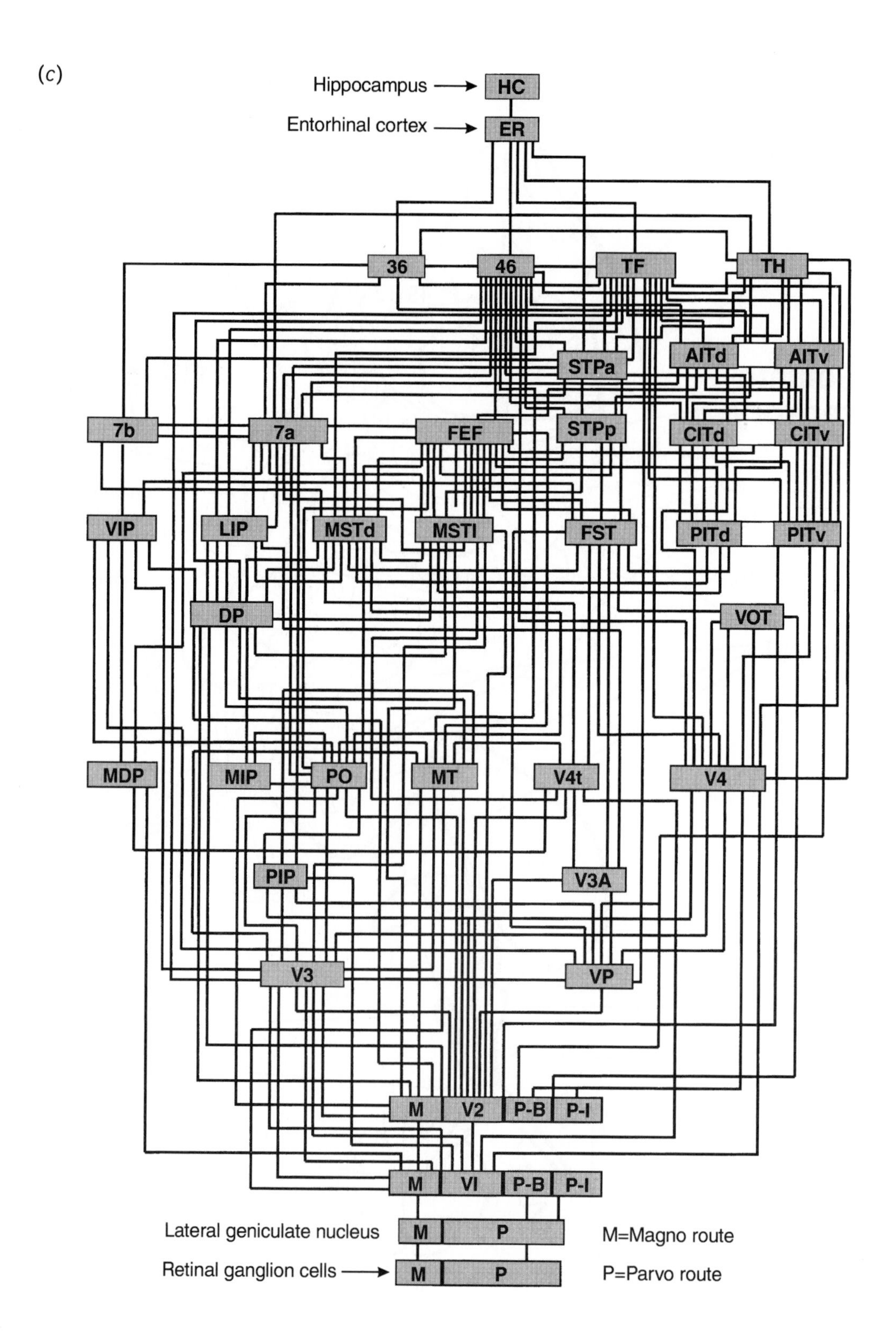
Hippocampus
HC
Entorhinal cortex
ER
36
46
TF
TH
STPa
AITd
AITv
7b
7a
FEF
STPp
CITd
CITv
VIP
LIP
MSTd
MSTl
FST
PITd
PITv
DP
VOT
MDP
MIP
PO
MT
V4t
V4
PIP
V3A
V3
VP
M
V2
P-B
P-I
M
VI
P-B
P-I
Lateral geniculate nucleus
M
P
Retinal ganglion cells
M
P
M=Magno route
P=Parvo route

see if the peroxidase has reached them. If that is the case, one may conclude that the two areas share white-matter connections.

The results of many different investigators of the monkey cortex have recently been analysed by David Van Essen, Charles Anderson and Daniel Felleman.[19] Their amalgamation of information from the various sources resulted in a circuit diagram that is complex and yet orderly. They find that it is possible to place the various cortical areas on a hierarchical grid, which has a number of different levels. The lowest strata of the scheme are close to the visual input, while higher levels correspond to the association areas (see figure 7.7). Still farther up in this picture are the regions that lie beyond the association areas. All the regions enter into multiple inter-area connections, although the degree of multiplicity is not uniform. Some areas have connections with less than ten others, whereas the number is closer to twenty for some of the other regions. It must be emphasized that the word connection, when applied in this way to the junction between two interacting areas, actually refers to millions of white-matter fibres, that is to say to the millions of axons which stem from large *pyramidal cells* in the sending area.[e]

Despite the complexity implied by the sheer profusion of its connections, this type of circuit nevertheless conceals one further degree of complication; the above-mentioned reciprocity is not indicated explicitly. Just as with the typical subway map, such as that of the London Underground, single lines are drawn to indicate situations in which the traffic actually travels in both directions. Jack Scannell and Malcolm Young have overcome this difficulty by use of a colour scheme that differentiates between reciprocal connections, forward-only connections and backward-only connections.[20] It is a pity that these beautiful diagrams cannot be reproduced here, for technical reasons.

So much (for the time being, at least) for the connections which lie roughly in the plane of the cortex. What about those that join neurons lying at different depths of the cortex's roughly three millimetres, within a single area? These too have been extensively investigated, particularly by Charles Gilbert and Torsten Wiesel in the case of the cat.[21] Once again, we find circuitry which at first glance looks forbiddingly abstruse. It is possible to make certain rationalizations, however, the main one of these being that the interlayer links appear to play excitation and inhibition off against each other. If one follows the various routes, and roughly estimates when signals would arrive at the various neuronal layers, one sees that in several cases an inhibitory wave would follow hard on the heels of a burst of excitation. Just why this should be useful to the system is not yet known, but it would certainly have the effect of breaking up continuous signals into bunched packets of activity. The frequency with which these packets would arrive at a given cortical layer can be shown to lie in what is known as the *gamma band*, which is to say 35–60 cycles per second.[22] The actual pattern of the inter-layer connections is rather more complicated than might have been imagined. It is not

e / Pyramidal neurons take their name from their shape, the vertices of the pyramids always pointing toward the cortical surface.

merely a case of adjacent layers interacting with each other. Some of the connections skip a layer, and there are links in either direction (see figure 7.8). It should also be noted that Gilbert and Wiesel found evidence of recurrent collaterals in several of the cortical layers in the primary visual area. This could mean that there are self-organization processes at work, of the type described by Kohonen.[f]

We have reached the point at which it would be useful to have a thumbnail picture of the main intra- and inter-area connections in the cortex and the underlying brain structures. The major *receiver* of signals is layer IV, irrespective of whether this concerns afferents from a sub-cortical centre, such as the lateral geniculate nucleus, or from another area of the cortex that is lower in the hierarchy. There are three major *senders* of signals from a given cortical area, namely layers II–III, V and VI. Layer II–III transmits signals to other cortical areas lying further up in the hierarchy, whereas layer V transmits signals in the reverse direction, which is to say that it sends to areas lower in the hierarchy.[g] In many areas, there are also projections from layer V to sub-cortical regions of the brain. In areas lying sufficiently close to the sensory input, layer VI sends signals to the thalamus (see figure 7.8).

The transmission of signals between different cortical areas, via the white-matter fibres, is the means whereby different levels of the hierarchy interact. Given the fundamental importance of this signalling, it is worth noting that the circuitry ensures that these signals do not collide with one another. If that happened, it is difficult to see how the system could avoid loss of information. In the ascending direction, layer II–III in one area sends to IV in another, whereas the reverse trip involves transmission from layer V in the higher area to layer IV in the lower area. Given that there are internal routes between layers II–III, IV and V in a given area, this would seem to indicate that signals can travel around a closed loop, from the lower area to the higher area and back again. There is a very important issue connected with the possible existence of such *closed loops*, and it is so major that a separate section will have to be devoted to it, later in this chapter.

Meanwhile, let us look at another aspect of cortical anatomy. It is known as the *minicolumn*, and its existence was first conjectured by Vernon Mountcastle, in the late 1950s, and independently by Janos Szentagothai.[23] The minicolumn lies roughly at right angles to the cortical surface, and it traverses the entire three-millimetre thickness (see figures 7.5, 7.6 and 7.8). In the cat, it has the approximate shape of a cylinder, the diameter of which is about 50 micrometres (i.e. about 0.05 millimetres, which means that the thickness of an ocular dominance column would be spanned by about ten minicolumns placed side by side). One should not gain the impression that the minicolumn is a neatly defined entity, with boundaries marked by special tissue. Its discoverers merely noted the tendency for bunching up of

f / Such self-organization was discussed in chapter 5.

g / The traditional distinction between layers II and III has become blurred of late, and they are now often referred to jointly.

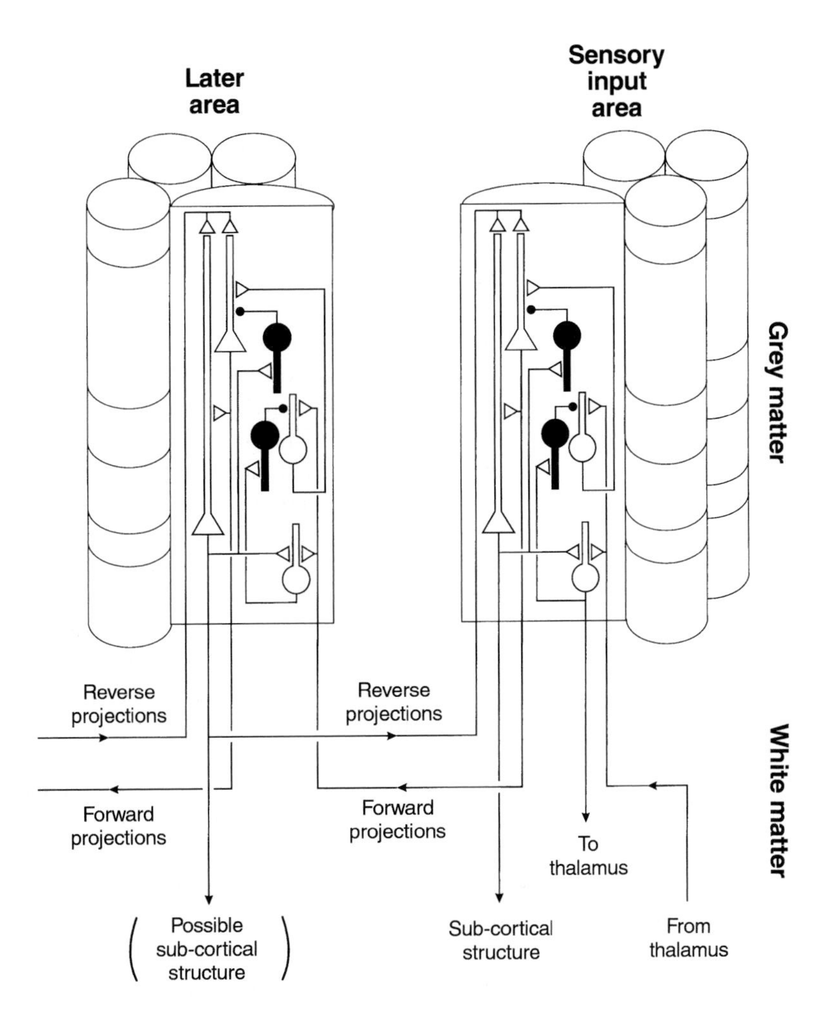

Figure 7.8
The work of Charles Gilbert and Torsten Wiesel has produced details of the neural circuitry within the primary visual area of the cat, and these have been used in this highly schematic diagram of the inter-layer connections in a given minicolumn. The picture also shows the main input and output routes. (No effort has been made to represent the multiplicity of connections of each type, and no indication is given of the vergences via which the neurons in a given layer send signals to neurons of other layers in other minicolumns.) In the case of the primary visual cortex, it is to be noted that layer II–III neurons send to later cortical areas (via forward projections); that layer IV neurons receive from both the lateral geniculate nucleus (LGN) of the thalamus and from layer V neurons of later cortical areas (via reverse projections – which synapse onto the distal dendrites of those receiving neurons); that layer V neurons send sub-cortical projections to the superior colliculus; and that layer VI neurons send sub-cortical (corticofugal) projections back to the LGN. There is evidence that this overall pattern, with minor variations, is seen throughout the cerebral cortex. The sub-cortical projections from layer V, for example, may be the manner in which the cortex generally exerts its influence on deeper-lying structures.

the neuronal processes that run between the various cortical layers, with thinned out regions lying laterally between. From its dimensions, one can calculate that a minicolumn contains about two hundred neurons, and if these are fairly uniformly distributed, each of the four layers II–III, IV, V and VI would have about fifty neurons.[h] These neurons will be of various types, which will in general be both excitatory and inhibitory.

We still do not have a definitive picture of a minicolumn's function, but some suggestive facts have emerged. The number of these units in the primary visual area of the macaque monkey (in one cerebral hemisphere), for example, is about the same as the number of fibres in the optic radiation. If each of these fibres carries a distinct signal, the implication would be that each minicolumn is capable of handling what could be called a unit of information. This naturally raises the question of why it should be desirable to have perhaps twenty to thirty excitatory neurons in each cortical layer of a minicolumn, all of them presumably serving the same function. It could be that this *redundancy* gives the brain a certain robustness against neuronal loss, but there might also be more subtle things afoot. Given the statistical nature of neuronal activity, there will be no guarantee that identical sensory inputs will produce exactly the same activity pattern in a given neuron. But if there are twenty or thirty neurons doing the same job, there will be a high probability that *some* of them will react in the desired manner. At a given moment, it will not matter which neurons are actually doing the work; it is merely important that some of them are functioning.

The column concept recently received a strong boost. Ichiro Fujita, Keiji Tanaka, Minami Ito and Kang Cheng have reported the presence of this type of structure in the anterior inferotemporal area in macaque monkeys.[24] This is one of the regions which contains neurons responsive to composite features in the visual input. They discovered groups of neurons in this area that are arranged in columns lying roughly at right angles to the cortical surface, all cells in the group being activated by the same visual figure. The relationship between input and activity was invariably complicated. There were neurons which showed maximal activity when the visual system was exposed to a circle joined lollipop-fashion to an elongated rectangle. The response was markedly diminished if either the circle or the rectangle was removed. In other parts of the same cortical area, the greatest reaction was elicited by composites of various colours, combinations of contrasting hues such as green and red, and blue and yellow being particularly common.

Of particular interest in this work were the observed dimensions of the columnar structures. They were about 0.4 millimetres in diameter, which makes them comparable to the ocular dominance columns that were described earlier, but considerably larger than the minicolumns present in the primary visual area. The total area of the anterior inferotemporal region is between three and four hundred square millimetres, and this indicates that it

h / In practice, the distribution is not so uniform, but such details need not concern us here.

comprises roughly two thousand columns. The most direct interpretation is that this is the number of distinct object features that can be represented by that part of the cortex, although Fujita and his colleagues gave reasons why the actual number of codable features would be somewhat smaller. That there are far fewer columns in the anterior inferotemporal area than there are in the primary visual area is in keeping with the fact that many component features must conspire to produce the type of composite figure detected by the higher region.

It does seem that this important discovery lends credence to the hierarchical principle of cortical processing, even though it might stop short of endorsing the existence of grandmother cells. The results are certainly interesting, but they do not in themselves tell us how the cortex gives us our cognitive abilities. Indeed *they do not even provide definitive proof that the hierarchical concept is correct*. There are in fact two rival ideas which have received a great deal of attention in recent years, and it is important that we consider both, and compare them with the hierarchy model.

The first alternative sees cortical processing as being carried out simultaneously in parallel streams of activity. Strong arguments in favour of such a view were put forward in Margaret Livingstone and David Hubel's classic paper of 1988. Their starting point was an apparent conflict: localized damage to a particular cortical area often leads to a specific cognitive deficit (we reviewed some of the evidence in chapter 3), and yet our perception of a visual scene seems to be totally unified, the tracking of coloured moving shapes presenting our visual systems with no difficulty.[25] This clearly raises the question of how the various attributes of an observed object can be so effectively bound together. Livingstone and Hubel took the evidence from localized damage to indicate functional subdivision in the visual pathway, and they supported this view with observations on the activity level of an enzyme known as *cytochrome oxidase*. We need not go into the details here, but this substance enables one to trace the paths taken by various signalling streams.

To appreciate what the Livingstone–Hubel work achieved, we will have to go into a little more anatomical detail. The lateral geniculate nucleus of the primate is a layered affair, the six strata being divided into the two *magnocellular layers*, which lie ventrally, and the four dorsally directed *parvocellular layers*. The ganglion cells which feed into these two types of layer are mutually distinct. Wavelength is predominantly handled by the parvocellular neurons, for example, while the magno system is essentially wavelength-blind. When it comes to contrast sensitivity, on the other hand, it is the magnocellular cells which carry most of the responsibility. Another important factor is the speed of a neuron's reactions, because this has a strong bearing on its usefulness in the service of motion detection. In this respect too, it is the magno system that carries the day. Now Hubel, together with Torsten Wiesel, had already established that the distinction between the two systems is maintained in the primary visual area of the cortex. Livingstone and Hubel's analysis of subsequent areas of the cortex convinced them that the separation is continued in those regions too, and the upshot was a model in which there is a major division of tasks between large sections of the cortex. In brief, the Livingstone–Hubel rationalization has the visual input coming into the primary visual area, whereafter there is a branching such that motion is handled in the (dorsal) parietal region while form and wavelength (and ultimately, colour) are the province of the (ventral) temporal region. Crudely speaking, one could say that *the temporal lobe tells us what*, while *the parietal lobe tells us where and when*.

But *how* do they tell us these things? Livingstone and Hubel did not essay an answer to that continuing mystery, but they did make a big assumption regarding the mediation of information, namely that the latter is contained in the signals that can be detected in the individual neurons they investigated. As Kevan Martin remarked, they were inferring that an individual cell can handle 'all the news that's fit to print'. Martin suggested that the more likely situation is one in which information is hidden in the activities of *combinations* of neurons.[26] Evidence has been accumulating which conflicts with the neat division envisaged by Livingstone and Hubel. For example, Vincent Ferrera, Tara Nealey and John Maunsell have shown that neurons in the visual area V4, which handles wavelength, are activated by both the parvo and magno systems.[27] And Jean Bullier and his group have shown that neurons in area V5 (also known as MT), which contributes to the perception of motion, continue to respond to visual stimuli even when transmission through the primary visual area is blocked. This area admittedly receives signals from a second pathway that involves the superior colliculus and the pulvinar, but neither of these structures contain magno or parvo neurons.

Martin's conclusion was that separation is not as crisp as it had seemed to be. He opted for the alternative that takes its inspiration from the parallel distributed processing (PDP) concept which we considered in chapter 6. In view of the harsh criticism of the artificial neural network in the later part of that chapter, this might seem to be a retrograde step, but we must bear in mind that there is more to the PDP idea than is embodied in that particular strategy. Accepting that signalling pathways can both diverge and converge, and that they can be superimposed for parts of their routes, does not have to imply that one is assuming the presence of what we referred to as *the teacher*. The need for such a surrogate homunculus is obviated by the brain's own reward system, but this still leaves the question of how the brain gauges the difference between what is desired and what is currently being achieved. In the artificial neural network, this mismatch determines how much adjustment must be made by back propagation. Most people would surely agree that the brain does not back propagate. What happens instead is still a moot point. I will propose an alternative in a later section of this chapter.

Although this requires a brief digression from our main theme, we ought to briefly consider the reward system that I just invoked. It was discovered in 1954 by Peter Milner and James Olds,[28] who noted that a rat with an electrode implanted into certain regions of the brain will continually press a lever that delivers electrical stimulation to that region. It is as if these regions act as *pleasure centres*. The phenomenon is now referred to as *intercranial self-stimulation*, and its basis is contingent association between the response and its consequences. Aryeh Routtenberg and Eliot Gardner have shown that this drive can be so powerful that higher primates, and perhaps also humans, will even indulge in self-starvation in order to obtain rewarding brain stimulation.[29] Several brain regions are involved, including the medial forebrain bundle, which is a group of nerve fibres passing through the hypothalamus, and also the amygdala and the hippocampus, both of which are known to implicated in the learning of goal-oriented behaviours. The brain thus appears to possess a mechanism that rewards the advantageous behavioural correlation.

Meanwhile, let us return to those cortical columns. The mini versions in the early visual system (and possibly in the corresponding areas of the other sensory modalities) comprise a couple of hundred neurons, as we have seen, and these units must be capable of more nuanced behaviour than that displayed by a single neuron. Perhaps these elements are somewhat analogous to the pseudoneurons of the artificial neural network, but with the additional characteristic of being able to generate periodic output from a continuous stream of random input pulses. If that is the case, the situation would be (very) loosely comparable to what happens in a radio receiver. Signals can be carried by direct current, to be sure, but the use of alternating current opens up more exciting possibilities, which exploit the potential inherent in temporal variation.

What advantage lies in parallel distributed processing, given that the brain has no bottom line of neurons? Perhaps the answer lies in the signals which travel through those axons which project *out* of the cortex from layer V, bound for sub-cortical regions of the brain (see figure 7.8). It could be that they are simply tapped off from those columns which are currently involved in the most useful activity. Let us take a closer look at one particular case: the projections from the primary visual cortex to the *superior colliculus*. The latter structure is located dorsally at the top of the brainstem, below the thalamus and above the cerebellum. Its structure is rather special. As in the primary visual cortex, it contains a map of visual space, and its map is also somewhat distorted; the foveal region of the retina is represented by the lion's share of the neurons in the superior colliculus.

There is another map in the superior colliculus, in the layer immediately adjacent to the one connected to the visual system. This other stratum serves the motor system, and the two maps are in mutual registry, an arrangement in keeping with the fact that the superior colliculus plays an important role in coordinating vision and movement. The function of that part of the visual system which serves the area beyond the fovea is to draw attention to activity of potential interest. If something moves in a region peripheral to our gaze, the superior colliculus turns the eyes so as to bring that activity into the fovea.

The superior colliculus is just one of the many sub-cortical structures guided by signals from appropriate regions of the cerebral cortex. In attempting to gain a picture of the overall strategy, we need not consider all the subsystems served in this manner. Their details will naturally vary because they have different tasks to perform, but the general principle will probably be one of tapping off signals from relevant cortical regions. There are two structures which do need further comment at this point, however, namely the hippocampus and the cerebellum, because the generality of their functions can easily be underestimated. Experiments in which rats learn to negotiate mazes have strongly implicated the hippocampus, so it would be easy to gain the impression that this structure exclusively governs navigation. But we learned in chapter 5 that the hippocampus mediates what we referred to as *declarative memory*. Indeed, we saw that people who have lost this structure are unable to store new long-term memories. How is this to be squared with the seemingly more modest navigational requirements of the rat? It would be tempting to conclude that knowing what is where is almost everything that a rat needs, whereas our own requirements are far more sophisticated. The connection between space and declarative memory might be more intimate than we suspect, however. We might be more spatial than we realize. One could wonder whether our syntax and semantics are stored in memory in a manner that relates their elements to a sort of generalized set of spatial coordinates.

Such a generalized set of coordinates may well be implicit in the interactions that occur in the cortex and related structures. John O'Keefe and Lynn Nadel have exhaustively reviewed the evidence that space is indeed represented in a relative manner in the brain, and they referred to the egocentric spatial system as *taxon space*.[30] They put forward a model based on the existence of place-coded neurons in the hippocampus, and the learning mechanism by which these become identified with particular generalized locations depends critically on the *theta rhythms* that have been recorded in the hippocampus during certain exploratory behaviours. The theta rhythm, which had long been known as a feature of the EEG, has a frequency of around 5 cycles per second, which is to say a periodicity of about 200 milliseconds. In the O'Keefe–Nadel scheme, this rhythm plays the key role of providing a phase against which the activities of the various groups of neurons (i.e. those of the dentate gyrus, the CA3, the CA1 and the subiculum) are to be gauged. Only when the phase relationship is favourable will the cells in those different regions be able to fire, and the theta wave runs through the system, in strict periodicity, acting like a neuronal counterpart of a conductor's baton.

There is another aspect of hippocampal function that remains a puzzle. How are things selected for their novelty? There must be a mechanism whereby familiar things do not become candidates for new declarative memories, but where is the interesting wheat separated from the routine chaff? Is it possible that not everything gets through to this structure? Perhaps when things are familiar, their short-term memory traces do not persist long enough to get handled by the hippocampus, because of the quenching influence of inhibition. If that were the case, there might be no winner in the winner-take-all process that we have considered several times earlier. Could it be that it is only when things are sufficiently surprising that they have a chance of getting through to the hippocampus, from which they can subsequently be rationalized into the cerebral cortex for long-term storage? How all this could be achieved is, of course, one of the big unknowns, but it might have something to do with the temporal integration that was mentioned earlier.

Fine tuning

The role of the cerebellum may similarly have been underestimated. It has been customary to see this structure (which contains just as many neurons as the rest of the brain combined) as the centre which permits movements to become second nature if they are repeated often enough. The consummate skills of musical and dance virtuosos are often cited as examples of what the cerebellum can do for those prepared to practice sufficiently. More recently, however, there has been the suggestion that the word *movement* should be replaced by the word *thing*. There is a growing suspicion that the cerebellum can mediate automation of virtually anything connected with the nervous system. Let us consider some interesting facts.

To start with, it must be emphasized that the cerebellum is not indispensable. Motor functions can certainly be executed in its absence, as clinical observations have amply demonstrated, but in those cases where the cerebellum has been extensively damaged limb movements are jerky, tremulous and inaccurate. This led to the suggestion, by Marie-Jean-Pierre Flourens in 1824, that the cerebellum coordinates movement. A little over a hundred years later, Gordon Holmes advocated a more detailed mechanism in which the cerebellum's role was that of a comparator. This is a term that is familiar in engineering, and it refers to a device for comparing the actual value of a system parameter with its desired value. To appreciate why this should be advantageous, we will have to consider a type of situation which is common in engineering practice.

A frequent requirement in engineering is the exercise of adequate control over processes that are to function automatically. The word process is used here in the general sense, and it includes the workings of machines of all types. The standard approach to this problem involves constructing a model of the process, and this word too is broadly interpreted; it does not refer to the sort of thing that can be purchased in a toy shop. The gist of an engineering model is frequently captured in a mathematical equation, which dictates the desired guidance. Such a strategy also requires knowledge of the values of the process's variables, at any instant, and these are fed into the appropriate places in the equation. The equation then yields the amount by which the control variable must be altered, so as to keep the process on schedule, or the machine on course. This common strategy usually goes under the name *predictor–corrector*, for obvious reasons. Such a method of automatic control is the best available at the present time, but that is not to say that it is totally satisfactory. On the contrary, it suffers from the serious limitation that only rather simple processes are amenable to modelling. When the underlying physics of the process is said to be non-linear,[i] and unfortunately most processes of engineering interest *are* non-linear, the exercise of control is very difficult.

The brain has a great advantage over the engineer. It does not have to account for the functions it performs by deriving the corresponding

i / In the present context, the word *non-linear* can be taken to mean lacking simple proportionality.

equations. It merely learns those functions through the gradual changes that occur in the synapses of the relevant neural network. As we have been discussing, that network lies in the cerebellum, and it is remarkable in several ways. For a start, the arrangement of its neurons is far more regular than is seen elsewhere in the brain, and this is especially true of the neuronal processes. Then again, four of the five most common types of neuron found in this brain structure are inhibitory, an unusually large fraction.

Very briefly, the layout is as follows[j] (see figure 7.9). One set of excitatory signals (from the spinal cord and cerebral cortex, via the *pontine nuclei* of the brainstem) enters the cerebellum via the *mossy fibres*, which activate the excitatory *granule cells*. The overall activity level of the latter is kept in check by the inhibitory *Golgi cells*, which possibly impose the sort of winner-take-all regime that we have considered earlier. The axons of the granule cells collectively form the *parallel fibres*, which are remarkable for their regularity; they resemble myriad telephone lines. These parallel fibres run in the direction of the numerous thin convolutions that run from side to side on the cerebellar surface. These *folia*, as they are called, are readily visible to the naked eye. The parallel fibres synapse onto the dendrites of the inhibitory *Purkinje cells*, the dendritic arborizations of which are unusually flat, these being deployed roughly at right angles to the parallel fibres. They look like the outstretched fingers of a hand, as they intercept the parallel fibres, except that they are far more numerous.

Two other types of cell are also activated by the parallel fibres, namely the inhibitory *basket cells* and the inhibitory *stellate cells*. The axons of both these types of cell synapse onto the Purkinje cells, and their role must therefore be one of *disinhibition*. The stellate-Purkinje synapses are made on the dendrites of the Purkinje cells, in the normal fashion. The basket-Purkinje synapses, on the other hand, are rather special because they are established at the very root of the Purkinje cell's axon (that is to say, at the *axon hillock*), where it emerges from the soma. This has led to the conclusion that basket cells have the power of veto over the Purkinje cell's output. The final component of this circuitry is the set of *climbing fibres*, which stem from the *inferior olivary nucleus* of the medulla (ultimately activated by the cortex), and which synapse onto the Purkinje cells, on a roughly one-to-one basis. To be more precise, each climbing fibre innervates up to about ten Purkinje cells, but each Purkinje cell receives its input from only one climbing fibre.[31]

There are other numbers which we should bear in mind. For example, each Purkinje cell receives input from upwards of 100,000 parallel fibres, while each parallel fibre contacts about 100 Purkinje cells. And each granule cell, which is presumably the source of just a single pair of counter-running parallel fibres, receives input from roughly 4 mossy fibres, while each mossy

j / This description of the cerebellum is considerably simplified. It ignores the fact that there are actually three distinct systems in this brain structure, each of which makes its own contribution to motor control.

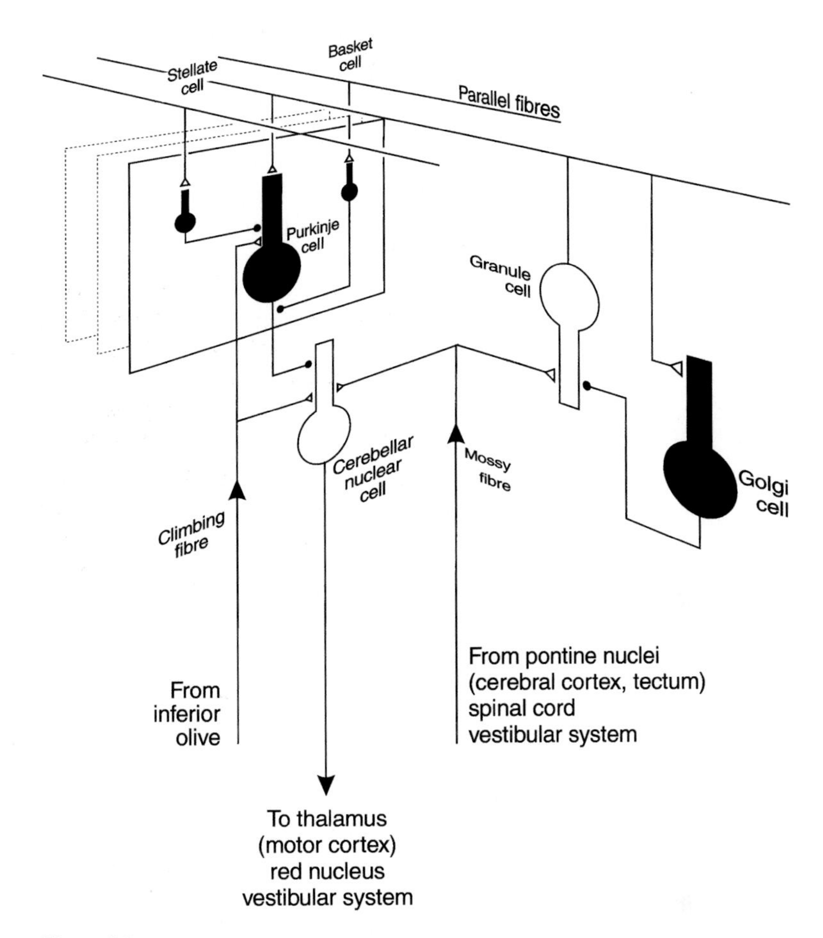

Figure 7.9
The five different types of neuron found in the cerebellum are all shown in this schematic diagram, and their relative positions and interconnections are also reliably depicted. But it is impossible to give the correct impression of their relative numbers because the granule cells outnumber the Purkinje cells by a factor of many thousands. The granule cells are the only excitatory neurons in the system, and the inputs to them are opposed by the inhibitory influence of the Golgi cells, which probably exercise winner-takes-all control. The granule cell axons form the parallel fibres, which feed excitation to the Purkinje cells, the basket cells and the stellate cells. The cells of these three types all have dendritic systems which tend to lie at right angles to the parallel fibres, as if in an attempt to maximize their interception of the latter. The basket cells and the stellate cells pass inhibition to the Purkinje cells, but in different ways. The stellate cell influence is of the dispersed type, whereas the basket cells veto the Purkinje cell output. Because the Purkinje cells are themselves inhibitory, the stellate and basket cells effectively contribute disinhibition to the overall functioning of the system. The climbing fibres activate the Purkinje cells, on a one-to-one basis. Although the anatomical layout of the cerebellum has been known for several years, there is still no universally accepted model of cerebellar function.

fibre innervates approximately 460 granule cells. All in all, therefore, a given Purkinje cell will have a catchment range which spans approximately 400,000 mossy fibres, and this is to be compared with its other excitatory input from a *single* climbing fibre. Offsetting these positive influences are the inhibitory inputs from the basket and stellate cells. John Eccles, Masao Ito and Janos Szentagothai have made a very interesting observation about the relative positions of these various inputs.[32] Because of the impressively tidy arrangement of the parallel fibres, a given Purkinje cell will receive excitatory input only from a bundle of parallel fibres having the width of its dendritic tree. Everything outside this strip will exert an inhibitory influence on that Purkinje cell, through the mediating agencies of the basket and stellate cells, both of these types of cell also receiving input from the parallel fibres, and both having dendrites which lie roughly perpendicular to those fibres. As Eccles, Ito and Szentagothai pointed out, this gives the areas of influence on a given Purkinje cell a centre-surround arrangement not unlike that we encountered in connection with the retinal ganglion cells. The difference is that in the case of the Purkinje cell, the pattern is symmetrical about a line rather than a point.

Despite the impressive regularity of its layout, the cerebellum is a reasonably complicated structure. What does it do for us? There have been recent suggestions that it functions in a manner reminiscent of the engineer's predictor–corrector mentioned earlier. In chapter 3, we noted that the nervous system has two different strategies at its disposal for steering muscular movements. There is feedback, which requires time to be effective, and there is the rather faster feed-forward mechanism. Let us consider what is involved in the seemingly simple act of hand–eye coordination. When humans track visual objects, their eye muscles can correct the direction of gaze up to about four times per second. (This rate is comparable to that achieved by the leg muscles of the best sprinters; these runners use 45–50 strides in a 10-second 100 metre race, so each stride consumes between a fifth and a quarter of a second.)

Christopher Miall, Donald Weir, Daniel Wolpert and John Stein have estimated the delays that would be incurred during the various stages of such visual tracking: 50 milliseconds for signals to travel from retina to primary visual cortex; a similar amount of time to get from the primary visual cortex to the association and motor cortices; delays in processing the motor command, and so on.[33] They find that visual feedback would require a third of a second or so, and the maximum rate at which the necessary muscular corrections could then be made would not exceed two per second. Miall and his colleagues concluded that the control must be of the feed-forward type, with the participation of the cerebellum acting as a predictor–corrector of the special type described by O. J. M. Smith in the late 1950s.[34] They proposed a mechanism in which the cerebellum forms linked pairs of neural models, the first of these providing a rapid prediction of the outcome of a motor command, and the second a delayed copy of that prediction. The latter would exactly match the outcome of the feedback arising from the movement if, and only if, the control was already ideally precise. By combining both these functions, Miall and his colleagues suggested, control could be achieved with the required speed and stability despite the long delays incurred in the neuronal loops.

More primitive ideas along similar lines had appeared earlier, and two of these must be mentioned here because their hypotheses appear to have been vindicated. They were made by David Marr and James Albus, who had been contemplating roughly coincidental arrivals of (excitatory) signals through a particular climbing fibre and through the relevant mossy fibres.[35] I use the word *relevant*, here, to mean those mossy fibres which activate the granule cells that give rise to parallel fibres which, in turn, contact the Purkinje cell in question. Albus suggested that the climbing fibre input would decrease the effectiveness of those mossy fibres (by weakening the synapses between the appropriate parallel fibres and the Purkinje cell), and he noted that this would tend to correct the mismatch between the intended and the actual movement. To see why this would be efficacious, we must note that the output of the (inhibitory) Purkinje cell connects to what are known as the *deep cerebellar nuclei*. These neurons are excitatory, and it is they which ultimately lead to activation of the relevant motor centres (by way of the thalamus). The Albus scenario is thus that the *reduction* of the influence of the mossy fibres on the Purkinje cell in question, *decreases* the activity of the latter, and this *increases* the activity of the appropriate deep nucleus. The net result is thus one of *facilitation*. It is intriguing to speculate whether Nature could have achieved a similar result via direct excitation rather than through a chain of inhibitions. When learning the English language we are told to avoid double negatives; Nature seems not to be impressed by such strictures.

The investigation that provided the critical endorsement of the Albus idea was carried out by Peter Gilbert and Thomas Thach.[36] They trained a monkey to grasp a handle and then maintain it in a new location, using the muscles of the wrist. A monkey has no problem in achieving this, provided the mechanical load on the handle is constant, but a period of rehabilitation is required if that load is suddenly changed. Gilbert and Thach monitored the activity level of the relevant Purkinje cell, and they could differentiate between the contributions of the climbing fibre and mossy fibre inputs. They did this during the various phases of the monkey's performance. Under constant load, they observed a regular series of single impulses due to the mossy fibre, with the occasional interruption by a more complicated response to the climbing fibre. When the load was abruptly changed, there was a sharp increase in the complicated variety, and this occurred at the expense of the single impulses. Gilbert and Thach discovered that the former pattern was gradually restored as the monkey became increasingly adept at the modified task. This crisp result had demonstrated that the climbing fibre is modulated during learning, and it gave credence to the view that such modulation *decreases* the influence of the relevant mossy fibres on the Purkinje cell in question.

These advances have given us an encouragingly clear indication of how the cerebellum goes about its job, but they leave us tantalizingly short of the complete picture. What we would now like are the details of how the synaptic facilitations in the cerebellum couple to the elements of motor control. Those elements are the firings of individual cells in the motor cortex, of course, but we still need to know how the rather broadly tuned neurons in that area conspire to give the precise control achievable by the performing animal. With the very large and very fast computers now appearing on the market, it may not be long before we see simulations of the activities of thousands of Purkinje cells. This will mean accounting for the activities of millions of granule cells, as a function of time, as well as their influences on cortical cells, but such a mammoth book-keeping exercise is now becoming feasible. Meanwhile, the work of Apostolos Georgopoulos, Andrew Schwartz and Ronald Kettner has provided a strong hint of what such simulations must build a bridge toward.[37]

They trained a rhesus monkey to reach out and push red buttons if these had been lit. The buttons were located directly in front of the animal, at shoulder height. There was a central button, and it was symmetrically surrounded by eight others equally spaced on a circle of 12.5 centimetres radius. A particular trial always started with the central button, whereafter one of the others was chosen at random by the experimenters. The goal of the investigation was to discover the neuronal correlates of the resultant arm motion, and an important preliminary was the locating of 224 neurons in the arm region of the motor cortex whose activity was found to be dependent on the direction of movement. The tuning of the individual neurons with respect to direction was rather broad. Subsequent monitoring of the activity of these neurons, as a function of various movement directions, revealed a remarkably logical relationship. Their activity levels always corresponded to what would be expected if they were contributing as *vector components* to the resultant movement.[k] If the resultant motion was aimed well away from the preferred direction of a particular neuron, its activity was very low. If the direction eliciting its maximal response almost coincided with the direction indicated by the lit button, on the other hand, it emitted a vigorous stream of action potentials. The satisfaction generated by this rational mechanism received an added boost, a couple of years later, when Choongkil Lee, William Rohrer and David Sparks found clear evidence of similar *vectorial addition* in connection with the activity of neurons in the *superior colliculus*.[38] As we have noted earlier, the role of this particular structure in the brainstem is to dictate the *saccadic eye movements* with which the visual system rapidly scans an object in the visual field.

To appreciate the significance of what Georgopoulos and his colleagues achieved in a follow-up investigation, we will first have to make a small digression and consider the pioneering work of Roger Shepard and Jacqueline Metzler in 1971. They asked a number of subjects to compare two mutually disoriented pictures of three-dimensional objects and judge whether they were the same.[39] The answers came progressively slower as the offset angle was gradually increased, and from their remarkably linear plot of time

k / Readers not conversant with vector summation may be familiar with one of its examples: the parallelogram of forces encountered in mechanics.

delay against angle of rotation Shepard and Metzler concluded that we can mentally rotate pictures at a rate of about 60 degrees per second. The linearity of their plot was a major revelation, whereas there is probably nothing fundamental about the numerical value of its slope; we would expect to be able to mentally rotate simpler figures are a greater speed, and Shepard later showed that this is the case (for a simple *two*-dimensional figure, the speed was 400 degrees per second). This is also what was observed by Georgopoulos and Schwartz, now in collaboration with Joseph Lurito, Michael Petrides and Joe Massey.[40] This time, they trained a monkey to move its arm to a location lying *perpendicular* to the illuminated button, and they again monitored the activity levels of the implicated neurons in the motor cortex. They were able to show that the various vectorial elements alter their activities in such a way as to rotate the resultant (appropriately summed) vector in the desired direction. Once again, therefore, they had revealed the existence of a remarkably logical mechanism. And given the fact that the monkey's mind was being required to rotate only a *one*-dimensional figure, namely a simple direction, we might not be surprised by their measured value of 732 degrees per second. I say *might*, here, because there is the question of whether monkey and human performances could be comparable; they indeed seem to be.

Returning to the cerebellum itself, from structures it merely helps to control, one of the most interesting things to appear in recent years has been the suggestion that this part of the brain may also contribute to the broader faculty of cognition. Amongst the observations which appear to warrant this expanded view have been direct visualizations of the cerebellum in action, through positron emission tomography (PET). (We recall that a PET scan reveals which parts of the brain are active under a given set of circumstances.) It was found, for example, that there is a significant increase in activity in the cerebellum when a subject is asked to generate a verb associated with a noun, such as ball/throw or weight/lift, and the increase prevails even when the subject is not immediately required to verbally communicate the word chosen. The converse type of observation, that is to say impairment of a cognitive function due to a cerebellar lesion, is also common. Richard Ivry and Juliana Baldo have recently reviewed the evidence for the cerebellar-cognition link, and it is becoming increasingly impressive.[41]

Something still more striking emerges if one regards the learning mediated by the cerebellum as a form of conditioning. In chapter 5, we saw that the classical conditioning epitomized by Pavlov's salivating dog has a counterpart in the adaptive behaviour of *Aplysia*. But if even the relatively primitive neural circuitry of that lowly sea snail admits of conditioning, why should we need a far more complex machinery to achieve much the same thing? The answer is of course that our repertoire of learned reactions has to be far more comprehensive and finely tuned. One could say that having to have a cerebellum is the price that more advanced species have to pay if they are to exploit the full potential of their sophisticated neural equipment. The conditioning analogy is a valid one, in any event, as Richard Thompson has stressed. He found that damage to the cerebellum disturbs acquisition of the conditioned eye-blink reflex in the rabbit.[42] Indeed, one can even identify which cerebellar components are implicated in the various aspects of conditioning. From what has been discussed earlier, we should recall that there are *unconditioned* and (to be) *conditioned* stimuli. It turns out that the unconditioned stimulus is conveyed to the Purkinje cells via the climbing fibre route, and thus stems from the olivary nuclei, while the conditioned stimulus enters through the mossy fibres, and thus emanates from the pontine nuclei (see figure 7.9).

One sees, then, that the situation regarding our understanding of the cerebellum is becoming more satisfactory. If I have seemed to labour the details, it is because the functions served by this major brain structure are fundamental to the optimal performance of its possessor. Amongst these functions is the mechanism which enabled the primate to gain one of its decisive evolutionary advantages, namely the delicate use of its four fingers and one opposing thumb. And with the strong hints now emerging that the cerebellum also serves cognition, the importance of this organ justifies the attention we have been giving it. It comes as something of a surprise to learn, as we are about to, that the brain has yet another system devoted to motor control. Let us pass straight on to it.

Subtler control

The *basal ganglia* collectively form a system consisting of four large sub-cortical structures which serve the control of movement. In this respect, they support the same aim as that of the cerebellum, but they accomplish this in a different manner. The cerebellum has direct connections to the spinal cord, whereas the basal ganglia interact only with the cerebral cortex. Their contribution to the control of movement is thus more subtle. As is the case with the cerebellum, the basal ganglia are also involved in processes not directly related to movement. Malfunction of this system thus manifests itself in two different types of impairment: motor handicaps and disorders of mood and thought.

The four major components of the basal ganglia are the *striatum*, the *globus pallidus*, the *substantia nigra* and the *sub-thalamic nucleus*. The first of these is divided into two anatomically distinct but nevertheless contiguous regions, namely the *putamen* and the *caudate nucleus*. The main flow of information in this system runs from the cerebral cortex to the striatum, whereafter there is a forking of the route into branches that variously go to parts of the globus pallidus and substantia nigra, as well as to the sub-thalamic nucleus, this part of the scheme also involving feedback loops. Finally, signals are dispatched to the *supplementary motor area* of the cerebral cortex, via the thalamus (see figure 7.10).

It is not appropriate to go into the minute detail of this system, so I will confine myself to some general comments regarding its structure and function, particularly where they contrast with that of the cerebellum. The latter receives signals only from that part of the cortex which is directly involved in sensorimotor function, whereas the basal ganglia are contacted by the

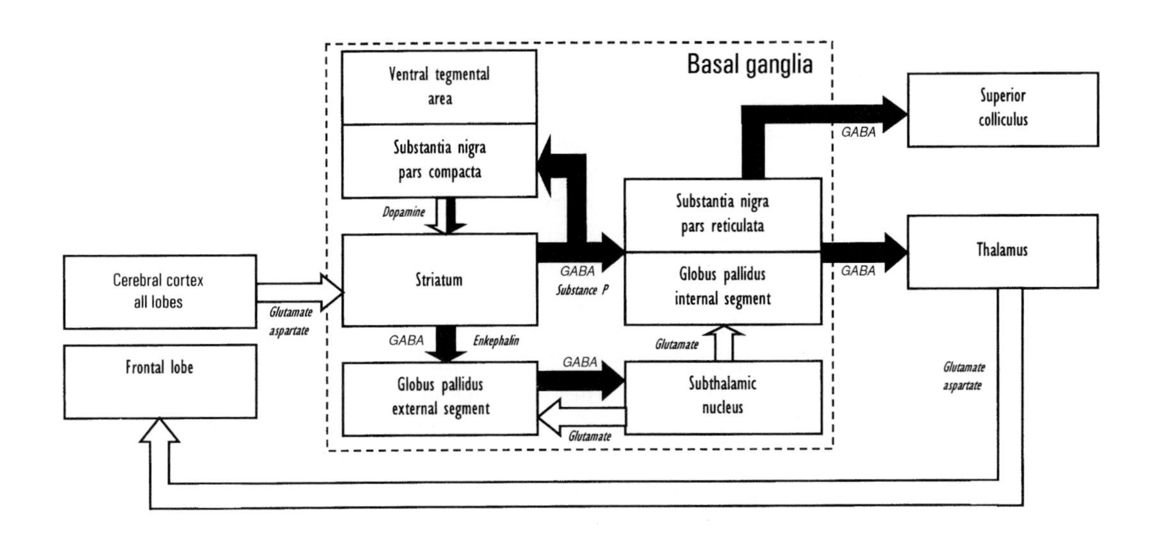

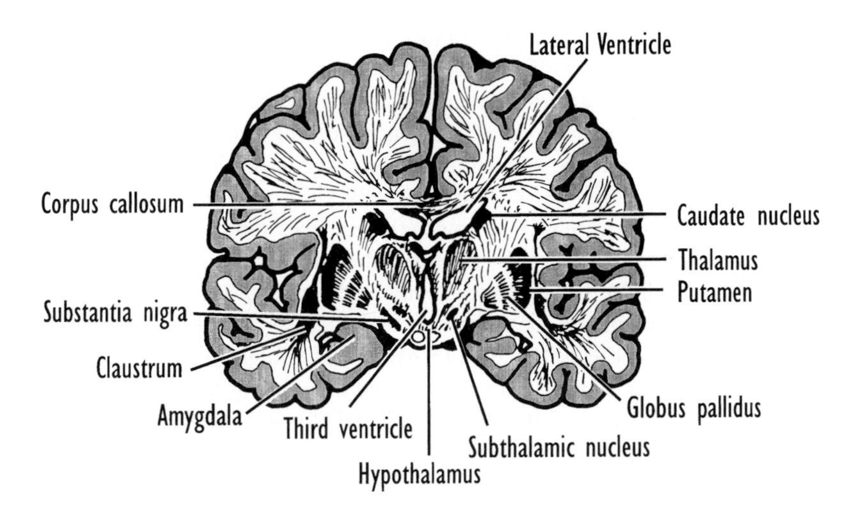

Figure 7.10
There are four main components of the basal ganglia, which collectively exercise secondary control on the motor system. These are the globus pallidus, the substantia nigra, the sub-thalamic nucleus and the striatum, the latter consisting of the putamen and the caudate nucleus. These structures employ an impressively broad selection of neurotransmitters, including glutamate, GABA, acetylcholine, dopamine, enkephalin and the so-called substance P. Unlike the cerebellum, the basal ganglia exert motor control indirectly; they have no direct contacts with the spinal cord. The lower diagram, inspired by a more detailed picture by R. Nieuwenhuys, J. Voogd and C. van Huizen, shows a coronal section through the basal ganglia, and locates the positions of the various components.

entire cortex. These contacts are nevertheless rather specific, and Ann Graybiel and Clifton Ragsdale Jr showed that the pattern of their targets in the striatum is consequently quite heterogeneous; one could loosely say that they resemble a patchwork quilt.[43] The domains in this pattern have a rather suggestive size; their diameters range from roughly a millimetre to about a third of a millimetre. This range is comparable to that of the diameters of the *columns* that we have encountered several times earlier in this chapter. The sensorimotor region of the cortex preferentially sends signals to the putamen, while the association cortex projects to the caudate nucleus. The feedback destinations show similar contrast with what is observed in the cerebellum. The output of the latter is exclusively directed toward the premotor and motor regions of the cortex whereas the output of the basal ganglia goes not only to those areas but also to the *prefrontal association cortex*. We should recall that the last of these is implicated in what was referred to in the preceding chapter as *the mind's blackboard*, which is to say working memory. It is deeply involved in the planning of actions. The cerebellum is implicated in the direct regulation of movement, whereas the basal ganglia participate in what could be called the cognitive dimensions of movement. Their dual role is the planning and execution of motor strategies.

One of the most impressive things about the basal ganglia is the number of different neurotransmitters they employ. These include glutamate, GABA, acetylcholine, dopamine, enkephalin and something known as substance P. This multiplicity is in keeping both with the underlying subtlety of the system, and with the wide range of faculties it serves. The basal ganglia may well be involved in virtually *all* the functions supported by the cerebral cortex itself. Quite how the system goes about its many tasks is still not totally clear, but there is evidence of the sort of interplay between excitation and inhibition that we have encountered earlier in this chapter, for example in the antagonistic arrangement that underlies the centre-surround response seen in the ganglion cells of the retina. This might permit comparison between signals being sent to muscles and those being returned from the sensory spindles in the latter. The basal ganglia appear also to participate in the initiation of movement. Another thing worth noting, because we encountered the same sort of thing when discussing the cerebellum, is that the basal ganglia employ the same double-negative strategy of disinhibition rather than direct excitation. Given that we will see much the same thing yet again in the next section, when we consider the brainstem reticular formation, it seems safe to conclude that disinhibition is a particularly effective way of exercising control in biological neural networks.

Certain diseases associated with the basal ganglia have been shown to arise at the molecular level. The affliction identified by James Parkinson in 1817 leaves its victims prone to tremor, muscular rigidity, a shuffling gait and difficulty in initiating movement.[44] In 1959, Arvid Carlsson reported that eighty per cent of all the normal brain's dopamine is located in the basal ganglia.[45] Neuromelanin, a dark polymer pigment derived from that neurotransmitter gives part of the substantia nigra the colouring from which it gets its name. Oleh Hornykiewicz showed that the trouble is linked to a dopamine deficiency in that region. This upsets the feedback from the substantia nigra to the striatum, a feedback that needs to balance excitation of that part of the striatum which forward projects inhibition to the substantia nigra, and inhibition of the other part, which forward projects inhibition to the external segment of the globus pallidus.

Another disease, recognized by George Huntington in 1872, initially merely produces clumsiness and absentmindedness in its victims, but these symptoms gradually give way to deteriorating speech, increasing muscular impairment, and growing mental decrepitude.[46] The ailment is genetically determined, and it is believed that several of the women burned as witches in Salem, Massachusetts, in the seventeenth century, suffered from the disease. In this case, there is a deficiency in the striatum of neurons which act through the neurotransmitters acetylcholine and GABA. The first of these is involved in the signalling within the striatum, while the latter is used in the extra-striatal projections. We need not go further into the mechanisms, and my only reason for giving this much detail has been to indicate that subtlety of function stems from the recondite nature of basal ganglia physiology.

The length of these last three sections has been such that we are in danger of losing sight of the goal they serve: the overall plan. Let me therefore reiterate what I see as the underlying strategy. It is a cerebral cortex that handles signals in a fashion which marries the ideals of specialized areas and parallel distributed processing. It is the numerous sub-cortical structures that tap off signals from the cortex and use them as their cues. It is a hippocampus that somehow sifts out the novel from the (current) routine, and stores both snapshots and sequences, later to transfer them to the permanent repository of the cortex. It is a cerebellum that tightens up the slack produced by inevitable delays in signal transmission. It is basal ganglia that exercise an even subtler control over movement. And it is the various other loops which we will examine in more detail later on. All these conspire to give us consciousness, attention and thought. But before we can make the final push toward explaining those attributes, we will have to take a closer look at three further things: the way in which the emotions are brought into the strategy; the actual nature of neuronal loops in the cerebral cortex; and ultimately what I tend to think of as *global closure*. These three items are covered in the following three sections respectively.

The power of feeling

When I get very angry, my pulse rate increases and my breathing becomes irregular. When I get upset in other ways, I can often feel the effects in my abdomen. My emotional response in such situations is entirely normal, and we considered such reactions when dealing with the autonomic nervous system in chapter 3. It would be wrong to conclude that the autonomic nervous system is the exclusive province of the emotions, however. After all, simple physical exertion also influences both one's pulse and one's breathing, and a bout of the collywobbles can be caused by something as innocent as over-spiced food.

It was Walter Cannon and S. W. Britton, in 1925, who started to put emotions on a neurophysiological footing when they gave the name *sham rage* to a group of symptoms in cats that included arching of the back, attempts to claw and bite, erection of the hair, and increases in blood pressure and adrenal secretions.[47] Cats deprived of their cerebral cortices display just such behaviour. Three years later, Philip Bard showed that systematic removal of increasing amounts of cortical tissue leads to disappearance of sham rage only if the hypothalamus has been excised.[48] Conversely, he found that large parts of the cortex can be cut away without sham rage developing. It was Heinrich Klüver and Paul Bucy who first demonstrated, in 1939, that removal of the temporal lobe, including the underlying amygdala and hippocampus, produced a dramatic change in experimental monkeys. They became much tamer, and yet reacted more vigorously to stimuli. Their memory function was clearly impaired, and their level of sexual arousal had risen so much that they even mounted inanimate objects and members of other species.[49]

Meanwhile, other types of experiment were giving information that supplemented these ablation studies. Stephen Ransom, and later Walter Hess, probed the reactions of experimental animals to electrical stimulation applied through fine electrodes implanted into different brain regions.[50] It transpired that the hypothalamus functions as a coordinator, gathering signals from various parts of the brain and dictating a set of appropriate responses in both the autonomic and somatic systems. Although much remains to be done concerning the neural seat of the emotions, it is now reasonably clear that James Papez had perceived the essentials when he identified the key structures in 1937, following the work of others just described.[51] These brain components are those now loosely referred to as the *limbic system*. They are highly interconnected, prominent members being the *hypothalamus*, the *cingulate gyrus*, the *fornix*, the *mammillothalamic tract*, and the *prefrontal cortex*. We cannot be sure that this list is complete. For example, Lawrence Weiskrantz has shown that striking behavioural changes can be produced if the midbrain structure known as the *amygdala* is deliberately lesioned.[52] Richard Bandler has seen similarly impressive changes in the behaviour of cats electrically stimulated in another midbrain region known as the *periaqueductal grey region*.[53]

One of the things we should be impressed by is the fact that the connections between the different structures in the limbic system are invariably reciprocal; there are usually nerve fibres that run in either direction when any two of these components interact. But this does not make the overall system easier to understand. One recent theory, published by Edmund Rolls in 1990, sees the amygdala as forming stimulus-reinforcement associations, and the orbitofrontal cortex as correcting behavioural responses when these are no longer appropriate because of changing circumstances.[54] The amygdala receives input from the hippocampus, so this view is not implausible. Indeed, when one considers the various routes taken by information entering the limbic system, it does seem that reinforcement of associations could be the underlying process. Various forms of sensory information receive preliminary processing in the early visual area, the somatosensory area, and the primary auditory area, while the relevant region for olfaction is what is known as the *prepyriform cortex*. The information then converges on the association cortex, where it is further processed, and the resulting signals are projected to temporal and prefrontal cortices. Both the latter have neuronal links with the amygdala. Just what happens thereafter is not entirely clear, but it seems that signals sent onward from the amygdala to the hypothalamus could be the ultimate mediators of autonomic control.

Let me switch now to another sub-cortical complex of structures which might seem to have rather little to do with emotions. One member of the group is already familiar to us, namely the thalamus. I have emphasized that this structure is more than just a relay station *en route* between receptor cells and the relevant primary region of the cortex (as identified above), but gave no justification for this attitude. The time has come to put this on a firmer footing, and the key word is control. The case of vision will provide us with a relatively straightforward illustration of the point. It is not difficult to work out, from the numbers given in the first section of this chapter, that the fovea covers a visual angle of about 5 degrees. This roughly corresponds to the angle subtended at the pupil by the width of one's thumb when it is held at arm's length. Let us carry out a series of small experiments.

Extend your arm, close one eye, and look at the back of your hand. You will find that you can take in the entire width of one finger at a glance. As long as you hold your gaze on that finger, the adjacent fingers will not be seen to the same degree of definition. Now notice that you can narrow down your region of attention so as to take in just one wrinkle on one of the knuckles on that finger; you will simultaneously lose the ability to discern detail elsewhere on the finger, if you hold your gaze quite fixed. Finally, note that you can focus even more finely, concentrating your attention on just one small part of that wrinkle; again, the other parts of the wrinkle will simultaneously become fuzzier.

Apart from the obvious significance of the control system that we have just put through its paces, this series of tests exposes something more general about the brain's ability to regulate its activities. It is revealed by the way we use the word *concentrate*. Note that it is employed in two senses, which are distinct but nevertheless related. We have the extent of concentration dictated by the focal diameter of our gaze (this sense being the same as that invoked when an artillery battery is said to concentrate its fire, for example), and we also have the degree of concentration measured by the focus of our thoughts and intentions. The latter connotation is not so easily defined, but it is the more common of the two. The exciting prospect is that these two meanings of the word may be related to the same process at the neuronal level.

Before going to that plane, however, let us consider examples that illustrate the point. One arose during a round of political campaigning in the USA, during the 1970s. A candidate said of his rival that the latter was unable to simultaneously walk and chew gum. I will not state names because the remark was singularly unkind, an unwitting gibe at the handicapped. But the biology was sound enough. The more we concentrate on one thing, the less attention we are able to pay to another. Conjurers and stage magicians exploit this when they distract our attention with their wand waving and their abracadabras, while surreptitiously setting up their next illusion. Similarly, every school teacher knows that pupils will absorb nothing so long as one of their number is distracting the rest.

The essence of these examples is that they expose a fundamental limitation to the amount of information that our cognitive systems can handle at any instant. And they raise two questions whose importance hardly needs stressing: what is the neuronal basis of this limitation, and what is the mechanism by which we are able to vary the degree of concentration of our attention? We do not have the definitive answers to either of these questions, but on the basis of what is known about the underlying circuitry, it seems that the control may be exercised in the thalamo-cortical region, while the limitation may be imposed by the extent or efficiency of the connections between different areas of the cortex. I will have to defer consideration of that latter issue until chapter 9, so let us confine our attention to that intriguing control mechanism. As usual, because it is the best understood of the sensory modalities, we will use the visual system as an illustration.

As we have already noted, the thalamus sends visual signals onward to the primary visual area of the cortex, via the optic radiations. The latter encounter, and pass through, an important nucleus on their way to the cortex. This is the *nucleus reticularis thalami* (the nRt for short), and it surrounds the top of the thalamus rather as the fingers of one hand can be made to cover the clenched fist of the other hand. We can imagine the optic radiations as passing through the spaces between those covering fingers. The nRt receives excitation from the optic radiations, through axon collaterals, and it returns inhibition to the thalamus. The term reticular refers to the fact that the surface of the nRt is patterned like a network.[1] It is believed that this pattern is related to the local nature of the nRt's feedback connections to the thalamus. In any event, the nRt's influence on the thalamus is one of controlling, via inhibition, the strengths of the signals being sent to the cortex.

The nRt does not act alone. A second structure lying at the top of the brainstem, caudal to the thalamus, appears to have dominion over both the nRt and the thalamus itself. This is the *brainstem reticular formation* (BRF for short), and it too exerts its influence by inhibition. It projects inhibitory axons to both the nRt and (it is believed, at least) inhibitory interneurons interdigitated amongst the excitatory neurons of the thalamus. Once again, therefore, we find the brain preferring to use the double negative of disinhibition rather than direct excitation (see figure 7.11). The BRF itself is actually the most ancient part of the brain, and it is the one brain component which is absolutely indispensable to the life of its possessor. It is located near the pons

1 / The reticulated python similarly gets its name from the network pattern on its skin.

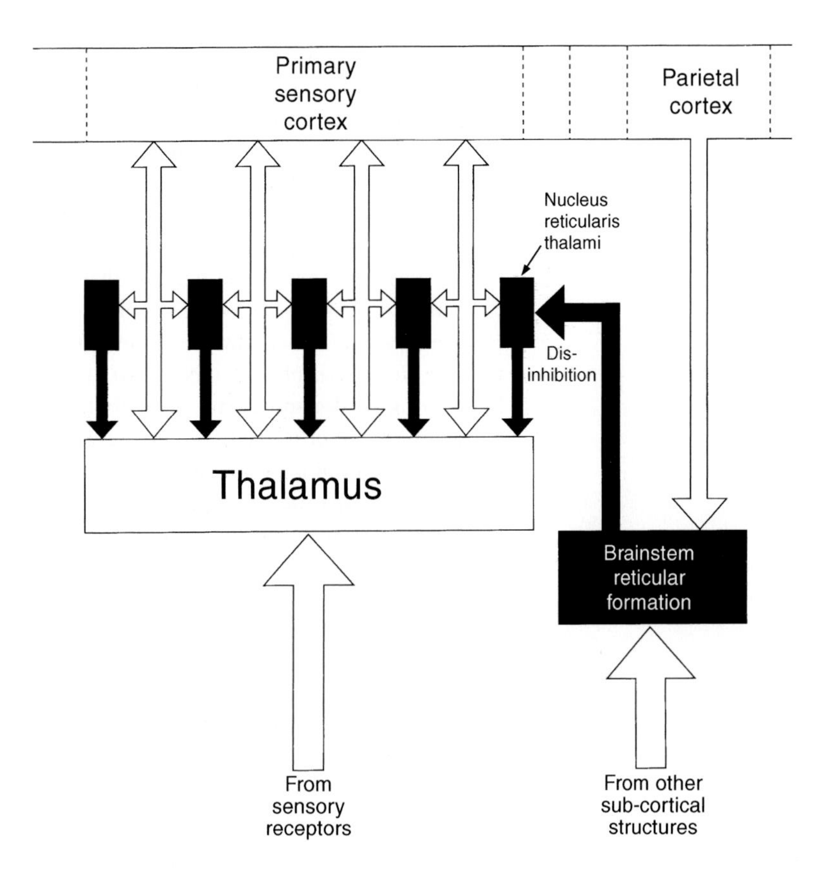

Figure 7.11
The excitatory pathway from a particular part of the thalamus (the lateral geniculate nucleus, in the case of vision) to the relevant part of the cerebral cortex (areas 17 and 18, in the case of vision) is modulated by the nucleus reticularis thalami (nRt). The inhibitory neurons of that structure form a layer that surrounds the thalamus, and the thalamo-cortical projections have to penetrate through this sheet. These projections activate the nRt neurons *en route*, through their axon collaterals, and the activated nRt neurons then return inhibition to the thalamus. The degree of this negative feedback is further controlled by the brainstem reticular formation (BRF), the neurons of which are also inhibitory. The BRF therefore exercises what could be called a gain control, and it does this by disinhibition. Finally, the BRF itself is controlled by, amongst other things, the parietal region of the cortex. It determines the state of arousal of the animal. In this highly schematic diagram, and in keeping with the other illustrations in this book, excitatory neurons are shown in white, while their inhibitory counterparts are coloured black.

and medulla, and the range and pattern of its projections attests to its enormous influence on the rest of the nervous system. There are connections which are directed caudally to the spinal cord, others run dorsally to the cerebellum, while the rest project rostrally toward the thalamus and beyond. Amongst the other brain structures which interact intimately with the BRF are the globus pallidus and the substantia nigra, both of which we encountered when considering the basal ganglia.

Around 1950, it was believed that the BRF regulated awareness, this conclusion stemming from experiments carried out by Giuseppe Moruzzi and Horace Magoun.[55] They stimulated the BRFs of anaesthetized cats and found that the electroencephalographs (EEGs) took on the active appearance they have during the dreaming state. It was for this reason that the BRF acquired its alternative name of *reticular activating system*. Their findings are best considered in the light of prior work on cats by Frederic Bremer.[56] His studies had attempted to verify the assumption that sleep is brought on by lack of sensory stimulation. Bremer cut right through the upper part of the brainstem and observed that his cats were thereafter perpetually drowsy. This, he rationalized, was due to the interruption of the flow of information passing up the spinal cord from the extremities of the body's surface. This simple picture was shattered by Bremer's own subsequent experiment. Making a through-cut at a slightly different level, that is to say between the medulla and the spinal cord, he found that his cats were this time wide awake. The only difference between the two planes of dissection was that the latter did not cut off input from the face and the surface of the head. Bremer realized that there must be a control centre located between the positions where the two cuts had been made. Remarkably, this was precisely the place chosen by Moruzzi and Magoun.

As study of the BRF continued, it came to light that this structure actually controls more than behavioural arousal; it also governs the regulation of muscle reflexes, the modulation of pain sensation, and coordination of the autonomic system. But let us confine ourselves to the question of arousal, and draw on a review of the BRF–nRt interaction published by Wolf Singer.[57]

As we have noted earlier, all regions of the visual field are not equally served by the primary visual system. The fovea is disproportionately well represented, and there are thus many more thalamo-cortical projections for that central region than there are for the more peripheral parts. As the BRF becomes more active, it will more strongly *dis*inhibit both the nRt and the (putative) inhibitory interneurons of the thalamus itself, and both of these processes will tend to increase the flow of information to the primary visual cortex. Conversely, as BRF activity decreases, the inhibitory effect of both the nRt and those interneurons will increase. But because the foveal path is most favoured, it seems plausible to suggest that it will be the *last* route to be quenched into silence by the inhibition. This could be the mechanism whereby we were able to focus right down on a small part of a single wrinkle on a single finger.

This ultimate control system really deserves more consideration than we have place to give it here. I will close this section by raising just one more point. It is, however, a major one. There is another structure that exerts control over the nRt, namely the *prefrontal cortex*. It does this by way of what are known as the *thalamic non-specific nuclei*, and this route represents a possible override mechanism which ultimately stems from the highest cognitive levels; we recall that the prefrontal cortex plays a vital role in working memory. (We recall also that malfunction in the prefrontal region is implicated in schizophrenia.) As Arnold Scheibel has noted, this additional route might provide a mechanism which allows not only selective focussing of the conscious state but also what he calls *superogative control* over body functions and pain sensation achieved in certain states of concentration, meditation and hypnosis.[58] There could hardly have been better justification for my placing a discussion of the nRt and its associated structures in a section declared to be dealing with feelings.

Vergence and reciprocity

This is an appropriate point at which to return to that question of loops in the reciprocal connections between different cortical areas. We are about to find that this issue is of major importance, but it will be necessary to go into a fair amount of detail before this can be appreciated. The first thing that deserves reiteration is that the word *reciprocal* is used only in a general sense; the counter-running axons do not maintain intimate contact, like a couple of courting snakes. And in any event, this could not possibly be true of the axon *branches*, since those arborizations are positioned in two different cortical areas, one upstream, as it could be called, and the other downstream.

Now the point is that the sort of arrangement actually present in these reciprocal projections *need not incorporate any closed loops at all*; if an axon emanating from a neuron in minicolumn number 527 of area gamma makes a synaptic contact with a neuron in minicolumn number 934 of area epsilon, say, there is no genetic constraint that requires an axon from a neuron in minicolumn 934 of epsilon to contact a neuron in minicolumn 527 of gamma. In practice, with upwards of ten thousand synaptic contacts being made by the branches of each neuron, the chance that there will actually be some of these closed loops is quite high, but they could nevertheless be in the minority (see figures 7.12 and 7.13).

It might seem that a paucity of closed loops would thwart the possibility of information flowing back and forth between the implicated cortical areas. Far from it. The areas can still exchange information, but they are forced to do it in a manner that requires a *high degree of coherence*. To see why this should be so, we have merely to recall three fundamental facts about the activation of a neuron: first, the typical resting potential lies around –80 millivolts; second, a depolarization to around –40 millivolts is necessary to activate the neuron; and third, a single synapse can only influence the voltage of the postsynaptic membrane to the tune of about 0.5 millivolt. Taken in conjunction, these numbers indicate that approximately eighty synapses will have to exercise their influence on a single receiving cell, *in or near unison*, in order to activate it. Given the statistical nature of the firing of individual neurons, perfect unison will not prevail, and the resultant

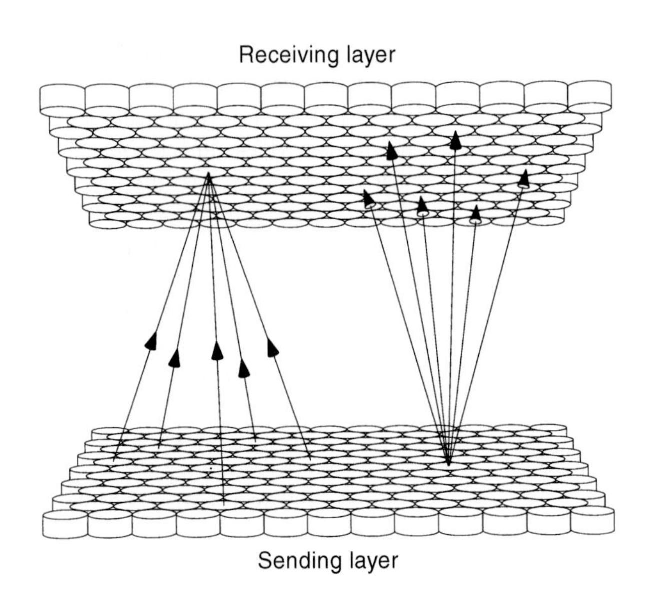

Figure 7.12
The reason why convergence and divergence always occur in conjunction, in the context of cortical projections, is illustrated by this schematic diagram. A neuron in a minicolumn of a given cortical area and layer projects to neurons in many other minicolumns of a different layer (either in the same area or in another area). This implies that a given neuron in that receiving layer receives signals from neurons in many different minicolumns in the transmitting layer.

wastage of effort will mean that several hundred other cells will have to pool their efforts if they are to activate a single recipient neuron. This might seem to put individual neurons at a disadvantage, but it is precisely this need to act in phase that enables the cortex to weed out signals that have less relevance. It may well be this fact which underlay the evolutionary pressure for neurons to develop such a large number of axon branches and dendritic ramifications.

There are two other features of cortical anatomy which we must also consider. One is that NMDA-type glutamate receptors are preferentially located in what are referred to as the *superficial layers* of the cortex,[59] which is to say layers I and II–III. These are the layers which receive the reverse projections, from higher areas. So if it is predominantly the NMDA-type receptors which are responsible for capturing correlations, thereby providing the basis for one aspect of short-term memory, then learning occurs on the *back*-stroke. This goes against what it seems natural to assume, but it is not difficult to conceive of a scenario in which learning by modification of the reverse-projection synapses would be an ideal situation.[60]

Let us see how this could work. A pattern of signals arrives at the primary cortex of the sense in question. It activates the corresponding minicolumns, and these activate numbers of minicolumns in higher cortical areas. The latter, in turn, send signals back to the primary area, through the reverse projections. The round-trip time for this out-and-back transfer of information would be very roughly 50 milliseconds.[m] The returning signals will reach the

m / I have arrived at this number by counting the minimum number of synapses that would have to be traversed, working from the circuit details given in figure 7.8.

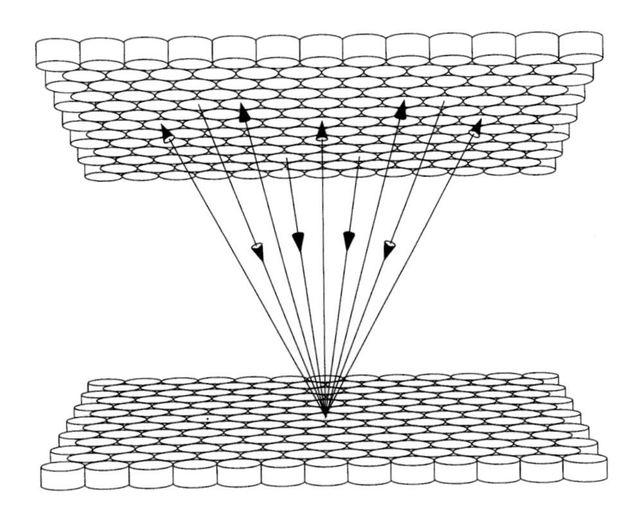

Figure 7.13
The fact that two different cortical layers (in either the same area or in different areas) share reciprocal projections does not have to imply that the reciprocity applies to all pairs of participating neurons; a neuron in one layer which receives signals from a neuron in another layer need not necessarily send signals back to that neuron, closed-loop fashion. There is merely reciprocity in the global sense, as indicated in this diagram (which grossly underestimates the number of synapses). Because a neuron will become active only if it simultaneously receives signals from many other neurons (around one or two hundred), this lack or scarcity of closed loops will mean that activity will be passed from one layer to another only if there is a high degree of coherence in the activity pattern in the transmitting layer. In vision, for example, this could contribute to the system's ability to separate an image from its background.

synapses that the reverse projections make with the neurons in the primary area. But given the fact that the duration of the input would normally be in excess of 50 milliseconds, there will still be signals arriving at the primary area from the sensory input, depolarizing the membranes of those same neurons. This is just the sort of situation that we earlier saw was necessary for activating the NMDA-type receptors, so the conditions for learning via the reverse projections are in fact fulfilled.

Assuming that such *back-stroke learning* can occur, let us consider what will happen when a pattern of signals *resembling* the familiar pattern subsequently arrives in that same primary cortical area. Many of the formerly activated minicolumns in the higher cortical areas will again be excited, and they will send back a reasonably familiar pattern of signals to the primary area, via the reverse projections. If the resemblance is sufficiently pronounced, the missing gaps in the pattern will be filled in, and what we have earlier referred to as autoassociation will have been achieved. The complete item will have been recalled, the agency which makes this possible being the cooperative action of the many feedback synapses that feed into each neuron. There would be something quite profound about back-stroke learning, if it does occur, because percepts gleaned from the sensory input would be learned only indirectly. The thing *directly* learned would be the *significance* of that percept for the organism, since it is this that the back-stoke flow of information would most immediately serve; that information would be returned to the only place where it can be of any use, namely the cortical area which possesses the appropriate *sub*-cortical projections.

In the case of vision, for example, it is the early visual areas which project to the superior colliculus, and it is therefore they which direct our gaze and help tell us what is where. It is difficult to believe that there would have been any evolutionary pressure to duplicate this machinery elsewhere in the visual system. It would be more efficient simply to refer the relevant signals back to this area, and let them serve the system in the usual way. That is what the reverse projections accomplish. This would presumably also apply to situations in which percepts are merely *imagined*; these too would be referred back to the appropriate area. It is difficult to see how any other strategy would avoid presenting the brain with insuperable logistical problems. In chapter 2, we contemplated what I called the suspicious accuracy of the imagination. I believe that this accuracy derives from the mechanism that has just been sketched.

Let us now return to that unexpected lopsidedness reported by Zeki and Shipp. It will be convenient to employ the same names as earlier, and we can use gamma for the area closer to the sensory input and epsilon for the area higher up in the system. Zeki and Shipp found that a neuron (or minicolumn) in gamma (forward) projects into a relatively small area of epsilon, whereas an epsilon neuron's (reverse) projections into gamma are far more widespread. Why should this be? In order to analyse the situation, we will need to appreciate something not yet touched upon concerning projections. If the epsilon projections into gamma are widespread, then it automatically follows that a given neuron (or minicolumn) in gamma *receives* signals from a wide area of epsilon. Similarly, the smaller area of projection from gamma into epsilon implies that an epsilon neuron (or minicolumn) receives from a relatively circumscribed region of gamma. The Zeki–Shipp result thus indicates that an area receiving reverse projections will normally draw its information from a relatively large area, whereas an area that gets its information via forward projections has a more limited catchment area. One could put it colloquially and say that the recipients of reverse projections are better informed.[n,61,62]

Why would this be advantageous, however? I believe that it provides the system with a means of *binding together* disparate parts of a perceived object. Two close, but not overlapping, parts of an image will send separate sets of signals to an upstream area, via forward projections, but the wider dispersion of signals in the reverse direction may perceptually link those parts together. This might be related to something that has been emerging in recent years, namely that what we perceive appears to be determined both by what we observe and by what we imagine we observe; *perception, it transpires, is influenced*

n / Mention should be made of the considerable amount of work carried out by Gerald Edelman and his colleagues, because it too has stressed the importance of reverse projections. They use the term *re-entry* to describe the multi-directional flow of signals, whereby concepts are gradually established by the integration of signals stemming from different cortical areas. In one of their particularly ambitious computer simulations, the Edelman group has demonstrated the consolidation of form, colour and motion into a coherent whole, this binding being mediated by the forward, reverse and lateral projections. Edelman has chronicled this work in the series of books listed in the bibliography.

Figure 7.14
In this illusion invented by Gaetano Kanizsa, one gets the impression that there is a single white triangle which is occluding parts of three black circles and a black triangle.

by consciousness.[o] A familiar example of this effect is provided by the ease with which we miss typographical errors when reading quickly. Once we get the gist of a passage of text, our anticipation of what is to come causes us to gloss over details, and errors become difficult to detect. This is the bane of proofreaders, who sometimes actually resort to reading galleys backwards, so as not to be influenced by the context of the printed words. Although I do not have direct knowledge of this, I suspect that children just learning to read would not be prone to such difficulties. Their reading skills would be too primitive to generate influential concepts.

A similar type of top-down process is probably afoot in the visual illusion invented by Gaetano Kanizsa, who built upon work by F. Schumann early in the twentieth century. Three black circles with missing sectors and three pairs of lines, when suitably positioned, give the impression that they are three complete circles and a single complete triangle which are being occluded by a white triangle. The illusion admits of numerous variations, but the original had the form shown in figure 7.14.

The white triangle is seen so vividly that it is difficult to believe that it

o / Semir Zeki has put the idea succinctly by paraphrasing a famous sentence: *Ask not what vision does for perception, ask what perception does for vision.*

does not exist. Even more impressive is the related figure involving six appropriately placed black circles, all of them again having missing sectors, because the illusory white object is now clearly three-dimensional, as can be seen by inspecting figure 7.15.

The figure appears to show a white cube partially occluding six full spheres. George Carman and Leslie Welch have recently taken this type of illusion one very significant step farther.[63] They exposed five subjects to sets of black circles from which parts more complicated than simple sectors were missing. As can be seen from the example shown in figure 7.16, these parts had *curved* edges, and the impression was one of full spheres being partially occluded by a curved white sheet. The novel aspect of the Carman–Welch

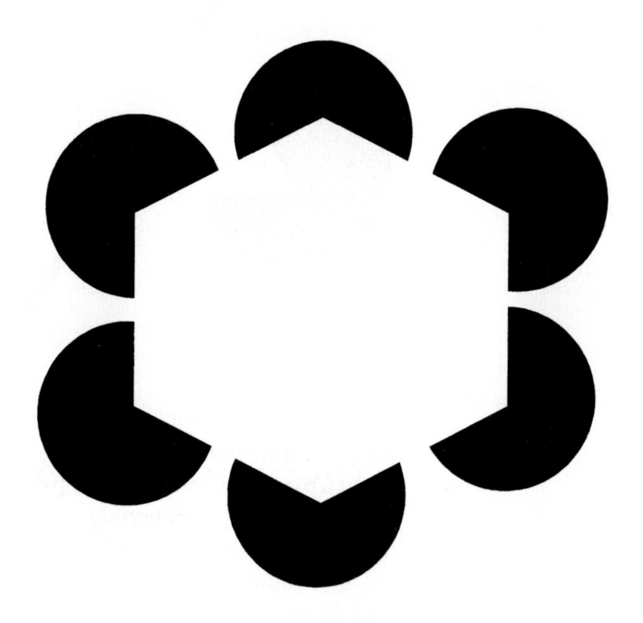

Figure 7.15
In another illusion of the type invented by Kanizsa, a white cube appears to occlude parts of six black spheres.

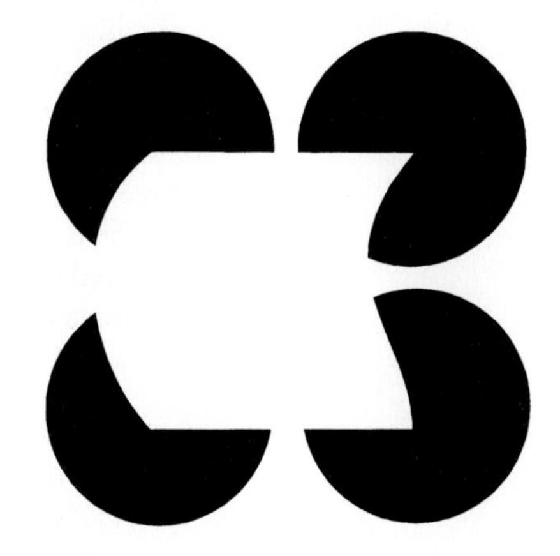

Figure 7.16
In an interesting extension of Kanizsa's work, George Carman and Leslie Welch have shown that the illusion of a curved white sheet can be gained from a set of suitably placed black circles from which regions more complicated than simple sectors are missing.

study was that they exposed their subjects to the same imaginary shapes under different viewing conditions. All the subjects reported seeing the same illusory objects, despite the changes in cue, position and orientation.

It had earlier been demonstrated that there are neurons in the brain which actually respond to illusory two-dimensional contours.[64] As Carman and Welch noted, there must also be neurons that react to the *attributes* of three-dimensional objects, whether real or imaginary. Indeed, there might not *be* any difference between real and imaginary, at the appropriate level of the system. This brings us back to our main theme. That deeper level where the real and the imaginary merge, in the province of the concept, will still have to make its conclusions felt in the cortical area possessing the appropriate sub-cortical projections. It will be the job of the reverse projections to ensure that the necessary information is returned downstream. In this respect, we ought to note that the *reverse* projections of one sensory modality are the *forward* projections of another sensory modality (see figure 7.17). Imagine two streams of sensory input converging on an association area, and indeed setting up remembered associations there. If the senses involved are audition and vision, for example, the association might be that of a name with a face. If one subsequently hears only the name, the auditory *forward* projections will carry the appropriate signals to the association area, and *reverse* projections will then permit the face to be conjured up in the relevant parts of visual areas.

But what exactly does that conjuring up of the face entail? I have been

Figure 7.17
In this highly schematic view of part of the cerebral cortex, each of two different sensory modalities, A and B, is served by two successive areas, arranged hierarchically. All inter-area links in both routes involve forward and reverse projections, and they ultimately feed signals into an association area. Each area also sends projections to a suitable sub-cortical structure, although in practice this might not be the case for all areas. The key point of this diagram is that the reverse projections of sense A appear to be forward projections for sense B, and vice-versa.

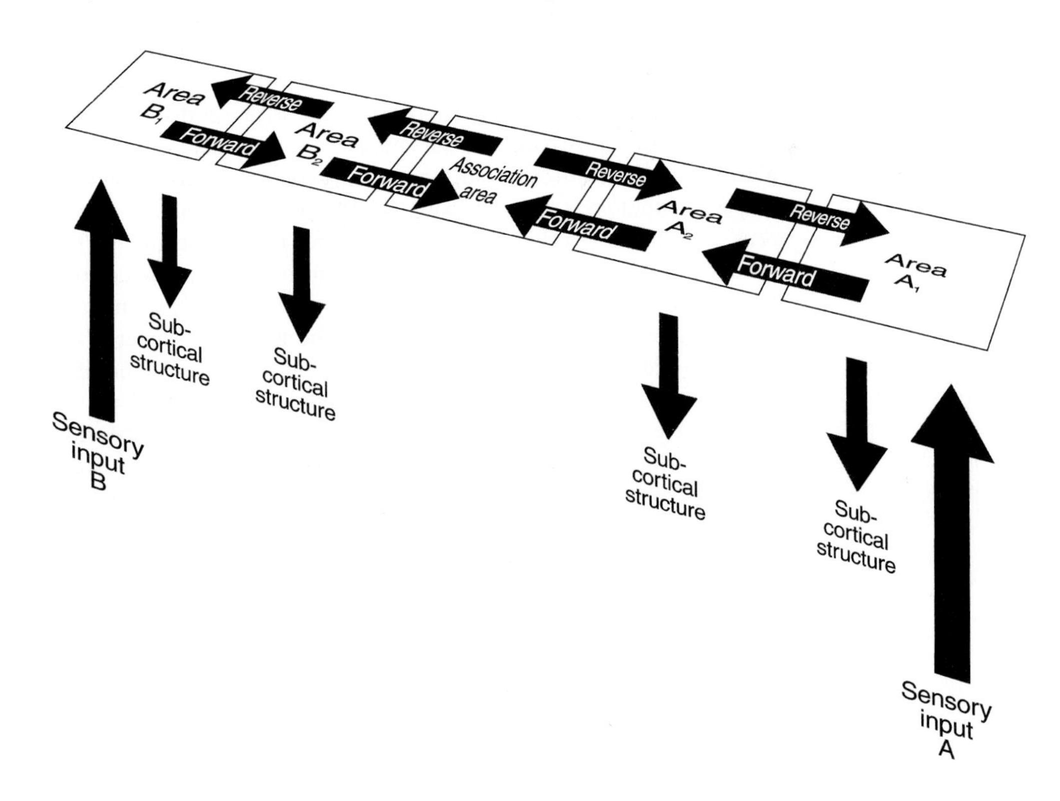

invoking the fact that layer V neurons in the primary visual cortex send sub-cortical projections to the superior colliculus. And we have noted that the superior colliculus dictates the direction of gaze. But there is surely more to imagining a face than merely gazing in a particular direction. This very important issue is still shrouded in mystery, but I would like to indulge in a bit of guesswork as to what the underlying mechanism might be. We know that there is no homunculus, inspecting the patterns of activity generated in the early visual areas by those reverse projections. What, then, are those patterns doing for imagination. Perhaps the key lies in that work of Georgopoulos, Schwartz and their colleagues, and in the vectorial mechanism that their studies revealed. Could it be that the memory of a face really involves storage of a set of vectors, rather than of a set of picture elements? The imagining of a face would then involve the activation of that set of vectors, and it might not be necessary for the centre of gaze to actually trace out the imagined face; perhaps the image of the face can somehow be 'felt', in its entirety, because of the influence of those vectors on the relevant motor elements. If this is not complete nonsense, it could suggest an explanation of that apparent duality of roles currently being seen in the cerebellum. The otherwise surprising finding that this brain centre mediates both movement and cognition might simply be a consequence of the fact that cognition, in a sense, involves *implied* movement.

By way of closing this discussion, we ought to compare the ideas that have just been expressed with what happens in a three-layer artificial neural network. Although the cerebral cortex clearly does not function like that particular network, there is a feature of the latter which loosely resembles the compartmentalization in the cortex that we have been considering. This is the hidden layer, and I am going to compare the *neurons* in that layer with the *areas* in the cortex. We recall that the various neurons in that hidden layer handle different aspects of the input, when the network has been suitably trained. A specific input gives a specific pattern of activities in the hidden-layer neurons, and this pattern changes if there is a switch to a another input. The distribution of activity amongst the hidden-layer neurons is, in turn, determined by the strengths of the synapses. Because the initial synaptic strengths were assigned arbitrary values, no significance can be attached to a neuron's position in the hidden layer. In the cerebral cortex, on the other hand, the different areas are *not* arbitrarily arranged; numerous lesion studies have amply demonstrated that position *has* significance. But it would be perfectly feasible to dictate which hidden-layer neurons in an artificial neural network are to do the lion's share of the work, for a given input, simply by giving the initial synaptic strengths the appropriate head start. Perhaps this is what evolution has done for the cortex. It could be that our genetic endowment produces a cortex with a predetermined distribution of tasks amongst the different areas.

Global closure

Of the many mysteries which still veil the brain, the one which most intrigues me is not often discussed. Indeed, it seems to have been rather ignored. It concerns the following situation. One goes to sleep with something on one's mind, a personal problem perhaps. It is often the case that this same problem is the first thing that one thinks about the next morning. What is the mechanism by which this recall occurs? Given that brain activity is markedly diminished during the night, except for the dreaming periods that is, memory must have something to do with it, but why should the problem in question be given such priority? It would not be difficult to conjure up an explanation based on a sort of last-in-first-out principle for memory, but this would have the unattractive feature of interfering with thought processes during the balance of the day; it seems like too good a prescription for obsession. In this final section of the chapter, I am going to suggest that this fascinating phenomenon is just one manifestation of what I tend to think of as *global closure*.

The overall time delay in the above example is about eight hours, although the actual figure may be only half of that if one allows for something to occur during dreaming which is beneficial for memory. I can think of a process in which a form of memory involves an even longer time delay. As we know, great anxiety over a sufficiently prolonged period can lead to the production of a stomach ulcer. When such a lesion has been formed, messages dispatched from the stomach to the brain convey a sense of pain. One could say that the ulcer is a form of record of prior mental events. This is a crude form of memory, admittedly, but it would be difficult to deny that it is nonetheless effective. Blessedly milder pangs can be generated by hunger, but this too could be called a form of memory. The point is that lack of sustenance in the stomach can be looked upon as the product of mental events and motor commands which did *not* lead to the consumption of food. The signals of discomfort that are later transmitted to the brain inform of the negative associations of this particular set of events and commands.

The point of these three examples is that they suggest the presence of memory that is not mediated by changes of synaptic strength in a neural network. There is a hint that there are more global things afoot, things related to the body's chemistry. This is hardly surprising, because one of the important messages of chapter 4 was that many neurons exert their influence by injecting substances into the blood stream, rather than by launching neurotransmitters across synaptic clefts. Given that diffusion and circulation are involved in this other type of process, it is not surprising that longer time delays are incurred.

What this is working up to is a more general view of memory, one that is not limited to synaptic changes in neural networks. This is not to say that these other memory factors can compete with the associative capabilities of the types of neural networks we have been considering in this book; they work at a much lower level of discrimination. But they augment the body's repertoire of responses, and in some cases they may even provide the dynamo of reaction. Keeping a pressing problem in centre stage throughout a night's slumber might be an example of this broader type of memory. Indeed, it might be generally true that the time scale of such a generalized memory will be longer the greater is the impact on the emotions. Seen in this light, it is more understandable that the ancients had difficulty in locating the seat of the mind. Perhaps we should be more inclined to cede Shakespeare the point when he lets Bassanio ask that question cited at the beginning of chapter 3.

What would be the use of allowing our definition of memory to take on this added breadth, however? I believe that there would be a philosophical advantage, if nothing else, and it concerns the way in which memory was introduced in chapter 2. We noted there that memory can result if a sequence of chemical reactions can be made to differentially influence the physical state of biological structures. In the examples that we subsequently considered, particularly in chapter 5, those changes were at the molecular level and they occurred in or around biological membranes. Our broader definition simply makes the point that *any* differential effects on chemical distribution, including those that act on a much larger scale, have the potential for serving memory. It is merely the case that the larger the scale, the lower is the degree of discrimination and the longer are the times involved. It is possible that this broader type of memory was acting in the case of Steven Rose's chicks, whose acquaintance we made in chapter 5.

One can regard all these mechanisms, independent of their individual scales, as forming loops. We have already considered the neuronal loops involved in muscle function, with the motor commands running one way and the signals that convey information about the state of a muscle travelling in the other direction (see figure 7.18). And in the preceding section, we contemplated the reciprocal loops of connections in the cortex. Prior to that, there was discussion of the looped neuronal circuits that incorporated the cerebellum, and there were also those of the basal ganglia. An important aspect of those other loops was that they frequently showed evidence of being interconnected. Taken together, these facts suggest that the various loops that we have been considering are simply parts of what in reality is one large and multiply connected network.

There is one final point to be made, and it concerns what we learned regarding the loops in the cortex. In the preceding section, we noted that the reciprocal nature of the forward and reverse projections does not apply slavishly to every single axon that joins two different cortical areas. The reciprocity merely refers to an overall trend; there need not be closed reciprocal loops. We also saw that this had a very important consequence, namely that distributed signals could pass from one cortical area to another only if their individual components arrived at their respective minicolumns in phase with one another. In other words, it is the *lack* of closed *individual* loops that forces two interacting cortical areas to function as a *globally* closed loop.

We have just seen that there are other systems of loops, and that they too are interconnected. They are not reciprocally connected with respect to all their individual regions, but from the above argument this does not prevent them from operating in a globally closed fashion. On the contrary, it will be their global mode of function which will be of greatest interest. The situation will be complicated by those different characteristic times, of course, but this will not prevent the system from functioning collectively. Far from it. There

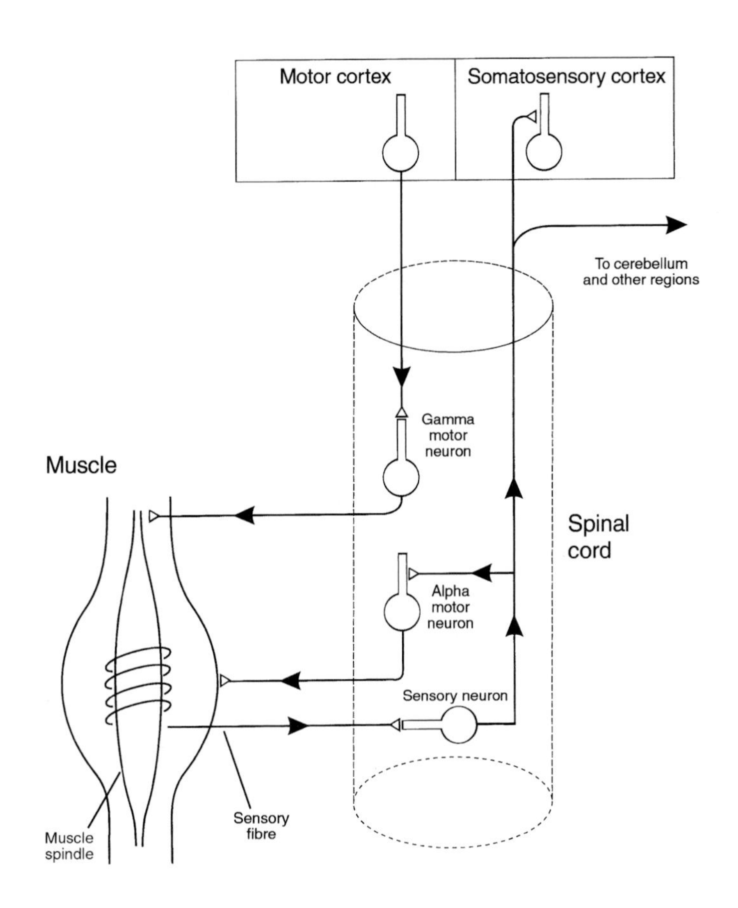

Figure 7.18
One of the most striking features of the central nervous system is the profusion of looped routes, in which pairs of regions are served by signalling paths in both directions. Good examples are seen in the interplay between the cortex and muscles. In the simplified case shown here, a pyramidal cell in the motor cortex contracts a muscle by first activating the intrafusal fibres of a muscle spindle (see also figure 3.8), via a gamma motor neuron, located in the spinal cord. This causes the spindle to contract, and afferent signals are thereby passed to a sensory neuron, which, acting via an alpha motor neuron, causes the extrafusal muscle fibres to contract. The gamma motor neuron continues to be activated, and this permits the spindle to monitor the state of the muscle, thereby allowing the muscle to adopt the required length. Signals from the muscle spindle are also passed to the somatosensory cortex, the cerebellum and other brain structures. The author of the present book believes that muscle spindles, together with their simulated versions – in efference copy loops within the brain – play a key role in consciousness.

will be an interplay between the processes occurring at different rates. A hint of the richness provided by these different time scales is probably to be seen in the typical EEG pattern. There are oscillations of several different frequencies, from the seventy or so cycles per second of the gamma waves, at one extreme, to the roughly one cycle per second of the delta waves at the other. And at even lower frequencies, there seems to be no reason why these waves should not feed their influence into oscillations that operate at the one cycle per day of the circadian rhythm.

This view has a respectable antecedent. Over a hundred years ago, in the 1884 Croonian Lectures, John Hughlings Jackson suggested that higher levels of the nervous system represent and re-represent all lower centres.[65] He saw the functions of lower systems as being represented and controlled by, but not replaced by, phylogenetically newer structures. The picture sketched here differs from this only in its perceived separation into different time scales for the different contributing processes. Things happening on the smallest time scale, that is to say those occurring most rapidly, take place in the cortex, which functions as a correlator, as a mediator for the smooth cooperation of sensory and motor functions, and as a distributor of tasks for sub-cortical centres. There was an evolutionary advantage in the development of such division of labour. We have noted that the sounds we hear do not appear to be located at our ears, in spite of the fact that this is the place at which sound waves are converted into nerve signals. Things heard are, on the contrary, heard *out there*. The same is true of things seen. We should be content to leave them out there; there is no need for them to be pulled inside the skull, where there would then arise the problem of enlisting some agency to scrutinize them for us. The brain is an impressive building, but there is no one at home!

8 The first half second

Is not vision perform'd chiefly by the Vibrations of this Eatherial Medium, excited in the bottom of the eye by Rays of Light, and propagated through the solid, pellucid and uniform Capillamenta of the optic Nerves in the place of Sensation?
Isaac Newton *(Opticks)*

Mechanics of the mind

It is not often the case that a qualitative picture of a phenomenon is sufficient for a full understanding of what that phenomenon entails. Far from it. If study of the phenomenon has not yielded the relevant numbers, and if these have not been accounted for in a quantitative manner, what appears to be understanding may well be misunderstanding. Before the early astronomers started to attach numbers to their observations, for example, interpretations of celestial phenomena tended to be subjective at best. There was insufficient contrast between astronomy and astrology. The situation changed dramatically when Kepler put the planetary trajectories on a firm mathematical footing. By deriving a reliable prescription for determining when a planet would reach a given position, and thereby adding a predictive dimension to the issue, he established the science of celestial mechanics.

During the hundred or so years that have passed since the discovery of the neuron, the phenomena associated with neural function have been studied with increasingly sophisticated techniques. And each advance has furthered our understanding. The invention of the cathode ray oscilloscope, and its use in conjunction with suitable electrodes, paved the way to elucidation of the microscopic events that underlie the nerve impulse. This was possible because experimenters had been given access to time intervals well below the millisecond duration of that impulse. A recent and further advance in that particular technology is currently permitting neuroscientists to simultaneously record from many neurons, and the fascinating things that have come to light will be concerning us later in this chapter. There is also still much to be learned from single electrode experiments. In this section, I am going to give an account of what they have recently revealed about the speed with which the cortex goes about its cognitive function. It is now possible to determine when signals reach a given position in the cortex, and this has given quantitative significance to the term *cortical mechanics*. Before starting on that, however, it would be advisable to reflect on what we already know regarding the characteristic times of the nervous system.

The nerve impulse occupies about a millisecond, so this is a natural choice for the unit of duration. Impulses are generated by a neuron at a rate determined by the amount by which its threshold has been exceeded, the frequency saturating at very roughly five hundred per second. The shortest interval between successive impulses is thus around two milliseconds. For a myelinated axon, the conduction speed can be as high as a hundred metres per second, whereas the figure is far more modest for unmyelinated axons, namely a couple of metres per second. If I prick my finger with a pin, the fastest signals can thus make the approximately one metre journey to my brain in about 10 milliseconds, via what is called the *A-delta route*, whereas the slower signals that convey the sense of pain travel through unmyelinated (*C-fibre route*) axons and they take about 500 milliseconds. These numbers obviously have a bearing on the speed with which one can respond to sensory input, but because reactions lead to muscular movement, other processes must be involved as well. And it is here that uncertainty creeps into the picture. My reaction to that pinprick would generally come before those 500 milliseconds have elapsed, and yet I am left with the impression that I have responded to the pain rather than to the piercing of my skin.

The situation is further complicated by what we have considered several times earlier in this book, namely that our nervous systems are able to anticipate the course of events through the use of internally generated models. We saw that this is probably one of the main functions of the cerebellum. When we undergo tests of reaction time, therefore, our responses are as much a product of this faculty for internal modelling as they are of signal transmission from our sensory receptors. Moreover, it is vital to distinguish between tests that involve simple reactions and those in which there is also an element of discrimination, or even choice. In 1868, Carl Donders determined a duration of 50 milliseconds for reactions involving discrimination, and 150 milliseconds when selection was added. Experimental psychologists now take these figures with a pinch of salt, because of those anticipatory factors. Indeed, tests of pure discrimination are looked upon as being more fruitful than those in which the subject is also being asked for a quick reaction.

One of the most interesting examples of pure discrimination that I have come across was published in 1988 by Hitomi Shibuya and Claus Bundesen. It was an example of what is referred to as a *partial report experiment*, the subjects being asked to discern between randomly arranged digits and letters, in displays which were presented on computer screens for different periods of time.[1] More specifically, the task was to report the digits (called the *targets*) from a circular array of digits and letters (called the *distractors*), the brief exposure of the arrangement being terminated by what is known as a *pattern mask* (i.e. replacement of the digits and letters by meaningless symbols of the same size). The number of correctly reported targets was analysed as a function of the total number of presented targets (2, 4 or 6), the total number of presented distractors (0, 2, 4, 6 or 8), and the exposure time.

Shibuya and Bundesen rationalized their observations with a theory known as *the fixed-capacity independent race model*, and the outcome of their work was a set of quite intriguing numbers. The visual processing capacity of normal human adults was found to be 45 items per second, (for targets and distractors of that particular complexity), which is equivalent to just over 20 milliseconds per item, while the short-term storage capacity turned out to be a mere 3.5 items. The other product of their work was a value for the longest ineffective exposure duration; this was 18 milliseconds. It seems reasonable to interpret this latter period as being the shortest time required for something in the visual image to make any impression at all on the cognitive system. From what we noted above regarding the frequency of emission of nerve impulses, this period would span just a few action potentials in a given nerve fibre. This does not mean that the act of cognition is *completed* in those 18 milliseconds, however, and we should bear in mind that the retinal receptors take some 30 milliseconds to integrate and respond to light stimuli (so bright, high-contrast scenes with lots of detail can be perceived with much shorter exposures than this). The number simply measures the *temporal span* of the signal which subsequently generates cognition higher up in the system. It is nevertheless fascinating that a few action potentials should provide the system with enough to work on.

Those few impulses subsequently travel from the retina to the lateral geniculate nucleus (LGN) of the thalamus, and thence to the primary visual cortex. They thereafter pass through the hierarchy of visual areas, in the parallel and distributed manner that we have considered earlier. They ultimately impinge upon the association areas, there to elicit the appropriate response from the motor system, be this a limb movement, a spoken word, a written word, or whatever. The question obviously arises as to *when* these signals reach the various cortical areas; we would like to know something about the underlying mechanics. The remarkable fact is that it is now possible to supply the numbers. The basis for this possibility has arisen from the fact that places have been located in the cortex in which there are neurons that respond to well-defined features in the visual field. An example that is rapidly becoming a classic is the presence in the inferior temporal visual cortex of neurons which are selective for faces.[2] This has been the remarkable achievement of Charles Gross and Edmund Rolls, and their respective colleagues, the studies having been carried out on monkeys.[a]

Once such a face-selective neuron has been located, it is a relatively

a / I have seen no reference to neurons specifically selective for the face of a *grandmother* monkey.

straightforward matter to determine the arrival time for signals at different intermediate points on the route from the retina. A picture is flashed in the visual field of the attending monkey, and the (electronic) stopwatch is started at the same instant. Electrodes are already in place at the positions of interest, it having been established beforehand that they are impaling neurons which are involved in the responses to the picture in question. The arrival time at the inferior temporal visual cortex itself lies around 100 milliseconds. The timetable for stations *en route* is 40 milliseconds for V1 (the primary visual cortex, which is the Brodmann area 17), 55 milliseconds for V2 (the secondary visual area), 70 milliseconds for V4, and 85 milliseconds for the posterior inferior temporal visual cortex.[3]

These numbers are quite enlightening. We see that the signals can progress from one cortical area to the next in about 15 milliseconds (see figure 8.1). Now from what was described in the previous chapter, concerning the neuronal arrangement in a typical cortical area, at least two neurons (that is to say, synapse–dendrite–soma–axon paths) would have to be traversed by the signals as they pass through the different layers of a single cortical area. It seems reasonable to add a further synapse for each transit from one area to another. This gives us a total of three neurons and their associated synapses to be crossed for each additional cortical area visited. Given that the travel time down a dendrite, across the soma, out along an axon (this latter part occupying an admittedly negligible amount of time), and across a synapse would not be less than about 5 milliseconds, we find that the measured transit time between cortical areas can just about be accounted for.

The fascinating thing about these transit times is that their duration roughly corresponds to those same few action potentials that emerged from

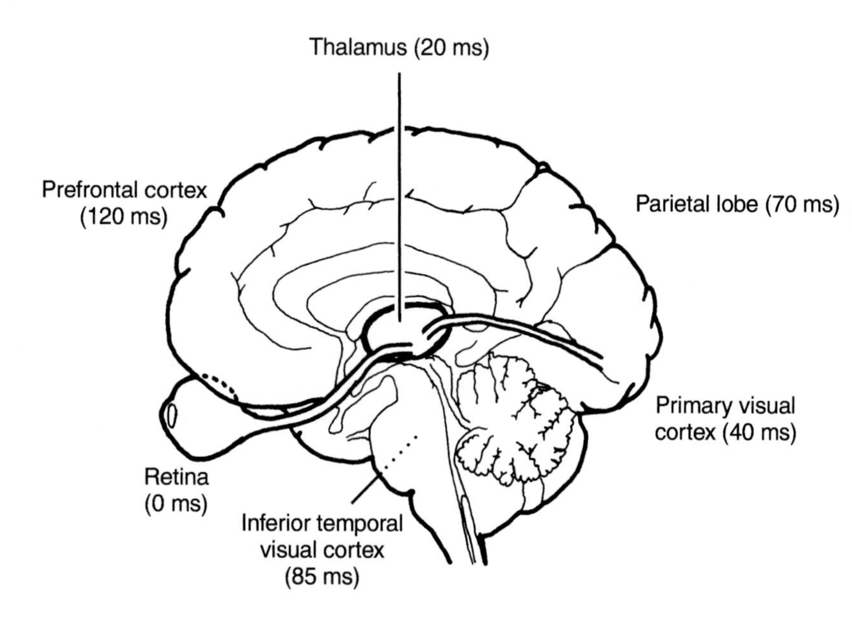

Figure 8.1
Because the various cortical areas form a hierarchy, as indicated in figure 7.7, and because sensory signals enter the cortex only at specific areas, the different cortical regions will not all react simultaneously to sensory input. This diagram indicates the time of arrival at different regions of the macaque monkey cortex when visual input to the retina occurs at time zero.

the work of Shibuya and Bundesen. Each cortical step has to make its initial contribution on the basis of this small number of signals. Just as interesting is the fact that the looped circuits made possible by the reciprocal projections, which were discussed at such length toward the end of the preceding chapter, do not participate in these *initial* stages of the cognitive process; there is simply not enough time available during the 15 millisecond inter-area transit time to fully activate them. This is not the same as saying that they do not *ultimately* play an important part in the cognitive process, however, because even the above-quoted 100 millisecond period is insufficient to establish conscious awareness. (This will become clear by the end of the present chapter.) In contemplating this disparity, we are getting the first inklings of the distinction between pre-attention and attention.

A further important point is that the two millisecond interval between adjacent action potentials applies only in the case of maximal output frequencies from the neurons involved. We can turn this around and say that *only* those neurons which are driven at their maximum rates can be contributing to the perceptual process. This provides a more quantitative criterion for the winner-take-all idea that has been invoked several times in earlier chapters.

In the human being, the times consumed by two further processes are of interest. The first is that of the onward passage of signals from the inferior temporal visual cortex to the appropriate association areas, which can be looked upon as the deepest recesses of the cortex, the regions most remote from the sensory inputs. This remainder of the afferent (inward) journey would take about 15 milliseconds if, as seems possible, the inferior temporal visual cortex and the relevant association areas are adjacent in the hierarchy. That gives a total of 115 milliseconds for the trip from the receptors to the association areas in the case of the visual system. Subtracting the 40 milliseconds it takes for the signals to travel from the receptors to V1, we get a time of 75 milliseconds from V1 to these particular association areas.

The second time-consuming process is that of the signals' subsequent progress from those association areas to the cortical regions responsible for generating the response, which is an efferent (outward) journey. To see what this involves, let us consider what happens after the signals have reached the association areas. By a mechanism which is still a mystery, and which we must therefore discuss later, a visual *percept* will somehow give rise to a *concept*. After suitable associations have been struck up, the latter could ultimately lead to the utterance of a word or a phrase. From what we have noted earlier, we know that this reaction involves both Wernicke's area and Broca's area, and it ultimately leads to activation of the relevant muscles underlying vocalization, including those of the diaphragm. The trouble is that we do not have direct measurements of the amounts of time these latter efferent processes require.

One can make a very rough guess at their duration, however, by comparison with a process that maximally involves both afferent and efferent signalling. When the hearing of a word conjures up a visual image, afferent signals travel from the auditory receptors to the association areas, and signals then travel from the association areas to the early visual areas.[b] If we assume that the number of synapses between the receptor cells serving *any* sense and the relevant association areas is pretty much the same, irrespective of the type of sense, it seems logical to conclude that our estimated value of a roughly 115 millisecond interval from receptors to association areas would *always* apply. But we have also noted that the forward projections of one sensory modality are the backward projections of another. If we further assume that the rates of progress of signals do not depend upon whether a forward or a backward projection is being used, the time lapse between sound waves striking the ear and a visual image being created would be roughly 180 milliseconds, this number being composed of the 115 millisecond afferent part and the 75 millisecond part. It must be emphasized that this is little more than a guess, but the value does happen to be in quite good agreement with the 200 milliseconds often quoted for a simple cognitive process. If we make the further gross assumption that all efferent target areas are roughly equidistant, in synaptic terms, from the relevant association areas, that value of 75 milliseconds might not be too bad an estimate for the missing efferent duration in the case of muscle activation.[c]

Even if this exercise in approximation can be looked upon as being satisfactory, it does not mean that our understanding of the cognitive process is adequate. That happy state of affairs is still some way off, for it is quite literally the fact that there is more to this issue than meets the eye. Let us return to the fascinating work of the Gross and Rolls groups. They reported the existence of neurons responsive to faces, amongst other things. They said nothing about neurons specifically responsive to vases. The question is, why on earth should I bring vases into the story. What I have in mind is the intriguing sort of picture that Edgar Rubin gave prominence to in 1921. One of the best known of these is shown in figure 8.2, and a little contemplation reveals its ambiguity; depending on what is regarded as the background, the figure shows either a vase or two counter-disposed faces. The way in which our visual systems distinguish between an object and the background was of great concern to *Gestalt psychology*, which was founded in 1912, the pioneers being Max Wertheimer, Kurt Koffka[4] and Wolfgang Köhler. The term is German, and the usual English translations *shape*, *pattern* or *form* fail to do adequate justice to what is really a rather subtle concept. The Gestalt is the *wholeness* or *configuration* of a figure, and it is this which enables one to separate it from the

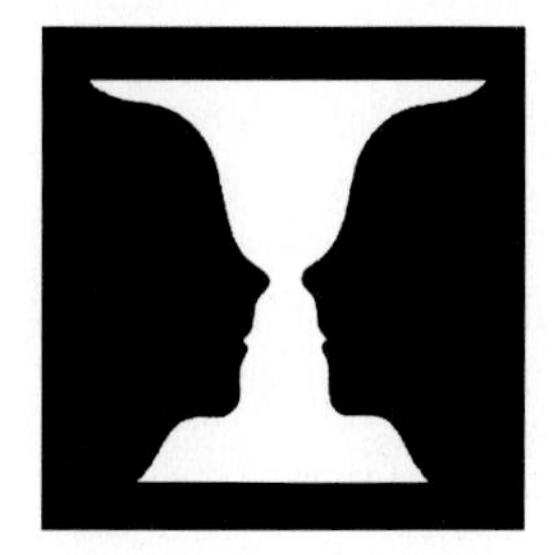

Figure 8.2
This celebrated picture first published by Edgar Rubin, in 1921, can be interpreted in two ways. It shows either a white vase against a black background or two black counter-disposed faces against a white background. It is impossible to perceive both alternatives simultaneously, but switching from one interpretation to the other can be achieved in as little as about half a second.

b / The involvement of the latter part of this route is implied by that result we considered earlier, namely that the early visual areas are involved in visual imagination.

c / Anticipating chapter 9, we should note that the areas in question do not make their *final* responses on the basis of the *initial* information; these responses evolve relatively slowly. And these responses are not the result of a series of individual decisions, but rather of a global, cooperative, rise-to-threshold process, culminating with a winner-take-all selection.

background. It is what would now be referred to as an *emergent property*, the idea being that the whole is more than the sum of its parts.

Let us return to those face-detecting neurons and vase-detecting neurons (if such exist), both presumably located in the inferior temporal visual cortex, and both therefore triggered around the 100 millisecond mark. We now see that it cannot be their mere activation which produces even a percept, let alone a concept, because we are aware of *only one* of the interpretations at any instant. We see the face *or* the vase, but not both simultaneously. There would thus have to be an additional passage of time, during which the cognitive process was completed. The duration of that supplementary period has not been measured, but there again appears to be the possibility of a rough estimate. Using a stopwatch, I have timed myself over ten reversals from face to vase, interspersed with ten reversals the other way. Making an intense effort to do this as rapidly as possible, and repeating the entire thing five times in order to achieve good statistics, I find that I can get the average reversal time down to 550 milliseconds.

During this test, I noticed that the centre of my gaze was systematically shifting from the region immediately between the two noses to a point somewhat higher up, around the centre of gravity of the vase, and back again. Such *saccades*, as these eye movements are called, have been measured to occur every 250 milliseconds or so when the eye freely scans over an object, and the actual movement occupies less than 50 milliseconds. This movement is therefore too rapid to have been a major factor in the 550 milliseconds of those cognitive reversals.

What is going on during that approximate half second? We cannot be sure at present, but we can at least make a simple calculation of what would be possible, using the rudimentary cortical mechanics that we have been establishing. In 550 milliseconds, there would be time for seven or eight of those 75-millisecond V1-associator or associator-V1 journeys, or nearly four V1-associator-V1 round trips.[d] If that is indeed what is happening, it would not merely be a case of the same signals passing backwards and forwards along this route; such ping-pong dynamics would achieve nothing, apart from possibly supplying the cortex with a *reverberation period* that could have relevance to *very short-term memory*. It seems likely, however, that the signals would be progressively modified as they run to and fro between the various cortical areas. Whether or not part of that modification has to *undo* the old concept before it can make a start on establishing the new one is a moot point.

Let us put these interesting issues on hold for a while, and return to the problem that so preoccupied the Gestalt psychologists. On the face of it, each of our sensory systems is confronted with a rather difficult task. Our

d / From what was stated in the preceding chapter, the recall of visual images would also involve early visual areas other than V1, and the return-trip transit times for these would be shorter. One could think of the early visual areas as functioning together in such a process.

retinas, for example, are under constant bombardment by the light rays (or at the finest temporal scale, by photons) reflected from the surfaces of the myriad objects in our field of view. Consider the complexity of the average country scene. How do our visual systems discover that the various parts of a single object, such as a grazing horse, belong together? What mechanism enables us to separate such an object from the background? Adherents of a rival theory known as *Structuralism* believe that each of our cognitive systems breaks down the thing being sensed into its component parts, and then synthesizes them, tinker-toy fashion, into a coherent whole. The Gestalt view is that things are perceived as wholes. This still leaves the question of how the parts of wholes are detected as belonging together. The Gestalt prescription was what were known as the *laws of grouping*, individual items on this list covering the contingencies of *proximity*, *similarity*, *closure* and *good continuation*. By this is meant that components will be perceived as belonging to the same thing if they are close together, if they are similar to one another, if they form a closed contour, or if they move in the same direction (see figure 8.3). Two new laws were later added to this set. Stephen Palmer noted that we tend to group components together if they are located within the same perceived region (the law of *enclosure*), and he and Irvin Rock discovered the law of *connectedness*, which states that the visual system tends to perceive as a single item any uniform and continuous region, be this a spot, a line or an area.[5]

What is the neuronal (or neural network) basis of these laws, however? The short answer is that we do not know. But it seems worthwhile to ask whether any of the processes we have been considering could supply a clue. The grouping appears to be accomplished so rapidly that it is difficult to see how saccadic eye movements could play a vital role. Of the things mentioned earlier in this section, the only process which occurs on a much briefer time scale is the one measured by Shibuya and Bundesen. We recall that their elementary events occupied a span of 18 milliseconds, which is well below a tenth of the time lapse between saccades. What could be happening at the neuronal level during such a brief period, given that the eyes are essentially frozen in position on this time scale? Perhaps the answer lies in the winner-take-all mechanism. Suppose that an item amongst several that obey one or more of the Gestalt laws manages to win in the race to establish a corresponding activity pattern in (say) the primary visual cortex. It does not seem far-fetched to suggest that this pattern could be extinguished by feedback inhibition from the surrounding inhibitory interneurons, within the 18 milliseconds measured by Shibuya and Bundesen. During that time, the pattern will have given rise to a further pattern farther on in the visual system, otherwise the effort will have been wasted. Once feedback inhibition has wiped the slate clean, as it were, another item in the Gestalt group will have a chance of making its impression on the primary visual cortex, and so on.

Indeed, the name *independent race model* given to their theory by Shibuya and Bundesen harmonizes nicely with the winner-take-all concept. But what about the other part of their title; what about the *fixed capacity*? We recall that their work also yielded a number for the short-term storage capac-

Figure 8.3
Although the dalmatian dog in this picture (by Ronald James) is well disguised, the remarkable thing is that we can perceive it at all, given so little information. According to the ideas of Gestalt psychology, we can do so because our perceptual systems avail themselves of a number of rules, including those that give priority to proximity, similarity, closure and good continuation. The current view is that these rules are a natural consequence of the influence that higher cognitive function has on elementary perceptive processes. That influence is presumably mediated by the reverse projections that are such a common feature of cortical circuitry.

ity of the visual system. According to their observations, it is 3.5 items. This is only an average, of course, and it surely would be difficult to attach meaning to a half item. The actual value is probably three or four. If we multiply 18 milliseconds by three, we get 54 milliseconds, and we get 72 milliseconds in the case of four items. This latter value is only slightly less than that 75 millisecond period that we operated with when discussing the forward and backward flow of signals between the early visual areas and the association cortex. Any sizable increase in the number of items would presumably put the system in danger of confusion because the late items still executing the forward sweep could run into the early items that had already started on the backward journey. Could it be this which fixes the capacity of the system?

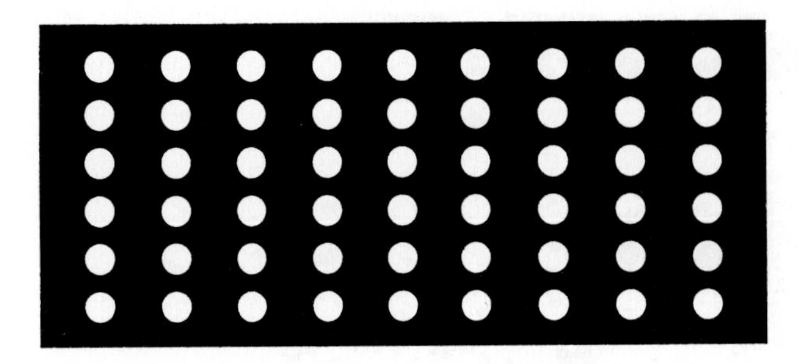

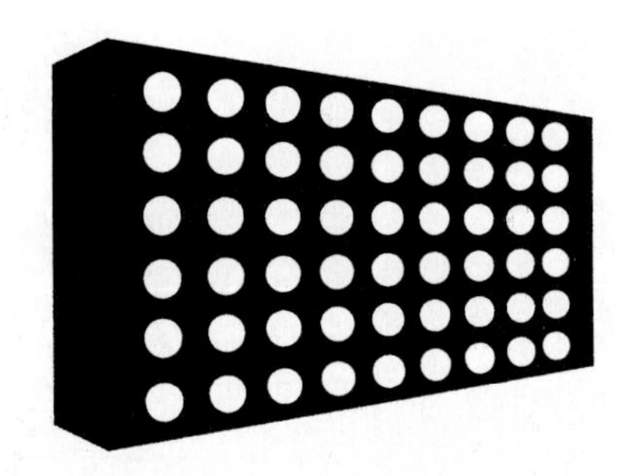

Figure 8.4
The beads in the left-hand diagram appear to form columns, rather than rows, because the spacing is less in the vertical direction than in the horizontal. The reverse is the case in the right-hand version, but the columns still take priority because of the other perspective clues. This emphasizes that the perception of proximity is three-dimensional.

In any event, this type of analysis does make the assumption that the laws of grouping are being implemented rather late in the visual system, and in that respect it conflicts with what Wertheimer originally envisaged. When he considered the law of proximity, for example, he saw this as arising from the closeness of points on the retina. More recently, however, experiments performed by Leonard Brosgole and Irvin Rock have raised doubts about the tenability of Wertheimer's view.[6] They suspended a number of equally spaced luminous beads on each of a series of parallel and equally spaced strings, such that each bead was closer to its neighbours vertically than horizontally. When viewing this assembly from the front, in the dark, an observer's primary sensation is one of columns rather than rows. If one end of the suspension frame is pushed backward, so as to present the observer with a sufficiently tilted (and thus foreshortened) assembly, the *apparent* horizontal spacing will be less than the vertical spacing. This can be appreciated by comparing the two parts of figure 8.4, which show the two situations. The beads are nevertheless perceived as forming a series of columns. This shows that the perception of proximity is three-dimensional, and that it is not merely determined by spacing on the retina. Brogsole and Rock had thus demonstrated that grouping by proximity must be achieved by a mechanism that operates after that of depth perception. Since the latter involves not only the retina but also the primary visual cortex, this was tantamount to showing that at least one aspect of Gestalt grouping is handled *higher up* in the system. This harmonizes with the analysis given in the preceding paragraph.

Showing where something is occurring is not the same as explaining how it is happening, however, and this brings us to a profound issue that was raised by Stuart Sutherland.[7] It concerns the difficulty that the nervous system would seem to have in keeping a check on what is where in visual space. (The problem probably exists in audition, to a lesser extent, but let us confine our attention to vision, in which it is particularly severe.) Stated simply, the trouble arises from two conflicting requirements. The things we see are usually built up of numerous components, the relative positions of which are clearly important. But the system also has to be impressively flexible regarding the way it handles those components. As Sutherland noted, the first of these demands arises because patterns have to be broken down into subpatterns, possibly through several stages. An example is seen in the sequence:

person, arm, hand, finger, joint. But the system has to classify these things despite variations in perspective, viewing angle, orientation, and the like. A considerable demand on the system is even made by the requirement of *size invariance*. Let us consider a simple example. Our visual systems must effortlessly detect the equivalence of o, o, O, O, O and O. Such a task is so commonplace that it is easy to miss the fact that the images of the closed circles in those six different sizes of letter are falling on different sets of retinal receptor cells.[e] This could be called the problem of *cognitive compression*; a thing perceived at different sizes, orientations and angles must nevertheless be properly assigned to the same cognitive pigeonhole.

The other side of this issue, the need to preserve juxtaposition, was nicely illustrated by an analogy due to Christoph von der Malsburg.[8] I will use the final sentence of the preceding paragraph to bring out the point. That sentence takes about eight seconds to read, and short-term memory can retain its sense (if not the exact wording) for several times that period. In that sentence, the letter *e* occurs 26 times, *s* 13 times, *i* 14 times, *o* 13, *t* 12, *r* 8, *a* 8, and so on. One could imagine such statistics influencing synaptic changes in some region of the cortex. Given that the sentence's 163 letters have been scanned at a rate of about one per 50 milliseconds, the exposure times are compatible with what we have considered earlier concerning the speed of synaptic modification, when NMDA-type glutamate receptors are involved. Now von der Malsburg's point is that the *sense* of that sentence cannot be reconstructed from such statistics. In addition, we would need to know how the various letters are arranged. As we noted earlier, one would have to know what appeared where.

It has been suggested that the brain could get around the problem of cognitive compression if signals are processed in such a way that their temporal relationships are significant. This proposal appeared in a paper by Peter Milner which has possibly not been given the attention it deserves,[9] and also in von der Malsburg's subsequent paper that was referred to above. The gist of the argument, in both cases, was that the individual neuron is ideally suited to play the role of correlation detector. The signals travelling inward along the dendrites of a neuron will have maximal influence on the somatic region of that cell if they arrive in mutual synchrony. As Milner put it, the visual system would be able to separate a figure from the background if *cells fired by the same figure fire together but not in synchrony with cells fired by other figures*. And von der Malsburg specified that the resultant discrimination between synchrony and asynchrony would have to be to a resolution of between 2 and 5 milliseconds. These suggestions acquired added significance when the multielectrode technique mentioned at the beginning of this section revealed suggestive correlations between the activities of cerebral neurons. I will suspend discussion of such things until we consider oscillations in general, later in this chapter.

e / This statement is true (when the page is viewed at the normal reading distance) even if the finite size of the individual receptive fields of the retinal ganglion cells is taken into account.

Meanwhile, let us close this section by returning to that apparent schism between Gestalt psychology and Structuralism. I say *apparent* because I believe that comparisons between them miss the important factor of time. The wholeness of the Gestalt approach would be valid only over the time span occupied by one of those V1-associator sweeps, I feel, whereas Structuralism refers to a rather longer period. My conviction on this point stems from contemplation of the fascinating type of *impossible object* conceived by the French artist Marcel Duchamp in 1916, and independently by the Swedish artist Oscar Reutersvard in 1934. It surfaced again in 1958, when it received the attentions of Lionel Penrose and his son Roger. The latter's first example, and possibly still the best known because of subsequent work by Richard Gregory,[10] had the appearance shown in figure 8.5. I do not agree with Gregory when he

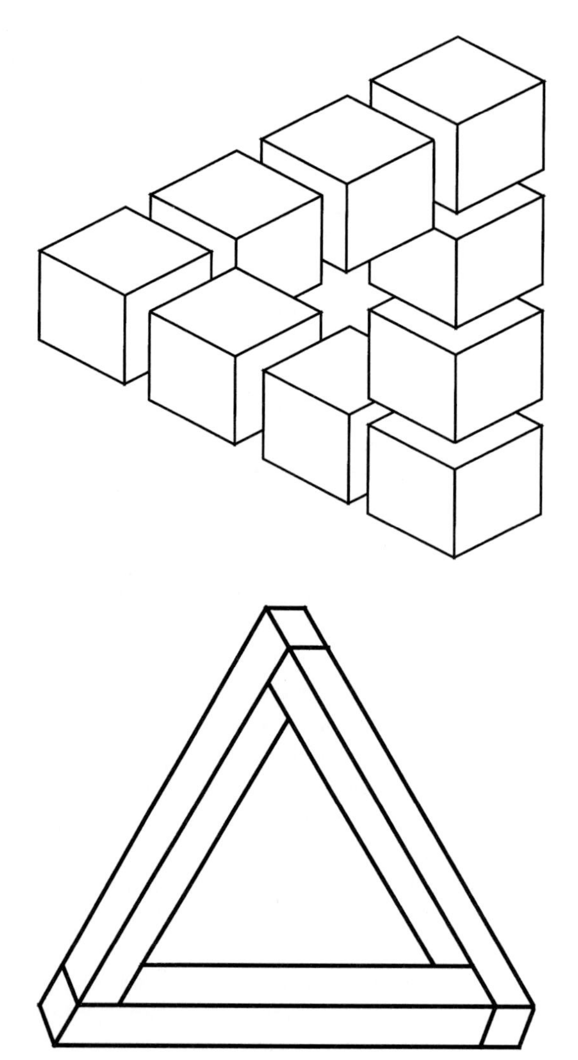

Figure 8.5
This famous arrangement of cubes (above), to form an impossible triangle, was devised by the Swedish artist Oscar Reutersvard in 1934. His work on such impossible objects was honoured by the issue of a Swedish stamp in 1982. The impossible triangle was later featured amongst the objects studied by Lionel Penrose and his son Roger, its impossibility emerging only through careful study (below). The visual system instantly accepts its local features, because they violate none of the rules of geometry that one has acquired through experience. The underlying conflict is revealed only through sequential (conscious) inspection of the various parts, together with brief retention of their mutual juxtapositions. The importance of this type of picture lies in its exposure of possible discord between percept and concept; the picture is not impossible, since it can be drawn, but the object it conveys is impossible. The first recorded drawing of an impossible object, by the French artist Marcel Duchamp, depicted an impossible bed frame, and the theme recurs often in the work of the Dutch artist Maurits Escher.

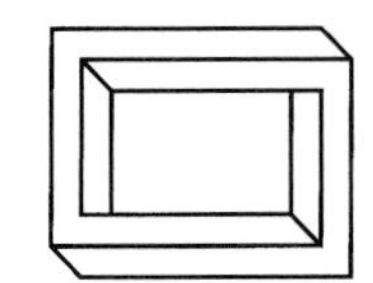

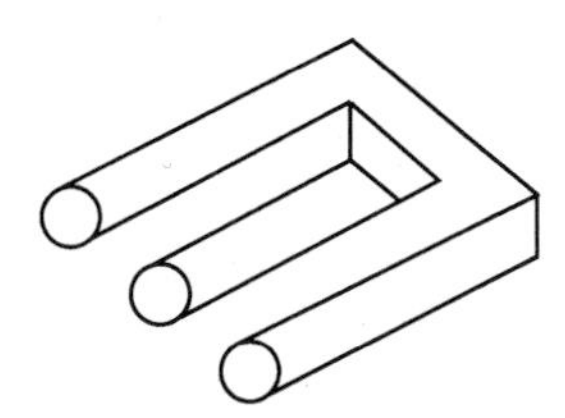

Figure 8.6
Two more percept–concept conflicts are illustrated in these diagrams. The ability to detect that three-dimensional embodiments of the diagrams would not be possible depends upon a facility for remembering the mutual juxtapositions of the component parts, and on the power of gauging the implications of those relationships. The ability to hold a number of facts in short-term memory, and grasp their consequences, might be the most satisfactory measure of intelligence.

regrets that the Penroses did not call this an *impossible figure.*[11] Surely, the fact that the figure can be drawn shows that it is not impossible. What *is* generally impossible, on the other hand, is fabrication of an object that complies with the figure. That Gregory ingeniously devised a way of constructing an object that is consistent with the above figure, when that object is viewed from *just one angle*, does not mean that the object implied by the figure is *generally* possible.[12] Indeed, Gregory's demonstration relies on an illusion. By the same token, the following figure is real enough

$$2+4+7-3+9-5=8-3+1+4-2-6+5$$

and anyone oblivious of the rules of arithmetic would happily accept it for what it is: a pattern of lines. It is only when one attempts to give it a *meaning*, in terms of what one has learned about arithmetic, that a difficulty arises.

Returning to geometrical things, our fleeting acceptance of the Penrose triangle, and its cousins, such as those shown in figure 8.6, is covered by the Gestalt principal of *Prägnanz*, which states that when stimuli are ambiguous, our perception will opt for the simplest interpretation consistent with past experience. The point is that in the short term, we can only scrutinize the individual *parts* of these figures, and they do not violate Euclidean geometry. At first glance, our visual systems do not have sufficient time to spot the global inconsistencies. It is only during the longer times which I believe apply to the realm of Structuralism that we discover the impossibility of rational interpretation.

Ripples in the cortex

In 1987, Charles Gray and Wolf Singer reported the discovery of oscillations in the cortices of anaesthetized monkeys for certain types of visual stimulation.[13] The observed frequencies were in the range of 35 to 70 cycles per second (i.e. 35–70 Hertz), which put them in what is known as the *gamma* band. They are now commonly referred to as 40 *Hertz oscillations*, frequencies around that value being particularly prominent. The oscillations were clearly important because they were simultaneously detected in electrodes that were impaling neurons in different cortical positions, the neurons having similar orientation preferences. In some cases, the electrodes were spaced as far apart as 7 millimetres, and later work by Gray and Singer, in collaboration with Andreas Engel, Peter König, Andreas Kreiter and Thomas Schillen, even picked up oscillations that were coherent across the corpus callosum, that is to say simultaneously in the two cortical hemispheres.[14] These discoveries were soon confirmed by the extensive studies of Reinhard Eckhorn and Herman Reitboeck and their collaborators.[15]

It is important to understand the significance of the special stimulation conditions which gave rise to these observations. To see what was entailed, the reader should cross the two index fingers and move the rear finger back and forth in the direction parallel to the front finger. There is a portion of the rear finger which is never seen during the execution of this motion, but one is in no doubt about that portion's presence. Indeed, there is no doubt about the fact that it is this hidden section which acts as the link between the two visible parts of the finger, permitting them to move in unison. That Gestalt grouping law which invoked similar motion helps us to draw such a conclusion. Now when we hold our crossed fingers thirty or so centimetres in front of us, the retinal images of those two visible portions of our rear finger are giving rise to signals in places in the primary visual cortex which are several millimetres apart. This is why the experimenters used such spacings between their electrodes.

They also tried several different modes of stimulation, each of which involved two *separate* moving bars of light. In one mode, the bars were moved in opposite directions. In another mode, they were moved in the same direction but at mutually varying speeds, such that sometimes it was the one bar that was leading, while at other times it was the other bar. Finally, measurements were made when the two bars were moved in unison, but separated in the same way as those two visible portions of the moving finger. It was only this third mode which gave the oscillatory response.

These were naturally not the first observations of oscillations in the brain. The occurrence of periodic variations of voltage in the visual system have been known at least since the studies of Francis Gotch in 1903, in which they were detected in the retina of the frog. The phenomenon became particularly familiar through the electroencephalographic (EEG) experiments of Hans Berger in the early 1930s.[f] Oscillations seen in EEG traces fall within a number of different frequency bands. The regular oscillations observed when a subject is at rest, with the eyes closed, have a frequency in the range 8 to 13 Hertz. These are known as the *alpha* oscillations. The fact that they are replaced by less regular responses with lower amplitude was earlier attributed to the fact that the observed alpha oscillations are really composites of many weaker alpha oscillations, all in step with one another, sensory stimulation causing desynchronization and thus irregularity. In reality, the situation is more complicated because oscillations of other frequencies are also present. *Delta* oscillations have frequencies in the range 1 to 4 Hertz. They normally occur during sleep, but are also seen during waking, in the vicinity of tumours and stroke damage. *Theta* oscillations, with frequencies in the range 5 to 8 Hertz, are prominent in the hippocampus. Precisely what they do for the system remains obscure, however. Finally, there are the *beta* oscillations. Their frequencies lie in the approximate range 15 to 30 Hertz, so they span the gap between the alpha and gamma bands.

The disappointment is that all the precise EEG measurements (and latterly MEG measurements as well) have not been rewarded by unequivocal

f / Electroencephalography actually dates from 1875, and the work of Richard Caton, but his studies were carried out only on animals. The original work of Berger was cited earlier in this book.

demonstrations of the origins of all these oscillations.[g] It is for this reason that the new observations described above are of such importance. It is true that the first work of this type was carried out on anaesthetized animals, and this has prompted some people to raise doubts as to their relevance, but the observations have now been duplicated in awake cats and even monkeys. They are certainly to be taken seriously. The question remains, however, as to the underlying mechanism. If these gamma band oscillations can be given a neuronal-level explanation, it is clear that we will have taken a significant step forward in the understanding of brain dynamics. Let us consider the theories that have been put forward, and begin with one due to Singer himself.[16]

When considering the responses of the various cells in the early visual areas, we noted that they show preferences with respect to eye (right or left) and orientation. We also saw that there is a reasonably systematic arrangement of these cells, such that the ocular dominances occur in alternating slabs, about 0.5 millimetres thick, while there is a continuous variation in orientation preference as one traces a line on the cortical surface, roughly at right angles to those ocular alternations. Things are actually somewhat more complicated than this, because the ocular dominance and orientation systems are superimposed on the (distorted) mapping of the visual field. Then there is the further complication that some neurons detect position while others pick up motion. The upshot of all this is that the early visual cortex is functionally quite patchy.

When Singer and his colleagues selected well separated neurons that had similar orientation preferences, they were picking out cells with particular specialties from something that resembles a patchwork quilt. The straight bars of light they used for stimulation were oriented to give a maximum response in the chosen cells. Those responses were elicited when the moving images of the bars passed across the corresponding receptors in the retina. This would not automatically give regular trains of action potentials that were mutually in step, however. Singer saw that something else would be needed to achieve that, and he invoked the action of excitatory *interneurons*. These, according to the Singer model, would receive excitation from the orientation selective cells, and return pace-making excitation to them. The interneurons would, in effect, be serving as the conductor's baton; they would orchestrate the coherence of the oscillations. The trouble is that such circuitry does *not* of itself guarantee regular oscillations. A single pulse disseminated by the pace-maker cell would naturally trigger the orientation cells in unison, but it would receive its prompting in an incoherent fashion, unless the motion detecting neurons were already firing in step. We must dig deeper.

The Singer model sees oscillations as arising from the to-and-fro of the signals between the orientation sensitive neurons and the interneurons. Let us start by checking that the frequency to be expected of this process

g / Certain features in the EEG and MEG responses have been reliably linked to brain processes, however, and we will be considering them in later chapters.

is in agreement with the observations. We have been quoting an all-inclusive value of between 5 and 10 milliseconds for the passage of signals from the soma of one neuron to the soma of any neuron that it interacts with. If we take the latter period and double it, because we need the round-trip time, we see that the corresponding frequency would lie around 50 Hertz. This is indeed in the gamma band. The next thing to check is whether interneurons have been found which could play the pace-making role. Again the situation is encouraging. Charles Gilbert and Torsten Wiesel,[17] and also Kathleen Rockland and Jennifer Lund,[18] have traced the connections of visual area neurons which have processes extending in the plane of the cortex, predominantly in layer II+III. The exciting thing is that these neurons provide lateral links between other neurons all having the same orientation preference. Such cells have not yet been given a name, but one could call them *cluster-binding neurons*.

This leaves the problem of *regular* pace-making. It could be that a subtlety in the dynamics enables the pace-maker and the orientation-sensitive neurons to mutually bring each other into phase, in a regular firing mode. Another possibility arises if the signal returned to the orientation-sensitive cells from the pace-maker is insufficient by itself to trigger those cells.[19] For the sake of illustration, let us say that it can only supply eighty per cent of the required membrane depolarization. If the orientation cells are in the process of being depolarized, but are not in step with one another, the arrival of that eighty per cent pulse will push all cells with a current depolarization in excess of the twenty per cent mark over the limit, irrespective of their actual depolarizations. And the receiving cells will fire coherently. Moreover, a sufficiently stimulated pace-maker will be emitting a regular series of pulses, so the orientation-sensitive cells will be induced to do the same thing. The frequency of such a series of pulses is indeed usually in the gamma range. There is one other observation that is nicely explained by this mechanism. It is frequently the case that the cortical oscillations disappear briefly and then return *at a different frequency*. The frequency of any cell, including a pace-maker, depends upon the strength of the signals with which it is being driven. But the pace-maker will continue to do its job, irrespective of the frequency at which it is currently firing.

Cortical oscillations have been observed in other contexts, examples being their detection in connection with olfaction by Walter Freeman and his colleagues;[20] in connection with tactile stimulation by Gregor Schöner, Klaus Kopecz, Friederike Spengler and Hubert Dinse;[21] and intriguingly in connection with the planning of movement by Gert Pfurtscheller and Christa Neuper.[22] There have also been reports implicating sub-cortical structures in oscillatory behaviour, most notably the thalamus since it figures so prominently on the route from the senses to the cortex.[23] The work on olfaction serves as a timely reminder that things might be rather more complicated than I have been giving them credit for here. The point is that there is no obvious counterpart in the sense of smell to those coherently moving bars of light in the visual experiments. And yet one observes oscillations, which are even in the same (gamma) band of frequencies as those seen in vision. There have been interpretations of the olfactory observations which invoke what is known as *chaos theory*. The underlying mathematics is beyond the scope of this book, but the gist of the conjectured mechanism is a high sensitivity to new sensory inputs, which violently throw the cortical dynamics from one mode to another.[24] Freeman and his colleagues see the relevant parts of the cortex as lurching from one spatiotemporal oscillation to another, every time the nose gets a whiff of a new scent.

When considering oscillations, therefore, we need to proceed with caution. Christoph von der Malsburg and Joachim Buhmann have pointed out that the presence of interconnected exciters and inhibitors in the cortex more or less guarantees that oscillations will occur.[25] Oscillatory behaviour is even seen in the neural circuits of *Aplysia*, which does not have a cerebral cortex.[26] But the fact that oscillations occur even when there is no obvious advantage of detecting spatial coherence, as there is when separating a figure from its background, does not mean that oscillations are never related to such coherence. I will therefore return to Singer's ideas, and attempt to put them in the context of the picture of the cortex that emerged toward the end of the preceding chapter. We recall that the cortex's role in global closure was mediated by those reciprocal projections. We also saw that such reciprocity is only an overall characteristic of the links between two cortical areas; it does not apply microscopically to every white-matter axon. Finally, we learned that this forces groups of activated minicolumns to act cooperatively in order to influence cortical dynamics. It is the paucity of microscopically closed inter-area loops which decrees that *inter-area oscillations* will primarily arise only when separated elements in a retinal pattern are moving in synchrony.

The Singer analysis was general in that it merely invoked the cohering influence of excitatory interneurons. We have already seen that cluster-binding neurons lying in the plane of the cortex, in a given cortical area, can fulfil this function. Groups of coherently acting neurons in *another* cortical area can play the same role, assuming that the other area shares reciprocal projections with the area in question. Olaf Sporns, Joseph Gally, George Reeke Jr and Gerald Edelman studied a model which endorsed this idea,[27] and a similar model investigated by Claus Nielsen and the present author produced much the same thing.[28] Both models invoked the minicolumn structure of the cortex. The former assumed the presence of orientation-selective cells, but did not specifically incorporate the known inter-layer circuitry in a given area. The latter work embodied those details, but made no specific reference to orientation selectivity.

It seems possible that two mechanisms could function in conjunction, one being mediated by cluster-binding neurons in a given area and the other by the reciprocal projections between areas. They would probably operate at two different frequencies. Perhaps it is this sort of situation that lies behind the multiplicity of frequencies seen in EEG and MEG traces. There might be a hierarchy of interactions involving the cooperation of increasing numbers of areas, and presumably producing progressively lower frequencies of oscillation because the signal transmission times would become systematically longer. It might even be the case that one should update the classical explanation of the alpha rhythm. The traditional picture, as I mentioned earlier, was one in which many small contributing waves were in phase with one another. Perhaps it is more a case of many, or even all, cortical areas passing pulses between each other that have no information content, pulses that would merely serve as a sort of temporal glue, supporting what could be called an idling oscillation. Then when sensory signals begin to arrive, the hierarchy is galvanized into activity, with first the intra-area oscillations, then the inter-area variety. Although it is only a very rough analogy, I am often reminded of the sailors' term *old sea*, which describes the waves that can still persist during a calm because of the remnant effects of a recent wind. They have much longer wavelengths than those of the waves that reappear as soon as the wind starts blowing again.

A recent investigation carried out by Katsuki Nakamura, Akichika Mikami and Kisou Kubota lends credence to this view.[29] They recorded the activity of neurons in the temporal lobes of monkeys while they were performing a short-term visual memory task. The monkeys were required to memorize specific visual stimuli, and sustained activity in the form of oscillations was detected in the ventral part of the lobe while this was being done correctly. When it was not, the activity was absent. Interestingly, in view of what we have been discussing, the frequency of the observed oscillations varied considerably, the range being from 1.7 to 22.2 Hertz. The indications are that a number of different inter-area modes are indeed involved when the visual system handles sensory input, and temporarily stores percepts. Finally, and returning to the work of the Singer and Eckhorn groups, the gamma oscillations were usually observed to persist for about 200 milliseconds, the word *spindle* being applied to such a brief rhythmic volley. It is interesting to note that this period is roughly comparable with the length of time that we were earlier associating with the cognitive process.

Early warning system

Having considered the useful role synchrony could play in neural processing, with its possible ramification into actual oscillations, we can return to the general question of sensory processing. As before, it will be expedient to focus primarily on vision, although we will make brief excursions into audition as well. The first point to be made is that the scenario invoked in the first section of this chapter was grossly simplified. The discussion concentrated on the forward and reverse flow of nerve impulses, without much thought being given to what information was actually being conveyed by those signals. There was a tacit assumption that the information-carrying capacity of the system is merely dictated by the number of neurons in each cortical area. Although this was not stated in as many words, the areas were being regarded as grids in which all elements were on an equal footing. Equal, that is, apart from the distortions arising from the fact that the various cortical neurons have different sizes of receptive field.

That this is too sweeping an assumption can readily be seen from the surprising fact we noted about our visual fields of view. We can sharply discriminate only with the foveal region of the retina, even though objects observed with that centrally located area fuse smoothly with the things seen with the less discerning extra-foveal part. So smooth is that fusion, indeed, that we get the impression that we are seeing equally well over the entire field of view. Two questions are raised by this non-uniformity: why has the system not evolved so that equal discrimination is possible over the whole field, or failing that, what use is the surrounding part, with its lower resolution? The answer to the first question is that the subsequent parts of the system simply do not have the information-handling resources to devote to all the extra information that would then be showered upon them. And the answer to the second question is that the low-resolution surround is a useful detector of things of potential interest. Something moving in the field of view, but not within the centre of gaze, may be of interest; it could pose a threat, or it might hold promise. In either case, it causes the eyes to swivel toward its direction, so as to bring it into the foveal region. More subtle forces are at work when the off-centre item of potential significance is not moving.

The mechanism whereby this low-resolution detection is accomplished is known as *preattention*, the existence of which was suggested by Ulric Neisser in 1967. Preattention has been studied particularly intensely, in recent years, by Bela Julesz and independently by Anne Treisman. I will return to the latter's work later on in this section. Julesz has investigated the *effortless* visibility of specific patches of larger patterned areas.[30] The word effortless can be quantified, because Julesz permitted his subjects glimpses of the pattern lasting 160 milliseconds or less; he found that the mismatched region displays immediate *pop out*. One refers to this as texture discrimination, and Julesz was

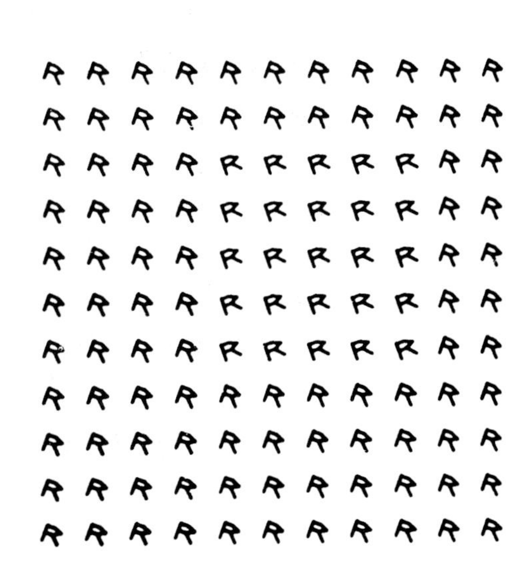

Figure 8.7
A quick glance at each of the two parts of this figure, due to Bela Julesz, is enough to show that there is in each case a patch just above and to the right of centre which is different from the remainder of the pattern. Such instant discernment of heterogeneity is known as ***pop out***.

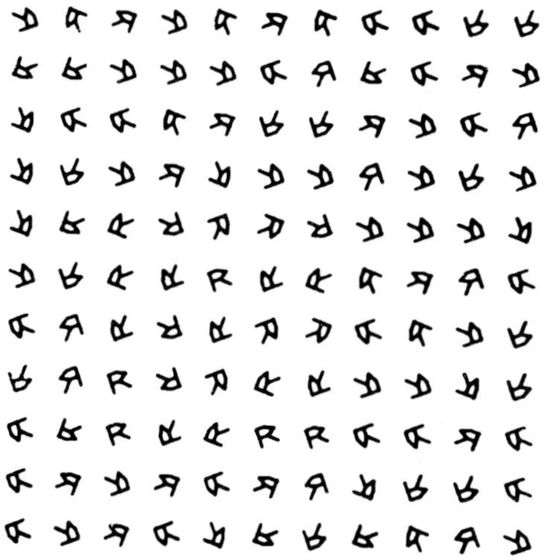

Figure 8.8
Although there is also heterogeneity in this diagram (again due to Julesz), it is of a more subtle type than that shown in figure 8.7. The heterogeneity is subthreshold for pop out, and its detection requires systematic inspection.

able to develop a set a rules which appear to govern such situations.[h] A rapid glance at each of the two parts of figure 8.7, for example, without studying their details, is sufficient to reveal that there is in both cases a patch just above and to the right of centre which is different from the remainder of the pattern.

Julesz found that this discrimination can be made in as little as 50 milliseconds, which is too short a period to have involved eye movements. In a subsequent investigation, he progressively decreased the mismatch until pop out no longer occurred. The subject then had to scrutinize the figure for a longer time to detect the subtle differences between the patch and the surrounding pattern. Julesz had thereby located the divide between preattention and attention, and he noted that visual processing in the former is *parallel*, while it is *serial* in the latter. An example related to the above diagrams which is sub-threshold for pop out is shown in figure 8.8. The figure has to be inspected rather carefully, if the deviant patch lying slightly below and to the left of centre is to be detected.

Julesz found that pop out discrimination is possible only for sufficiently low values of what is referred to as the *statistical order of the pattern*. In a figure with uniformly first-order statistics, the probability that randomly thrown *dots* will land on a certain colour (black, for example) is the same in all areas. Uniform second-order statistics applies to figures in which the vertices (i.e. the end points) of randomly thrown *lines* have the same probability of landing on a certain colour, irrespective of position in the diagram. In general, *n*th-order statistics obtain when this uniformity proposition is true for randomly thrown *n*-sided *polygons*. In the case of the three examples shown in figures 8.7 and 8.8, we can simply regard the first, second and third orders as

h / One could loosely say that the Julesz rules are the opposite numbers of those encompassed by Gestalt theory.

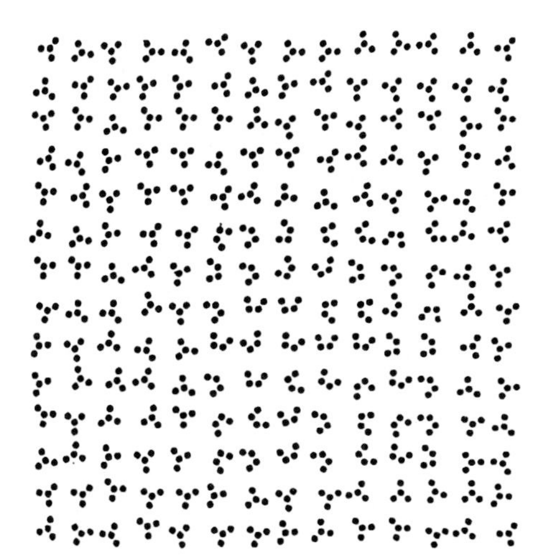

Figure 8.9
Julesz has derived rules for the preattentive detection of heterogeneity. They are related to the statistical order of the elements of the pattern, and pop out usually occurs if there is violation of second-order statistics. Some patterns in which there is no violation of second-order statistics nevertheless show pop out, however, the extra condition that must then be fulfilled being the presence of certain local features. Julesz called these textons, and they are conspicuous because of the magnitude of their third-order violations.

referring, respectively, to the size, orientation and handedness of the figure's elements. Julesz observed that patches which violate the uniformity of first-order or second-order statistics can immediately be distinguished. This explains why we have no difficulty in noticing the offending patches in the first pair of diagrams, shown in figure 8.7; they respectively violate first-order and second-order statistics.[i] In the case of the third picture, which does not produce pop out, the violation is in the third-order statistics.

The fascinating question naturally refers to the origin of the pop out, when it occurs. Is the visual system functioning piecewise or globally when it permits such discrimination? Are there basic elements in perception out of which higher percepts are constructed, or are such higher percepts indivisible? This is, of course, the Structuralist–Gestaltist dichotomy in its starkest form, and Julesz discovered the answer through cases which failed to comply with the above rationalization. It turns out that there are patterns in which patches pop out even though there is no violation to second-order statistics, the extra condition that must then be fulfilled being the presence of conspicuous local features. Julesz called these *textons*.[31] The crux of the matter lies in the magnitude of the third-order violations in such cases. This is well illustrated by the pair of diagrams shown in figure 8.9. In the diagram on the left, there is a patch near the centre, seven columns wide and seven rows deep, in which the dots are arranged on a curving line; elsewhere in the diagram, the dots form a forked motif. If this entire pattern is briefly viewed, a few degrees away from the direction of foveal gaze, the patch does not pop out. The right-hand diagram has a similarly sized and centrally located patch which immediately

i / It is of no consequence that the patch in the left-hand example of figure 8.7 happens also to violate third-order statistics. The lowest-order violation will always be the determining factor.

pops out under those viewing conditions. Inspection of its details reveals that it differs from the first case in that the curved line has been made somewhat straighter, while the forked motif is more irregular. As Julesz noted, quasicolinearity makes for preattentive discrimination in such cases.

This work had given a strong indication that local processing plays a part in the preattentive mechanism, and it obviously became relevant to speculate about its neuronal basis. A vital further clue was provided by pairs of figures such as those shown in figure 8.10. Only the right-hand member gives pop out. Close inspection reveals that the elements of the left-hand picture come in two varieties: there are S-elements and there are IO-elements. Now despite their differences, these two types of element both possess just two line-ends. By contrast, the elements in the readily discriminated central region of the right-hand figure have five line-ends, whereas the balance of the figure has exclusively two-ended elements. Julesz discovered that pop out occurs merely on the basis of whether or not there is violation of the first-order statistics of textons. Patterns conforming with the requirement of exclusively first-order statistics are said to possess *ideal camouflage*.

What is the neuronal foundation of this early warning system? In the 1960s, David Hubel and Torsten Wiesel had extended their earlier work on orientation-selective neurons in the primary visual cortex, and discovered cells that responded not only to lines lying at a particular angle but also to a specific geometrical feature.[32] Amongst the features that gave marked reactions was indeed the abrupt termination of a line. Following the discovery of textons, David Marr incorporated detectors of edges, elongated blobs and end-of-lines in a theory of early vision which he called the *primal sketch model*.[33] Since then, further refinements have been made to the observations. It has been discovered, for example, that texture discrimination depends upon the inter-

Figure 8.10
Only the right-hand diagram in this pair gives pop out. There are two types of element in the left-hand diagram, but they both have only two line-ends. The pop out elements in the central region of the right-hand diagram have five line-ends, while the remainder have two. Julesz discovered that pop out occurs if there is violation of the first-order statistics of textons. Patterns which do not comprise such violation are said to have ideal camouflage.

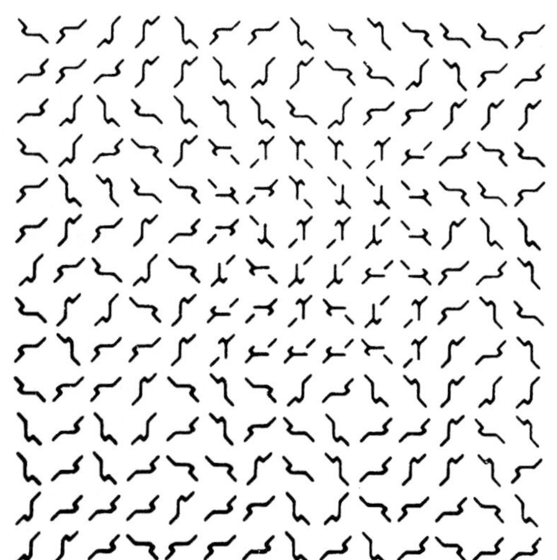

element distance, there being a requirement that the packing density is sufficiently high.

When the distance between elements exceeds about twice the diameter of an element, the visual system automatically goes over to a scanning mode, in which there are small-angle shifts of what has been referred to as the *searchlight of attention*, every 20–50 milliseconds. This is about ten times more frequent than the normal saccadic rate, and it therefore cannot involve movements of the eyes. The durations of these small shifts are in fact comparable to the minimum time required to make an impression on the cognitive system, as discussed in the first section of this chapter. Outside what could be called the beam of the searchlight, the preattentive system can only detect where differences in texton quality occur, not *what* those differences actually are. But this is sufficient for the redirecting of gaze, by a mechanism that involves the superior colliculus.

Texture discrimination tasks are capable of telling us more about the way the visual system handles information, as the work of Anne Treisman has shown.[34] She investigated pop out in patterns of coloured letters, such as Os and Vs coloured red and blue. A patch of Os will pop out from a background of Vs (and vice-versa) even though both colours are represented in both populations of letters. Similarly, a patch of blue (or red) consisting of a random mixture of Os and Vs will pop out from a red (or blue) background that also randomly comprises Os and Vs. Conversely, a patch consisting of blue Os and red Vs will not pop out from a background comprising blue Vs and red Os. As Treisman noted, the *parsing*[j] of the field of view by the visual system takes heed of individual features, not combinations of features. The consolidation of the various attributes into a coherent whole must be accomplished higher up in the system. (A rather more complicated picture that illustrates this principle of late consolidation is reproduced in figure 8.11.)

This led her to predict that errors of consolidation would occur under favourable circumstances, and that is precisely what was observed. In one set of tests, subjects were shown collections of V shapes and I shapes, mutually oriented so that the Is lay parallel with the direction of the lines that would have been required to transform the Vs into triangles. Following brief exposure to such collections, subjects will sometimes report having seen some complete triangles, even though none were present. If the Is did not lie in the appropriate direction, no complete triangles were reported. Treisman discovered that if the collection also contained closed symbols, such as Os, the erroneous observation of triangles increased considerably. The presence of one type of closed symbol apparently biased the visual system toward fallacious detection of other types. This indicated that early visual processing also includes analysis for closure.

j / *Parsing* is the process of analysing the structure of something. The something could be an object or pattern in the visual field or a train of sounds, such as the succession of tones in a piece of music or a spoken sentence.

Figure 8.11
A cursory glance at this picture is enough to identify it as Leonardo's famous Mona Lisa, replete with her elusive smile. If the page is rotated to recover the more familiar orientation, however, the picture is found to illustrate something else, namely that the individual elements of an image are handled separately by the early part of the visual system, and that they can thus have a life of their own. (After Julesz, B. (1986). Stereoscopic vision. *Vision Research*, **26**, 1601–12.)

A particularly revealing test required subjects to detect symbols which *lacked* a feature, such as an O amongst Qs; the former lacks the small straight line at the lower right. It was found that this required much more time (for a given total number of symbols) than that needed to find a Q amongst Os. This result clearly shows that the early visual system reacts to the *presence* of things, but not to their absence. It is suggestive of a mechanism in which each elementary attribute gives rise to independent signals, which are subsequently added to those emanating from other attributes. The resultant composite is then subjected to a winner-take-all filtering process. In the case of the Os and the Qs, the elementary attributes are circles and small straight lines. In the winner-take-all competition, the Qs will outstrip the Os because they possess that extra attribute.

It was through ingenious little tests like this that Treisman and her colleagues were able to compile an inventory of individual features that the early visual system can cope with. The list includes colour, size, contrast, tilt, curvature, line-endings and closure. All these simple characteristics can provide the basis for preattentive detection and discrimination, whether it be by the foveal region for very brief exposures or by the extra-foveal region. How are the various components properly assembled higher up in the system, however? The answer is that they are *not*, if they are located extra-fovealy, irrespective of how much time is available. And if they are positioned in the fovea, assembly takes time, and also something to which we must now turn, namely *attention*.

Paying attention

Anne Treisman's work provides a good starting point for considering the very important phenomenon of attention. She again used patterns that were briefly exposed to subjects, but in this case the patterns were somewhat more complicated. In one of them, the target was a red or blue H, say, amidst a number of red O and blue X distractors.[k] Interest centred not only on the subject's ability to identify the target, after brief exposure to the display, but also to properly locate it once it had vanished from the screen. It transpired that errors in remembering the target's location were more frequent than mistakes in simply identifying it. In a second set of displays, the target was a blue O or a red X amongst a number of red O and blue X distractors. This task was fundamentally different from the previous one, because the target had something in common with each of the two types of distractor: form with one and colour with the other. In such a display, one speaks of features being *conjoined*. The subjects' scores for this variant of the test were quite different from those for the previous challenge: correct identification was now strongly correlated with correct localization. This demonstrated that attention has to be focussed on a specific location if the features of a target are to be mentally combined, and thus properly mimic the actual conjunction in the display.

These studies were obviously related to what was discussed in the preceding section. The various elements of the displays were naturally being handled preattentively, and they were then being combined further up in the visual hierarchy. One refers to such consolidation by the term *bottom up*, and because conjunctions of letters with colours is an abstract exercise, bottom-up consolidation was probably the only process operative in these tests. As Treisman emphasized, many conjunctions have the additional dimension of *prior knowledge*. Blue bananas and red eggs are not things that one often encounters, and if such items occur in test displays, one could imagine that they will meet with a certain amount of resistance, from the cerebral powers that be. One speaks of *top-down constraints* in such cases. Irving Biederman has shown that identification of objects occurs more rapidly if top-down strictures are not being defied.

Anne Treisman and Deborah Butler have explored top-down influences further. They used displays in which three target figures, equally spaced along a line, were flanked by two distractors. The distractors were digits, while the targets were simple geometrical shapes which could nevertheless be linked with familiar objects, the idea being to thereby introduce the possibility of top-down influence. An elongated isosceles triangle will be reminiscent of

k / The words *target* and *distractor* have the same meanings as those employed in the first section of this chapter.

a carrot if it is coloured orange, for example, while an ellipse will resemble a lake if it has the appropriate shade of blue; a thickly drawn circle could be an automobile tyre, if it is black. The entire five-member display was obliterated by a pattern mask after exposure to the subject for 200 milliseconds, a pointer simultaneously indicating the position previously occupied by one of the three targets. The subject was required first to report which two digits had been used as the distractors, and then to identify the object that had lain in the position indicated by the pointer.

The subjects in these experiments were divided into three groups, which were required to perform under different test protocols. Members of the first group were told that they would be seeing carrots, lakes and tyres, but that a quarter of the objects would appear in the wrong colour, to preclude short-cut identifications through colour alone. The second group was told that the objects would be orange triangles, blue ellipses and black circles, again with a certain fraction of deviant figures such as orange ellipses. Members of the final group were made to expect only the natural colours; their operational paradigm was the same as for the first group, except that they were not informed beforehand that false colourings might occur. Not surprisingly, the first group scored better than the second; their reports included fewer erroneous identifications. The real interest centred on the third group. Treisman and Butler discovered that the subjects did not generate illusory conjunctions to fit their expectations. In cases where the pointer indicated the position of a triangle that was unexpectedly blue or black, the likelihood that they would erroneously report it as having been orange was quite independent of whether that colour had appeared elsewhere in the display.

Treisman and Butler drew two fundamental conclusions from this test: prior knowledge and expectations aid attention in properly conjoining features of a target figure, but they do not promote illusory exchanges of features in order to make unexpected combinations revert to what was anticipated. Illusory conjunction thus appears to arise at an earlier stage of visual processing than that which has semantic access to knowledge of familiar objects. Conjunction of features apparently occurs in the bottom-up manner characteristic of preattention; it seems not to be subject to top-down interference. But how does the bottom-up mechanism feed into the upper echelons of the hierarchy? I used the term cognitive compression, earlier, when describing what happens to the inward-travelling information. Let us consider an example cited by Treisman herself. A bird perched on the limb of a tree is perceived to have a certain size and shape. As it flies away, these attributes change, and yet we still see it as the same object. We can even cope with the radical change in colour that can occur as the bird unfolds its wings. How does it maintain its perceptual integrity in the face of such variability? What mechanism ensures that the cognitive compression will nevertheless funnel these changing sensory inputs into the same conceptual pigeon-hole?

In collaboration with Daniel Kahneman, Treisman has suggested that the intermediate consolidation process involves what they called a *spot-*

light of attention.[1] The crux of their model is the construction of a temporary representation that is specific to the object's current appearance, this being continually updated as the object changes. The temporary representation's contents are then passed up the hierarchy, for comparison with what is already stored in memory. Treisman compared this latter part of the process to the scrutinizing of a file. Continuity of the perception of an object will be guaranteed if the contents of the spotlight continue to fit one particular file better than they fit any other. In order to test this idea, Treisman and Kahneman, in collaboration with Brian Gibbs, devised a test in which identity had to be robust against change of position. Two different letters were briefly flashed on a screen, each surrounded by a square that remained visible when the letters disappeared. The empty squares then moved to new positions, and a letter was briefly flashed in one of them. This second singleton letter that fleetingly appeared in the randomly selected square could either be the same letter that had been there the first time or it could be different.

When there has been prior exposure to the same letter, that letter's first appearance is referred to by the term *priming*. It is well known that priming speeds up recognition; if the second letter is not identical with the primer, recognition takes slightly more time. The key point in the experiment was whether priming is effective only if it has occurred in the same square. This is just what was observed; subjects recognized the match an average of 30 milliseconds faster if it was taking place in that location. This confirmed that the conjunction of the letter's identity and its link to a particular square was being maintained despite the movement of the square. Indeed, it is as if the visual system is attributing movement to the letter as well, by association.

As a result of these experiments, Treisman proposed a scheme for visual processing which accorded the spotlight of attention a decisive role. She saw the presence of individual features of a stimulus as being signalled without this necessarily including an indication of their whereabouts. She also conceived of a *master map of locations*, one area of which was instantaneously highlighted by the spotlight. According to her theory, only that area currently being highlighted can be the goal of instantaneous links from the various features. It is the momentary establishment of those links which gives the spotlight its content, and it is that content which is of relevance for the above-mentioned file. Finally, Treisman suggested, the contents of that file are compared with descriptions stored in a recognition network, that is to say in an appropriate part of memory. This network specifies the defining characteristics of the things involved, and it supplies their names, their likely behaviour, and their current significance. According to Treisman's view, *it is the interactions between the file and the memory bank which are the basis of conscious awareness*.

In view of what was described in the preceding chapter, concerning the layout of the cerebral cortex, the Treisman hypothesis would seem to be

1 / Some authors prefer the term *searchlight of attention*. This conveys the impression of a slightly more active role, which may or may not be appropriate.

grist to our mill. The various stages of the hierarchy invoked in her model might appear to be identifiable with the various stages of the visual system (if it is vision that is implicated, of course). In her scheme, the feature modules come before the map of locations, that map comes before the object file, and the object file comes before the recognition network. Could it be that the feature modules are simply in the primary visual cortex, that the map of locations and the object file are in the subsequent visual areas, and that the recognition network constitutes the appropriate association cortices, functioning collectively? I believe that this would be too facile an interpretation of her ideas. In my view, there are a couple of things which militate particularly strongly against such direct equating of her levels and the chain of areas that undoubtedly exists in the cortex. For a start, such a scheme leaves no role for the reverse projections. Two-way traffic of neural signals occurs in the Treisman strategy only between the uppermost two levels, that is to say between the object file and the recognition network. Two-way traffic in the cortex, on the other hand, is seen between many of the areas, even in those which receive the sensory input. Even more seriously, Treisman's conclusion that the feature modules might not be place-specific is difficult to reconcile with the actual situation in the cortex; patches earmarked for colour, orientation and direction coexist side by side in the quilt arrangement referred to in earlier chapters.

There is a third, rather more subtle, objection that can be levelled against too direct an interpretation of the attractive model that Treisman proposed. As a matter of fact, its roots can be traced back to 1910, and that paper by Cheves Perky that I cited in chapter 2. As we saw, plenty of corroborating evidence has emerged since that time, most notably the recent extensive work by Semir Zeki on the various visual areas, which was also cited earlier. When we imagine something, we often speak of seeing something *in our mind's eye*. The point of Perky's investigation, and of the experiments that followed in its footsteps, is that it showed that the mind's eye and vision's eye have a lot in common; we apparently use the same areas when imagining as we use when seeing. The cortex is apparently parsimonious with its areas; it so likes to get good milage out of them that it uses them for both tasks. It is this economy which necessitates a more subtle interpretation of what Anne Treisman was postulating. It means, I feel, that her feature modules and her map of locations might share the same part of the cortex. If this is so, the unidirectional conduits which she conceived as passing signals from the former to the latter might in fact be two-way connections. They might be a combination of the forward and reverse projections that we know are a very common attribute of the cortex's hard wiring. Such a scheme would permit us to use the fact, emphasized earlier, that things like colour, orientation and direction *are* spatially mapped in the primary visual cortex. Before continuing with this alternative picture, however, we ought to consider other evidence that has a bearing on this quite central issue. There is certainly plenty more evidence to review.

We could hardly choose a more appropriate starting point than the mid-1950s, and the pioneering contributions of Donald Broadbent.[35] He investigated a number of situations which strongly pointed to essential limitations in the nervous system's ability to handle signals. We instinctively allow for these limitations, hardly realizing that we are doing so. We sometimes close our eyes in order to concentrate on a piece of music, or on something being said. Conversely, we turn off the radio when we wish to give all our attention to something we are reading. The distrac*tors* used by the experimental psychologist and the distrac*tions* of everyday life are, of course, close cousins. Broadbent's early work in this area was motivated by the type of practical problem that accompanied the development of communication networks and the growth of mass transport. A typical example is seen in the case of the flight controller, who must make rapid decisions while being bombarded with numerous streams of information.

Experimental work in this area benefited from the invention of the tape recorder and the development of stereophonic techniques. These advances gave Broadbent the means of varying the rate at which competing flows of auditory information are fed to the two ears. He used what is known as a *split span protocol*, three (prerecorded) spoken digits impinging on the left ear while a different set of three is simultaneously fed to the right. The task of the subject was to recall as many of the six digits as possible, and it was found that the optimal rate of delivery was two digits per second. Despite the fact that the subject was being presented with three synchronous pairs of digits, Broadbent found that they were being recalled in two groups of three: the entire set for one ear, followed by all three digits for the other ear. His conclusion was that the subject attended to one ear first, and that the information simultaneously reaching the other ear could be retained in an unattended form of short-term memory for a few seconds. Attention was then switched to that other information, while retention of the first three digits became the responsibility of short-term memory. These observations formed the basis of Broadbent's *filter theory*, the gist of which is that the brain constitutes a single-channel information processing system with limited capacity. This channel, according to his findings, could select only one sensory input at a time, the switching of selections occurring at a maximum rate of about two per second. This is about the same as what I was able to achieve with that faces–vase picture that we encountered in the first section of this chapter.

Limitations of information-carrying capability are often encountered in signal transmission technology, the measure of capacity usually being what is referred to as the *bandwidth*. Broadbent's conjectures are particularly reminiscent of a technique known as *time-division multiplexing*, in which several information streams can be handled simultaneously, if it is permissible to pay full attention to any one of the streams only periodically.[36] Brief attention is rapidly and cyclicly shifted from one signal to the next, this mechanism being known as *gating*. The strategy works only if the persistence of any stream is longer than the time lapse between attentional shifts. If that is not the case, information will be lost. In time-division multiplexing, the bandwidth can be roughly defined as the ratio of the period over which the signals persist to the time lapse between shifts; this determines how many signalling streams can be coped with by this approach. In the Broadbent scheme just described, the persistence of a signalling stream was simply that of short-term memory. We will have to suspend, until later, discussion of what it is in the brain that determines the rate of gating, but from Broadbent's observations we can see that the rate is about two per second; the minimum time that must be given to any one thing is the period that seems to be cropping up all over the place in this chapter, namely 500 *milliseconds*.

Broadbent's early work was followed up by studies of what is known as *speech shadowing*, the pioneer of this technique being Colin Cherry. Anne Treisman herself was active in this area, as was Neville Moray.[37] The subject is asked to repeat aloud a piece of text that is presented simultaneously with one or more distractor passages. A number of different factors were found to aid concentration on the correct signal, including tone of voice, the direction and distance from which the various signals were arriving, and the cadence of the delivery. Treisman discovered attributes of a distractor which enable the unattended passage to impose itself, at the expense of the correct text.[38] One of these factors is emotional content, an example being the sound of one's own name. Another potent distractor is text that is closely related to what one is supposed to be concentrating on. This is rather revealing, because it touches on that issue of bandwidth. When two signals bear a close mutual resemblance, it might no longer be correct to regard them as being independent; the routes that they take through the nervous system might have a lot in common. This particular distractor phenomenon is quite logically known as *breakthrough*, and we have good reason to be grateful for its existence. There are occasions when breakthrough can be vital to life and limb, indeed. Let us consider an example.

I am often shocked, when travelling in my car, to discover that my mind has not been on my driving. There are occasions when I am lost in thought, even though the car is moving at considerable speed. It is almost as if the vehicle were driving itself, with me as a passive passenger. There are other occasions when an impatient driver immediately behind me has to sound his horn, to shake me out of my reverie, and draw my attention to the traffic light that long since turned green. One would think that such absent-mindedness would put me at constant risk. Yet the fact is that in forty years of driving, I have never had an accident. What guardian angel has been watching over me all this time? The benevolent agency has probably been a combination of Donald Broadbent's bandwidth and Anne Treisman's breakthrough. The bandwidth is probably greater than he initially envisaged, however, and the threshold for breakthrough is probably higher in the driving situation than it is when two streams of speech are vying with each other.

Day-dreaming when driving is especially interesting because the latter enterprise makes considerable demands on one's sight. It is thus mysterious that one can nevertheless see things with *the mind's eye*, at the same time. I say this because it has been emphasized several times already that the visual component of imagination appears to use some of the same cortical areas as those employed in sensory vision. How can it be that this double usage does not lead to confusion? What prevents scenes perceived with the real eyes and scenes imagined with the mind's eye from getting all mixed up? One factor must be the difference between what is being seen and what is being imagined, because that experiment performed by Cheves Perky with the real and the imagined tomato (as described in chapter 2) showed that a mix-up *does* occur when the difference is inadequate. One could say that her experiment explored the limiting case, in which breakthrough was mandatory. When there *is* a difference, the possibility of keeping them separate must stem from the fact that the items are being handled by different sets of neurons, in the appropriate area of the cortex. We saw earlier that there is a region where faces are processed, and it might be the case that there are analogous regions that handle roads, fields and trees. It is this separation that provides the safety factor when one imagines someone's face while driving through the countryside. (Breakthrough is impressively illustrated in the type of test first described by J. Ridley Stroop in 1935; an example is shown in figure 8.12.)

Distribution of perceived items amongst different cortical regions can only be part of the story, however. There remains the important question

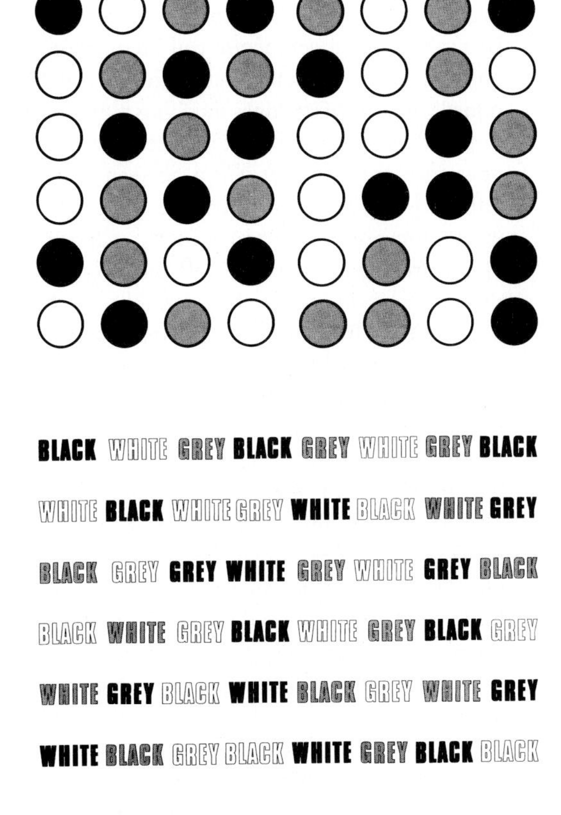

Figure 8.12 Breakthrough, the interference of one percept by another, can be demonstrated with what is known as a Stroop test (Stroop, J. R., *Journal of Experimental Psychology*, **18**, 643–62). In the example shown here, the perceptions are of colour and shape. When the shape is a circle, there is no conflict between the two percepts, and breakthrough does not occur. If shape is used to convey meaning, however, this may well clash with the perception of the colour, and breakthrough will be possible. That this happens in the case shown here can readily be demonstrated by comparing the times taken to read aloud the **actual colours** in the sequences above and below.

of when priority is being allocated during the processing of information; is selection made late, after the stimuli reaching the senses have been fully operated upon, or much earlier? If we could answer that question with confidence, we might be closer to an understanding of how the brain really handles sensory input. A strong hint is provided by situations in which the system is not supplied with opportunities for selection. In such cases, interference is found to depend on individual features of the stimulus and not on the responses they provoke. We saw examples earlier, when considering Anne Treisman's work. If the subject knows what part of a stimulus field to select, there is no interference between the various features of the input. The interference must thus occur before the choice of response has been made; selection is made early. We might have anticipated this by contemplating the cortical map reproduced in the preceding chapter. We recall that in vision, for example, the number of alternative paths confronting signals as they proceed up the hierarchy steadily increases. Early in the system, therefore, there must be more sharing of neuronal resources than there is later on. If we accept that there will be more interference between two tasks in regions where they are competing most vigorously, we see that the case for early selection is quite persuasive.

Francis Crick and Christof Koch have described a delightful illustration of interference.[39] For reasons which will become apparent shortly, it is known as the *Cheshire Cat experiment*, and it stems from what is known as *binocular rivalry*. The phenomenon can be observed with the aid of a convenient object, a mirror, and a featureless area such as a white wall. The mirror is positioned with one of its edges touching the tip of the nose and the middle of the forehead, as if it were about to slice vertically through the head. An object is now positioned so that it can be seen only by the eye lying on the non-reflecting side of the mirror. In the relaxed state, the other eye will also point in the direction of the object, but it will actually be receiving the uniform illumination from the blank wall; the subject will not be aware of the fact that the object is being seen only with one eye. If the hand on the reflecting side is now moved in front of the wall, so that it apparently passes through the part of the object actually being attended to by the other eye, that part of the object will briefly vanish. If the object is the face of a cat, and the moving hand passes across the appropriate portions of it, only the smile will not be briefly obscured. Hence the allusion to Lewis Carroll's famous animal. In this fascinating experiment, the hand's movement momentarily captures the brain's attention, apparently because motion is accorded the higher priority. The missing parts of the image rapidly reappear, once the distraction is gone. I have timed this restoration to occur within as little as a couple of seconds.

The question naturally arises as to the nature of the neuronal interactions that underlie binocular rivalry. The phenomenon has yet to be explained, though I shall be putting forward a theory of my own in the next chapter. Meanwhile, let us consider the more limited issue of how the system gives priority to motion. It seems reasonable to assume that inhibition must play a role. Perhaps the inhibition is of the *en passant* (or *spoiler*) type that I described in chapter 4. The neuronal pathway that carries the signals for motion detection might dispatch axon collaterals toward the pathway that handles the signals for stationary objects, and it might inhibit the latter through the mediating agency of inhibitory interneurons.

There is some anatomical evidence, admittedly tentative, that could endorse this idea. The point is that investigations of the cat's visual system have revealed the presence of two separate pathways which have been designated the X and Y sub-systems.[40] Both X and Y pathways project to the primary visual area, which seems to be mostly concerned with the stationary attributes of an observed object, whereas only the Y pathway projects to the secondary visual area, which apparently lies on the main motion-detecting route. Now the spatial response of an X-route neuron displays high resolution, whereas it is temporally sluggish; it is well suited to stationary objects. Conversely, Y-route neurons show good temporal resolution but poor spatial discrimination; they are ideal for the detection of motion. The anatomical scheme just described therefore makes good sense, apart from that fact that the primary visual area also has Y-type neurons. They would seem to be unnecessary and misplaced. What are they doing in the primary visual area? Could it be that their mission is *en passant* inhibition? Are they there as potential spoilers, waiting to give motion priority if it occurs?

Much painstaking work will be required to settle that particular issue, but if such *en passant* inhibition does play a role, it would have important repercussions for theories of attention. In that binaural set-up employed by Broadbent, for example, it could explain how first one ear and then the other is attended to. To explain breakthrough, however, we would have to postulate that the inhibition can fail if the competing signals are too similar. Just how this would happen is not clear, but it might involve those reverse projections that were discussed at such length in the preceding chapter. If this is the case, it would be an example of a higher cognitive process influencing the early stages of signal processing. As we have seen earlier, there is independent evidence for this sort of thing. Perhaps it is the occasional intervention of such a top-down mechanism which has made it difficult for Broadbent and his peers to settle the early selection *vis-à-vis* late selection issue.

The introduction of *en passant* inhibition into the story raises another interesting possibility. It could even force us to adopt a different attitude toward the sort of hierarchical cortical diagram that was given prominence in the preceding chapter. We were impressed by the welter of inter-area routes, many but not all of which are reciprocal. Could it be that a sizable fraction of these routes arise to serve the needs of *en passant* inhibition? If that is so, it could mean that the hierarchy actual accomplishes for *groups* of neurons what McCulloch and Pitts tried to attribute to single neurons. Perhaps there really is a sort of logic operating in the cortical hierarchy, but one which permits different cortical *areas* to variously compete or join forces. Being based on large groups of neurons, such a scheme would not be subject to the catastrophic vulnerability that so blighted the McCulloch–Pitts single-neuron model.

I mentioned, in connection with the Cheshire Cat experiment, that restoration of the disrupted image takes a couple of seconds. This puts that process on the outer reaches of our main interest in this chapter; we are considering what happens in the first half second. However, the work of Broadbent suggests that other *mutually* inhibiting processes (as opposed to *en passant* inhibitory processes) may well operate around the half-second time scale. We saw that his experiment involved first listening to a train of sounds impinging upon one ear, and then switching attention to a different stream arriving at the other ear. We must now enquire as to what mechanism permits identification of the sounds being attended to by a given processing route. In the first section of this chapter, we were identifying visual events that apparently occupy as little as 18 milliseconds, but nothing was said about the process which conspires to produce cognition from such brevity. When discussing the work of Shibuya and Bundesen in that section, I concluded that a winner-take-all mechanism could govern things in the visual domain, because the time scale seems to fit this possibility. Could a winner-take-all strategy also apply to hearing? Before attempting to answer this question, let us check whether that brief time scale is also relevant to hearing.

Spoken words are composed of phonemes. At what rate can the ear intelligibly receive those atoms of articulation? The highest speed recorded in a public speech was achieved by the late President John F. Kennedy,[41] in December 1961. In one particularly rapid burst, he reached a rate of 327 words per minute. This corresponds to just under 200 milliseconds per word. Allowing an average of two syllables per word, this gives just under 100 milliseconds per syllable. If we further allow for an average of two phonemes per syllable, we arrive at the final figure of just under 50 milliseconds per phoneme. Although this is slightly longer than the elementary time that emerged from the visual studies by Shibuya and Bundesen, it is nevertheless similar to the value we have encountered earlier for those switches in visual attention that do not involve eye movements. It does seem possible, therefore, that the underlying mechanisms rely on similar neuronal processes. Just what these are, and whether they do indeed involve the winner-take-all strategy, is not yet known. But there is increasing evidence that an important role is played by the *pulvinar*. In order to appreciate how this sub-cortical structure comes into the story, we should first consider some important studies of eye movements carried out during the past few years.

Eye movements are an important exception to the fact that attentive processes are usually hidden from us. They are necessitated by that lack of uniformity in the retina which gives priority to the centrally located fovea, with its high acuity. The oculomotor system swivels the eyes in their sockets, so as to point the fovea in the direction of an item of potential interest. As we have already noted, the swivelling is not smooth and continuous. It proceeds in the jumps known as saccades, which occupy a mere 50 milliseconds. It is between such saccades that the system inspects the details that can be encompassed within the foveal region. Alfred Yarbus established that these momentary fixations focus predominantly on the most salient features of an object.[42] It is worth emphasizing, however, that attention can also be paid to an object not lying in the direction of gaze. We humans can master such surreptitious attention, concentrating on things seen out of the corner of the eye. Monkeys can be trained to do the same thing, the incentive to remain fixated on a particular spot being a suitably edible reward.

Robert Wurtz, Michael Goldberg and David Lee Robinson have studied the responses of neurons in various parts of the cortex, and also in the superior colliculus, in monkeys that had been trained to fixate.[43] It is worth recalling that the superior colliculus controls the direction of gaze, and that it too receives input from the retina, even though the bulk of the retinal signals are passed to the cortex via the thalamus.[m] In addition, we should recall that the receptive field of a neuron anywhere in the visual system is simply that portion of the total visual field to which that neuron can respond. Receptive fields for the foveal region are typically four degrees, whereas extra-foveal regions may give rise to receptive fields four or five times larger than this. Wurtz and his colleagues made a serendipitous discovery. They were studying the response to a spot of light lying at an angle to the direction of gaze, by probing the activity of a particular extra-foveal neuron. Suddenly, their monkey inadvertently released its fixation and glanced at the spot, about 200 milliseconds after it had appeared on the screen. The investigators observed a surge of activity in the neuron they were monitoring, starting a mere 50 milliseconds after the spot had appeared. Subsequently establishing that there was no such response if the eye movement was made in darkness, Wurtz and his colleagues concluded that this activity guided the swivelling of the eyes, the location of the probed neuron relative to the fovea defining the direction of movement. The surge in activity was clearly the precursor to a saccade. Conversely, as the investigators later established, a response is not seen in the appropriate collicular neurons when an extra-foveal spot is merely being attended to by the monkey, without subsequent shift in gaze.

When Wurtz and his colleagues made a similar investigation of the primary visual cortex, they obtained results which differed from those seen with the superior colliculus, in virtually every respect indeed. A slight rise in activity of an extra-gaze neuron could be detected, admittedly, but this did not appear to define the direction of a saccade. It seemed that the increased activity was related not to attention but rather to the general state of awareness. The conclusion was that the primary visual cortex and the superior colliculus have widely different significances for the visual system. The response of that region of the cortex is highly discriminative with respect to the input, but its resultant activity pattern is not directly related to any immediate reaction by the oculomotor system; the primary visual cortex appears to be oblivious of significance. The response of the superior colliculus is crude, on the other hand, but it is eminently purposeful; for this brain structure, significance is of the essence.

Wurtz and his colleagues went on to show that there are two other cortical regions which behave more like the superior colliculus than the

m / The situation is just the opposite in more primitive species, such as frogs and fish, the main target of retinal output being the *optic tectum*, which is the counterpart of the superior colliculus.

primary visual area. These regions are known as the *frontal eye field*, which lies in the frontal cortex, and the *posterior parietal cortex*. The former had been shown by David Ferrier to provoke eye movements when it is directly stimulated, while damage of the latter, in one cerebral hemisphere, is known to lead to neglect of parts of the other side of the body.[44] A person who has sustained a lesion in the posterior parietal cortex may leave one side of the hair uncombed, and, if the person is male, one side of the face unshaven. Vernon Mountcastle showed that cells in this cortical area display a rise in activity immediately prior to an eye movement, just as Wurtz and his colleagues reported for the superior colliculus.[45] Working with Catherine Bushnell and Gregory Stanton, Wurtz, Golberg and Robinson were able to show that there is nevertheless a difference from the collicular response.[46] The appropriate neurons in the posterior parietal cortex begin to react even when an object in their receptive field is merely being attended to. The corresponding neurons in the frontal eye field and the superior colliculus come into play only some tens of milliseconds later when a saccade is imminent.

We should pause here and briefly contemplate the broader significance of what this work was establishing. One of the key messages of the preceding chapter was that the organism can exploit the differential routing of the various characteristics of sensory input. We likened this apportioning of attributes to what happens because of the parallel distributed processing (PDP) in a multi-layer artificial neural network of the feed-forward type, with different aspects of the input being handled by different pseudoneurons in the hidden layer, even though the analogy with such a network must perforce be a loose one. And it was also emphasized that such PDP, in the case of the cortex, does not mean that the whole sensory message can be regarded as a conglomerate of independent parts. Far from it, indeed, for the brain is so constructed that this wholeness is carefully protected by the binding mechanism, which we have also discussed. Differential routing certainly does occur in the cortex, however, and we have considered the good example seen in the visual system, with the temporal region handling the *what* of visual input while the parietal region looks after the *where* and *when*. Now let me come to the point of this little digression. If the designer of the brain wished to tap off from the cortex information that would be useful for attention, which region would be the favoured choice? Bearing in mind that the items being attended to might not be stationary, the parietal region is seen to be a particularly strong candidate. Viewed in this light, the results described in the preceding paragraphs have a satisfying logic about them.

We have not yet done with the parietal region, and its significance for attention, and there are two other parts of the brain that we will also have to take a close look at before the end of this section, namely the *nucleus reticularis thalami* (nRt), and the *brainstem reticular formation* (BRF). But before returning to our main thread we must clear up another piece of unfinished business. It concerns those 50 milliseconds that are consumed by the eye movement of a saccade. Had it not been for what was described in the first section of this chapter, we might have been inclined to look upon this as a period too brief to be worthy of further consideration. But that visually significant interval of 18 milliseconds, as measured by Shibuya and Bundesen, suggests that we might overlook the period of movement at our peril. Indeed we would! As we are about to discover, what happens in those 50 milliseconds is a piece of biological orchestration that is of the utmost importance. It is also possessed of a compelling beauty. And the conductor of the piece is the *pulvinar*. Let us start by taking a closer look at this centrally located component of the brain.

The pulvinar is located in the thalamus, partially sandwiched between the lateral geniculate nucleus (which serves vision, of course) and the medial geniculate nucleus (which plays an analogous role for audition). Its size has been found to be an excellent indicator of the evolutionary rank of a species; the more advanced the animal, the larger is the pulvinar with respect to the rest of the brain. Although we do not need to go into all the details, it should be noted that it is a composite structure, several parts of which have been found to comprise *retinotopic maps*. The latter term refers to the spatial arrangement of the receptive fields, which are congruent with the arrangement present in the retina. Several of the pulvinar components receive strong projections from the superior colliculus and various early visual areas of the cortex, and they dispatch reverse projections to those same regions. It is particularly noteworthy, in the present context, that the cortical areas involved in this exchange of signals include the posterior parietal area and the parieto-occipital area (both of which are on that *where–when route*), and the middle temporal area (on the *what route*). Given the posterior parietal area's involvement in spatial attention, which we have just been discussing, its inclusion in this list might make us respond with a *Aha! The plot thickens!* It certainly does.

Earlier in this long section (and we are far from finished yet), the term *filter* cropped up in connection with the pioneering work of Broadbent. Now is the time for it to make an encore, for it transpires that vision-related pulvinar neurons carry out a filtering which discriminates between movement of the retinal image caused by movement in the external visual field and movement of the retinal image caused by movement of the eyes themselves.[n] A sizeable fraction of these neurons respond to movements in the visual field during fixation, but their activity falls sharply during a saccade. As David Lee Robinson and Steven Petersen have noted, it is as if these neurons do not see, during eye movement, an object which they see readily during periods of fixation.[47] Moreover, these pulvinar neurons receive projections from the superior colliculus, and Robinson and Petersen have postulated that the latter modulates the activity of the former. The *visual characteristics* of these pulvinar neurons are nevertheless dictated by what they receive from the early visual areas rather than by what comes from the colliculus.

We can now begin to glimpse the reason why the wiring of the brain

n / The reader may be familiar with a simple demonstration of this difference. Stationary objects observed as the head is rotated are perceived as being motionless. But if the corner of the eye is gently poked, the entire visual field appears to move.

has to be so complicated. The mechanism we have just been discussing involves the superior colliculus, the early visual areas and the pulvinar, with each of them having responsibility for a different aspect of the overall process. But what a complex process it is. In fact, there is evidence that the pulvinar can accomplish things that are still more subtle. When pulvinar neurons are monitored immediately after an eye movement, it is found that their activity level takes a brief amount of time to establish itself again. There is even evidence of certain pulvinar neurons being able to filter out extraneous information when there is movement in the background, during a period of fixation. This would heighten the salience of the object being attended to. Finally, experimental results are accumulating which suggest that other parts of the pulvinar can achieve analogous filtering for auditory and tactile signals. The pulvinar appears to be the brain's conductor *par excellence*.

I am going to add my own piece of conjecture to this story, and attempt to go beyond what has already been established experimentally. We have seen that the pulvinar essentially switches vision off during the 50 milliseconds of eyeball movement, and we also know that the eyes stay stationary for about 250 milliseconds thereafter, before the next saccade. The question arises as to what happens to the visual snapshot gained during the preceding 250 milliseconds, when the eyes shift to a new target. It cannot already have disappeared from the system, because very short memory lasts at least a couple of seconds or so. I wish to suggest that the system not only retains that snapshot but that it can accurately fit it to the new snapshot that is being acquired. I believe that this could happen because the superior colliculus is able to keep track of the direction and magnitude of each eye movement. If this guess is correct, it would mean that the visual system is able to build up a composite picture, jigsaw fashion.[o] Ultimately, of course, the individual snapshots of this composite array will fade, and they will do so in the order in which they have been acquired. The visual system will thus be in a continuous race against time; it will be as if the motifs on the individual pieces of the jigsaw puzzle gradually fade away.[p] This is no doubt the reason why the eye returns again and again to the same areas in the visual field. Wurtz and his colleagues were easily able to establish this, and they noted that features of particular interest were visited most often by the centre of gaze. An emphatically schematic illustration of this principle is shown in figure 8.13.

o / The mechanism also bears a certain resemblance to the technique of *aperture synthesis* in radio astronomy, in which a number of well-spaced and small antennae are electronically linked up in such a way that they become equivalent to a single much larger radio telescope.

p / The use of the term *jigsaw* should not be taken to imply any significance for the concept of interlocking, in this context; one should rather think of the simplest type of jigsaw puzzle, composed of a few simple squares, which are used for very young children. In the balance of this book, I will prefer the more technical term *aperture synthesis*, sometimes more generally applying it to situations that are not limited to the visual domain.

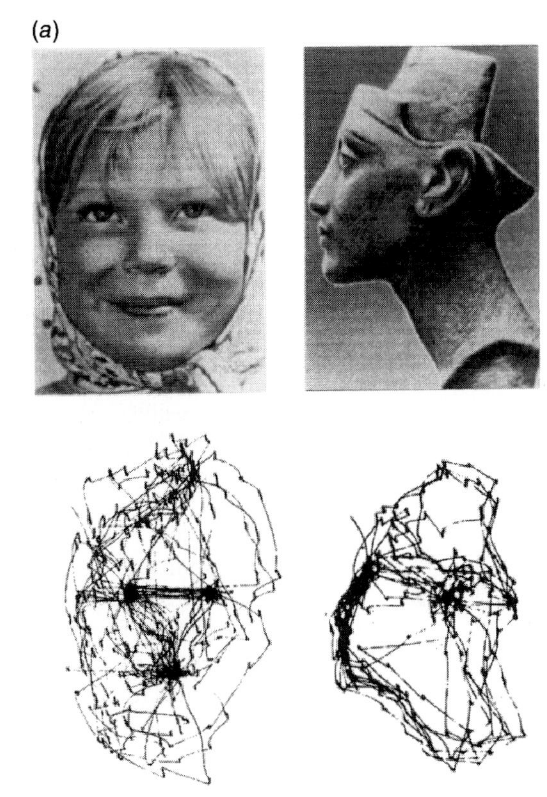

Figure 8.13
When the eyes inspect an object, they move position about four times every second, the small movements being known as saccades. Their scanning of the object is not haphazard. The eyes tend to spend proportionately more time on features of greater potential interest. In the examples shown in part (*a*) of this illustration, the straight line segments trace the route taken by the direction of gaze. There is a mechanism which suppresses the formation of a visual image during the saccadic movements, and evidence suggests that the system pieces the various quarter-second snapshots together, jigsaw fashion. This process is referred to as aperture synthesis in the present book. Because each snapshot starts to fade from short-term memory the moment the eyes have moved to a different position, the visual system has to cope with fleeting information. Part (*b*) of the illustration shows, in a highly schematic fashion, such jigsawing when the fading away is slower (left) and faster (right). As discussed in chapter 10, it is possible that a more general form of too-rapid fading away underlies the truncated working memory believed to underlie some forms of mental retardation. (Part (*a*) after Yarbus, A. L. (1967). *Eye Movements and Vision* (New York: Plenum Press).)

An analogous mechanism could operate for the other senses. Let us briefly consider hearing, for example. Although that sense has not been treated to the same degree of detail in this book, we can quickly lay an adequate foundation. Just as there are no lights flashing around inside the head, so is that inner sanctum devoid of all sound. In both cases, the information has been converted to trains of action potentials. In vision, it is the position and shape of what is being perceived (not to mention colour and movement) that are of the essence, and we have seen that these are analysed through the receptive fields of the relevant cortical neurons. In audition, the important attribute is frequency content. But the system converts frequency to position, in what is known as the *tonotopic map*, and this too can be analysed through the receptive fields of the appropriate cortical neurons. This means that the requisite machinery for *aperture synthesis*, as I have decided to call it, also exists in hearing. Given that the pulvinar is also implicated in the tactile sense, I will guess that the same sort of thing can even apply to touch.

It must be emphasized that this is mere speculation on my part, but the various components required for such a consolidation process have been shown to exist, so the idea is not particularly far-fetched. If this picture is a reliable one, it could provide a mechanism for the bandwidth limitation in Broadbent's filter process. We recall that the three pairs of digits fed simultaneously to the two ears, in his classic experiment, were attended to as two sequential triplets; the three digits impinging upon one ear were stored first, and the three reaching the other ear were stored thereafter. If the mechanism just described is correct, it would provide the means for storing each of the triplets. And because the span of very short memory is naturally limited, we can readily see that the number of pairs of digits fed simultaneously to the two ears could not be made arbitrarily long. The two trains of three digits could not be replaced by two trains of fourteen (say) digits – not if the subject was to have any chance of remembering them all.

There is circumstantial evidence which supports the aperture synthesis idea, at least in respect of eye movements, which are fundamental to an important aspect of perception. The synthesis mechanism would require that the system maintains a record of the most recent saccades, and automatically shifts the image fragments to their correct positions in the overall picture. There are several independent studies, including those by David Sparks and others by John Schlag and Madeleine Schlag-Rey,[48] which strongly suggest that the system does indeed possess this capability. And it is important to note how vital the concept of position *is* to perception, for it is also implicated in the observation of shape. As Jacques Paillard[49] reminds us: '... an object contains its own network of spatial relationships. Thus each channel has to compute a topographic representation of spatial features but differs in its basic encoding modes. One concerns the space of shape and the other the space of locality, thus reflecting two different processing modes'.

The superior colliculus plays a central role in eye movements, of course, and Daniel Guitton and his colleagues[50] have suggested that it functions like a *virtual arena*. It is located on a feedback loop, which returns signals to the brainstem. The colliculus itself comprises several neuronal layers, the superficial ones displaying purely visual responses while many cells in the deeper layers, which are also organized into a map, discharge high-frequency bursts of nerve impulses immediately before the initiation of a saccade. They conclude that the superior colliculus provides the brainstem premotor circuitry with a topographically coded command specifying initial eye motor error. Guitton and his colleagues note that the locus of maximum neuronal activity passes across the surface of the colliculus as the gaze shift runs its course; whence the concept of the virtual arena.

We could go on to further fascinating speculation along this road. Does aperture synthesis also operate at the higher levels of cortical function, for example, and are there conditions which limit the number of pieces that can fleetingly be held in very short memory? I will be returning to such issues in chapter 10, when we consider the depth of reasoning that any given brain is capable of. For the time being, therefore, let us return to those other brain components that we put on temporary hold. Let us get back to the parietal lobe, and consider what happens when it has been injured. Michael Posner, John Walker, Frances Friedrich and Robert Rafal have emphasized that the cognitive act of shifting attention from one place in the visual field to another can be accomplished covertly, that is to say without movement of the eyes.[51] They have rationalized such shifts as comprising three distinct stages: first there is *disengagement of attention from its current focus*, then there is *transfer to the new target*, and finally there is *engagement of the new target*. From their studies of human patients, they were able to conclude that damage to the parietal region causes a defect in the disengagement stage, if the target happens to lie to the side of the head opposite to that which has the lesion. The researchers also studied patients with damage to other parts of the brain, including the frontal and temporal lobes, and also the midbrain. None of these other damage sites were found to cause difficulties with disengagement, even though it is known that they all receive projections from the parietal region.

In this respect, other projections from the posterior parietal cortex may be rather more significant. As Vernon Mountcastle has stressed, there are particularly strong sets of connections to the superior colliculus and the BRF (brainstem reticular formation).[52] He has described evidence that these components somehow manage to gauge errors in the momentary alignment of the eyes, so that this can be reduced to zero. It was not possible to construct details of the mechanism from the observations, but there are a number of other pieces of work which, if taken together, might provide a clue. I will mention them in the order which best reveals the anatomical loop made up of the structures they invoke. Let us start by tracing the loop itself. The first link is the above-mentioned path from the posterior parietal cortex to the BRF. Secondly, there is the influence that the BRF exercises on the nRt (nucleus reticularis thalami). Thirdly, there is the fact that the nRt appears to act as a gate between the thalamus and the cerebral cortex. The projections from the thalamus to the cortex do not include direct lines to the posterior parietal cortex, which would close the loop. Instead, the line goes from the thalamus (more specifically, the lateral geniculate nucleus, of course) to the early visual areas. These, in turn, feed signals up the hierarchy, and it is in this way that the loop is ultimately closed, when the line reaches the posterior parietal cortex.

The case for a closed loop thus appears to be a good one, but what would be its significance? We can gauge this by considering each of the links in turn. Let us start with the BRF. Wolf Singer has reviewed the anatomical evidence, and he suggests that the BRF controls the activity level of the nRt, through a process of disinhibition.[53] As was discussed in chapter 4, this mechanism appears to be capable of finer control than that of direct excitation. We must then ask what the nRt achieves for the system. Francis Crick has suggested that it functions in conjunction with the thalamus to intensify a particularly active thalamic input to the cortex, thereby providing a *spotlight of attention*.[54] We recall that the existence of such a spotlight was postulated by Anne Treisman. Let us consider how its structure would make it suitable for such a task. The nRt surrounds the thalamus in much the same way as the fingers of one hand can be made to surround the clenched fist of the other. Nerve fibres, that is to say axons, stemming from the thalamus must therefore penetrate past the neurons of the surrounding nucleus, on their way to the cortex.

Now it turns out that these nRt neurons are all inhibitory (the neurotransmitter being GABA), so they oppose the activity of the thalamic neurons via feedback inhibition. Although Crick invoked an observed peculiarity of the thalamic neurons which has not stood the test of time, a spotlight mechanism can still be salvaged from the observed anatomy. The point is that the inhibitory neurons of the nRt will subject the oncoming signals to a winner-take-all process, and this will automatically provide a salience filter; it will function like a selective gate. The beauty of the system, if this scenario is correct, lies in the circular nature of the control. The BRF controls the nRt, which controls, as would a gate, the information passing from the thalamus to the cortex, some of which, no doubt in a modified form, helps to control the BRF, via the posterior parietal cortex. The key phrase here is *in a modified form*, because if it were not for that possibility of modification, the closed loop would seem to have little purpose. The modification in question will be the result of cortical processes, which is to say, ultimately, of thought. Some time will pass, surely, before all the details of this picture are fully worked out. When they have been, we will be much the wiser regarding the mechanism that facilitates a search when we know what we are looking for, and we will probably be closer to an understanding of what enables us to pick out a familiar voice through the din of a cocktail party.

Recent discoveries regarding the anatomical structure of the nRt have removed what was earlier seen as a difficulty for Crick's ideas. His theory required that the nRt be composed of well-defined topographical maps, one for each sensory modality, but evidence of such maps was not available at the time his original suggestion was published. It has now been shown that these maps do exist, however. The pathways connecting both the thalamus and the cortex to the nRt establish distinct reticular sectors that relate to functionally distinct parts of the thalamo-cortical pathway.[q] The Crickian view has thus been vindicated.[55]

A relevant limitation in colour vision

The idea of a limitation in the signal-handling capacity of the brain was a central theme in the treatment of attention given in the preceding section. But although a good case was made for such a restriction, direct evidence was lacking. In the case of colour vision, the evidence for a restriction is quite compelling, and it is thus appropriate to make a small digression and consider the facts. I must emphasize, however, that I am not going to go into the philosophical question of what it means to perceive colour. That issue belongs to the broader realm of consciousness, which will be occupying us for most of the following chapter.

There are three types of wavelength-sensitive (cone) receptor in the normal human retina, and they display peak responses to what we know as the colours red (R), green (G) and blue (B). It is not the case that these three different types of cone give rise to three separate sets of nerve fibres running from the retina to the thalamus, and thence on to the cortex. Far from it, because there is already interaction in the plane of the retina, between the various types of cone. This gives rise to signals that derive from additions and subtractions of the individual cone responses. Thus neural pathways have been found which signal R–G, G–R, (R+G)–B, and B–(R+G). The sum R+G in the latter two combinations is equivalent to the colour we know as yellow. This is probably the reason why that colour is the fourth psychological primary, even though we do not have receptors for it.

Now how does the visual system exploit the possibilities inherent in these combinations to derive the maximum efficiency from the available resources? It turns out that it uses an approach analogous to that devised around 1950 by television engineers, when they were attempting to produce the first colour receivers. These designers naturally wanted to restrict the amount of information that had to be transmitted, and there was also the requirement that owners of black-and-white sets should nevertheless be able to receive colour programmes. The simplest approach would have been to record scenes through three different filters, namely R, G and B, and arrange for them to be recombined in the colour receiver. But this was found to give a black-and-white picture that was too biased toward what was being detected by the G channel. Another drawback was the trebling of the amount of information that had to be transmitted.

It was then realized that a better alternative would be to let one carrier convey *luminance* information, since this is all the black-and-white sets had use for.[56] They could then simply ignore whatever else was being

q / We recall that the word *reticular* refers to a net-like structure. These new results indicate that the structure stems from the distinct anatomical divisions.

transmitted in the composite signals. The question then arose, however, as to how colour and luminance were to be separated in the colour receivers. The solution lay in determining the relative sizes of the contributions to luminance made by the three primaries, and the upshot was an equation that read $L=0.3R+0.59G+0.11B$, where the three letters now stand for the actual *magnitudes* of the respective signals. The three numbers (referred to as coefficients, or weightings), reflect the contributions to luminance made by the corresponding colours. The engineers then derived the two difference equations $R-L=0.7R-0.59G-0.11B$ and $B-L=-0.3R-0.59G+0.89B$, with L representing luminance. These latter two equations are said to convey *chrominance* information, and it turns out that less bandwidth has to be devoted to its transmission, because human visual acuity is less for colour than it is for luminance. It can be demonstrated that the employment of one luminance signal and two chrominance signals leads to optimal use of transmission resources.

The fascinating thing is that the visual system apparently avails itself of the same possibilities inherent in the luminance–chrominance strategy, thereby making the most of its limited capacity for information transmission.[57] This has been a protracted digression, but it has served to demonstrate that bandwidth limitation is both real and important in the visual system. It does not seem like a dangerous generalization to assume that such a limitation would apply to *all* parts of the nervous system. Nor does it seem excessive to assume that the entire system has evolved so as to make the most out of the available resources.

Bandwidth, breakthrough and working memory

Despite the clear evidence of limitations on signal-handling capacity, as described in the preceding two sections, this chapter has not yet dealt with the neuronal nature of the bottleneck. The impression could easily be gained that this is essentially dictated by the span of short-term memory, the determinants of that attribute being uniform throughout the system. Results that have been accumulating during recent years suggest that this view is too simple.

As has so often been the case with issues dealt with in this book, evidence supporting this change of paradigm has come from observations on people with brain injuries. In chapter 5, we learned of patients suffering from the classic amnesic syndrome, in which short-term memory is unimpaired despite an inability to lay down new long-term memories. The situation became more complicated when Tim Shallice and Elizabeth Warrington identified a second class of patients whose trouble appeared to be just the opposite; they possessed normal long-term memory but had a short-term memory span limited to one or two items.[58] This observation was difficult to reconcile with the accepted picture of a short-term store acting as working memory, and thereby mediating long-term accumulation. Patients with this deficit are relatively rare, so Alan Baddeley and his colleagues tested normal subjects with a dual-task technique in which they were required to remember digits while performing other tasks. Investigation of a subject's ability to retain a string of digits is referred to as a *digit-span test*, and the investigators first determined the span of each subject. Subsequent observations revealed that the subjects could still reason and learn even when their full digit-span was being required of them.[59] This result, and related findings, led to abandonment of the idea that a single short-term memory serves as the working memory.

It was replaced by a tripartite scheme comprising a *central executive* and two subsidiary *slave systems*. The executive was seen as including an *attentional controller*, while the two slave components were envisaged as an *articulatory or phonological loop* and a *visuospatial sketch pad*. The former of these subsidiary components was assumed to maintain speech-based information, while the latter was looked upon as being responsible for setting up and manipulating visuospatial imagery. Figure 8.14 was inspired by Baddeley's schematic representation of this system.[60] The various parts of this composite have been investigated by different, and complementary, techniques. The central executive is probed by what is known as the *psychometric approach*, which investigates the extent to which performance on working memory tasks can predict individual differences in the relevant cognitive skills. The slave systems, on the other hand, are best examined by the *neuropsychological approach* and the *dual-task approach*, which collectively demonstrate the separability of the memory systems which serve vision and audition.

A typical example of the psychometric approach is the test in which subjects must retain the last word from each of a series of sentences. For example, a person being presented with *The sailor sold the parrot; The vicar opened the book; The chicken crossed the road*, should respond with *parrot, book, road*. The *working memory span* is deemed to have been reached when the number of sentences is so large that the subject can no longer recall all the terminal words. If subjects are subsequently tested on standard reasoning tasks, like those used to assess intelligence, it is found that working memory capacity shows a strong correlation with reasoning skill.[61] The two things are not actually equivalent, however, because reasoning shows the greater reliance on prior knowledge, whereas working memory span is more dependent on sheer speed of processing.

It has been found that *the key function of working memory is coordination*

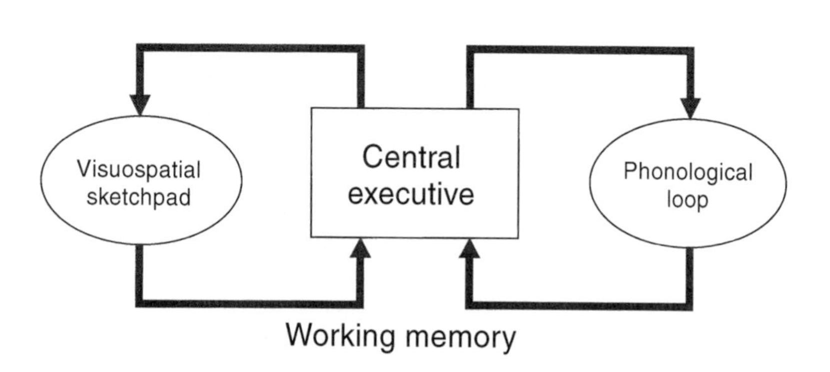

Figure 8.14
According to the model of working memory advocated by Alan Baddeley and his colleagues, there are three components, as indicated in this highly schematic diagram. The central executive, which is an attention-controlling system, might be located in the prefrontal cortex. It marshals reverberatory processes in the two slave systems, namely the visuospatial sketchpad and the phonological loop. These might each be a series of successive cortical areas, their interactions being mediated by the forward and reverse projections that are known to be a common feature of the cortex.

of resources.[62] Strong endorsement of this proposition comes from observations on patients with Alzheimer's disease. When required to simultaneously perform two tasks, one visual and one verbal, these people manage very badly, despite the fact that they can handle either task individually just as well as their healthy counterparts. Moreover, although the patients show little deterioration in their ability to handle those individual tasks, as the disease progresses, their facility for combining tasks falls away markedly; it is clearly the ability to coordinate their mental resources which is suffering.[63]

Our understanding of the peripheral slave systems has progressed even more rapidly, and the dual-task approach has confirmed the separate nature of visuospatial imagery and verbal repetition. Imagery is interfered with if the subject is concurrently required to perform another visuospatial task, whereas no such clash occurs if the secondary task is a verbal one. A similar lack of interference is observed if the primary task is verbal and the intended distractor is visuospatial. Moreover, there is further separation between positional and pattern processing in the visual domain, and the corresponding neural mechanisms have been causally linked to the occipital and parietal lobes, respectively. Such separation is endorsed by neuropsychological data. There are patients who experience difficulties in recalling the shape of a spaniel's ears or the colour of a pumpkin, and yet can still easily navigate their way through a labyrinth. Conversely, there is another type of patient whose deficits are just the opposite.[64]

The phonological loop is the easiest component of working memory to investigate, and it is consequently that part which has been studied most extensively. It appears to comprise two parts: a *phonological store*, capable of retaining acoustic information for one or two seconds, and an *articulatory control mechanism*. The latter plays a mediatory role, and it is this part of the system that permits us to register visual information in the phonological store, by sub-vocalization. This model of the loop explains observations such as the *acoustic similarity effect*, the *irrelevant speech effect*, the *word-length effect*, and *articulatory suppression*. Let us briefly consider these effects, in that order. We find it easier to recall a list of dissimilar words, such as *pit*, *day*, *cow*, *pen*, *rig*, than a sequence of sound-alikes such as *man*, *cap*, *can*, *map*, *mad*. Baddeley attributes poorer retention of the latter group to the fact that similar items have fewer distinguishing cues, and are therefore more easily forgotten. I will suggest an alternative explanation shortly, but let us note here that similarity of *meaning* does not produce this effect; this sub-system appears not to reflect *semantic* coding.

The irrelevant speech effect refers to impairment of recall through interference of irrelevant spoken material. As we noted earlier in this chapter, the pioneer in this type of study was Anne Treisman. Here too, the semantic character of the material is not important. The word-length effect was discovered when it was found that memory span for words is inversely related to the spoken duration of the words. Finally, articulatory suppression occurs when subjects are required to include a vocalized gratuitous sound, such as the word *the*, in a word sequence that is being sub-vocally rehearsed. This effect is also found to disrupt the registering of visually presented material in the phonological store.

The consensus view of the phonological store is that it serves as a backup system for comprehension of speech. The definitive tests leading to this conclusion have involved repetition of non-words, the subjects usually being children, so that their subsequent development could be gauged. It is found that children who encounter difficulty in repeating non-words increase their vocabularies at a much slower rate than those with a facility for recalling gibberish. Short-term phonological memory is apparently crucial to the acquisition of new words.

Returning to the system in its entirety, working memory is best regarded as a mechanism that permits performance of complex cognitive tasks through its ability to temporarily store information related to the various sensory modalities, particularly those of vision and audition. In Alan Baddeley's view, working memory stands at the crossroads between memory, attention and perception. As such, its role in cognition could hardly be exaggerated, and any damage to this system would obviously imply serious disadvantage for the sufferer. In chapter 10, we will be contemplating the plight of people who appear to have such a deficit.

What bearing do these findings on working memory have on some of the things discussed earlier in this chapter? For a start, they are obviously related to the bandwidth concept embodied in Broadbent's filter theory. They have served to fill in more of the detail regarding the various components of the system; they have shown that we must differentiate between the active executive function and the passive reverberatory mechanisms, that is to say the visuospatial sketch pad and the phonological loop. The question remains, however, as to where these various components are actually located in the brain. It seems likely that the reverberatory loops are simply the relevant regions of the cortex that are interlinked by those forward and reverse projections, which we contemplated when discussing cortical mechanics in the first section of this chapter. And as for the central executive, which must clearly be accessible to signals emanating from different sensory modalities, we should recall the work of Patricia Goldman-Rakic and her colleagues which was cited in chapter 5.

They made a very strong case for what they were calling working memory being located in the prefrontal cortex, and we learned in that chapter that this region has connections with many other parts of the brain; it is ideally connected to play a marshalling role, organizing and coordinating information from various other brain centres. Armed with what the work of Baddeley, his colleagues, and his contemporaries has revealed, it is tempting to reappraise that other evidence and conclude that the prefrontal cortex specifically provides only the central executive, rather than serving as the entire working memory.

How does it go about its job of steering the traffic, at the neuronal level? We do not know. The fact is that we know very little about such traffic steering in *any* part of the cortex. But I believe there may be a hint at this mechanism in two different aspects of the work of Anne Treisman and her colleagues: breakthrough, and the fact that it is difficult to detect an O amongst Qs, but easy to find a Q amongst Os. Breakthrough, I feel, is suggestive of inhibition that is momentarily being overcome. And we have already noted that it probably underlies the acoustic similarity effect. Moreover, we saw earlier that the results of the O–Q study indicate that the cortex reacts to the presence of things, rather than to their absence; the cortex works through neuronal activations, not through a lack of them.

Let us assume that the circle and the short straight line which constitute a Q, in the visual domain, can be compared with the phonemes that constitute words in the auditory domain. When the cortical region corresponding to a given phoneme is activated, it has presumably become the momentary winner in a winner-take-all process. In chapter 4, I suggested that one inhibitory factor in that type of mechanism may be supplied by *feed-forward inhibition*, and I likened the process to one in which a pattern of lights is trying to penetrate a fog; only the brightest parts of the pattern succeed in getting through. Having been thus excited, the neurons associated with a particular phoneme will maintain their activity only for a limited period of time. Trouble for the system would surely ensue if this were not the case. What ultimately curtails the activity, however? It could simply be the leakage of charge that occurs in any electrical system. But a crisper termination of activity will be achieved if that activity is quenched by inhibition. And the natural candidate for this *second* opposing factor is *feedback inhibition*, because it will arise just where it is most needed, namely near the site which has just been activated; that is to say, at the neural correlate of the phoneme in question.

Unlike the feed-forward variety, however, feedback inhibition involves a time factor. Activation of the relevant inhibitory interneurons, and the consequent return of inhibition to the activators, takes time. That out-and-back trip traverses two sets of dendrites, as well as the related synapses. The net transit time could be at least 20 milliseconds, and complete quenching might take several times that period. We saw earlier that phonemes can follow on the heels of each other at about 50 millisecond intervals during rapid speech. This means that the terminal phoneme of the previous word might still be activated when the leading phoneme of the following word is becoming activated, and the upshot could be a false conjunction further up in the hierarchy. Perhaps this type of process underlies breakthrough in general.

This is a rather qualitative argument, admittedly, but I will be attempting to build upon it in the next chapter, when we contemplate the question of how we know something. Meanwhile, let us once again subject ourselves to that self-imposed restraint implicit in this chapter's title, and return to the trail of those first 500 milliseconds.

Dissection of sequence

A most intriguing series of experiments has been carried out in recent years by Benjamin Libet and his colleagues Bertram Feinstein, Dennis Pearl and

Elwood Wright. Their first report appeared in 1979, and subsequent papers consolidated the impression that they were revealing things of paramount importance to the understanding of brain dynamics.[65] The work continues to be the focus of great interest (and its interpretation has, for example, been questioned by Ted Honderich[66]). The subjects were human patients who were to have electrodes surgically implanted into certain cortical regions, in connection with a programme of therapy. The clinical situation offered two major advantages, which Libet and his colleagues exploited with the permission of the patients. The exposed cortical surface could be directly stimulated with suitable probes, and the effects of this provocation could be directly reported by the subjects because they were fully conscious.[r]

Direct stimulation of part of the somatosensory cortex gives the subject the impression that the corresponding part of the body's surface is being touched. It is not surprising that such bypassing of the normal signalling route should be possible; a related mechanism underlies the phantom limb effect, in which an amputee feels sensations in an arm or leg that has been removed. In the case of subjects who still had their limbs, the exciting prospect lay in being able to compare the sensations resulting from stimulation of the limb with those elicited by direct stimulation of the somatosensory cortex. Access to two alternative routes provided Libet and his colleagues with the chance of investigating the logistics of sensation; in effect, they were given the opportunity of dissecting sensory sequence. The experimental results produced some surprises.

Broadly speaking, the investigations concerned two issues: firstly, the temporal relationship between the occurrence of a stimulus and subjective awareness of the resulting sensory signals, and secondly, the initiation of a voluntary act relative to the time of conscious intention to act. Let us consider them in that order. The limb stimulation, denoted by the letter S, was a half-millisecond electrical pulse to the skin at the back of one of the hands. The cortical stimulation, referred to by the letter C, was a train of half-millisecond electrical pulses, administered at a frequency of 60 Hertz (which is equivalent to a lapse time of about 16 milliseconds between pulses) to the region of the somatosensory cortex corresponding to one of the hands. The stimulus had an intensity about fifty per cent higher than the *liminal level*.[s] Each experiment involved an S stimulus and a C stimulus. They were given at a variety of relative times, and the subject was requested to report on the perceived sequence, that is to say *left first*, *right first* or *together*.

Before considering the results of these experiments, we should note that the subjects could easily distinguish between the S-generated and C-generated sensations, when they were administered in isolation. The S pulse is experienced as a localized tap, and it can readily be detected despite its brevity.

r / It is worth repeating here that brain tissue has no pain sensors, so the subjects suffered no distress during these investigations.

s / The *liminal intensity* is the level of peak current required by repetitive pulses in a one-second train to elicit a threshold sensory report.

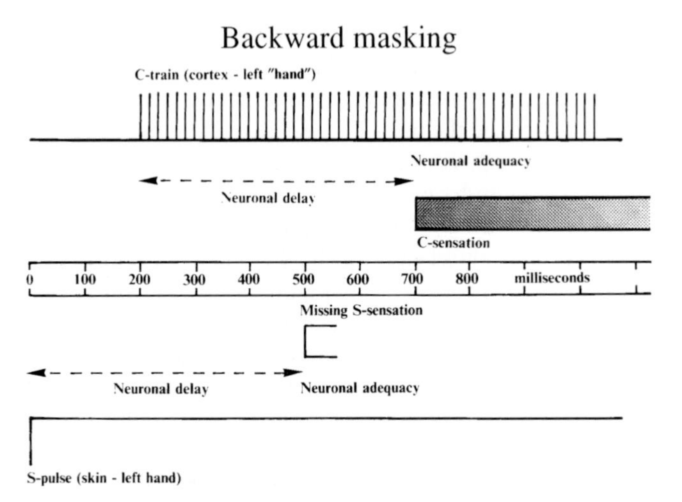

Figure 8.15
In the experiments of Benjamin Libet and his colleagues, stimulation (C) was applied directly to the somatosensory cortex while the subject was conscious. It was first established that such stimulation must be continued for at least half a second if it is to be consciously experienced, that condition being referred to as **neuronal adequacy**. A stimulus (S) was also given to the skin of one hand, either before or after the cortical stimulation, and the subject was asked which one was given first. The results were surprising. If both stimuli were applied to the same 'hand', and S occurred less than about half a second prior to C, the subject did not experience S. This is known as **backward masking**.

The C train induced a tingling or wave-like sensation, and it had to be continued for 500 milliseconds to give any sensation at all. This 500 millisecond requirement also applies if the stimulus is applied to sub-cortical cerebral pathways, for example to the relevant part of the thalamus or the *medial lemniscus*.[t] Libet and his colleagues coined the term *neuronal adequacy* to describe the point during the subsequent neural process at which the subject becomes aware of the sensation (see figure 8.15). Now let us turn to the experiments that involved both the S pulse and the C train.

If an S pulse is followed by a C train to the same 'hand', after a delay of about 200 milliseconds (or even somewhat more than that), the subject is generally not conscious of the S pulse. One would expect the subject to report having felt two stimuli, but it transpires that only one of them makes an impression. The C train is somehow able to erase or block the effects of the S pulse that preceded it, and this remarkable phenomenon is now referred to as *backward masking* (see figure 8.15). An obvious question concerns which of the words *erase* or *block* is the more appropriate. As we noted in the preceding chapter, the fastest signals travel from limb to cortex within 15–20 milliseconds, and it is they which carry the signals emanating from the S pulse. This means that they arrive a good 180 milliseconds before the onset of the C train, so the word block is *not* suitable; the masking appears to result from erasure. Because the C train starts 180 milliseconds or so after the signals from the S pulse have arrived at the cortex, its ability to erase the *effects* of the latter must be related to their duration; the indication is that these effects last more than 200 milliseconds. The effects clearly need a considerable amount of time to

t / The *medial lemniscus* consists of axons arising from the dorsal column nuclei in the spinal cord, their target being the thalamus, (and ultimately the somatosensory cortex). It lies on the principal pathway for somatosensory perception, and its axons are somatotopically organized.

make an impression on conscious awareness, presumably because they involve a chain of neural events.

A more recent experiment by Libet and his colleagues has shed further light on this fascinating result. They refer to it by the term *retroactive enhancement*.[67] The experimental protocol again called for an S pulse to precede the C train, but this time by as much as 400 milliseconds, or even more. Moreover, *two* S pulses were used rather than one, the interval between them being a standard 5 seconds. The pulses actually had the same strength, but this was not divulged to the subject. The subject's task was to report the perceived intensity of the second S pulse, relative to the first, so the possible reactions were *stronger*, *same* or *weaker*. Control runs were made first, without administration of the C train, and the averages over four subjects and many repetitions were 15% for *stronger*, 65% for *same* and 20% for *weaker*; the sameness was clearly being properly evaluated. The C train was then added, at various delays after the second S pulse, and there was a marked shift in the distribution of answers. For delays in the range 300 to 400 milliseconds, the averages were now 68% for *stronger*, 29% for *same* and a mere 3% for *weaker*. Even for delays of 500 milliseconds or slightly longer, the three percentages were 45%, 45% and 9%, respectively. As Libet and his colleagues concluded, the use of the C train enhances the perception of the second S pulse.

An explanation of this new effect is at hand if one accepts the interpretation of the backward masking experiments: the arrival at the cortex of the signals generated by that pulse is followed by neural processes which require time to fully develop, the end result of those processes being awareness. In these enhancement experiments, application of the C train apparently does something to those processes to increase their effectiveness. It is roughly analogous to giving a swing an extra push, to make it reach greater altitude at its highest point; this is much easier if the swing is already in motion. Moreover, the results of this second investigation indicate that the duration of the processes touched off by the arrival of the S-generated signals must be as long as about 500 milliseconds.

When we compare the two experiments, however, there seems to be one serious flaw. *Why is that second S pulse not backwardly masked?* Libet and his colleagues concede that the difference could arise from the use of somewhat different electrode configurations in the two experiments. The electrode in the enhancement experiment was a fine wire, one millimetre long, while the masking runs employed a metal disc ten millimetres in diameter. But it is possible that it is the addition of that (five seconds) prior S pulse which is responsible for the unexpected observation. Since the subject *knows* that there are going to be two pulses, the arrival of the signals associated with the first might initiate processes related to anticipation of the second. This sort of thing is quite familiar in experiments in which EEG traces are recorded from subjects being exposed to series of stimuli.

A further surprise awaited Libet and his colleagues when they applied the S and C stimuli to different 'hands', so as to permit greater

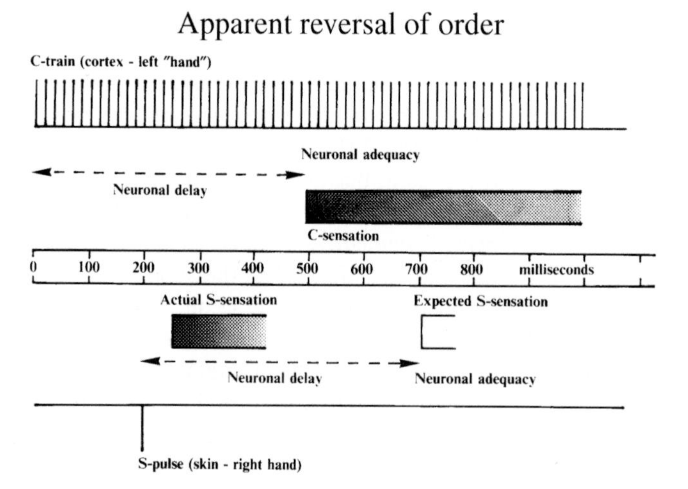

Figure 8.16
Benjamin Libet and his colleagues also studied situations in which stimuli were applied to different 'hands', and they discovered that a subject will report having felt S before C even though S was actually applied after C, but within about half a second of the latter. There is thus an apparent reversal of order of the stimuli, in the subject's mind.

flexibility regarding sequence. They discovered that their subjects consistently reported the S sensation as coming *before* the C sensation even though the S pulse was actually administered as much as 400–450 milliseconds *after* the C train commenced (see figure 8.16). This was certainly a remarkable finding. The earlier experiments had shown that neuronal adequacy requires about 500 milliseconds for completion. This new result showed that the subject does not associate the sensation thereby developed with that completion time. Libet and his colleagues had demonstrated that the *content* of experience is influenced by events occurring well before the 500 millisecond mark. It is this which causes the subject to associate the S-generated sensation with a time much closer to the administration of the S pulse. The subjective experience thus involves a *retroactive referral of sensation*, and the researchers made a suggestion as to what the sensory apparatus uses as a cue, namely what is known as the *primary evoked potential* (see figure 8.17). This is something that is seen in the EEG trace, and it will be worth our while to consider it briefly.

Although interesting EEG traces can be obtained from subjects when they are merely resting, most recent work in this field has probed the responses to various stimuli. The average of many individual EEG traces must be determined, in order to obtain reliable statistics. The resulting plot of amplitude as a function of time is known as the *average evoked response* (or AER), and this comprises a number of standard features for all normal subjects. Around 20 milliseconds after the stimulus is applied, the AER shows a distinct swing in the positive direction, and this trend peaks (actually with two maxima) around 50 milliseconds. The peak is known as the *primary evoked potential*. This is followed by a larger negative swing that reaches a maximum (again with a split-peak structure) around the 100-millisecond mark. After further swings in the positive and negative directions, the amplitude decays to zero around 500 milliseconds. Libet and his colleagues identified this final stage of the AER with neuronal adequacy, and they made a good case for the cue-giver of backward referral occurring around the time of the primary evoked poten-

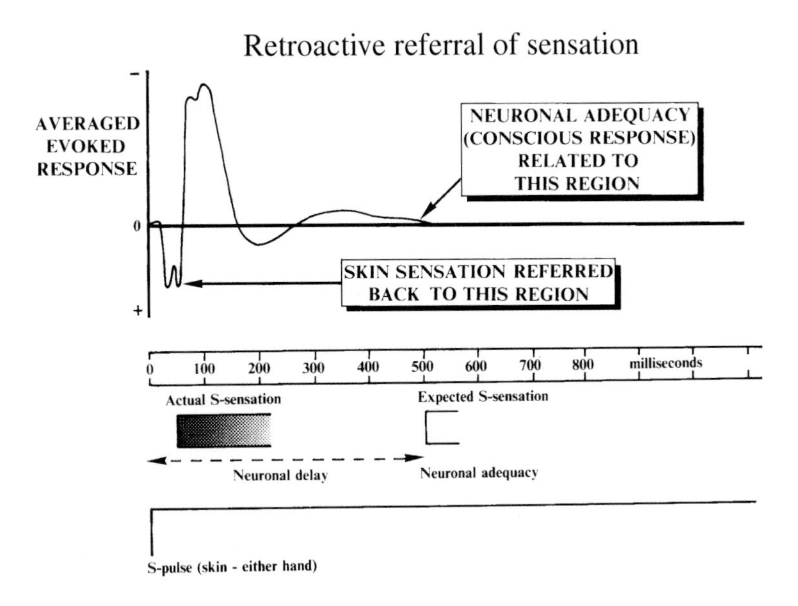

Figure 8.17
The significance of the approximately half-a-second period was explained by Libet and his colleagues through reference to what is seen in the typical electroencephalogram, when this is averaged over many individual measurements. The average evoked response, as it is called, first shows a positive sweep around 50 milliseconds. It then develops both negative and positive trends, and finally decays away to zero at around half a second. The conclusion was that full conscious awareness does not develop until about half a second has elapsed, because this amount of time is required for the cortex's appropriate nerve cells to fully process the incoming information. Libet and his colleagues gave the term **neuronal adequacy** to the fully developed response. The surprising experimental observations can be explained if the brain refers its conscious response back to the initial positive sweep around 50 milliseconds.

tial. Once neuronal adequacy has been established, there is a *backward referral* to the vicinity of the 20–50 millisecond range. So although this enables the subject to make a quite good estimate of an event's timing, an error of about 20–50 milliseconds is nevertheless incurred.

Everything that has been described thus far required only the passive participation of the subject. We must now turn to the subsequent studies which addressed the very important issue of *volition*. Libet and his colleagues appeared to have uncovered some fascinating new facts concerning the timing of sensation. It now remained to see whether there were further surprises in store concerning the timing of cortical signals related to initiation of a voluntary act. Are such signals detectable before one becomes aware of making the decision to act, or is the reverse the case? Through their observational technique, Libet and his collaborators had placed themselves in a position to address the question of *free will* itself.

Before considering the experimental observations, we ought to take note of some preliminaries. Firstly, there is the operational term *unconscious*, which is applied to mental functions that are not reportable as introspective subjective experiences. This use of the term should not be confused with the one invoked in connection with sleep and coma.[u] I agree with Libet when he suggests that the terms *pre-conscious* and *sub-conscious* could serve as useful alternatives.[68] From the observations made by him and his colleagues, one can conclude that cortical processes having a duration less than that required for neuronal adequacy could nevertheless mediate unconscious mental function,

u / Such confusion would be excusable, given the inadequacies of terminology. I will be attempting to correct this situation in the next chapter.

without accompanying subjective awareness. The second point concerns an important discovery made by Hans Kornhuber and Lüder Deecke in 1965. Studying the event related potentials in subjects who were making what are called *self-paced voluntary movements*, they found significant features in the experimental traces at 800 milliseconds or more before the actual movement. This *readiness potential*, as they called it, is a sort of advanced neuronal notice of impending action.[69]

The Libet group further investigated the readiness potential, and discovered that it actually comes in two versions.[70] If there has been mental preparation, or preplanning, for the voluntary act, the readiness potential can occur as much as 1050 milliseconds before the movement. If the act is spontaneous, on the other hand, the readiness potential precedes the movement by about 550 milliseconds. The arrival of the movement command at the relevant muscle can be detected electrically, through what is known as an *electromyogram* (EMG), and the signals can be shown to be dispatched from the motor cortex about 50 milliseconds before the EMG picks them up.

The prior work had given Libet and his collaborators a useful tool for establishing the timing of the subjective awareness of the wish to move, this instant being given the label W. That tool was the S pulse, about which so much information had been accumulated during the earlier studies. In effect, S could be used as a sort of stopwatch; during the experiments, S pulses were delivered at random moments, and the subject had to report when they were experienced, relative to W. As was noted above, a subject can temporally place the S pulse to within about 50 milliseconds of its actual occurrence, so the surrogate stopwatch was a rather reliable one. The difference was thus a quite convincing one when the W was perceived to occur at about 350 milliseconds *after* the onset of the readiness potential, for the case of spontaneous movement.[71]

In view of this result, we must conclude that a voluntary act is initiated unconsciously. The requirement of a substantial period of development, during which neuronal signals build up cascade fashion, must thus be extended to include the important case of awareness of volitional mental processes. As Libet himself emphasized, this seems to deny free will the chance of initiating actions. But he also noted that significant things can happen in the interval between W and the dispatch of the commands from the motor cortex. As we have seen, these occur at −200 milliseconds and −50 milliseconds, relative to the detection of signals by the EMG. Libet and his colleagues showed that their subjects were actually able to *veto* the consummation of the volitional process during that temporal no-man's-land.[72] The will could thus exercise a control function over the performance of a motor act.[v] Whether this makes it worthy of the label *free* is a moot point.[w]

v / The more recent work of Roger Carpenter and Matt Williams sheds further light on such veto effects. I shall be reviewing their very important findings in the next chapter.

w / One could say that these latter experiments demonstrate that even if we do not have free will, we do at least have free won't!

Let us leave that philosophical issue for the time being and briefly turn to the equally difficult question of consciousness. This, after all, was the vital product of neuronal adequacy, so it is relevant to ask whether the experiments of Libet and his colleagues were providing new insights into its nature. Perhaps they were, indirectly at least. We could compare the 500 milliseconds required for neuronal adequacy with the observed frequency of automatic saccades. We have earlier noted that this is about four per second, on average. (In practice, there is a statistical distribution of inter-saccade intervals, and many of these are in fact below 100 milliseconds.) The 200–250 millisecond inter-saccade interval is briefer than the time required for neuronal adequacy. It has to be admitted that there may be a significant difference between consciousness of a tap on the hand (which was the stimulation favoured by Libet and his colleagues) and consciousness of an image presented to the retina. If the comparison is nevertheless valid, the relative magnitudes of these two intervals are quite suggestive. They indicate that the eyes would be moving more often than neuronal adequacy could keep up with. But this would be the case only if a new drive toward neuronal adequacy has to be initiated every time the eyes move. In practice, one might expect some sort of dynamic equilibrium to be established, with the processing of one inter-saccade snapshot starting before the processing of the previous snapshot has been completed.

The recently established physiological facts surrounding the pulvinar, discussed at length earlier in this chapter, lend support to this idea. It was there suggested that one role of this structure is to filter out irrelevant visual information during eye movement, and to permit production of what could be called a scissors-and-paste job on our visual inputs. The upshot would be continuity salvaged from eyeball observations of the world that are, perforce, fragmentary in the temporal domain. I have also argued that the pulvinar may provide us with the vital glue in the spatial domain, and I referred to this as aperture synthesis (or jigsawing). If there is any truth in this, we see that the surprisingly long time required to establish neuronal adequacy is actually a blessing, for it obscures the seams in space and time.

Finally, let us return briefly to that question of unconscious mental events, which has received such a boost through the work of Libet and his colleagues. The idea that our decisions are mediated by processes lying below the level of awareness is emphatically not a new one. This was a subject of considerable debate around 1900, and it gave rise to experiments that were remarkable when one realizes how primitive equipment was at that time. This chapter has been littered with references to timings in the millisecond range, and it is all too easy to take such measurements for granted. One has to admire the ingenuity of the experimental psychologists of that era. It is impressive to read the accounts of how they went about the task of timing events on the brief scale that we have seen is relevant for the processes of interest.

One such early artisan was Henry Watt, and his efforts were directed at measuring the time taken by the individual stages of association.[73] His apparatus, known as a chronoscope, included a mechanical device which rapidly raised a card that had otherwise been hidden from the subject's view, and a microphone for detecting the very start of the subject's verbal response. The tests were of the type referred to by psychologists as *partially constrained association*, and they required the subject to associate another word with the one displayed on the card, and to speak that other word. Various classes of word-pairings were employed, including those in which the second word was subordinate, for example *oak–beam*, coordinate, as in *oak–beech*, and superordinate, an example of the latter class being *oak–tree*.

The subject was told which class was to be used, before each trial, and the card was then displayed, whereafter the idea was to respond as quickly as possible with the second word of the pair. The individual stages of the overall association were first the presentation, then the mental search for a suitable association, and finally intonation of the chosen word. The surprising result of Watt's investigation was a demonstration that associations were being made automatically, without the need for introspection. One sees evidence of this sort of thing in games which require completion of words when only a fraction of their constituent letters are known, as in the popular television game *The Wheel of Fortune*. Successful players in this competition cannot explain the instantaneous flashes of insight which provide them with the missing consonants and vowels.

The work of Watt and his successors can be added to that of Libet and his colleagues, to bolster the case for unconscious mental processes, taking place before neuronal adequacy brings their products into conscious awareness. The question remains, however, as to what conscious awareness actually is. The point has been reached at which that question must now be squarely faced. This is the aim of the following chapter.

9 Midwives of reflection

Millions of items...are present to my senses which never properly enter my experience. Why? Because they have no *interest* for me. *My experience is what I agree to attend to*....Everyone knows what attention is. It is the taking possession by the mind, in clear and vivid form, of one out of what seem several simultaneously possible objects of trains of thought. Focalization, concentration of consciousness are of its essence. It implies withdrawal from some things in order to deal effectively with others.
William James (*Principles of Psychology*[1])

Man might be described fairly adequately, if simply, as a two-legged paradox. He has never become accustomed to the tragic miracle of consciousness.
John Steinbeck (*Log from the Sea of Cortez*, 1941)

An inadequacy of colloquialisms

It is a characteristic of the scientific endeavour that it seeks to define the objects of its curiosity. We saw in the first chapter that no satisfactory definition of consciousness has yet emerged. Some people believe that attempts to find one would be futile, because too little is known about the phenomenon. This is frustrating. It is also rather surprising, given the central role that consciousness plays in our very existence. How can it be that something so intimately familiar still defies adequate delineation? Dictionaries appear comfortable with descriptions that merely see consciousness as the totality of one's thoughts and feelings – a general awareness of one's existence, sensations and circumstances. It is sometimes suggested that brains are conscious if they can examine their own thoughts.[2] William James, toward the end of the nineteenth century, essayed a precise characterization along these lines when he suggested that the essential element of consciousness is self-reference.[3] That would seem to be a beneficial faculty, but it says nothing about the neuronal processes involved in such introspection. If self-reference is taken to include reference to one's thoughts, rather than merely to one's body in its immediate environment, this definition could also be in difficulty with respect to humans with certain brain deficits. There might be people incapable of reading their own thoughts.[a] Would we have to regard them as not possessing consciousness?

a / In the next chapter, I will be discussing the possibility that autistic people might not be able to examine their own thoughts.

In earlier times, it was tacitly assumed that consciousness is the brain's only waking state, and that the inner workings of the mind would eventually be laid bare through the investment of sufficient contemplation. Hermann von Helmholtz, in the middle of the nineteenth century, appears to have been the first to realize that there are mental processes which are not part of conscious experience, and Sigmund Freud subsequently argued that conscious thought can be, and invariably is, influenced by unconscious mechanisms related to an individual's past experiences.[4] Freud interpreted the working of the mind in terms of confrontations between impulses at the unconscious level,[b] the product of their resolution being passed to the conscious level of thought. At the end of the preceding chapter, we considered the strong experimental evidence for unconscious mental states that has been steadily accumulating during the present century; it stretches from the early efforts of Henry Watt, and continues today with the work of Benjamin Libet and his colleagues. Later in this chapter, I will be discussing studies by Roger Carpenter of eye movements, which are even more revealing.

One should be careful to distinguish between the unconscious mental events considered by von Helmholtz and Freud and unconscious *reflexes*. Amongst the best examples of the latter are those seen in the fluent performances of trained musicians and circus artists. Once frequently repeated movements have become sufficiently ingrained, they can be carried out on cue without conscious effort, and the performance even becomes awkward if an attempt is made to concentrate on what is being done. It is rare for one of these virtuosos to be able to describe the mechanics underlying the skill in question. To these more overt examples may be added the myriad lesser routines that relieve the burden on our consciousness, the things that we describe as being *second nature*. Consider, for example, how difficult it would be to drive a car if every control had first to be subjected to conscious attention. Practice relegates the individual manipulations of the pedals and steering wheel to habit, and these aspects of driving become unconscious.

Gilbert Ryle stressed that colloquial usage of the word consciousness has tended to blur its meaning. In his *The Concept of Mind*, published in 1949, he listed several distinct applications of the term.[5] It will be useful to enumerate them here. To begin with, consciousness may be invoked to convey the idea of being generally aware of, an example being the phrase *I was conscious that the furniture had been rearranged*. This has little relation to any specific sensory mechanism, and the same is true of cases in which the word is used in the sense of being self-conscious (i.e. shy), which is more a quality of character. Yet another sense of the term is that of heeding, and again one sees that the relationship to specific brain mechanisms is rather tenuous.

b / It is perhaps unfortunate that the word *unconscious* is used to describe both lack of consciousness, as during sleep or a coma, and mental processes that lie below the level of consciousness during waking. The word *subconscious* offers an alternative for the latter connotation, although its use seems to be frowned upon in some circles.

Ryle ultimately identified just two applications of the concept which *must* be related to processes occurring at the neuronal level. The first of these is consciousness as the opposite of unconsciousness, that is to say when conscious appears as a synonym of sentient. This form of consciousness would occur, by definition, in creatures that sleep, so all mammals would possess the capacity for it. Ryle's second example is more restrictive, because it requires the ability to think of oneself thinking. Thus whereas the drawing of inferences need involve only thought, the awareness that one is drawing inferences might also require consciousness. Ryle's example of the latter is a person saying to himself *Here I am deducing such and such from so and so*. He came to the conclusion that such a definition of consciousness would be too limited, because it would lead to an infinite hierarchy of levels, the next member being *Here am I spotting the fact that here I am deducing such and such from so and so*. In rejecting this definition, Ryle argued that a portion of mental activity must *always* lie below the level of consciousness, a conclusion that was in accord with what von Helmholtz and Freud had surmised earlier.

Ryle's point about difficulties stemming from common usage was well taken, and we could extend the list to include awareness, attention and thought. This chapter is only five paragraphs old, but these three travelling companions of consciousness have already managed to get into the act. It *is* confusing to have such a variety of terms, with *attention*, *awareness*, *consciousness* and *thought* all vying for our attention, awareness, consciousness and thoughts. It seems clear that there is a certain amount of redundancy in the way we use these terms. For example, the expression *conscious awareness* is frequently used when the single word *consciousness* would have sufficed. The practice is confusing, indeed, because it gives the dual impressions that consciousness sometimes needs a qualifier and that there could be consciousness without awareness.[c] It could naturally be the case that both of these propositions are true, but in that case we should need definitions of the words that are far clearer than anything we have at present. Indeed, if one looks up the dictionary entries for these two terms, one usually discovers that they are cross-referenced, as if they were equivalent. It is probably such uncertainty which leads us to use the words in conjunction.

At the beginning of this book, I emphasized the need to differentiate between the products of consciousness and consciousness itself.[6] Despite the appeal of what von Helmholtz, Freud and Ryle had to say on the subject, I am not sure that they succeeded in paring consciousness down to its barest essentials. In this chapter, I must attempt to home in on what consciousness actually *is*. Reference will naturally be made to what consciousness *mediates*, but my goal will be a definition of the phenomenon itself.

c / When the term conscious awareness is used in contrast to *un*conscious awareness, on the other hand, it *is* useful; we will shortly be encountering plenty of evidence that the latter exists.

Further probes of the unconscious

Not long after the advent of commercial television, there was discussion about advertising by *subliminal perception*, that is to say by flashing commercial messages on the screen for such brief periods that the viewer would not be conscious of seeing them. Not surprisingly, many looked askance at this suggestion, and some found the idea quite ominous. There has always been something sinister in the suggestion that we can be brainwashed, and the spectre of subliminal advertising soon gave rise to mumbling about the brave new world. The debate appears to have subsided, and one trusts that the proposal was never put into practice. The possibility of subliminal perception is nevertheless interesting, because of its potential for shedding light on brain processes that lie below the level of consciousness. We would do well to consider some of its ramifications.

The word *subliminal* is related to the psychological term *limen*, which means the limit below which a given stimulus ceases to be perceptible. It thus indicates the existence of a threshold, and it is natural to assume that this is related to the minimum nerve excitation required to produce a sensation. Just how the latter is related to the threshold excitation for a single nerve axon will depend upon the relevant neuronal circuitry, this relationship presumably being rather complicated. Formally, the limen is defined as the smallest stimulating intensity that can be detected with a fifty per cent probability. Signals with intensities below that level are too weak to be consciously detected, but they presumably still have some effect on the relevant neural assemblies. We should consider the evidence supporting their existence.

One fruitful technique involves what is known as *masking*. The approach involves confronting subjects with groups of letters displayed on a screen, the letters being arranged so as to produce either an actual word or gibberish. The standard procedure calls for a *priming stimulus* to be displayed before the *target stimulus*, and the subject is asked whether the target is a meaningful word. It is invariably observed that the subject's evaluation is facilitated by the use of primes that are meaningful words. The interesting thing is that when a third group of words, known as a *masking stimulus*, is inserted between the prime and the target, the prime–mask interval being such that the subject would not have sufficient time to become consciously aware of the prime, the expected influence (positive or negative, depending upon the arrangement of the letters) is nevertheless still observed. The conclusion must be that the unconscious processing of the prime (the term *preconscious* is favoured by some people working in this area) still influences the subject's decision.

Another source of evidence is provided by the phenomenon of *hypnosis*, even though care is needed in the evaluation of this phenomenon because of its vulnerability to charlatanism. As is well known, the process involves the response of a subject to verbal suggestions or physical stimuli made by the hypnotist, following the subject's insertion into an altered mental state best described as wakeful unconsciousness. A typical example is the failure of a subject to register pain when in the hypnotised state. It can readily be shown that this is not attributable to a breakdown in the normal pain-generat-

Figure 9.1
The inkblots shown here are typical of those used in the Rorschach test. The subject being tested is asked to identify the items suggested by the various shapes, immediately and without contemplation. The test is probably a good example of the influence of concept on percept, and of the importance of reverse cortical projections. Because it thereby accesses memory, through unconscious processes, it is capable of revealing some of the personality traits of the subject, although such interpretations require skill on the part of the tester.

ing mechanisms, because these are amenable to physiological monitoring. The heart rate, for example, becomes elevated as a consequence of the painful stimulus, even though the subject is not conscious of it. The indications are thus that the sensory machinery is able to provoke the appropriate physiological response, even though the fact of the stimulus is not communicated to consciousness.

Amongst the hypnotist's standard bag of tricks, none is more familiar than the phenomenon of post-hypnotic amnesia; the resuscitated subject is unable to recall events that took place during the hypnotic trance. An embellishment, particularly popular when hypnosis is being used as entertainment, is the reversal of the amnesia by some prearranged signal, such as the utterance of a cue word or even the mere snapping of the fingers. The fact that this can be accomplished demonstrates that the subject's initial inability to recall reflects an underlying disruption of memory retrieval, rather than a diminished capacity for storing memories.

Hypnosis becomes a particularly potent source of information on the unconscious state when it also involves the subject's reactions to orders given either during the trance or when the normal state has been re-established. In the former case, because the orders can be subsequently recalled upon administration of the prearranged cue, the subject's actions must have been steered unconsciously by directives that have nevertheless been stored in memory. Compliance, by the subject, with orders given post-hypnotically is suggestive of an even richer chain of unconscious responses, because it implies a twofold lack of consciousness. The subject cannot anticipate which tasks are to be performed on the hypnotist's cue, because the directives were imprinted under hypnosis, and the reason for performing them will be equally hidden from the subject's conscious awareness.

Amongst other interesting manifestations of unconscious processes, we have the celebrated inkblot test devised by Hermann Rorschach[7] (see figure 9.1). The subject is asked to inspect the irregular shapes that constitute a typical group of blots, and to state without contemplation what the various shapes suggest. Rorschach himself proposed that people who were inclined to 'see' humans in the blots, and particularly humans in action, were more likely to have rich powers of fantasy, and to indulge in day dreaming. According to what has been discussed in the preceding chapters, this test probably demonstrates the influence that concept has on percept, the influence being mediated by the reverse cortical projections. And the concepts in question would be those laid down in long-term memory, as a consequence of previous experience. There is a close relationship between the Rorschach test and fortune telling by those who 'read' tea leaves. Such 'predictions' reveal more of the teller's past than they do of the tellee's future.

Finally, there is the growing body of evidence concerning what is referred to as *blindsight*.[8] Testing of patients who have lost their primary visual cortex, or part of it, have been found to perform better than randomly in identifying objects that they should not be able to see. The identified objects do not enter such a patient's consciousness, because they cannot immediately influence behaviour. These remarkable feats of recognition must thus be labelled *unconscious awareness*.[9] They are presumably made possible by the fact that there are several different routes from the retina to the various visual areas of the cortex, even though these are not equivalent in their signal-handling capacities. One such secondary path passes through the superior colliculus. It has recently emerged that some of the data on blindsight are not as unequivocal as one might have hoped,[10] but the phenomenon nevertheless provides a valuable reminder that there is more to vision than meets the primary visual area. And indeed that there can be awareness without consciousness.

The philosophical nub

Having considered an important aspect of the conscious state, namely the unconscious mental process, we must now turn to our main theme. And let us immediately draw a distinction between consciousness in general and consciousness of one's own thoughts. Despite occasional redundancy in the two-worded form, as noted earlier, let us call the former state *conscious awareness*. We can then reserve the term *introspection* to describe thought processes that involve awareness of one's own thoughts. The possibility that there could be individuals who are capable of conscious awareness yet incapable of introspection is an intriguing one. Indeed, if our species acquired the former before the latter, as some have suggested, the world must once have been peopled by such creatures. There has been the suggestion that autistic people might not be aware of consciousness in others. Whether this would have to imply that they are not aware of their own consciousness remains to be seen. Equally interesting is the question of whether some of the other species are conscious of consciousness. I will be discussing both of these challenging issues in the next chapter.

The philosophical difficulty surrounding both conscious awareness and introspection lies in their subjectivity. They are *first-person* phenomena, but we would like to explain them in a *third-person* and objective manner. Unlike anything else that has come into the scientific focus, thoughts are private possessions, inaccessible from outside the head. It is true that mental activity can be detected externally, and we have considered the techniques that have made this possible. But until now at least, the nature of thought has remained a mystery. It might seem that we could take the vicarious view and, by invoking the similarity between members of our species, simply dispense with thinking about the workings of other people's minds and draw the necessary inferences from observations of our own. There is of course the objection which claims that such deliberations would disturb the very thought processes they were attempting to shed light on. This does not seem to be a serious threat, however, because one may still contemplate thought processes one has experienced in the recent past, and probably in sufficient detail to make the exercise useful.

This sort of approach simply will not do, however, because the consciousness issue is not confined to our own species. We will have accomplished precious little if we are unable to gauge the likelihood of consciousness in other animals. And in the context of the present book, it would be a crushing disappointment if we were unable to say anything about possible consciousness in machines. We must therefore return to the line taken in chapter 2, and couch our discussion in terms of those black boxes. In particular, we have to consider cases in which exclusively internal processes are at work, independent of any input or output. How would we decide that something could not function as such a box? Are there constraints which we could use to narrow down the field? Let us start by considering an extreme case.

A metal rod has internal states. There are (negatively charged) electrons which are perpetually on the move through the lattice of heavier (positively charged) ions, dashing this way and that, in a manner that requires quantum mechanics for an adequate description. Granted, the relevant motions within our own nervous systems are of ions rather than electrons, but the physics of the two situations would not seem to be so different. Could a piece of metal be conscious, therefore? If our answer is *no*, it is probably because we would be loathe to accord what appears to be one of the main products of our own consciousness, namely feelings.[d] But how are we to take

d / More formally, discussions of this issue are usually couched in terms of *qualia*. These refer to what it *feels like* to be conscious of a particular set of circumstances. I will use the terms feelings and qualia interchangeably in the balance of this book.

an objective position on something so subjective as feelings? And it is no use passing off feelings as a mere embellishment to something we need only account for in its basic form. How can we be sure that feelings are not the *essential ingredient* of conscious awareness?

There is something much worse that has been waiting in the wings. It has been stressed by Thomas Nagel, amongst others.[11] He framed it in terms of a question that has become a classic in the field: *What does it feel like to be a bat?* Walt Disney made a spectacularly successful career out of persuading children that all manner of creatures feel pretty much as we do. It is rather unlikely that this is the case. We can appreciate the gravity of Nagel's question by considering another, seemingly more circumscribed, philosophical issue: *What does it mean to say that one is observing the colour red?* If we can answer that, we will be well on the way to appreciating why Nagel's bat question is such a thorny one. Let us make a first stab at it.

We say we are seeing red when light with wavelengths lying in a particular range falls on the retina. This light interacts with certain protein molecules in the membranes of the relevant receptor cells. This, in turn, gives rise to neural signals which travel to the various visual areas, and thereafter to the association regions of the cortex. If we have previously been exposed to light with the appropriate range of wavelengths, and have at the same time heard the word *red*, associations will be generated and we will either verbalize the word or merely have it in mind. The interesting thing is that this response is *subjective*, whereas the physical attributes of the light are *objective*. Wavelength and intensity are amenable to measurement, by methods which can be made quite independent of vision. And it can easily be established that the sun emits light having a continuous spectrum of wavelengths. There is nothing subjective about such facts. But these emissions have relevance for us only because we have those protein molecules to make the necessary conversion to neural signals. In a very real sense, therefore, red exists only in the mind, and the same is naturally true of all the other colours. The association of light with wavelengths in certain ranges and words such as red, green and blue is in reality nothing more than a convention. And for a person suffering from colour blindness, the convention has a more limited relevance.

We have to be vigilant with these conventions of ours, because they have a way of being modified from time to time. Until around 1970, one would have defined a piece of glass as something that is transparent, brittle and electrically insulating. Ten years later, all three of these hallmarks had been swept away by the advent of metallic glass. This remarkable material is opaque, ductile and electrically conducting. All that is left of the original characterization is that the atoms in a glass are arranged in a much less orderly fashion than they are in a crystalline solid, and we now realize that *this* is glass's one defining trait. The degree of atomic order may be monitored by the use of X-rays, which have wavelengths outside the response ranges of the proteins in our retinas. If there were creatures which could detect X-rays directly, they would observe the world in a quite different way. Solids which we call opaque would appear transparent to them, so window glass and metallic glass would not produce the sense of contrast that they do for us.

What, then, *is* it like to be a bat? The answer is that we simply cannot tell. And our ignorance on this point actually stems from a number of factors. For a start, we do not have adequate knowledge of the physical responses of all the bat's receptors, though this could presumably be rectified through a concerted research effort. There would then be the question of how the various receptors feed signals into the higher reaches of the bat's brain, not to mention the establishment of associations. It is unlikely that these associations would have any relevance for us – we do not employ sonar, for example – so our chances of appreciating what they mean in the bat brain would be pretty remote. That being the case, the task of finding parallels between bat feelings and human feelings appears insuperable. In the case of the bat, therefore, and for all other animals, we are limited to objective considerations, such as their overt behaviour and their apparent motivations.

If the bat–human comparison, feelings-wise, is fraught with difficulties, what could we usefully say about mechanical devices *vis-à-vis* our species? Very little, one would think. If feelings are an integral part of consciousness, therefore, we are going to be hard put to understand the ramifications of consciousness in other types of creature, even if they do possess consciousness. The best that can be said of this impasse is that it shifts the focus back to the workings of our own brains. Let us return to human perception, therefore, and start with our senior sense, namely vision. We can gainfully begin by contemplating something that might seem rather esoteric, and of only peripheral significance to our theme. It is referred to as *filling in*. Daniel Dennett, with whom I took issue in my opening chapter, does not believe that it exists.[12] Let us consider the evidence.

The reality of filling in

An optician, inspecting the back of a person's eye, through the pupil, sees a fine network of blood vessels which feed oxygen and nutrients to the retina's neurons. The widths of these vessels are considerably larger than the diameter of a retinal receptor cell. Yet even when we look at a featureless expanse, such as a cloudless sky or a whitewashed wall, we are quite unaware of these capillaries.[e] The visual system somehow compensates for those parts of the field of view that they would otherwise obscure. The blind spot[f] represents an even

e / A newly ruptured blood vessel does in fact give a visible irregular patch, but this rapidly disappears again, as the receptive fields of the surrounding neurons adjust to the changed circumstances. This mechanism lends support to the argument given in this section.

f / This region is properly referred to as the *optic disc*, and it is the place at which the axons of the retinal ganglion cells leave the eye to form the optic nerve. The term *blind spot* is more appropriately applied to the part of the visual field that corresponds to this region of the retina.

Figure 9.2
This particular demonstration of the blind spot apparently supplied King Charles II with much diversion; he used it to temporarily behead his courtiers! The reader will be able to accomplish the same effect by closing the left eye, by then focussing on the black spot, and finally by adjusting the distance between the eye and the printed page.

larger potential obstruction, but we are normally oblivious of this too. Again, the system appears to make the necessary adjustments, automatically and without our intervention. This is not to say that objects whose retinal images fall upon that area will nevertheless be seen, when the other eye is closed and thus not 'covering' for the missing receptors.[g] On the contrary, such occlusion is readily demonstrated, and I imagine that the majority of readers will have observed it for themselves (see figure 9.2). It has been related that King Charles II was fascinated by the effect, and that he exploited its potential for diversion, temporarily beheading his courtiers.[13]

The question arises as to why special steps have to be taken to reveal the presence of the blind spot, even for monocular viewing, and why we are normally not aware of it. The answer appears to lie in the compensation mentioned above, the phenomenon being known as *filling in*. If the area around the blind spot is uniformly and regularly occupied by a series of lines, for example, the brain will fill in the missing region with similar lines, which, moreover, will be in perfect registry with the originals (see figure 9.3). Charles Gilbert and Torsten Wiesel exposed an important component of the underlying mechanism through an experiment in which a small region of a cat's

g / The offsets of the optic discs from the foveas, in the two eyes, are in opposite directions, so one retina has neurons in the zone occupied by the optic disc in the other retina, and vice-versa.

retina was destroyed with a laser beam, so as to artificially produce a second blind spot.[14] They located the neurons in the primary visual cortex immediately adjacent to those corresponding to the damaged retinal area, and discovered that their receptive fields grew within a few hours, following production of the lesion. This expansion permitted these cortical neurons to take over the role of those that had been rendered inactive. Cortical neurons corresponding to the retinal regions adjacent to those capillary blood vessels must have similarly enlarged receptive fields.

Normally, when both eyes are open, there will be a contribution from the other retina, of course. The region of the primary visual cortex corresponding to one eye's blind spot nevertheless has neurons which receive signals from the other retina. The activity of these monocular neurons has been investigated for a situation relevant to the demonstration shown in figure 9.3. When two horizontal (and collinear) lines are presented on either side of the right eye's blind spot, for example, the monocular neurons in the relevant cortical area respond vigorously, whereas their reaction is much weaker if either line is presented by itself.[15] So the neural machinery that mediates filling in is now amenable to direct investigation.

How does filling in occur, however, and why does it not always dominate? The answer to the first of these questions appears to lie in the feedback processes mediated by the cortex's reverse projections. We recall from the description of these projections, given in chapter 7, that they pass signals from the higher visual areas (in the case of vision, that is) back toward the primary areas. But the information on which these backwardly travelling signals are based must have originated from the observed scene itself; it must have been derived from inwardly travelling signals, routed via the cortical forward projections. In the vicinity of the blind spot, that information will have been gathered by the enlarged receptive fields. If the information is *sufficiently definitive*, it will have the decisive influence on what the brain interprets as being

Figure 9.3
The important and revealing phenomenon of 'filling in' can be demonstrated with the aid of this figure. If the procedure adopted with figure 9.2 is repeated, the reader will find that the missing portions of the lines are supplied by the visual system.

the missing object. Failing this, nothing will be seen at that position. The lines in figure 9.3 apparently produce such sufficiency, for the result is indeed filling in. In the case of the heads of Charles II's courtiers, however, the surrounding pattern was too irregular, and the brain opted for decapitation!

Further exploration of these effects can be quite enlightening. A series of horizontal and closely spaced parallel lines, each drawn with a single and similarly positioned break, appears to be traversed by a vertical blank strip. The visual system creates this impression by producing *illusory contours* that span the gaps between the broken ends of the lines. If one arranges for a region containing the breaks to fall on the blind spot, the blank strip will appear to be completed across the occluded area (see figure 9.4). If the parallel lines are drawn sufficiently far apart, however, the brain opts for the other alternative and it is one of the broken lines that appears to be continuous.[16] Which type of illusion one experiences is apparently determined by those cognitive processes occurring further along in the visual system, processes which then return their verdict along the reverse projections.

The most intriguing effects of this type do not even require participation of the blind spot. Filling in can happen in a wide variety of situations provided that the viewer's attention is not focussed on the region where the compensation is to occur. As Vilayanur Ramachandran and Richard Gregory have demonstrated, a strong effect of this type can be obtained with two adjacent areas that have different pastel colours or slightly different degrees of grey shading.[17] The dividing line can be readily discerned if one looks directly at it, but it appears to evaporate when one stares intensely at a spot located somewhat away from the division. Rather unusual hues can be produced in this manner, including a quite fascinating pinkish green. This book is printed in monochrome, unfortunately, so figure 9.5 merely shows the grey version of this important example.

An important aspect of this latter type of illusion is its dependence on attention. With a little practice, one can make the illusion appear and

Figure 9.4
In this variant of figure 9.3, it is the portion of the white occluding strip which falls on the blind spot, rather than the missing parts of the parallel lines, which is filled in. If those lines are drawn much farther apart, however, it is the missing part of one of the lines which is filled in.

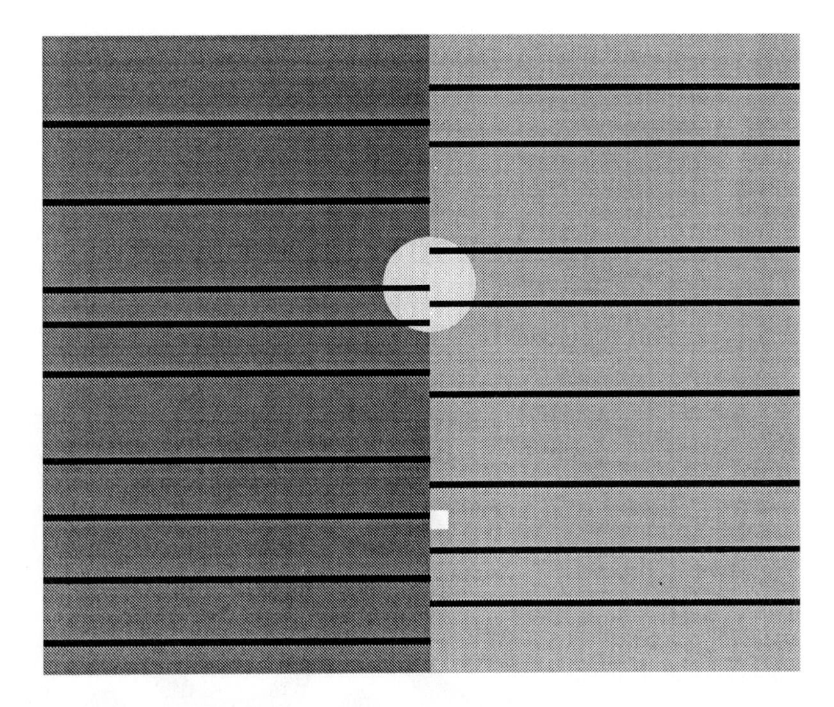

Figure 9.5
This diagram, which is a monochrome version of one due to Vilayanur Ramachandran, illustrates a form of filling in. It is of particular interest here because it shows that filling in does not require participation of the blind spot. If one stares intently at the small white square for several seconds, deliberately attempting to confine one's attention to that rectangle, the two different shades of darker grey either side of the small light grey circle will be observed to fill in that feature, gradually obliterating it.

disappear at will, the trick being to control the *degree* of attention on the offset spot. Our ability to impose such control is in itself important, and we discussed this in chapter 7, using the wrinkles on the back of a finger as convenient targets of varying degrees of attention.[h]

Daniel Dennett's dismissal of filling in, although questionable, would have been less important had he not tied it so strongly to his views on the nature of consciousness. And because consciousness is our central concern in this chapter, it obviously behooves us to consider Dennett's line of argument. This occupies many pages of his (entertaining) book, and it is naturally not practicable to repeat it *in extenso* here. Suffice it to say that he accounts for our normally being oblivious of our blind spots by assuming that the brain notes what lies in their vicinities and concludes that the missing parts of the images are simply 'more of the same'; that he believes the visual system simply *ignores* what actually falls on the blind spot; that he postulates there is an affinity between what happens in connection with the blind spot and the blindsight which was alluded to earlier in this chapter; and finally that he uses these arguments to justify the view that qualia do not exist.

Dennett illustrated his first point through reference to the artist Andy Warhol's celebrated picture which consists of a repeating pattern, wallpaper fashion, of Marilyn Monroe motifs. When one inspects this diptych

h / The reader may find it useful to repeat the brief exercises described in chapter 7, before continuing.

monocularly, one's blind spot does not cause awareness of any break in the pattern. Dennett believes that the brain simply assumes that the missing part would be 'more of the same', and that it can thus safely be ignored. This is related to his next point, to which we pass immediately.

The second of Dennett's four beliefs can easily be shown to be fallacious by inspection of the two parts of figure 9.6, which was inspired by, and builds upon, the work of Ramachandran.[18] The upper part consists of a regularly drawn series of annuli (i.e. doughnut – or bagel – shapes), each of which would cover an area *somewhat larger than the blind spot*. If one of them is made to fall on that region, the compensatory mechanism completes the element as if

Figure 9.6
As noted in figures 9.2, 9.3 and 9.4, the more familiar example of filling in involves the blind spot. The first part of this figure is due to Vilayanur Ramachandran. If one closes one's right eye, and stares intently at the small white square, and suitably adjusts the distance between the eye and the page, the black centre of one of the white annuli will suddenly fill in, causing the resultant full circle to pop out of the background of unmodified annuli. The second part of the figure displays the same effect, despite its irregularity. This demonstrates that the filling in phenomenon has nothing to do with the arrangement of such a figure's elements.

it were a full circle rather than an annulus. Because the completed element is now *different* from the others, it pops out. According to Dennett's 'more Marilyns' argument, the completed element should have resembled its brethren, so it should *not* have popped out. Our conclusion must be that *the brain does not ignore what falls on the blind spot*. The lower part of the figure is similar, but this time the arrangement of the annuli is *irregular*. The fact that the same pop-out effect is nevertheless observed shows that regularity, such as that employed in Warhol's picture, has nothing to do with the issue.

Dennett's motive for believing that there is no filling in is revealed by the following passage in his text:'... The absence of confirming evidence from the blind spot is no problem for the brain; since the brain has no precedent of getting information from that gap of the retina, it has not developed any epistemically hungry agencies demanding to be fed from that region ...' And he later continues:'... In other words, all normally sighted people "suffer" from a tiny bit of "agnosia"...'. In short, Dennett dismisses filling in by saying that the brain has no need for such a construct because there is no internal presence to look at the result, no 'little man watching a screen'. This is flagellation of the proverbial horse that already expired; I know of no person working in neuroscience who seriously entertains the concept of a homunculus.

Dennett essays a demonstration of continuity between what are known as scotomas, that is to say interruptions of normal visual processing. We have encountered three examples in this chapter: the natural blind spot; the artificial blind spot created by Gilbert and Wiesel; and the damage suffered by those who have blindsight. The first two of these are associated with irregularities of the retina, while the latter is caused by a lesion to the primary visual area. To these physical cases can be added the (short-lived) *artificial scotomas* created in Ramachandran and Gregory's ingenious experiments.

Dennett's argument runs as follows. Because visual information nevertheless falls upon the retina of the blindsight patient, just as information falls on the blind spot of the monocularly viewing normal person, both situations could be said to involve the *ignoring* of information. He then describes a thought experiment in which a blindsight patient gradually acquires the surprising cognitive skill mentioned earlier. One could say that the patient thereby becomes conscious of the items correctly identified. He then asks whether the patient would concurrently experience sensory qualia, and concludes that this would not be the case. Finally, he asks what the blindsight patient lacks which normal people do not, and his answer is *nothing*. Although I have had to telescope Dennett's argument, I believe this to be a fair representation of the crux of his case, a case which aims at undermining sensory qualia.

Robert McCauley has published a rebuttal of Dennett's claimed continuity between the blind spot and blindsight (a progression which he refers to as the blind leading the blind).[19] The thrust of this rejoinder uses Ramachandran and Gregory's observation that some of their artificial scotomas take several seconds to establish themselves in a subject's consciousness, and also that they often persist for several seconds after the presented picture has been changed. Dennett himself rationalizes these findings by noting that 'there is competition between two sources of information, and one gets overruled (gradually)'. He uses this as justification for discounting the relevance of the artificial scotoma work. But as McCauley points out, blindsight involves just such competition, namely that which takes place between the malfunctioning route via the lesioned primary visual cortex and the well-functioning secondary routes. This, McCauley shows, is ruinous for Dennett's claimed continuity, and it thus challenges his grounds for dismissal of qualia. And as Ramachandran himself has noted, regarding Dennett's linking qualia to the homunculus issue, belief in qualia does not make one a dualist.[20]

Before leaving the subject of visual filling in, and its dependence on attention, we ought to take a brief look at some relevant experiments with rhesus monkeys reported by Jeffrey Moran and Robert Desimone.[21] These studies provide a strong hint as to the underlying mechanism. As a preliminary, we should remind ourselves that the occurrence of vergence between interacting and hierarchically arranged cortical areas leads to a systematic increase in the sizes of receptive fields, as one progresses deeper into the hierarchy. That is to say, neurons respond to stimuli within an increasingly large portion of the visual field. Thus whereas a typical receptive field of a neuron in area V1 measures about a degree in any direction, the average neuron in area V4 responds to stimuli within a field that is typically four times larger along each of its edges. In the inferior temporal area, which is even farther up in the hierarchy, the receptive field of the typical neuron is actually not much less than the entire visual field. Moran and Desimone's most interesting results were obtained by investigating responses in V4.

They made a monkey fixate on a certain spot, much as we have been required to fixate when observing some of the illustrations in this chapter. But instead of having it attend to a patch surrounding the fixation spot, they made it attend to a patch somewhat offset from that point. In a preliminary investigation, they had identified a conveniently located neuron whose receptive field was also offset from the fixation spot, and they charted that neuron's responses to various visual stimuli in or near this field. They then measured its response under various conditions, all of which involved one favourable and one unfavourable stimulus, in or near the receptive field. Let us first consider the tests in which the favourable and unfavourable stimuli were *both within* the field. If the monkey was made to attend to a patch centred on the favourable stimulus, the response was strong. The two stimuli remained in their respective positions while the monkey was made to transfer its attention to the unfavourable stimulus. The response was markedly attenuated. So when both stimuli were within the receptive field, the response of the neuron was determined only by the characteristics of the attended stimulus. As Moran and Desimone remarked, it is *almost as if the receptive field has contracted around the attended stimulus* (my italics).

They then investigated the situation when the unfavourable stimulus was outside the receptive field of the neuron, the favourable stimulus still being inside. It now transpired that the response was always strong, regardless of which stimulus was attended. Moran and Desimone concluded that when attention is directed outside a receptive field, the latter appears to be unaffected. And because they observed similar responses for this arrangement, irrespective of whether attention was directed inside or outside the receptive field, the indications were that attention does not serve to enhance responses to attended stimuli.

It is perhaps a pity that Moran and Desimone did not investigate the remaining combination, namely that in which attention is directed outside the receptive field, when favourable and unfavourable stimuli are both present inside the receptive field. But based upon what had already been observed, the indications are that the responses would have been exclusively determined by the types and arrangement of the stimuli within the receptive field. This has a bearing on those filling-in experiments which we performed for ourselves earlier in this chapter. When we are not attending to an area in the visual field, the response is automatic, and determined entirely by the stimuli which happen to be present within that area. When we subject that area to attention, on the other hand, the receptive field effectively shrinks around the attended stimulus, and the features of the latter have a better chance of making a strong impression.

Benjamin Libet has suggested that filling in is a general phenomenon, and that versions of it occur in senses other than that of vision. He calls it *subjective referral of sensory representations*, and noted that the bizarre distortions of sequence observed in the tactile domain are just one of its manifestations.[22] Although this will be something of a digression, it is worthwhile to consider some of the other ways in which our senses can be misled. Let us look at a few examples.

The gullibility of senses

Filling in is obviously something that happens because of processes going on in the brain. The objects that appear to be filled in do not exist in reality. This puts the filling in phenomenon into the category of a visual illusion. Moreover, unless there is something special about this particular example, the study of filling in might shed some light on illusions in general. I believe this to be the case, and I suspect that many illusions stem from a common origin. It is important to remember that although visual illusions are particularly common, and indeed popular as diversions, vision is not the only sense that can be fooled into making mistakes. Other senses are just as gullible. The manufacturers of all manner of artificial tastes and fragrances are well aware of the tricks that can be played on our gustatory and olfactory senses, for example. Let us take a look at how two other senses, those of hearing and touch, can be similarly hoodwinked.

A fascinating auditory illusion can be produced by presenting one ear with a descending musical scale while the other ear is fed the same scale but as an ascending sequence. The individual notes are presented as staccato tones, and the presentations to the two ears are in phase with each other. But this is not what the average subject perceives. Instead, one ear appears to hear a sequence that first descends from the highest tone, then reverses itself half-way down and thereafter ascends again. The other ear, conversely, hears notes that first ascend from the lowest note, only to descend again after the half-way point has been reached. In other words, one ear appears to hear only tones in the high range, whereas hearing in the other ear is restricted to the low range. The interpretation of this intriguing phenomenon, as given by Diana Deutsch, is that similar sounds are likely to be emitted from the same source, while different sounds usually stem from different sources.[23] This is analogous to several of those empirical laws of visual perception that the Gestalt psychologists discovered, as discussed in the preceding chapter. The laws of grouping, proximity and similarity would all seem to be relevant, and the fact that there is a multiplicity of appropriate laws in the visual domain probably attests to vision's greater sophistication; I have several times in this book referred to vision as the senior sense.

Another auditory illusion that is even more amazing to the listener was reported in 1964 by Roger Shepard.[24] Using computer-generated tones that comprised the overtones as well as the fundamental, he repeatedly played a sequence which moved up an octave in cyclic fashion. That is to say the sequence ran C, C#, D, E♭, E, and so on up to B, whereafter it returned to the *lower* C and started over again. But the typical subject does not perceive it this way. Instead, an eternally ascending scale is heard, even though there is a vague feeling that such an endless sequence must be impossible. When Shepard played the sequence of notes in the reverse order, his subjects perceived an endlessly descending sequence. The resolution of this illusion had much in common with the one described above, because it again invoked a sort of Gestalt assumption on the part of the brain.

We need to note two facts before getting down to the explanation. The first is the difference between what are known as *pitch height* and *pitch class*. Pitch height simply refers to the absolute frequency; one can think of it as being given unequivocally by the position of the corresponding note on the piano keyboard. Pitch class is fundamentally different, because it is cyclic; it merely defines a tone's position within an octave. The difference between the two concepts is revealed by the fact that the question *which is higher C# or D?* cannot be answered unless one is also given the two pitch heights.

The important thing about both these illusions, in the present context, is that they arise from the influence that higher brain processes have on lower ones. Such higher processes are syntheses of many factors. In the Shepard illusion, for example, they can avail themselves of the overtones, to anchor the fundamental in its correct position within the pitch class. One could say that the concept is influencing the percept, in both the above illusions. And this is made possible by those reverse projections that were given so much prominence in the latter part of chapter 7.

Turning to the sense of touch, there are again some quite startling deceptions. These appear to bamboozle in the most innocent of ways. A good example was first described by Frank Geldard and Carl Sherrick in 1972, and it has been given the nickname *cutaneous rabbit*.[25] The resting arm of a subject is given a rapid series of taps at several different positions along its length, and in a systematic fashion. The time lapse between taps usually lies within the range of 50 to 200 milliseconds. Four taps are delivered at the wrist, for example, followed by four at the elbow and four at the upper arm. But this is not what the subject perceives. Instead an evenly spaced series of taps is felt running systematically up the arm, as if a little rabbit were hopping along it.

In trying to rationalize this illusion, we must take note of two important facts. One is that not all stimuli delivered to the skin surface necessarily cause excitation in the somatosensory cortex and an associated sensory awareness. There are control systems which can modify transmission of activation from the skin, as it makes its way to the cerebral cortex. These systems can be very potent, and they usually have their base in the brain itself. The other significant thing is the frequency of the taps. At the beginning of the preceding chapter, when we were contemplating the issue of cortical mechanics, we saw that signals take about 250 milliseconds to make the round trip from sensory input to the appropriate association area and back again to the relevant early cortical area. Taps administered at intervals less than this are going to be fair game for distortion, and the higher cognitive function will rationalize the sensation to the closest perception compatible with what it has experienced before. It would probably be no exaggeration to classify the hopping rabbit as another product of Gestalt justification.

Phantom sensations

People who have had arms or legs amputated will frequently complain of pain in the missing limb. These troublesome sensations were not well known around the time of the American Civil War, and when the prominent neurologist S. Weir Mitchell published his first account of the phenomenon, he submitted it as an anonymously authored short story to the *Atlantic Monthly*, rather than to a scientific journal; he feared the professional criticism of his peers. A large body of clinical evidence has now established the various characteristics of what is referred to as the *phantom limb effect*, and one of the most striking of these is the way in which the missing appendage even seems to participate in normal body movement. Its coordination with the surviving parts of the body invariably appears to be perfect. When its owner sits down, for example, a phantom leg will be felt to bend in unison with its physically present counterpart. The eerie reality of these ghostly members is further reinforced by their ability to mimic the feelings of heat and cold, and even of itching.

An early theory of the phantom sensation attributed it to signals emanating from the surviving nerve endings at the point of amputation. The severed nerves grow into nodules called *neuromas*, and these are known to send impulses to the somatosensory cortex, via the spinal cord and the thalamus. This idea led to curative surgery in which incisions were made at various places on this route, including the location of the neuroma in the stump itself.

Although this does bring a measure of relief to the amputee, sometimes even for a period of years, the cure is not permanent. A later theory placed the seat of the trouble in the spinal cord, the suggestive observation being that neurons in this structure spontaneously generate volleys of impulses. But paraplegics who have suffered total disjunctions high up in the spinal cord nevertheless feel pain in the legs despite the fact that the nerves transmitting signals from the latter are thus isolated from the brain.

Yet another idea, due to Frederick Lenz, is that the ghostly feelings stem from structures even farther up in the system, such as the thalamus and the somatosensory cortex itself. One could think of this as a last-resort theory, because the latter structure is the terminus on the tactile route. Even this extreme view was not entirely supported by clinical evidence, however; in some cases, removal of the affected parts of the thalamus and somatosensory cortex *still* fails to alleviate the pain.

It was in the face of this impasse that Ronald Melzack[26] proposed a radically new approach. He advocated a holistic view of the sensory system, and postulated the existence of what he called a *neuromatrix*. This was perceived as a network of neurons which has the dual task of responding to sensory stimuli and continuously generating impulses. The latter are to be distinguished from those that arise as reactions to incoming signals. Melzack saw these other impulses as being spontaneously emitted, their task being to inform the body that it is intact and unequivocally its own. He coined the term *neurosignature* to describe the characteristic pattern of impulses which carry the body's information about itself.

Melzack's neuromatrix would have to be a rather extensive affair. It would need to incorporate all the structures invoked in the earlier theories, and it would also have to include centres that are involved in emotion and the sense of identity. This would implicate the limbic system and parietal lobe, the latter being known to give an individual the perception of self. Although plasticity is one of the important characteristics of at least part of any network of neurons, as I have stressed several times, Melzack put forward sound arguments for a strong genetic component in its basic structure. He and his colleagues have examined many patients born without one or another limb who nevertheless experience pain in the phantom member.

The neuromatrix bears a striking resemblance to the global closure model that was discussed in chapter 7, but it has an important additional component, namely the spontaneous signals that collectively constitute the neurosignature. In its original form, on the other hand, the neuromatrix was perceived as being less extensive than that other model, because it was primarily addressed to the sense of touch. The global closure model envisaged all the senses, the entire cortex, the limbic system and the brainstem as forming a single network, the counter-running projections and the closed loops having the responsibility of broadcasting messages from any given area to the rest of the system. (That other model also included loops mediated by diffusing molecules rather than nerve impulses, and in this respect too it was broader than Melzack's neuromatrix.)

Recent studies have indicated that the two models may be even more intimately related because phantom effects have also been seen in vision and hearing. In 1769, Charles Bonnet wrote of the remarkable images reported by his cataract-plagued grandfather, Charles Lullin. Many accounts since that time have described the figures and objects paradoxically witnessed by others with the same affliction. A noteworthy aspect of the typical phantom episode is that it involves people never before encountered, and events never before experienced. The audible counterparts of these apparitions are every bit as disturbing, the most distressing being the incessant whistling sound that torments many people with impaired hearing. This is known as *tinnitus*. Some sufferers of this affliction experience a more structured series of sounds, which may even resemble music and speech. Perhaps the most dramatic report on this subject was one that described the spurious visions of a woman who had lost a large portion of both her vision and her hearing. One incident conjured up a circus, and she was able to give a lively account of the various acts and the accompanying music.

Although these occurrences are not pleasant for the patient, they do reveal important facts about the nervous system. It seems that the brain spontaneously generates its own experiences if it is denied sensory input for too protracted a period; the conscious brain appears to abhor idleness. And the fact that these experiences can be quite novel indicates that the brain is able to piece together its own scenarios, as if drawing on a set of basic elements that can be arranged in many different ways. This is reminiscent of what happens during periods of sensory deprivation. It also bears a marked similarity to the dreaming state.

I will close this section with two final suggestions, one made by Melzack and one of my own. He advocated the view that sensory inputs merely modulate that experience of the body which the brain is actually capable of providing for itself. I would like to add that the observations made by him and his colleagues might have a connection with what we considered in the previous section, namely the phenomenon of filling in. This connection would be a natural consequence if it is *mandatory* for the brain to produce a response to any stimulus, irrespective of how ambiguous or incomplete that input might be. If the connection that I am suggesting is a valid one, it raises an important question because we saw that filling in can be interfered with by attention; those dividing lines between the different shades of grey evaporated only when we were not concentrating on them. There is thus a strong suggestion here that *attention interferes with a process that would otherwise proceed automatically*. What this does not explain, on the other hand, is the actual nature of awareness. There are other aspects of the story that we still have to consider, however.

Deferment of automation

If filling in and phantom perceptions are related, it becomes important to ask how they are associated with the minimal condition discussed at the beginning of the chapter. Let us assume that the neuromatrix is present in all organisms that possess a nervous system. In simple creatures, whose systems are apparently *always switched on*, it could be that this is essentially all there is. Sensory input would superimpose itself on the internally generated pattern of signals, as in all species. This would create a fleeting disturbance to the neurosignature. The result would be an appropriate response on the part of the input's recipient, irrespective of how much that reaction was genetically wired into the system and how much it was learned from past experience.

What such a response might not include, however, is the *ability to anticipate*. The possessor of this apparatus would not necessarily be able to project into the future, and use its predictions to figure out before hand its best tactic. This is an obviously desirable faculty for any animal, and it is interesting to contemplate its equivalent in the provinces of mathematics and engineering. In the former, it goes by the name *extrapolation*. If the trend of a variable can be sufficiently well described over a given interval, it is possible to predict its behaviour in regions outside that interval. The better the characterization, the more reliable is the prediction and the greater is the extended interval over which it can be applied. This procedure is very common; one plots a graph using a given set of points, draws a smooth line through them, and then uses the extension of that line to make one's prognosis. The forecasting of weather, although far more complicated because it involves many variables, is merely another example of this same strategy. Those who play the stock market dream of being able to make reliable extrapolations.

Likewise, in engineering practice, control often has to be of the feed-forward type, and one could say that the designer attempts to give his machine the power of extrapolation. In chapter 7, we saw that this is embodied in the strategy known as *predictor–corrector*, and we learned that its biggest difficulty arises from non-linearity, which essentially means lack of simple proportions between stimulus and response. This book is concerned with a machine that can be dauntingly non-linear if it is owned by a sufficiently advanced creature. This complexity notwithstanding, the numerous processes that are carried out within the device, and indeed *by* it, have to be controlled. As we have just been discussing, the nervous system appears to be forced to respond to every stimulus; it has to come up with a reaction,[i] no matter what is thrown at it by way of sensory input.[j] It also has one considerable advantage over the engineer: it does not have to work via mathematical equations. Its underlying structure is genetically dictated, and it can build upon this endowment by learning from experience. These twin factors determine the system's modelling capabilities.

An internal model can have considerable predictive utility. There is one which gives the frog its ability to shoot out its tongue at just the right moment to catch a passing gnat. And there is another which gives the bat its

i / I am including in this repertoire of possible reactions the decision to remain immobilized.

j / The one proviso that ought to be added here is that the brain seems to be able to spare itself the bother if the input is merely random noise.

even more remarkable proficiency at intercepting a moth in flight, despite the fact that its victim is itself no mean aerial performer. Yet these impressive feats of anticipation may nevertheless be nothing more than sophisticated examples of automation. Is it possible that some modelling systems are much more impressive than the above examples? In particular, are possessors of conscious awareness capable of more complicated anticipatory feats?

I believe that this is the case, and I will start my argument by recalling some of the things that have already been described. One of them may have appeared to be peripheral, but I believe that it is very important. Phantom visual sensations were reported as frequently involving things that had never been experienced, and this was taken to indicate what could be called an innate modularity in human imagination. (Whether it could also be present in other primates, for example, is something that will have to be considered in the next chapter.) Let us also recall the conceptual linking of obligatory responses to the phenomenon of filling in. Finally, let us remember that filling in is *not* a feature of something that is being attended to. Putting these items together, we arrive at the conclusion that the benefits of attention may include deferment (or possibly even suspension) of automation and exploitation of the advantages inherent in modularity.

There are still several other major issues to be contemplated before we can consider drawing conclusions as to the nature of consciousness awareness. It is tempting, however, to take stock at this point, and put what has just been discussed into a broader perspective. In so doing, we will be using our power of projection, the very faculty that we have just been deliberating. We have seen that things outside our immediate attention appear to be handed over to automation. This suggests an automatic nature for the unconscious. Then again, because the neuromatrix must have its own rate of self-influence, dictated by the speed of signalling between its component parts, conscious awareness may provide the system with its only means of learning things that have novel timings and sequences. This is a more formal way of putting a point stressed by William James. He noted that a major task (and possibly *the* major task) of consciousness awareness is the handling of surprise.[k] It is the unexpected event which is invariably of the greatest interest to an organism. If that organism is primitive, unusual events may be able to impress themselves only by their alteration of the statistics of the conditioning stimuli. And because these statistics will change only gradually, the creature will not be able to respond immediately to the novel event. In more advanced species, attention, and the associated conscious awareness, may accelerate the rate at which the unforeseen occurrence can be assimilated into the repertoire of situations that must be handled. If the surprise is great, and

k / Surprise provides the basis for a large part of humour, of course, and it stems from the conflict between the actual and the anticipated. Anticipation is probably an inevitable product of unconscious mental processes, and is thus a close relative of filling in. It seems possible that it cannot be avoided even through conscious effort.

of sufficient significance to the animal, its assimilation may even be immediate. As we have already seen, it is the emotions which serve to adjudicate significance.

The performance of a frog in catching a gnat was cited earlier as a good example of the type of prediction that can be mediated by a nervous system. It is an accomplishment which we ourselves would not be able to match, and we thus find it impressive. It is not difficult to expose this creature's limitations, however. If a bead is swung before it, on a thread, the frog will go for this too, subsequently spitting it out only when it has proved not to be the assumed tasty morsel. Although we are not as quick as a frog, relatively speaking, we seldom make this sort of blunder. Our superiority stems from our more advanced cognitive abilities, and although the implementation of these consumes a certain amount of time, they nevertheless give us a decisive edge.

What is the nature of this advantage? Broadly speaking, it is the ability to distinguish between stimuli that are more complex than those that can be handled by simpler animals. Such distinction requires sophisticated machinery, however, and we have already had hints that its specialty is modularity. There is in fact plenty of evidence for this, on a gross scale at least. Our command of spoken language stems from our ability to string together successions of phonemes, the variety of which is relatively limited. Similarly, the thousands of written words that comprise even the more modest personal vocabulary are constructed from a much smaller selection of individual letters. And these letters themselves have a modular construction that is based on a series of line elements. The principle is reminiscent of the (metallic) Meccano construction sets that were very popular during my schooldays; given a few pieces of each basic element, and a copious supply of nuts and bolts, one could build all manner of interesting structures. Today's (plastic) equivalent is called Lego; putting the pieces together is easier in this system, but the underlying idea is the same.

A prerequisite for filling in is lack of focussed attention, and the process appears to be accomplished by what has been referred to as the neuromatrix or globally closed network. This system has characteristics which are acquired by the dual agencies of genetics and experience. Filling in thus appears to be automatic, and steered by what the system already comprises. This raises the question of whether the only things that can be remembered, and thus added to the system's repertoire of automatic responses, are those that have been the subject of attention. The answer must be *no*. If an animal does not possess an attention mechanism, conditioning would be the only way of altering the system. When such a mechanism is present, it would at least be expected to take preference over unconscious learning. Unconscious learning will still be possible, however, as those subliminal television commercials demonstrate only too well. But it seems probable that things learned unconsciously will also influence our reactions in an unconscious (or sub-conscious) manner. The surreptitious commercial might steer us to a particular brand of goods in the supermarket without us ever being aware of the fact.

All this has a familiar ring; it is evocative of the discussion of memory mechanisms that appeared in chapters 5 and 7. Let us recall the facts. Removal of the medial temporal lobe, including the hippocampus, in humans leaves both short-term memory and long-term memory intact, but it severely impairs the transfer of items from the former to the latter. Although it is still not known how many different forms of memory there are, it does appear that this removal (surgically or by injury) seriously handicaps those forms that require a conscious record. These latter are called *declarative* or *explicit memory*. Forms of memory that do not require the participation of consciousness are said to be *non-declarative* or *implicit*, and such classes are not affected by temporal lobe lesions. Finally, explicit learning is fast, while implicit learning is slow.

Adding these facts to what we have just been discussing, it seems that a synthesis is possible. Implicit memory looks very much like what we have been attributing to the neuromatrix or globally closed network. It has the required slow learning speed, as well as the independence of attention (if we loosely link this to conscious awareness, that is). Explicit memory, likewise, bears a striking resemblance to what we have suggested is the simultaneous handler of surprise, exploiter of modularity, and relier on attention. Those animals which possess only the machinery for implicit memory thus begin to be seen as the victims of circumstance. Their future is determined exclusively by their past conditioning and by happenstance; if something unexpected crops up, they will be unable to find a new strategy on the spur of the moment. They are rather disadvantaged.

A consciousness agenda

We are beginning to home in on conscious awareness. It is not that the discussion has produced a crisp definition, but the concept is becoming delineated by its beneficial consequences. We will consider more of these later. It would be a good idea to halt at this point, however, and ask whether it is known just what one would have to explain in order to produce a successful theory of consciousness. We are going to have a hard time trying to construct a hypothesis if we do not know what we are aiming at. So is there an inventory of such things, is there what could be called an official agenda? Insofar as there is not universal agreement on the issues, the answer is a regrettable *no*. But it is possible to compile a reasonably full list of items to be elucidated, even though it is difficult to say whether this would be exhaustive. One of the most active in this respect, in recent years, has been John Searle, whose writings on the subject are extensive.[27] Let us contemplate the characteristics that he has identified, commenting upon them as we go.

The most important attribute of consciousness, in Searle's opinion, is *subjectivity*. We must explain how processes at the neuronal level collectively give us our sense of awareness, a sense which we privately experience. A subsidiary issue concerns the *qualia* that often accompany this awareness; we would also like a neuronal explanation of these feelings. Max Velmans has addressed the apparent difficulty of explaining subjective experience in what are necessarily objective terms.[28] In what he calls a *reflexive model*, he proposes that the external phenomenal world be viewed a *part of* consciousness rather than *apart from* it. Observed events are then seen as being public only in the sense that they are shared private experiences, and scientific observations are objective only in that they are inter-subjective. Velmans believes that the gap between the physical and the psychological domains can be closed by noting that observed phenomena are repeatable only in that they are sufficiently similar to be taken as tokens of the same class of event.

Another item on Searle's shopping-list is the *unity* of conscious experience, what Immanuel Kant referred to as 'the transcendental unity of apperception'. In keeping with the currently favoured terminology, I have been calling this the *binding problem* in the present book. It refers to the fact that our experiences occupy a single conscious field, irrespective of whether we are sensing external events or are occupied by our thoughts. A related issue concerns the *continuity* of the stream of consciousness. Our experiences and thoughts are retained for a few seconds, as if they were being recorded single-file on tape, and this enables us to build upon them.

Searle also stresses the importance of *intentionality*, by which he means that mental states are usually related to, or directed toward, external situations and circumstances. He cedes the point that this is not always the case, however, notable exceptions being the states of anxiety, depression and elation that are merely nebulous moods. One aspect of intentionality concerns *choice*, irrespective of whether this implies the exercise of free will. Even if choice were not really free, the fact that we are able to handle it would still warrant contemplation. Searle's point is well taken. In the science of thermodynamics, there is no place for the word *purpose*, systems merely being treated on a statistical basis. In the eyes of the thermodynamicist, no arrangement of a given set of atoms is any more significant than any other arrangement. By highlighting intentionality, Searle has identified one of the defining characteristics of the higher organism.

Then again, as Searle points out, there is *the distinction between the centre and the periphery of our conscious awareness*. He cites the example of not being conscious of the tightness of one's shoes when something else is being attended to. But we can easily switch our attention to our shoes if we so desire, at the expense of ceasing to concentrate on what occupied us earlier. There is again a corollary issue; our perceptions form natural hierarchies, some things occupying centre stage while others are relegated to the background.

Searle also identifies *familiarity* as something that requires explanation, his point referring to the fact that we will usually not have seen recognized items from precisely the angles they are currently being viewed from. Likewise, we understand words spoken in our native tongue even when they are intonated in an unfamiliar accent, and we recognize melodies even when they are played by unfamiliar combinations of instruments. As Searle puts it, we seem to have prior possession of categories for all the things we comprehend. Later on in this chapter, I shall be attaching particular significance to this issue, when I discuss things referred to as *schemata*.[29]

Finally, there are what Searle calls *boundary conditions*. He notes that conscious states are embedded in what one might call their '*situatedness*'. Although we do not constantly have such things in mind, we are nevertheless always aware of where we are, of roughly what time it is, and of the season and the (at least approximate) date. Conscious states are thus experienced in the context of a situation.

Searle's compilation is valuable in that it serves as a guide to what a theory of consciousness must account for. But it does not provide us with a definition. Indeed, Francis Crick and Christof Koch suggest that it might be better not to seek a definition of consciousness at this time.[30] They make their point by noting how difficult it would be, for example, to give a neat definition of a gene. That analogy is not an ideal one, however, because a gene is not a process whereas consciousness appears to be one.

In default of an actual definition, we might note that a successful theory of consciousness will have been found when the top-down and bottom-up approaches to the issue have been fully reconciled, that is to say when the phenomenological aspects of consciousness have been provided with explanations in terms of the underlying processes at the neuronal level. A particularly lucid formulation of the top-down approach has been provided by Gerd Sommerhoff.[31] As in the strategy advocated by Gerald Edelman,[32] he specifically set his sights on what can be called *primary* consciousness, that is to say on the mechanism that produces the basic, sub-verbal awareness we must share with the blind deaf-mute, presumably also with the neonate, and possibly also with some non-human species. In adopting this strategy, Sommerhoff aspired at least to *tame* the problem of consciousness, just as some still-unsolved problems in particle physics and cosmology could be said at least to have been tamed.

Sommerhoff notes that the dictionary consensus on the subject defines consciousness as *an awareness of the surrounding world, of the self, and of one's thoughts and feelings*, and he concludes that these attributes can be jointly accounted for by the presence of a mechanism that integrates three main categories of internal representation. These categories are: representations of actual objects, events or situations, jointly amounting to a comprehensive representation of the current structure and properties of the environment (Category A, in Sommerhoff's nomenclature); representations of hypothetical objects, events and situations (Category B); and finally, representations which relate to internal representations or individual stimuli, and represent the fact that these representations or stimuli are part of the current state of the organism (Category C). The Category C representations thus play the vital role of combining the current representations of the other two categories. The upshot, according to the Sommerhoff scenario, is an integrated global representation (IGR), which operates as a coherent functional unit in the organism's transactions with the environment. We will do well to keep these top-down considerations in mind, and I shall attempt to marry them to the bottom-up mechanisms that have been this chapter's primary concern.

Perhaps the difficulty in finding a suitable definition of consciousness is related to something that was emphasized in the first chapter, namely that one must distinguish between the underlying mechanism of consciousness and its mere consequences. I will later be making a guess as to the nature of that mechanism. Meanwhile, let us return to those beneficial consequences of consciousness. Another of these is thought. Let us give it our attention.

Trains of thought

William James was in no doubt as to the primacy of two of our faculties, namely *discrimination* and *association*, and his analysis of brain function hinged mainly on these attributes. Indeed, he saw discrimination and association as serving all cogitation through a sort of relay race,[1] in which the concept currently at the centre of one's conscious awareness is continually passed back and forth between these two endowments. And James noticed two quite profound things about association: items in memory tend to be recalled in sequences, but each element is not merely cued by the thing that immediately precedes it. In other words, we tend to experience *trains of thought*, and these trains are dictated by forces that are not purely local. James concluded that things stored in memory are multiply connected, and that the context for recollection stretches over many other items in the sequence.

Regarding the actual mechanism of associative recall, James was intrigued by the fact that this sometimes takes a considerably length of time, even though one has the distinct feeling that the missing item is *on the tip of one's tongue*. His conjectures about the underlying dynamics came close to invoking the sort of mechanisms that are now so common in writings on neural networks. But the interacting units in his analysis were clearly larger than individual neurons, because they were able to store entire concepts. In the James theory, the way in which the focal point of nervous activation shifts from one item to another, as one strains to dredge up the missing information, bears more than a superficial resemblance to modern ideas about associative recall. And James saw habit amongst the neural elements as being the chief agency that dictates the course that a train of thought takes. *When two brain processes are active together or in immediate succession* he wrote, *one of them, on reoccurring tends to propagate its excitement into the other*. If one replaces the term *brain processes* by *neurons*, one obtains a rule for learning that is strikingly reminiscent of the Hebbian mechanism that has been invoked several times in this book.

Although the writings of William James were, perforce, couched in terms that are primitive by today's standards, he was in no doubt as to the lack of anything mystical in the mechanics of the mind. *The order of presentation of the mind's materials*, he wrote, *is due to cerebral physiology alone*. He used the term *irra-*

1 / A slightly better analogy would be the less common type of race known as a *parlauf*, in which just two runners alternate in carrying the baton around a closed circuit.

diations to describe what would now be called sets of nerve impulses, and he believed that it would be a long time before physiologists would be able to follow the route taken by these packets of activity, under a given set of mental conditions. Indeed, he expressed the opinion that this might never come to pass. But *never* is a dangerous word in science, and James would no doubt have withdrawn this pessimistic prognosis had he been able to witness first hand the sophisticated and precise types of monitoring that are routinely carried out nowadays. Just as importantly, access to the sort of detail that we now have of cortical anatomy would probably have put James on the trail of the mechanics of conscious awareness. Let us see how far one can now get with that quest.

It seems safe to say that all objects of conscious awareness can be divided into two broad categories. There is awareness of things in our environment, and there is awareness of things that are exclusively in our thoughts. Awareness of the external world would appear to implicate a greater share of the brain's resources, because sensory input is involved. A fuller picture will emerge, therefore, if we consider that particular manifestation of the phenomenon. In the preceding chapter, we saw that the neural activity pattern corresponding to a particular percept is generated within 50–100 milliseconds of the sensory stimulus. We also learned that as the percept gives rise to a concept, there is a backward influence on perception itself, mediated by the cortex's reverse projections. After the initial brief period, therefore, the neural activity pattern is gradually modified, increasing degrees of feedback exerting an ever stronger influence.

We should bear in mind the interesting finding by Semir Zeki and Stuart Shipp, that the reverse projections tend to be more spread out than their forward counterparts. We noted that this would promote the capturing of correlations between various features in the sensory input. A good example of this process at work is seen in Oliver Selfridge's famous two words

THE CAT

There is no physical difference between the sloppily drawn H and the equally careless A, but the contexts are sufficient to remove the ambiguity in each case. And those contexts are themselves the product of concept-directed perception.

The influence of concept on percept causes us to miss even the most glaring of typographical errors when reading; once the gist of a text is grasped, one starts to form mental images that can actually influence the way in which the printed words are scrutinized. This effect must be general; there is no reason why it should be limited to the written word. The mental images are provided by the associations that are already resident in memory, and their conjuring up has all the signs of being automatic. They can be regarded as a dynamic form of filling in. A proof reader wishing to avoid missing typographical errors will often resort to a tactic that defies automation, and aids awareness, namely reading the text backward. When that method is used, the incorrect spelling of a word can more readily be detected. But what is the actual mechanism of that detection? Why does it have conscious awareness as a prerequisite? I believe that it all has to do with timing, and with the duration of very short memory, and with something that I tend to think of as the device of *unfinished business*. Let us see how this works.

Unfinished business

In the first chapter of this book, and also in a previous section of the present chapter, I wrote of top-down and bottom-up approaches to the brain. The problems we are addressing here have no monopoly on such division. On the contrary, diversity in the possible lines of attack crops up quite frequently in science. A current goal of physics, for example, is the complete description of nature's four fundamental forces, and a full account of the relationships between them. If recent indications are reliable, there appear to be grounds for cautious optimism that this will be accomplished. Much of the recent work in this area has focussed on the microscopic realm of sub-atomic particles, which is an example of the bottom-up approach *par excellence*. But the history of fundamental forces has also enjoyed its top-down successes. Newton's falling apple and Oersted's deflected compass needle heralded the greatest of these. The top-down and bottom-up approaches will have met when a sub-atomic explanation is found for the force of gravity, and for its relationship to the other three fundamental forces, which have already been brought under one roof.

Where will the top-down and bottom-up approaches to the brain find their meeting place? I believe that it will be in *the mechanism which enables us to know*. One could call this mechanism *the comparator*, and we obviously would like to know how it works at the neuronal level. I believe that an important consideration here is *time*, and that this relates to that other great schism that we have been identifying, namely the one that separates unconscious awareness from conscious awareness. Unconscious processes are parallel and automatic, conscious processes are sequential and negotiable.

This is all very well, but we still have to ask how the comparator actually works, and why it needs the time dimension. I believe that the various bits and pieces of the mechanism are already to hand in what has been described in the preceding chapters. Let us take a whirlwind survey of them, and attempt to synthesize an overall picture: the brain in a nutshell, as it could be called.

Information emanating from the environment makes impressions on the various types of receptor cell, and thereby causes signals to be dispatched from the different senses toward the deeper recesses of the nervous system. Most of these signals are not passed to the brain, travelling instead to neurons which are able, unaided, to elicit responses. These responses occur unconsciously; they are automatic. As the evolutionary rank of the species in question becomes progressively higher, more and more of the signals are transmitted to the cerebral cortex, however, many of them passing through the thalamus *en route*. In the awake state, the signals from the various sensory modalities enter the cortex at different places, and they do so in versions that are modified by the vergence (di- and con-) that always occurs when signals are projected from one group of neurons to another. Each modality is served by a number of cortical areas, vision claiming the majority share in our species and in our close cousins. The signals, continually modified as they travel, progress via the forward projections through the various areas as through a hierarchy, and they do so in a parallel distributed manner, ultimately to arrive at the appropriate association areas of the cortex. Further modified by the associations that lie collectively stored in those areas, they provoke signals that continue onward toward the early areas of the other sensory modalities, this later part of their journey being mediated by the reverse projections in those target senses. Signals in one sensory modality are thereby able to provoke signals in other modalities, as when vision conjures up auditory associations, and vice-versa.

At appropriate positions in the array of cortical areas, signals are tapped off for projection either to sub-cortical regions of the brain or to the motor cortex.[m,33] The sub-cortical targets are also the sources of signals sent *to* the brain via separate projections, and this arrangement establishes loops. Some of these loops are special in that they permit control over the sensory input, a particularly noteworthy example being the thalamus, which is therefore seen to be much more than a mere relay station between sensory input and the cortex. Indeed, the thalamus is under even further control in that there is a more circuitous route that runs from the cortex via the brainstem reticular formation (also known as the reticular activating system) to the thalamus, and then back to the cortex, through the gating control of the nucleus reticularis thalami. This route is special in that it can even control the degree of arousal of the animal.

The importance of precise timing to the overall system is apparent in the structure and functioning of several of these loops. The one that involves the cerebellum controls the timing of movements. Finer control is served by the basal ganglia which, unlike the cerebellum, do not send signals directly to the spinal cord. Timing is also of the essence in the hippocampus, which, through its connections with many cortical areas, imbues the animal with a sense of its own personal space, but which also mediates remembering that is contingent on conscious awareness. Because all events relevant to the animal

m / One important role for those sub-cortical projections is to hold the reflex routes in check, by inhibiting them, and to permit those reflexes to act automatically only when cortical processing indicates that it is desireable, the 'permission' taking the form of disinhibition. Direct evidence for this type of mechanism is provided by the functioning of the superior colliculus, though this brain component is itself sub-cortical rather than cortical.

are related to its personal space, these two things are intimately connected, and the role of timing may be akin to that familiar in radar.[n]

The various signalling processes consume a certain amount of time in reaching their destinations, primarily because of the slowness of signals in dendrites, and modification of the signals during early sensory processing thus occurs unconsciously. The signals also have a certain duration or lifetime in the system, this *reverberation* appearing to last several seconds; it can be regarded as a very brief form of memory, though it is not clear whether it involves synaptic plasticity.[o] Conscious awareness of a sufficiently strong stimulus develops after about half a second, which is the time required for neuronal adequacy, as measured by Benjamin Libet and his colleagues.

Although we still have to decide what conscious awareness actually is, let us consider a consequence of these temporal delays and reverberations in the system. Action potentials arriving at a given cortical area will produce two main responses. There is the excitatory response, by which activation is passed on to subsequent areas in the cortical hierarchy, and there is an inhibitory response which causes the activity in the given area to diminish, and ultimately to die out altogether. This inhibition will in general arise through two familiar mechanisms, namely the feedback and the feed-forward varieties. We have already seen that the latter type can set up a cloud of inhibition which will then exercise a winner-take-all constraint on the arriving signals. Those signals that succeed in getting through this cloud, by dint of their strength and coherence, will proceed up the hierarchy and ultimately give rise to return signals, which travel via the reverse projections. They will also leave behind them a residual pattern of inhibition which arises from the feedback mechanism.

The information carried in the reverse direction will be the product of automation. Indeed, it will be a generalized form of filling in. If the original input is fully recognized by the system, the activity pattern set up in the original area by these return signals will not differ from that set up by the original input. If we assume that the pattern of feedback inhibition faithfully reflects the originally injected pattern of activity, and that seems like a reasonable assumption, this means that the return pattern will precisely match the inhibitory cloud, and the result will be instant extinction. One could say that there will be no *unfinished business* to attend to. If, on the other hand, the returning signal does not match the residual inhibitory pattern, there will be some remnant activity. This will be concentrated around the mismatch between the input and return patterns, and it would be available to provoke the appropriate responses by the system. It seems likely that one aspect of this response would involve continued attention, so as to give the system a better chance to

n / This analogy does not refer to the emission and reception of radiation, of course. It is the mechanism of precise timing which is important.

o / If it does involve synaptic plasticity, the important receptors may well be the NMDA-type glutamate variety.

examine the origin of the discord. Conversely, the lack of mismatch would presumably be registered by the brain's reward system, which was discussed in chapter 7. I believe that this is the way in which the comparator functions, and its ability to detect novelty is clear.

It would be easy to underestimate what possession of this comparator mechanism would do for an animal. As an example of its usefulness, let us consider the theorem of Pythagoras. The procedures required to implement its logic are not wired into the brain's neural networks, waiting merely for the lengths of the triangle's sides to be fed into the system. What *is* in the system? Let me reminisce for a moment on the time when I learned the theorem in school. I sat there in the class-room, surrounded by my fellow pupils, and the teacher informed us that we were going to be shown a piece of mathematical elegance. She started by *reciting* the proposition: *the square of the hypotenuse of a right-angled triangle equals the sum of the squares of the other two sides.* Then she took the 3–4–5 triangle as a specific example, and demonstrated the correctness of the proposition by adding 3 times 3 to 4 times 4, and noting that the sum equalled 5 times 5. My *whole body* registered the equality of the two 25s, arrived at by their separate routes. Numerous loops within my nervous system contributed to the feeling of well-being at the successful outcome of the teacher's little calculation. My muscles relaxed; I felt at one with the entire atmosphere of the occasion; no doubt, my various glands were registering the harmony of the situation. The teacher then used the 5–12–13 triangle as an encore, and my body became even more relaxed. *I was being told that there was no unfinished business, and my synapses were adjusted accordingly.* And what did I take with me from the class-room on that auspicious day? It was the memory that there was someone called Pythagoras, and that his proposition had something to do with right-angled triangles and the squares of their sides. A few days later, when the teacher touched on the subject again, I actually learned to recite the proposition, word perfect. It has lain in my memory ever since, and its insertion into that repository was mediated by my brain's reward system, which was steered by that lack of unfinished business.

In view of what has been said about the way the system could function, with the comparator checking reversely projected signals against those of the input, how should memories be characterized? It seems logical to conclude that they are *stored associations between stimuli and consequences.* Moreover, from the work on filling in, it seems that the comparator mechanism is functioning most effectively when attention is present. When attention is absent, the ability of the system to discern unfinished business is apparently diminished, and there is a tendency toward automation. The idea of stored associations is in tune with a more limited suggestion by Rodolfo Llinás and András Pellionisz concerning the coupling of sensory input to muscular output. They were able to demonstrate that such coupling can be rationalized in terms of the mathematical concept known as a tensor, and they successfully applied their theory to the specific coupling of the vestibular sense to the movements of neck muscles, in the cat.[34]

Another idea somewhat related to the one I am advocating is seen in the neural network strategy known as the *Boltzmann machine*, which was first described by David Ackley, Geoffrey Hinton and Terrence Sejnowski.[35] The gist of their approach is that the network is able to explore the vast number of possible couplings between input and output, in a random manner, small adjustments being made to the tuning of the system (which is to say the synaptic strengths) every time a better fit is found. In practice, to stop the system getting bogged down, a trade off was permitted such that a temporary worsening of the input–output match was tolerated in order to ultimately reach a more satisfactory solution. Finally, mention should be made of the work of David Zipser and Richard Andersen, in which they asked where the equivalent of the output units of a perceptron could exist in the brain.[36] After considering several possibilities, they concluded that the final spatial output might exist only in the behaviour of the animal; it may be found only in the pointing of the eye or finger accurately to a location in space.

The theatre that never closed

The most difficult problem facing anyone set on explaining brain function is to expunge the idea of what Daniel Dennett has called *the Cartesian Theatre*.[37] One laughs, these days, at the concept of the homunculus: that little being in the head who was supposed to play the triple role of observer, arbiter and activator. But the apparent independence of the imagination, which I wrote about at such length in the second chapter, still fools us into believing that mental events are somehow played out on an internal stage. This idea was particularly prominent in the writings of René Descartes, and it was for that reason that Dennett facetiously named the edifice after him, skilfully showing it to be a theatre of the absurd.

And yet the idea stubbornly persists, by implication at least, in much of the writing on the subject. It is still common to read of attempts at explaining how the line-detecting neurons discovered by David Hubel and Torsten Wiesel could feed into neurons that would be able to detect more composite aspects of a visual scene. That path leads to the grandmother cell, amongst other things, and to the question of what is to observe it in action.[p] In this section, in which I will start my long final build-up to an explanation of conscious awareness, I am going to suggest that there *is* a theatre that is relevant to brain

p / A better indication of the way in which the visual system handles information is provided by the caricature, which enables us to recognize the face of a familiar person, even though no facial features are drawn in their proper physical locations. Indeed, the more grotesque the artist's distortion of the facts, the easier it is to identify the person being represented. Recognition clearly depends on relative rather than absolute position, and this must be gauged deeper in the visual hierarchy. Another important factor is cognitive compression, in which different inputs produce the same effects deeper in the hierarchy. This may contribute to the phenomenon of *déjà vu*, incidents apparently experienced before merely resembling a previous episode.

function, and that Descartes merely made an error regarding its location. It is, I would like to suggest, a theatre that has never closed.

Before saying where this arena actually is, let me recall some things that have been stressed in earlier chapters, and construct upon them. Firstly, there was that fascinating analysis in which Philip Johnson-Laird and Keith Oatley developed their cognitive theory of the emotions, showing that all our feelings can be pigeon-holed into a large but not unlimited set of categories. Secondly, there were a number of points regarding the hippocampus. There was, for example, the idea that this part of the brain provides the nervous system with a cognitive map, thereby giving the animal a sense of its personal space. I suggested that it is able to do this because of its sophisticated way of handling the timing of action potentials, perhaps in a manner roughly analogous to the time-processing techniques that underlie radar. There was also the impressive evidence that the hippocampus mediates remembering of events specifically related to conscious awareness. Above all, we must note that the hippocampus is a prominent member of the limbic system, which is known to be involved in the emotions.

We are far from done with this synthesis, but let us note in passing how spatial our metaphors become when we speak of mental processes. As the reader may have noticed, I have a weakness for the word *approach*, when referring to theories, and one often speaks of *advancing* a hypothesis and *retreating* from a belief, or *position*. We talk of *grasping* the point, of *seizing* the initiative, of *holding* an opinion, and of *dropping* a hint. One has an idea at the *back* of one's mind, one considers the evidence *in front* of one, one contemplates it from all *sides*, and comes to the conclusion that it is either *above* one's understanding or *beneath* contempt. And when we are sufficiently aroused, we start to gesticulate, using bodily positions and movements that are the physical counterparts of these metaphors. We literally *throw up* our hands, or get *hopping* mad. If Johnson-Laird and Oatley ever turn their attention to our spatial metaphors, they may discover patterns that are just as systematic as those that they discovered in our emotions. And it would not be surprising if the two systems were actually connected.

I believe that there is something quite fundamental about all this, and that it is directly related to conscious awareness. It indicates, I feel, that mental events are indeed played out in what could be called a theatre, namely *the theatre of our personal space*. I believe, moreover, that this gives us a strong hint as to what conscious awareness actually is. It is simply *a mechanism which enables us to relate to our personal space*. That might sound like a rather modest attribute, until one considers how complex an environment our personal space really is. What is more, and unlike *Aplysia*, say, we have a personal space which extends well beyond our bodily surfaces. With the most highly developed of our senses, those that claim the lion's share of the cortex, we can see and hear things at considerable distances. Indeed, our personal space includes places that are not always in our immediate vicinity, such as our places of work and leisure, the surgery at our local practitioner, the supermarket and the pharmacy, the homes of our friends and relations and, if we are religious, our churches. Even more importantly, our personal space includes people: our spouse, our children, our parents and in-laws, our friends and our colleagues. It is not surprising that all these places and people are the greatest source of stress when they change. Bereavement, divorce, change of house, change of job: these are the things the doctor looks for when a patient complains of tension.

Julius Fast has written about subtle aspects of personal space in the workplace.[38] An administrator will sometimes intimidate an employee by invading the latter's domain, the violation usually being no more intrusive than the placing of a hand within the victim's portion of a temporarily shared piece of furniture, such as a table or a couch. Robert Ardrey's concept of the territorial imperative deals with the same issue, but on the level of the species.[39] The recent preoccupation with personal roots, that is to say with one's geographical origins, is yet another facet of the story.

But the aspects of personal space of greatest interest to the present discussion concern the body's more immediate environment. And here we see parts of the nervous system which go about their tasks in a quite unobtrusive manner. One of these is constantly monitoring bodily posture with respect to the surroundings. It ensures that we do not spend so much time in a given position that blood flow would be impaired. One part of this system has what could be called shock detectors in the ball of the heel, and it increases blood flow in proportion to the cadence of walking. It can be tricked by a device which administers the occasional light tap to the heel, and such gadgets are used to ensure adequate circulation in bed-ridden hospital patients. In another vein, I am often intrigued by my nervous system's ability to keep a check on the amount of coffee remaining in my cup. Without giving the matter any thought whatsoever, and without ever looking into the cup, I always seem to know whether or not it is empty (perhaps a signal is given when my proprioceptive system detects that a certain angle of my hand – and the cup – has been reached). A related mechanism may underlie a squirrel's relationship to its hidden nuts. Finally, it might be useful to consider some of the spatial phobias, including claustrophobia and agoraphobia, as impairments of the neural machinery associated with the sense of personal space.[q]

This brings us back to the question of mechanism. As we have seen, the hippocampus plays the definitive role, both in the precise timing on which our personal radar is based and in the temporary storage of memories that require the participation of conscious awareness. The importance of timing suggests that coherence in the brain's neural activity is significant. The word coherence is not used merely in its colloquial sense, however; in this case, it is meant to imply synchrony, or near synchrony, of the activities of considerable numbers of individual neurons. Both Peter Milner and Cristoph von der Malsburg have emphasized that neurons are detectors of temporal correlations, since whether or not such a cell emits a nerve impulse is determined by

q / Claustrophobia and agoraphobia are opposites, the former being dread of enclosed spaces while victims of the latter abhor open spaces.

the instantaneous sum of the inputs it receives. Similarly, Horace Barlow has written of neurons functioning as coincidence detectors. All these contributions to the story were discussed in the previous chapter.

Coincidence is a word that must be used with caution, however, for it requires specification of degree. Pulses spread over several milliseconds will not be coincident on the scale of microseconds, for example. This might seem like splitting hairs, but the point becomes a fundamental one when we compare learning in the neural networks of *Aplysia* and learning in our own cortico-hippocampal system. Aplysia's networks can capture correlations if the relevant coincidences are within half a second or so, but the synaptic modifications occur gradually, and require several repetitions of the paired stimuli. As we have seen, our own systems are capable of a rather higher degree of discrimination, and we can learn things we are exposed to just once, if they are sufficiently important. But is this difference sufficient justification for denying *Aplysia* the possibility of conscious awareness? As Thomas Nagel would remind us, there is no way of finding out directly whether or not that sea snail also possesses that attribute. We can only compare the two sets of neural circuitry and try to draw the correct inferences. Let us now attempt to do so, and let us start by returning to those black boxes we contemplated in chapter 2.

We can begin by recalling that the really significant box, for a given species, is the one that is complicated enough to represent the animal's full repertoire of responses. This is what places the creature on the ladder of abilities. An animal which displays behaviour analogous to that of a given box will also be capable of emulating the behaviour of simpler boxes. Our own reflexes are pretty simple, for example, even though our box-wise aspirations clearly lie higher up. *Aplysia* can certainly function as a fairly sophisticated box, because it has a memory system. But can it function as we do? Are there ever times when *Aplysia*, lying quiescent and unprovoked by any sensory stimuli, simply thinks things over and changes its attitude toward the outside world? Does *Aplysia* ever imagine? Does *Aplysia* possess the neural machinery that would permit it to do this?

We have strong indications as to what imagination involves, because we noted (in chapter 2) that it is suspiciously accurate. We came to the conclusion that it uses the same mechanisms and even the same neural circuitry as perception. Imagination may thus be regarded as being *simulation* of the outside world, and we have seen that our version of it relies on the presence of those reverse projections in the cortex. Now in one important respect, imagination appears to be on an equal footing with normal perception. Evidence for this comes from observations similar to those made by Anne Treisman on breakthrough. We recall her demonstration that competing trains of perception can mutually interfere if they become too similar. I once witnessed a good example of this, in a restaurant that had a resident pianist. I noticed that he could carry on a conversation without impairing his performance in any way. My curiosity aroused, I wondered if he could even do this when the conversation involved music. I went over to the piano and asked him if he knew a particular tune. Rather mischievously, I started to hum this other melody before he could react. Sure enough, it so put him off that he had to stop playing, much to the irritation of the other guests.

I apologized, of course, and we later enjoyed a discussion, during one of his rest breaks. It was particularly interesting to learn that he was not even capable of playing when merely *thinking* of a rival tune. Just imagining other melodies was sufficient to cause breakthrough. This shows, surely, that imagination and perception are not basically different from our neural networks' point of view. The competitive mechanisms that act on perceptual inputs work just as effectively on internally generated percepts. We might close our eyes and ears to improve concentration on our thoughts, but thoughts themselves can interfere just as effectively with the perception of our environment. We have all laughed at stories of the absent-minded professor; they are not exaggerations. There appears to be complete reciprocity between thought and perception. Experimental verification of this equivalence was obtained by Sydney Segal and Vincent Fusella, who found that a conjured-up image acts as an internal signal which can become confused with the external signal.[40]

At the risk of irritating the reader, I will interject a brief autobiographical note at this point, to introduce what I feel is an important and relevant principle. Many decades ago, when I was so young that the content of my Christmas stocking was still a major annual issue, I was surprised to find that my gifts one year included what appeared to be a leather-bound book, embossed in gold leaf with a small legend. Surprise gave way to disappointment when I discovered that it was not a book at all, but rather a book-shaped money box. It had a slot at the top, optimistically dimensioned to accept coins as large as a half-crown, which was about 3–4 centimetres in diameter, and there was a lock at the side, where there should have been the edges of pages. My sense of outrage was compounded by the fact that Santa Claus's generosity had not extended to inclusion of the key. I immediately saw this pseudo-book as part of a plot to get me to pay for my own presents in future years. But I was also intrigued to note that Santa appeared to share my mother's weakness for mottos, homilies and the like.[r] And that legend has remained in my memory over the decades, whereas all the other items in the stocking that year have long since been forgotten. The legend read: *Tall oaks from little acorns grow.*

I believe that the difficulty we encounter when trying to define consciousness arises from our inability to differentiate between the tall and richly varied oaks of our conscious experiences and the little acorn of the underlying mechanism. As I mentioned earlier, Francis Crick and Christof Koch have suggested that we temporarily shelve the issue of definition, and they cited the difficulty of saying just what a gene is, for example. By way of coincidence, Crick has written of the choice he once faced between researching into consciousness and researching into the borderline between inanimate molecules and the chemistry of life.[41] As is well known, he opted for the latter, and dis-

r / This occurred during the Second World War, and my father was away from home on active service.

covered (together with James Watson) that the crux of genetic inheritance lies not in the rich variety of organisms, but rather in the humble fact that a coplanar hydrogen-bonded arrangement of adenine and thymine has exactly the same overall length as a similar combination of guanine and cytosine. These equally long base pairs can thus be equivalently positioned as the rungs of the twisted ladder polymer that we now know as the famous double helix, and genetic variety stems from the variety of possible arrangements of those bases.

What is the gist of consciousness? What is the little acorn? What is the counterpart of those humble base pairs? I believe that it will turn out to be equally humble, and equally mechanistic. And I believe that the key factor is *time*. If the organism is to have the ability of responding to the temporal texture of its environment, on the time scale inherent in that texture, it will have to be able to retain a temporary record that spans a sufficient amount of that texture. And it will need cognitive mechanisms which extract relevant information from that texture, in the time available. Only then will a response be possible which exploits the choice implicit in the existence of that texture. Failing this, the information in that texture will be lost, and the resulting synaptic changes (if any) will merely reflect the statistics of the texture.

After such a build-up, I must admit that this idea sounds rather tame. But exploration of its implications will reveal that it could indeed give rise to those metaphorical tall oaks. Let us begin with that word *texture*. It conveys the concept of unevenness, without which there would be little information content. In the spatial domain, there is more information in a series of lines than there is in a blank page, and the line-sensitive neurons in our visual systems contribute to the extraction of that information. One can generalize the idea of space, and say that different combinations of molecules represent different coordinates in olfactory and gustatory space, and for those cases too, we have neural networks capable of extracting the relevant information. But the information of greatest potential interest to an organism concerns changes in those generalized coordinates. That series of lines will ultimately lose its interest for us, once its message has been assimilated, and smell and taste will cease to serve us if our nostrils and tongues are permanently subjected to the same molecular species. For our nervous systems, it is change which is of the essence, and because we are also seeking to explain that facet of consciousness which only involves imagination, we must extend the concept of texture to include graininess in the actual patterns of neuronal excitation.

Change will be wasted on us, however, if the system is not able to draw conclusions from it in the time available. What we need, therefore, is what could be called *a mechanics of consciousness*. In the opening section of the preceding chapter, we saw that experimental information on transit times for signals in the cortex permitted us to think in terms of the underlying mechanics. We must now try to extend those ideas to consciousness itself. It would seem that *conscious awareness is possible if temporal changes in sensory input (or in internal patterns of signals) can be detected while the signals resulting from that input are still reverberating in the system*. And we could go on to note that *such detection will permit the system to capture correlations between cause and effect so rapidly that their significance for the organism can immediately be appreciated, and also immediately used to modify the organism's repertoire of responses.* By this principle, detection of correlations can be exploited sequentially. It is this, I feel, which provides the basis for the continuity we invoke when we refer to the stream of consciousness. The reactions implied in the definition may be internal, external, or a combination of both. And in this respect, the equal standing of the imagination and perception, in neural terms, is of paramount importance. One consequence of this is that it enables the system to compare sensory input with previous experience, and thus to gauge novelty *immediately*. We do well to bear in mind the importance that William James attached to both the continuity of consciousness and the element of surprise.[42]

The abilities provided by the above principle make considerable demands on candidate neural networks. There must be the possibility of internal simulation, that is to say of imagination, so a system of reverse projections is presumably a prerequisite. Then again, there must be the possibility of handling simultaneous streams of information, and of keeping those streams separated. And possession of these pieces of circuitry would be futile if the corresponding signals were not in the system long enough to permit comparisons, by that *unfinished business* mechanism. The upshot of these various requirements is that there are fairly severe constraints on the underlying time constants. Conscious awareness is predicated on there being an adequate very short-term memory – the neuronal reverberation time – as has indeed been stressed by Crick.[43] It also requires that the transit times of the signals in the separate processing streams have the right order of magnitude.

If these ideas about the mechanics of consciousness are on the right track, they raise the obvious question about where the underlying process is implemented. Where does the unfinished business mechanism operate? Does it take place in the cortex, or in a sub-cortical structure, or in a combination of both those locations? The short answer is that we simply do not know. But there is experimental evidence that appears to be homing in on promising candidate regions of the brain. Before we can appreciate their full significance, however, we will have to put up with a digression, a rather protracted one, indeed, for it will occupy several additional sections. And the digression will start with a surprise, because we are going to return to something that we may have thought had already been discussed to exhaustion!

Black boxology revisited

In chapter 2, we considered the familiar black box, using it to represent an individual's nervous system and the structures it activates. The environment provides the box with various stimuli and these produce a range of responses. The challenge lies, of course, in explaining how the former lead to the latter in those cases where consciousness intervenes.

Roger Carpenter[44] has given a particularly succinct discussion of the various possible scenarios. In the case of the reflex, stimulus automatically

(a)

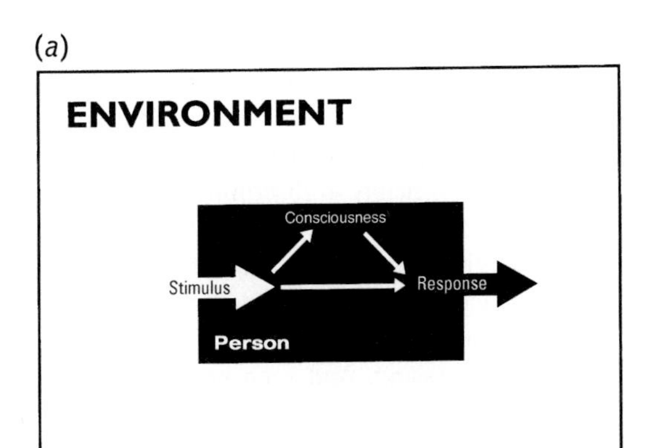

(b)

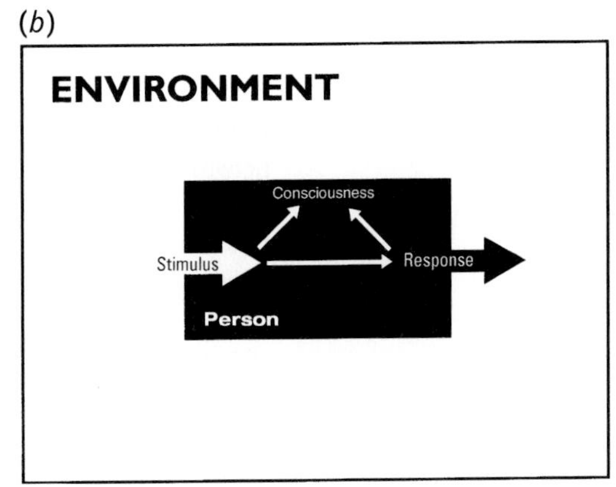

Figure 9.7
(*a*) The customary view of consciousness regards the nervous system and the associated muscles as a black box, which responds to stimuli emanating from the environment. Reflex responses are indicated by the direct route, while consciousness intervenes in the indirect path. This picture is appropriate only for the third-person view of consciousness. (*b*) The idea that consciousness plays no essential role for its possessor is captured in this version of the black box, in which one of the arrows in version (*a*) has been reversed. This theory regards consciousness as providing a spectator view of the system's inputs and outputs, without permitting participation in either, and it has been championed by Roger Carpenter.

and rapidly produces response, and consciousness is not required (though, as we saw earlier, reflexes can nevertheless be under the dominion of the higher brain centres). In slower and more deliberate situations, we get the impression that the stimulus provokes consciousness, and that it is the latter which then mediates the response (see figure 9.7*a*). Carpenter argues that this might be illusory, and he cites various situations in which responses of considerable complexity are actually carried out automatically and unconsciously. This leads him to suggest that even in these cases consciousness might merely be an ephemeral bystander, enabling us to observe our responses without really participating in them (see figure 9.7*b*). According to this view, consciousness is just an epiphenomenon, with no essential role to play. He suggests that one should not be disappointed by the idea that one's consciousness might be carried round in one's body as a kind of perpetual tourist, a mere spectator of the world's stage. Indeed, he recommends that we enjoy the trip!

I must admit that I dismissed this idea out of hand when I first read it. Surely, the usefulness of consciousness is manifest; it must be nonsense, I felt, to suggest that it is not serving some vital purpose. It is true that conscious awareness can be dispensed with even in impressively complex tasks, such as driving in unfamiliar territory and sight-reading novel musical scores, both while being preoccupied by thoughts of other things. A strong hint is provided by that word *preoccupied*, however, for the fact is that thought is always given priority in such situations; one cannot simultaneously think and perform a motor task and be preoccupied by the latter. If something occurs which shifts concentration to the motor task, the train of thought is inevitably suspended. My conclusion was that consciousness is for[s] some aspect of that which preoccupies us, and I believed the aspect in question to be *evaluation* and *choice*.[t]

s / More correctly, one should ask why consciousness confers an evolutionary advantage.
t / I also use the term *adjudication* in this chapter, to denote essentially the same thing.

Evaluation and choice would seem to be a rather paltry accomplishment, however. Even an artificial neural network of the feed-forward multilayer type can evaluate and choose, as we saw in chapter 6, when such a network was shown to encounter little difficulty in discriminating between the letters T and C. And no-one would claim that such a network is conscious. It is well to bear in mind that we do give these networks quite a bit of help, however. For example, as Francis Crick has noted,[45] their input and output 'neurons' have been *grandmotherized*; they have been relieved of the burden of considerable amounts of cognitive processing. And such networks do not have to choose between temporally extended events, which are characteristic of animal behaviour; grandmotherizing a *sequence* of neural signals is not an easy task. Finally, such a feed-forward network gets its information served *gratis*, data being fed passively to its input layer, and its output is read off by an intelligent agency; such networks usually do not acquire experience of their own 'volition'.

Now it might seem that we acquire information in a similarly passive manner. After all, we sit motionless, most of the time, as we attend the various news media which nowadays bombard our senses. There is very little evidence of activity, apart from the private workings of the mind. Many of those currently struggling to understand the phenomenon make just that assumption, indeed; they take consciousness to be an essentially passive mechanism. I take the diametrically opposite view. I believe that consciousness is always, perforce, an *active* process. I believe, moreover, that the realization of this fact can lead one to understand the phenomenon in all its beauty. The analysis requires one to work through several reasonably protracted steps, however. Let us make an immediate start on the first of these, which, it must be admitted, looks like an unnecessary digression.

Let us start by asking what is happening in the brain when our motor machinery is doing one thing, such as driving a car, while we are thinking of something else? How can the system simultaneously handle both processes, and avoid getting them confused? To start with, we should note that concurrent performance of different motor tasks presents no problem; we can all simultaneously walk and chew gum. Putting it in terms that we must now familiarize ourselves with, the two tasks involve different sets of muscular movements, actual or merely simulated, and their successful separation reveals that (in the sufficiently trained animal, at least) the important process is the selection and triggering of the appropriate sets of motor sequences (as opposed to the more onerous job of directing every individual muscle). It must be emphasized, however, that the two tasks must both implicate the same part of the brain, which will be identified later. It would otherwise be difficult to explain those interference phenomenon first explored in the auditory domain by J. Ridley Stroop,[46] and to explain the related effect known as *breakthrough*,[47] which occurs when competing signals in the brain display too much mutual similarity. These were discussed in the previous chapter.

The roots of these ideas are certainly not new; as we will see, they have been implicit in the writings of many authors on the subject. Indeed, in his obituary of Roger Sperry, Colwyn Trevarthen[48] mildly rebukes those who seek to explain consciousness while turning their backs on its relationship to muscular output. Sperry's views on the subject[49] are worth quoting in detail:

An analysis of our current thinking will show that it tends to suffer generally from a failure to view mental activities in their proper relation, or even in any relation, to motor behavior. The remedy lies in further insight into the relationship between the sensori-associative functions of the brain on the one hand and its motor activity on the other. In order to achieve this insight, our present one-sided preoccupation with the sensory avenues to the study of mental processes will need to be supplemented by increased attention to the motor patterns, and especially to what can be inferred from these regarding the nature of the associative and sensory functions. In a machine, the output is usually more revealing of the internal organization than is the input.

Sperry's point regarding one-sided preoccupation does not enjoy universal support, though emphasis on the brain's output can point to august antecedents. Charles Sherrington's seminal 1937–8 Gifford Lectures, published under the title *Man on his Nature*,[50] includes the passage: 'The importance of muscular contraction to us can be stated by saying that all man can do is to move things, and his muscular contraction is his sole means thereto'. This is quite similar to the statement he made over a decade earlier, which is quoted at the start of chapter 2 of this book. Sherrington was obviously fixated on the idea that despite the richness of its cognitive capabilities, our species has at its disposal only one type of external response: activation of appropriate muscles.

In the same vein, a passage in Edgar Adrian's equally celebrated *The Mechanism of Nervous Action*,[51] transcribed from his 1931 Johnson Foundation Lectures, states: 'The chief function of the central nervous system is to send messages to the muscles which will make the body move effectively as a whole'. His use of the word *chief* reminds us that the cognitive machinery produces a second type of output, namely that embodied in the functioning of the endocrine system. This is a purely internal matter, however, and it works over a rather longer time scale. Galvanizing the body's muscles into action is the nervous system's only response on a time scale corresponding to that of very short-term memory, which is to say within a few seconds.

Those distancing themselves from the philosophy of the above passages might stress that the brain is far from inactive even when one is immobilized, and merely thinks. The Sperry, Sherrington and Adrian statements do seem to gratuitously denigrate the brain's mediation of pure thought. Arthur Ritchie,[52] in his Tarner Lectures delivered at Trinity College, Cambridge, in 1935, had provocative things to say about that side of the issue. A paragraph in his section *Thought and Muscular Action* (page 127) reads:

At the suggestion that muscular movement may be intellectual I feel that there will be a stirring of indignation among the highbrows. They will say, "Why all this talk about acrobats and muscular activity? Granted that what the acrobat does is perfect of its kind it is not properly an intellectual activity, like the mathematician's for instance". In reply to this I should admit at once that what the mathematician does is much more useful than what the acrobat does; but is there any reason apart from snobbery for saying it is more intellectual? In effect the acrobat thinks with the muscles of his whole body while the mathematician thinks with – well, whatever it is he thinks with. Of course it may be that the mathematician does not think with anything but just thinks. Even if this were true, which is doubtful, I find it hard to see why thinking with nothing should be more truly thinking than thinking with your muscles. Because thinking is mental it does not follow that it is not bodily too.

Why have these ideas not been able to achieve a lasting foothold in the mind–body debate? Why, to use Sperry's own words, have they merely become 'the old motor theory of thought, now largely abandoned'? It is a tribute to the objectivity of Ritchie's book that his text provides a reason why the motor theory may not have earned the support I believe it deserves. Later in the above-cited section, he writes of a man who uttered the continuous sound '-E-E-E-E-' while doing mental arithmetic, and notes that this person therefore could not have been making 'implicit' mouthed movements while thinking about the task. (Returning briefly to the above-mentioned interference phenomenon known as breakthrough, Ritchie's man's mental arithmetic would have been made far more difficult if that uttered '-E-E-E-E-' had been replaced by a loudly articulated recitation of, say, the seven times table.)

It is high time we resuscitated the old motor theory. The nervous system could get around the difficulty posed by that uttered '-E-E-E-E-', if thought is not independent of the body's muscles. But the latter would have to play their role in a rather sophisticated and subtle manner. I shall later argue that this involvement is indeed of the above-mentioned *implicit* type, and attempt to demonstrate that thought proceeds through the *simulation* of muscular movements. In doing so, I shall be trying to provide an actual mechanism for what Antonio Damasio[53] referred to as the brain's *'as if' processes*.

Returning to our black boxes, but having become more aware of the central role played by the muscles, we might begin to suspect that something vital is missing from the situations depicted in figures 9.7(*a*) and 9.7(*b*). Perhaps the traditional view has put things back to front, and the time might be ripe for a radical change of paradigm. To begin with, Sherrington's statement indicates that the word *response* in those figures should be replaced by the word *movement*. One could then go on to ask whether the system would ever need to remember anything *not* related to movement. If we take Sherrington literally, the answer must be *no*.[u]

Now a particularly effective way of ensuring that only movement-related memories are stored by the organism would be to let movement play a conditioning role in the memory process. To see how this could be achieved, we will later need to consider certain anatomical and molecular facts. Meanwhile, let us note that if all we can do is move, movement might be our

u / We should here include memories related to decisions *not* to move.

(a)

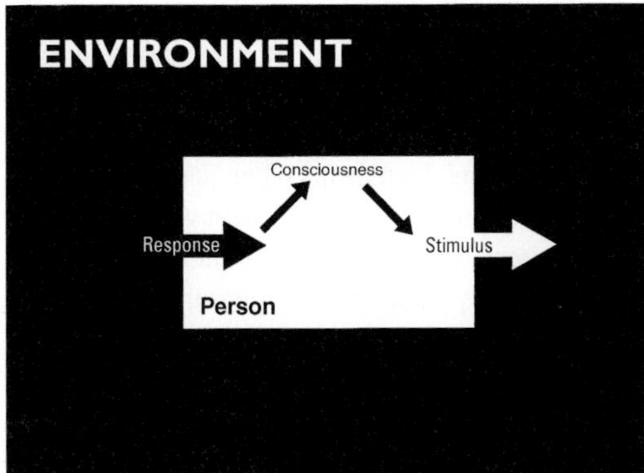

(b)

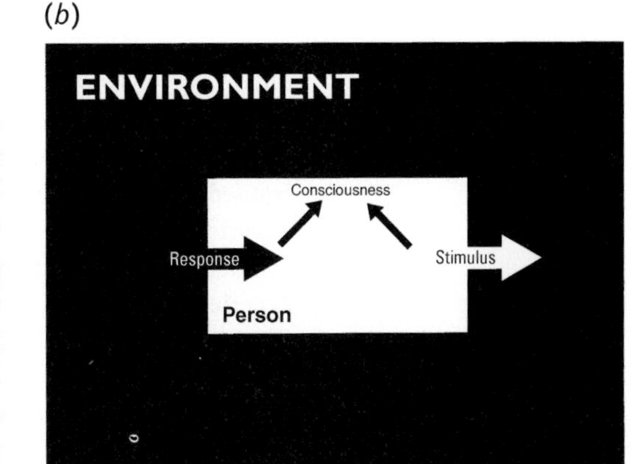

Figure 9.8
(*a*) A more realistic picture of the first-person perception of consciousness reverses figure 9.7(*a*), and views the environment as the black box. The plausibility of this idea can be verified through the expedient of closing one's eyes and reaching for an object in one's vicinity. Such reaching, and indeed all other motor acts, can be regarded as questions put to the environment. But it is important to bear in mind that not all such questions are 'asked' consciously. (*b*) The present author believes that this diagram is the most realistic representation of what actually happens during a conscious motor act, the affinity with Roger Carpenter's view – see figure 9.7(*b*) – being apparent. (It is to be noted that the direct route corresponding to reflex reactions is absent in this version, but its presence can be inferred by visual observations of one's own body.)

only means of gaining knowledge about the environment. This would certainly require a drastic change in the way we usually view the voluntary acquisition of information; it indicates that such appropriation is necessarily muscularly active, never muscularly passive. It would also suggest one critical modification to what Sperry stated: the organism's response might, viewed externally, appear to be an *output*, but to the organism itself it is an *input*, a stimulus whereby the environment is probed.[v]

The resulting change in the traditional diagram is indicated in figure 9.8(*a*), and indicates that movement *is the mechanism whereby an organism actively acquires information about its present and past environment*. I include the word *past* here because detection of the environment's response to the organism's stimulus could include the recall, and possible revaluation, of past experiences. The world *actively* is also important because this eliminates from the definition the passive (and non-selective) conditioning by the environment to which all organisms are subject.

It is to be noted that the direct route signifying unconscious reflex is absent from figure 9.8(*a*). But we do well to remind ourselves that our bodies are not the rigid rectangles that these black-box figures might suggest. Our physical flexibility permits us to scrutinize large parts of ourselves as if from the outside, and observation of our own reflexes must play an important part in the formation of our self-image. Mankind's discovery of reflections from water surfaces, and the later fabrication of mirrors led to further admixing of first-person and third-person views.

Before moving on, we ought to enquire whether there is any supporting evidence for the idea that memory is conditional on muscular activity. It turns out that there indeed is such backing, and that it is provided by the classic result reported by Richard Held and Alan Hein.[54] As described in chapter 7, two newborn kittens were confined to a cylindrical room, the walls of which were decorated with exclusively vertical stripes. The active kitten was

v / Those who feel that this suggestion is far-fetched should try closing their eyes and reaching for an object in their vicinity.

free to move about whereas its passive partner merely rode around in a gondola. The goal was to show that the acquisition of appropriate motor responses depends upon conditioning during early development, and the post-training tests included a visual cliff trial. In this, the kitten was confronted with a choice between shallow and deep drops when descending from a platform, the entire environment again being covered with vertical stripes. The actively trained kitten always chose the shallow drop, whereas its motor-experience deprived sibling could not appreciate the danger on the deep side; the vertically striped objects in its environment were invisible to this disadvantaged animal.

This has been a long digression, so it would be a good idea to remind ourselves why we embarked upon it. The point to be made was that acquisition of information is always an active process, never a passive one, and the arguments given above will later be supplemented by further evidence, described in a separate section. But we have *not* established that such active accumulation of information necessarily involves consciousness. It certainly does not, indeed. Take walking, for example. Each stride we take could be regarded as a question put to the environment: *is the ground still there?* Yet we are not aware of every individual step, and the act of walking would be awkward if we had to concentrate on such detail. It is only if we encounter an irregularity, such as a hole, that our motor explorations suddenly become conscious ones. Meanwhile, let us note that such active procurement of information will inevitably take time. Signals will have to be passed to the muscles, and signals being returned from the environment must pass through the relevant sensory modalities. Proprioceptive information must simultaneously return from muscles, related joints and body surfaces.

Time was the significant ingredient in the work of Benjamin Libet and his colleagues, of course, and we earlier encountered their 500-millisecond requirement to attain neuronal adequacy. This alone suggests that Roger Carpenter might be right when he suggests that we are informed too late of our actions, and that consciousness thus makes us mere observing bystanders of our own actions. The moment has come to take a look at his own work on eye movements, which lends further strength to this proposition. But it will be useful to make a brief digression, by way of introduction.

At the 1996 Olympic Games, the then current 100-metre champion, Linford Christie, was disqualified from the 100-metre final for twice jumping the gun. Many of the spectators must have been puzzled by this elimination, because the slow-action replays of both starts, displayed on giant screens in the stadium, clearly showed that the starter's pistol had been fired *before* Christie started to move. Why were the officials nevertheless justified in ousting him? The answer was provided by another technical development which is rapidly becoming standard practice at major athletic events, namely an analysis of the competitors' reaction times. This is usually given in milliseconds, typical values for top-class performers falling in the range 125–250. If, for a particular start, a runner is found to react within 50 milliseconds, say, the starter can safely conclude that there had been a deliberate attempt to jump the gun, because the above range shows that it takes a minimum of about 75 milliseconds *more* than this in order to react to the sound of the pistol. One could say that the offending sprinter has been felled by a combination of technology and statistics. In earlier times, there must have been many who jumped the gun and went undetected.

It might seem that such considerations would not be particularly relevant to the investigations of eye movements carried out by Roger Carpenter and Matt Williams, but let us consider their experimental set-up. The subject was asked to fixate on a suddenly appearing, centrally located spot of light, which thereafter persisted throughout a trial, and then to transfer the gaze to a second spot appearing sometime between 0.5 and 1.5 seconds later, either to the left or the right of the central spot. The subject was not told where this second (target) spot would show up, nor when during the one-second time window it would occur. Both of these variables, position and time, were randomized. The experimenters could control their overall distributions, however. The proportion of target spots appearing on one particular side was varied between 50 per cent (no bias) and 95 per cent (strong bias).

Carpenter and Williams were able to remotely monitor the time it took for the subject to make a saccade (that is, divert the eyes) in the direction of the second spot, after its sudden appearance. Not surprisingly, the stronger the bias to a given side, the shorter was the observed average time-lapse (or latency); the subject was better able to anticipate where the spot would materialize. Conversely, the average latency for saccades to the other, less likely, side was found to increase. But because of the randomization of the times of appearance of the target spot, the latencies naturally displayed a certain amount of statistical variation.

Carpenter[55] had earlier considered the processes which underlie the situation in this type of experimental arrangement, and he suggested that: '... the presentation of a target causes some kind of decision signal to rise steadily towards a threshold that triggers a saccade, with competing targets running a sort of race to see which gets there first'. (A similar type of race was assumed in Claus Bundesen and Hitomi Shibuya's analysis of visual discrimination in the presence of distractors, as described in chapter 8, and competitive neuronal processes in general have been analysed by John Taylor.[56]) The results of this study by Carpenter and Williams nicely supports that contention. They found that they can model their results, to a remarkable degree of precision, with such a race model – a race between a rising signal representing the original command to execute a saccade, and another later one representing the cancellation of the command. They were able also to establish that for very low prior probability, that is to say almost pure spontaneity, the latency lies around 300 ms, which is in excellent accord with the work of Libet and his colleagues discussed above.

Subsequent joint work by Carpenter and Doug Hanes[57] consolidated the earlier findings. They used an experimental paradigm in which the subject executed a saccade to a suddenly appearing target, just as in the earlier study, but in some of the new trials, after a certain delay, a stop-signal was presented that told the subject to suppress the eye movement. The average delay before administration of this veto command was systematically increased, and it was found that this progressively diminished the subject's ability to inhibit the movement. From measurements of this proportional relationship, and from control assessments of saccadic latency, Carpenter and Hanes were able to estimate the stop-signal reaction time. This varied little between different subjects, the observed range being 125–145 milliseconds. Here too, it was found that the data could be fitted to a race model, with impressive precision, indeed.

It is important to note that there is no essential difference between a stop signal administered externally, as part of an experimental trial, and one that arises internally, as a consequence of an unconscious thought process. In this vital respect, therefore, the various studies carried out by Carpenter and his colleagues are directly comparable to those of Libet and his associates. And the studies bear profoundly on the entire body/mind issue. They are evocative of the final two sentences of the epilogue of Charles Sherrington's Rede Lecture,[58] delivered in Cambridge on 5 December 1933: 'It has been remarked that Life's aim is an act, not a thought. Today the dictum must be modified to admit that, often, to refrain from an act is no less an act than to commit one, because inhibition is coequally with excitation a nervous activity.'

Before returning to the results obtained by Libet, and to his interpretation of them, let us briefly explore the similarities between these eye movements and those of the sprinter in his starting blocks. The latter, if he does not deliberately intend to cheat, will be simultaneously experiencing conflicting urges: to react as rapidly as possible, but also to avoid initiating his movements before he is sure that the starter's pistol has sounded. One could look upon the latter tendency as a sort of self-imposed veto. Similarly, in an eye-movement trial that has a clear bias to one side, the subject's tendency to anticipate the target's appearance will be opposed by the knowledge that there is nevertheless a finite chance of its showing up on the less-likely side. So again, there will be a tendency toward a self-disciplining, or vetoing. Intriguingly, Carpenter and Williams actually observed the saccadic equivalent of jumping the gun, there being more short latencies than there should have been.

Libet felt that his discovery of the veto function salvaged free will (or at least free won't!). But this conclusion was thoroughly undermined by the Carpenter–Williams and Carpenter–Hanes studies, which demonstrated that the countermanding process is just as automatic as the original movements themselves. If neither the initiation of a motor act or its last-moment veto are freely willed, however, what could possibly be the benefit to their possessor? In what way could unfree choice be any more use than the mechanical discrimination performed by a feed-forward neural network or, for that matter, by a conditioned *Aplysia*? The benefit is in fact a profound one: *the power of real-time choice (or veto-on-the-fly, as it were), even in the absence of free will, would still be an enormous advantage to the animal.*

That advantage can be expressed in a single word: *navigation*. Dictionaries usually give several meanings for this term, the one now most common dating from 1559, and which refers to '... the methods of determining a ship's position and course by the principles of geometry and nautical astronomy'. These methods are, of course, applied by a navigator. On board a ship or in a plane, the navigator is subordinate to the captain or pilot. What I have in mind here, however, is the other meaning of the word, the one more often applied to the person in charge. When we refer to the explorers of bygone days, figures such as Magellan, Cook and Columbus, we call them navigators in this second, more senior, sense. And when I invoke the word navigation, here, I too am referring to exploration. In this sense, conscious animals are navigators, probing their environments through the agencies of their (sequential) muscular movements, including (in advanced creatures) those belonging to the vocal apparatus.

What about a creature such as a bee, however. Doesn't it also navigate, as it flies around seeking nectar? Doesn't it too have to make choices? Yes, of course it does. But there is something that it does *not* do. When it is on the point of making one of its moves, it *never* suddenly realizes the significance of its intended action and then vetoes it.[w,59] It does not ask much of the imagination to see that this ability could sometimes make all the difference between life and death. I submit that the bee does not even have a mechanical (and deterministic) version of this capability, that is to say one in which the 'realization' does not involve consciousness. We, and our sufficiently close evolutionary relatives, I believe, do possess such a mechanism. Indeed, the work of the Libet and Carpenter groups demonstrates that this is the case, for our own species at least. The latter studies merely go that one vital step further, by showing that no freedom of will need be invoked.

But, and here is the biggest question of them all, *why should such a deterministic veto mechanism require consciousness*? Why could this unfree mechanism not simply be carried out unconsciously? Why did we have to become a captive audience of our own motor acts? The answer lies in that word navigation. As those intrepid mariners of yesteryear made their tentative ways around the globe, their every change of direction was dictated by the environmental feedback provided by their explorations; encountering an obstacle – impenetrable isthmus, unfriendly natives – they changed course, as recorded on their navigational charts. Likewise, we are able to switch tactics on the spot, which in this case means within the span of very short memory, and although no freedom of will is involved the process does rely on feedback, either from the current environment or from our past recollections of it, or indeed from both simultaneously.

But where is our system's counterpart of those indispensable navigational charts; where is *our* record of the muscular route recently taken? It is

w / The neural circuit underlying bee behaviour is now known, and reverse projections and efference copy routes are conspicuously absent.

present, though only fleetingly, in the muscles themselves. Their states, and the rates of change of those states, are perpetually monitored by the tendon–organ / muscle–spindle apparatus, and although the mechanism is surreptitious, the overall system *is* aware of these parameters. (I shall be going into this aspect in more detail, when we later consider qualia.) The record persists over the span of very short memory (Gerald Edelman's *remembered present*;[60] Nicholas Humphrey's *thick moment of time*;[61] and Robert Efron's *duration of the present*[62]). In other words, the mechanism requires that our systems are constantly being apprised of our surroundings, and of their significance for us. It requires, in other words, that we are consciously aware!

It would be difficult to exaggerate the repercussions that this story could provoke. It is telling us that exploitation of an unconscious veto mechanism, no doubt mediated by specific brain circuits, and made possible by an adequate span of short-term memory, paradoxically requires conscious awareness (see figure 9.8*b*). I suppose that one could call this *quasi-behaviourism*, in that consciousness merely emerges as an indispensable servant of the unconscious. But in contrast to the old behaviourism, we now have a picture in which there is room for significant internal processes, albeit deterministic ones.

Put in this way, consciousness seems like something of a gratuitous gift, even though it is clear that the system could not have functioned without it. This is the gist of what Carpenter tells us, of course, when he writes of being carried around like a spectator of his own actions. He seems to have reconciled himself to the fact of being thus relegated to the role of the proverbial fly on the wall. I must say that when this awful truth dawned on *me*, I was overcome with a feeling of nausea.

Is this strange story true? Is this the solution to the venerable mystery of the mind? I am convinced that it is, but I am also sure that the reader will require further proof before the idea can be accepted. The remaining sections of this chapter therefore seek to bolster my case. Some of this material will simply add to the amount of circumstantial evidence. But there will also be an explanation of the functioning of the underlying neuronal circuitry that is actually present in our brains. I find this evidence particularly compelling. There will also be an explanation of how the mechanism provides us with those all-important qualia, and how it meets those criteria enumerated by John Searle, by Gerd Sommerhoff and by others. Let us start by taking another look at that issue of active *vis-à-vis* passive.

Active acquisition of information

The view that the senses do not passively serve perception has been expressed by a number of authors. Jean Piaget,[63] for example, tells us that: '... all knowledge is tied to an action, and that one can only know an object or an event by using them through assimilation into schemes of action'. One encounters the same idea in the work of Marc Jeannerod[64] and Andy Clark.[65] And when James Gibson[66] discussed the manner in which the environment is perceived, he noted that '... the causal link is from response to stimulus as much as it is from stimulus to response'; he understood that we seek information through movement, irrespective of which sensory modality is being used to monitor the resultant feedback from the surroundings. It is tempting to suggest that the famous dictum of Descartes ought to be rewritten as: *Moveo ergo sum*.

Bruce Halpern[67] considered the olfactory and gustatory systems in particular, and concluded that the corresponding sensory processes are highly dependent on active exploration. As he put it, '... active sampling not only produces systematic changes over time in chemosensory stimulation but also, for olfaction at least, is coupled with CNS [central nervous system] processing of the peripheral chemosensory responses.'

The importance of self-paced activity in the acquisition of information is perhaps most subtly illustrated by the work of Yoshiaki Iwamura, Michio Tanaka, Okihide Hikosaka and Masahiro Sakamoto.[68] They looked for functional differences between the three Brodmann areas 1, 2 and 3, which together constitute the somatosensory region of the cerebral cortex. They discovered that whereas area 3 comprises a map of the body surface, area 2 contains cells that have very large receptive fields, indicating convergence over large regions of the body surface. Even more suggestively, they found that these cells are activated only when the animal is performing complex movements, such as grasping. Area 2 thus appears to be a site for integrating complex information concerning both tactile objects and behaviour. (Cells in area 1 appear to function in a manner intermediate to that observed for cells in the other two areas.)

Jim Bower and Mitra Hartman[69] have demonstrated a connection between motor and sensory functions, using microelectrode probes of the cerebelli of awake and behaving mammals. They observed the greatest cerebellar activity in rats when these creatures are exploring with their whiskers. In cats, on the other hand, the greatest activity corresponds to probing the environment with the paw, which is the preferred exploratory component for that animal despite the fact that it too possesses prominent whiskers.[x] Their observations are *precisely* those one would expect for the active acquisition of information.

Jia-Hong Gao, Lawrence Parsons, Jim Bower, Jinhu Xiong, Jinqi Li and Peter Fox[70] have recently published the results of functional magnetic resonance imaging (fMRI) investigations of the cerebellum, under certain circumstances involving human subjects, and they find evidence for a role in *acquisition and discrimination of sensory information*. Miguel Nicolelis, Luiz Baccala, Rick Lin and John Chapin[71] have detected a synchronizing of alpha-band oscillations in the cortex, thalamus and brainstem of rats indulging in

x / One might guess that in our own species, similarly, even those with handle-bar moustaches would generate greater activity in their cerebelli when exploring with their finger-tips. Our species is probably unique in the amount of cerebellar activity associated with its vocal apparatus.

tactile exploration. And the investigations of self-paced finger movements undertaken by Jonas Larsson, Balázs Gulyás and Per Roland,[72] with the help of positron emission tomography (PET), revealed simultaneous activity in the primary motor area, the premotor area, the supplementary motor area and, suggestively, the cingulate motor area. We are later going to see how important this latter area is in the overall scheme of things.

As revealed by the studies of Ian Darian-Smith and his colleagues, the vital role in perception played by movement is particularly prominent in the way textures are perceived with the finger tips,[73] as when blind people read braille. They find this virtually impossible if the pattern of raised dots is moved across their passive fingers. Only when the fingers themselves move does reading become possible.

Persuasive evidence of a different type comes from investigations of memory performance when a distinction is being made between line drawings of possible and impossible objects, like the ones we encountered in the preceding chapter. If the idea being advocated here is correct, our memories of objects unavoidably include stored prescriptions for handling them (and also impressions of the associated qualities gleaned from the corresponding sensory inputs). If a line drawing represents an object that cannot be fabricated, the brain will encounter difficulties in attempting to store such prescriptions. This is just what Daniel Schacter, Lynn Cooper and Suzanne Delaney[74] observed. Their results indicate that implicit memory depends on encoding of, and access to, structural descriptions of objects.

A somewhat similar situation arises in the case of ambiguous figures, of the type first used by Edgar Rubin[75] (see figure 9.9, and also figure 8.2). Each of these is capable of two different interpretations (for example, vase or two counter-disposed faces, duck or rabbit, and outward-pointing or inward-pointing cubes and staircases), but it is impossible to perceive both alternatives simultaneously. As I mentioned in chapter 8, the average minimum reversal time for one example of such an ambiguous figure is around 550 milliseconds. The related case of binocular rivalry will be discussed in a later section of this chapter.

Having briefly reviewed evidence that supports the idea that active acquisition of information is an important aspect of the story, let us now turn to the vital anatomical and physiological issues: where in the brain do the critical events occur which imbue us with consciousness?

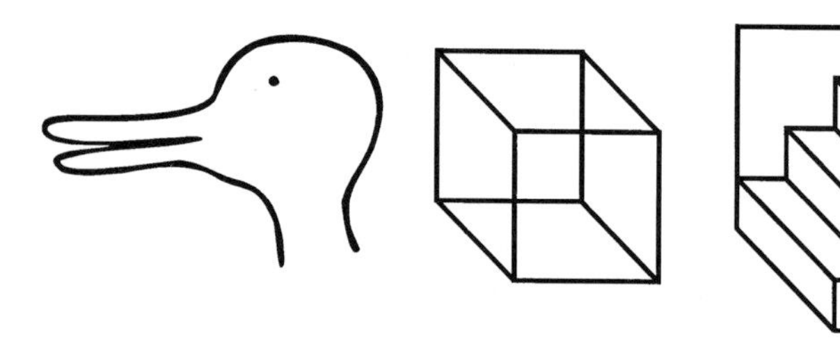

Figure 9.9
The competition between the rival interpretations of ambiguous figures, such as the one in figure 8.2, and the additional examples reproduced here, has been analysed by Hermann Haken (as described in this chapter), and shown to be a consequence of a self-organization process. The duck/rabbit was originally produced by Ludwig Wittgenstein, while the reversing cube was first described by the Swiss naturalist and crystallographer L. A. Necker, in a letter to the Scottish physicist Sir David Brewster, in 1832. The present author has been unable to trace the origins of the reversing staircase.

The core circuitry of consciousness

Without further ado, I am going to suggest that the above-mentioned veto function is mediated (in each hemisphere) by a centrally located circuit that incorporates the sensory cortex (that is to say the occipital, parietal and temporal lobes), the premotor area of the frontal lobe, the thalamic intralaminar nuclei,[76] the nucleus reticularis thalami and the anterior cingulate (see figures 9.10*a* and 9.10*b*). This is quite a large group of structures. Sceptics might be tempted to note that the list includes a considerable fraction of the entire brain; they might feel that I am guilty of hedging my bets. It is true that the inventory is a long one, and still other structures no doubt participate in providing consciousness with its content. The fact that so much of the brain is involved attests to the non-localization of consciousness; one cannot pinpoint the phenomenon to a neatly circumscribed region.

In defence of this essential circuit, let me point out what it does *not* comprise. It leaves out the cerebellum, the hippocampus, and that part of the frontal lobe lying rostral to the premotor area. Neither does it include the basal ganglia. This core circuit is more than just the sum of its parts, however. The key to its mediation of consciousness is to be found in its connective anatomy, as well as in its physiology. It has long been known that the sensory cortex feeds signals to various parts of the frontal lobe, of course, and it is equally well known that the premotor area of that lobe serves the planning of sequences of muscular movements, which are activated via the motor cortex. There has also been some prior speculation about the possible role of the thalamic intralaminar nuclei in consciousness. What has been lacking, until recently, has been a clear picture of the more global scenario in which these structures participate.

Planning of movement was invoked in a discussion of visual awareness by Francis Crick and Christof Koch,[77] in 1995. Indeed, they predicted which cortical areas would be associated with awareness: if an area does not project to one of the frontal lobes, they maintain, we are probably not aware of processes mediated by its neuronal activity. The corollary implication, although not specifically stated, was that if a region of the brain does not receive projections from the sensory cortex, it is probably not directly involved in consciousness. This appeared to be Koch's inference,[78] at least, for he argued against Joseph Bogen's suggestion[79] that consciousness involves the thalamic intralaminar nuclei, which do not receive such projections.

Subsequently published experimental evidence endorsed Bogen's suggestion. Shigeo Kinomura, Jonas Larsson, Balázs Gulyás and Per Roland[80] used positron emission tomography to study the changes that occur when otherwise relaxed human subjects are presented with an attention-demanding reaction-time task. Activation of the thalamic intralaminar nuclei was observed, and also of the midbrain reticular formation (and, indeed, the premotor areas). These structures have long been known to be involved in control of the level of arousal.[81]

The thalamic intralaminar nuclei are the target of dense projections from the frontal lobes,[82] and they themselves widely project to the cerebral cortex, with no discernable topography.[83] Moreover, activity in these latter

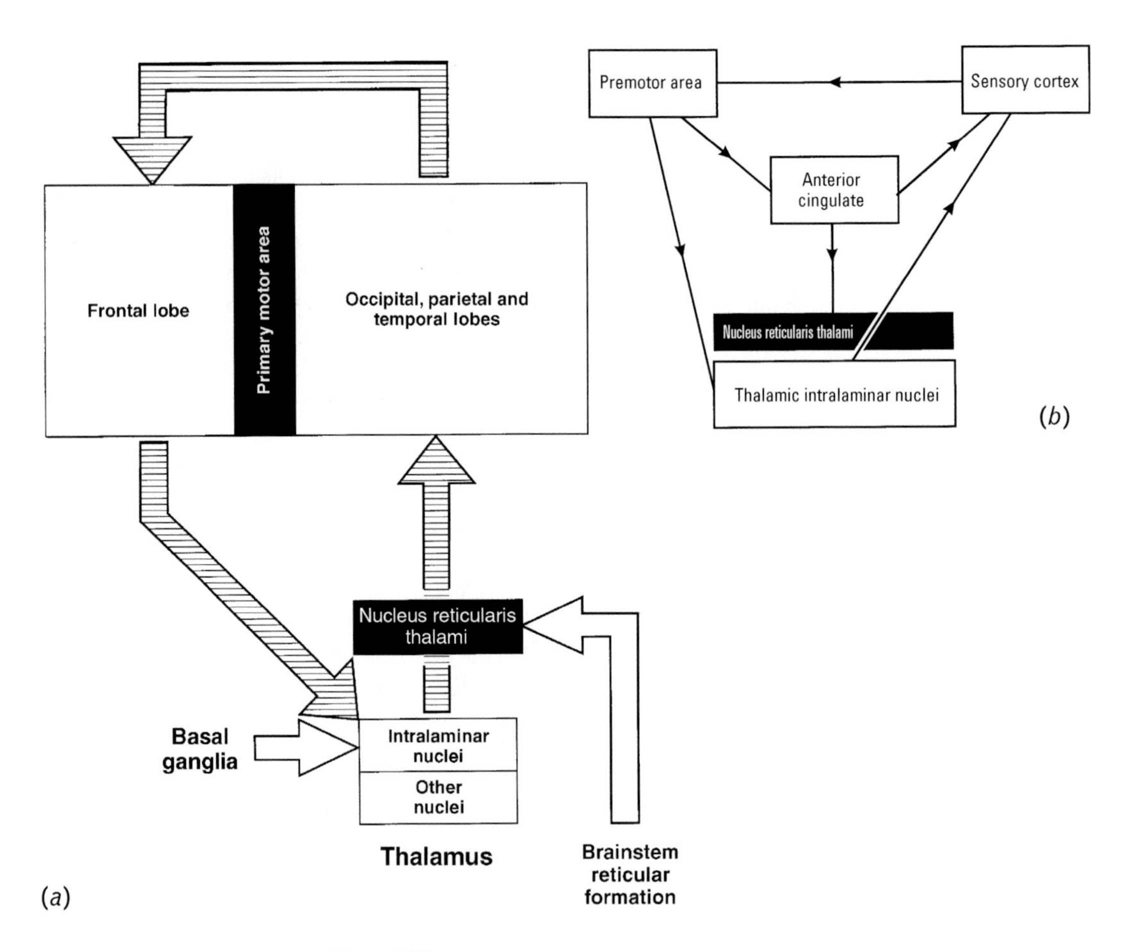

Figure 9.10
(*a*) The basic circuit underlying awareness and thought, in the present author's view, is conjectured to run from the three posterior lobes of the neocortex to the frontal lobe, and thereafter back to those posterior lobes via the thalamic intralaminar nuclei. This route is indicated by the striped arrows in this highly schematic diagram. The final branch of the route is under the control of the brainstem reticular formation (see figure 7.11), which exercises its influence through the nucleus reticularis thalami. According to the present theory, consciousness acquires its vital emotional dimension through the interaction of the thalamic intralaminar nuclei with other thalamic nuclei, and through the internal reafference from the muscle spindles (or its simulated counterpart, in the sufficiently experienced animal). (*b*) The deterministic and unconscious real-time-choice mechanism (or veto-on-the-fly), which the present author believes was the decisive evolutionary advantage that (paradoxically) requires conscious awareness, is mediated by the direct coupling between the anterior cingulate and the nucleus reticularis thalami, as shown here. It is interesting to note that removal of the anterior cingulate does not preclude consciousness, although it has a profound influence on the perception of pain. Indeed, patients who have had this brain component removed appear to be oblivious of pain.

projections appears to desynchronize the cortical EEG,[84] a reliable signature of conscious activity.[85]

Collectively, these facts suggest that the frontal lobes are not merely the goal of signals arising from earlier-established consciousness, as the Crick–Koch model appears to propose. On the contrary, they indicate that for consciousness to be created, the activity generated in the frontal lobes, by events in the more posterior part of the cerebral cortex, must produce further activity elsewhere: first in the thalamic intralaminar nuclei and subsequently, as a consequence, in that latter part of the cortex itself. The upshot would be a closed loop of sequential activity[86] (see figure 9.10*a*).

I had earlier conjectured that such a pattern of circulating activity must underlie thought, surely the hallmark of consciousness, and one of this pattern's appealing aspects would be the return of activity (though presumably in a modified form) to the region(s) that originally provoked it.[87] This harmonizes with the discovery that an area which serves perception of a given feature of sensory input is also used when that feature is merely imagined.[88] Further support comes from the observation that the frontal lobe does indeed play a role in mental imagery.[89]

The round-trip time for this activity could be estimated if one knew the number of synapses (and postsynaptic neurons) to be traversed during each circuit. It is possible that this time has been measured implicitly through the recent demonstration by Daniel Barth and Kurt MacDonald[90] of gamma-band oscillatory activity in the intralaminar nuclei during wakefulness (and spasmodically during sleep[y,91]). Mircea Steriade[92] has argued that these nuclei coordinate the oscillations, rather than drive them, which is in accord with the mechanism being considered here. The observed period is around 28 Hz, which gives a round-trip time of about 35 ms. This would roughly correspond to between four and eight synapses traversed per circuit, which does not seem unreasonable for a route that links three different brain regions.

We have not yet touched on all the components identified at the start of this section. How do the others fit into the story? In particular, what role is played by that structure that appears in the above list but which has hardly been mentioned so far in this book, namely the anterior cingulate? A strong hint of its significance was provided by PET studies reported by José Pardo, Patricia Pardo, Kevin Janer and Marcus Raichle.[93] Monitoring changes of blood flow in the brains of people subjected to a Stroop interference situation, they observed the largest increase in flow in the anterior cingulate region. They noted that their results: '... suggest that the anterior cingulate is involved in the selection process between competing processing alternatives on the basis of some preexisting internal, conscious plans'. Endorsement was

y / Munglani and his colleagues (see note 91), during a study of the efficacy of an administered anaesthetic, have demonstrated that detection of gamma-band oscillations is a reliable indicator of the presence of consciousness.

subsequently provided by the work of Michael Posner and Marcus Raichle,[94] who showed that it is the region where *executive attention* operates.

Brent Vogt, David Finch and Carl Olson[95] wrote, of this region: 'Anterior cingulate cortex may provide a meeting place for interactions between cognitive and motivational processes, particularly in relation to the generation of motor output'. Jeanne Talbot and her colleagues have shown that this structure plays a central role in the making of decisions when pain is a possible concomitant,[96] as have Pierre Rainville and his colleagues.[97] It has indeed been suggested that the anterior cingulate serves an important function in the avoidance of danger.[98] Danger and pain are not unrelated, of course, so this brain area might always have to be on duty during the exploratory movements (actual, or merely simulated) that I am postulating to be central to consciousness. This suggests that the anterior cingulate might function in parallel with the route that passes through the thalamic intralaminar nuclei (see figure 9.10*b*, which derives from a diagram by Keith Purpura and Nicholas Schiff,[99] but with the addition of several extra connections).

I mentioned earlier that consciousness arises not only through participation of these vital structures but also because of the way in which they are interconnected. We are now going to see a particularly important example of this principle. What could be called the missing link, literally indeed, is provided by a connection that has been found to run from the anterior cingulate directly to the nucleus reticularis thalami[100] (see figure 9.10*b*). This link was found in the rat, and it seems not unlikely that it would be present in all mammals. (Indeed, Nicholas LaMendola and Thomas Bever have recently shown that rat whiskers, which are the exploratory counterparts of our hands and fingers, display a left–right asymmetry in their cortical representations, quite like handedness in humans.[101])

Now what could be the use of such a linking between the two parallel routes? I believe it has to do with that veto function explored by Libet and Carpenter, and their respective collaborators. Indeed, as long ago as 1948, Arthur Ward wrote of the anterior cingulate as being one of the most powerful of the cortical suppressor regions.[102] With the anterior cingulate implicated in the avoidance of pain while the nRt is involved in attention, the link between the two structures indicates that evasion of danger is handled by them collectively. Already in 1971, Björn Merker[103] had speculated that the nRt might control attention, this work building on the anatomical studies of Arnold Scheibel.[104] Francis Crick[105] later conjectured that the nRt might provide the basis for the searchlight of attention.

It is not yet clear how the cingulate–nRt interaction actually functions. It could be that impending danger, gauged by the anterior cingulate through remembered prior experience, causes signals to be passed to the nRt which focus the animal's attention and, in effect, suggest that a planned action might not be prudent. That the system is able to plan ahead, and predict an action's possible outcome, must follow from the involvement of the frontal lobe. The precise mechanism is not yet known, of course, but it seems likely that it does not incorporate an emergency switch (a stay of execution, as it could be called). The work of Roger

Carpenter and his colleagues, discussed earlier, indicates that it is more a question of signals not acquiring sufficient strength to activate the relevant muscles.

The mechanism I have been sketching here would be unable to function if the various characteristic times were not of the correct magnitude. It is clear, for example, that the above-cited cognitive process leading to the recognition of danger would have to occupy less time than the overall procedure, the latter clearly including the planning of the possibly dangerous action as well as the subsequent (conditional) diversion of attention. The durations over which we can remember fleeting impressions have been the subject of numerous investigations, the work of Alexander Luria[106] being regarded as seminal.

His lead was followed by George Sperling.[107] Employing a five-by-five array of 25 random target letters (that is to say, letters which did not spell words) and varying the time between exposure and the cueing of the subject's report, he was able to show that very short photographic memory, now also referred to as *iconic memory*,[108] persists for no longer than about 200 milliseconds. We may take this as the above-mentioned time for cognitive recognition. Determination of the persistence time of short-term memory is made difficult by the need to eliminate any effects arising from inadvertent rehearsal. This is most readily achieved by asking the subject to repeat the remembered items in reverse order. Lloyd and Margaret Peterson[109] established that this longer component of memory lasts for about 10 seconds.

These, then, are the magnitudes of the two persistence times referred to earlier, and we see that they do indeed display the sort of mismatch that would be required of the conjectured veto mechanism. The neural signals would be reverberating around the 'primary' intralaminar-activating loop long enough for the 'secondary', cingulate-activating, loop to exercise its attention-directing influence on the nRt. As I noted earlier, this mechanism might appear modest, and of rather limited scope, but the danger-avoiding successes of its possessors guaranteed that it would score handsomely in the Darwinian stakes. And the fact that it just happened to imbue the animal with conscious awareness, even without free will, opened the flood gates for all manner of embellishments that Nature could not anticipate. This was the mechanistic little acorn, from which the tall oaks of human culture and technology have grown.

The internally simulated navigations (Antonio Damasio's *as if* processes[110]) made possible by the functioning of the core circuit and its interactions with other parts of the brain, draw upon the predictions provided by the cerebellum (the *what if* processes conjectured by Chris Miall and his colleagues[111]) and the planning capabilities provided by the prefrontal cortex (David Ingvar's *memories of the future*[112]), while automatically notifying the body of their significance. Seen in this light, there is no *problem* of body/mind, only a *situation* of body/ability-to-simulate-body's-interactions-with-environment.

Schemata

The time has come to introduce another technical term: schema (plural: schemata). It will be prominent in the balance of our story. Let us start by asking what the signals must accomplish which arrive at the premotor area. Amongst other things, they must set up patterns of motor acts which prevent the various muscles from coming into mutual conflict. When the animal is perceiving, this amounts to policing the various possible sequences of motor response appropriate to the stimulus. Frederic Bartlett[113] referred to such a sequence as a schema, the origin of this concept being traceable to the writings of Immanuel Kant.[114] The premotor area, presumably in conjunction with the anterior cingulate, must ensure that only viable schemata are made available.

The exact nature of schemata is still the subject of debate,[115] the uncertainty being in part due to the fact that they have yet to be unambiguously associated with specific brain regions. Kant saw schema as structures of the imagination that connect concepts with percepts, and he described them as procedures for constructing images. This is too vague to be of much help here. One of the most useful definitions has been given by Ulric Neisser,[116] according to whom:

A schema is that portion of the entire perceptual cycle which is internal to the perceiver, modifiable by experience, and somehow specific to what is being perceived. The schema accepts information as it becomes available at sensory surfaces and is changed by that information; it directs movements and exploratory activities that make more information available, by which it is further modified. From the biological point of view, a schema is a part of the nervous system. It is some active array of physiological structures and processes: not a centre in the brain, but an entire system that includes receptors and afferents and feed-forward units and efferents.

Caution is required, however, in connection with his phrase 'part of the nervous system', because this appears to indicate that specific neural circuits exclusively serve a single schema. It is more likely that schemata are embodied in synaptic modifications, in a distributed and superimposed fashion, as is the case for memories in general. A better definition would be:

A schema is a reproducible coactivation of neurons linking a specific pattern of motor-planning activity in the premotor area to relevant activity in the sensory areas, the reproducibility stemming from the fact that schemata are laid down in memory.

My aim here is to put the schema on a more secure neuronal footing. During perception the premotor area participates in the provision of schemata appropriate to a given input.[z,117] Moreover, the time scale of the planning provided by a given region of the frontal lobe appears to increase the more rostral that region lies,[118] so the schemata would be expected to become progressively more sophisticated as increasingly anterior regions of

z / The planned muscular movements associated with a given schema are presumably embodied in specific synaptic connections to neuronal *pattern generators*.

the prefrontal cortex are recruited.[aa,119] In addition, the reciprocal interaction between the ventromedial frontal region and such emotion-linked structures as the amygdala and hypothalamus would be expected to give these schemata personal and social dimensions.

The important (though distressing) case of Phineas Gage[120] lends weight to this argument. Hanna Damasio and her colleagues[121] have used his skull in a computer-aided reconstruction of the actual damage to his cortex, and find that this was extensive in the prefrontal regions, whereas the premotor area in each hemisphere was spared (as were all regions lying posterior to these). Following his accident, Gage's ability to make rational social decisions was clearly compromised, even though he retained his facility for basic logic.

Further insight into the manner in which the frontal lobes[122] function has been provided by recent experiments with monkeys. Masataka Watanabe[123] has recorded from single neurons in the forebrain as the subject performed a delayed-reward task. He observed that some cells fire selectively in anticipation of a particular type of reward, whereas some are selective for the reward's location. A third group of neurons were found to be influenced by whether the reward had previously been seen or merely inferred through prior experience. As Charles Jennings[124] remarked, when commenting on these results, '... the combined activity of many such neurons could in principle encode a detailed description of the expected reward, which could then be used to control goal-directed behaviour'. This formulation emphasizes that a schema includes implicit information about the goal of a planned sequence of motor acts.

The specificity of planned actions is exposed by the recent experiments of Vittorio Gallese, Luciano Fadiga, Leonardo Fogassi and Giacomo Rizzolatti.[125] For example, they have found neurons in the frontal lobe which increase their firing rate when the monkey is grasping a small morsel of food, such as a raisin, and also when the animal observes a person doing the same thing. But these neurons do not react to the mere sight of raisins, and neither do they respond to the animal's act of grasping in the absence of the raisin, or indeed to the sight of a person doing the same thing. Gallese and his colleagues concluded that the activation of these neurons contributes to the matching of observed movements to their actual execution. Again, therefore, one glimpses the frontal lobe in action, selecting an appropriate schema.

Reza Shadmehr and Henry Holcomb[126] have used PET to endorse the idea that acquisition of a motor skill involves learning an internal model of the dynamics of that task, and they found that within six hours after completion of practice, while performance remains unchanged, the system engages new brain regions to perform the job. They showed that there is a shift of activity

aa / It seems possible that the critical determinants of intelligence are associated with the frontal lobes, and that this region of the brain is particularly important for generation of Sommerhoff's category B representations.

from the prefrontal region to the premotor area, posterior parietal area and the cerebellar cortex. A quite suggestive augmentation of this result was provided by the work of Gregory Berns, Jonathan Cohen and Mark Mintun,[127] using the same technique. Their subjects performed a simple reaction-time task, the various stimuli being equally likely and apparently occurring randomly. In practice, however, the stimuli covertly followed a complex pattern. Subsequent behavioural tests showed that the pattern had been unconsciously learned, the acquisition being accompanied by increased blood flow in the premotor and anterior cingulate areas.

The proposed connection between schemata and consciousness is illustrated by the following simple example. If the back of my left hand itches, I might scratch it unconsciously with the fingers of my right hand, as a reflex while my attention is elsewhere. Alternatively, I might simply contemplate the itch. But what then informs me of the itch's location? It might seem that the answer lies in the presence, in the brain, of a map of the body's surface, and there are indeed several of these.[128] But what agency inspects the activity levels in the various regions of the map, and thereby discovers where I am itching?

Consciousness of the itch must involve something quite different. Knowing its location entails having access to a schema that would bring my right-hand fingers to the appropriate place for scratching. And this schema will have been the product of past experience, which may have had a component of random chance; volition need not have been implicated.

Access to a suitable schema permits us to use it reflexly, as in the first case of the above example. And its availability also serves perception, as in the second case. But the logic seems incomplete, because the mechanism still appears to require an agency, to witness the use of the schema. We have overlooked a vital factor, however, namely the efficacy of the scratching. The reflex version of this act remains unconscious only because it succeeds in removing the itch. The cessation of the itch therefore produces information: the scratching was effective. And this principle can be generalized, because all the body's needs make themselves felt through the production of signals. It is remarkable that we thus acquire information through the quenching of signals as well as through their generation.

It would be easy to underestimate the power of this argument. A schema provides not only a prescription for a coordinated motor act; it also carries implicit information related to the goal of that act. Indeed, a particular schema is not incorporated in our repertoire of schemata unless it has previously produced a successful outcome. Schemata therefore simultaneously provide us with our means of perceiving and our mechanism of knowing.

Why do I feel that I am in my head?

Ever since I can remember, I have relied on the working assumption that I am in the middle of my head. My me-ness, if you will, seems to be a few centimetres symmetrically behind my eyes, mid-way between my ears, and under what is left of my hair. I get the distinct impression that I am slightly above my nose, and most decidedly above my mouth[bb] and the rest of my body. Likewise, when I visually observe things, I feel that I am seeing them *out there*, in the environment. I do not get the impression that such observed objects are being presented, second-hand, to an inner observer, as a consequence of some intermediate process.

John Eccles,[129] Eric Harth[130] and Max Velmans[131] have written quite eloquently on this theme, and Velmans listed a number of authors who had discussed it even earlier. The concept of schemata removes any mystery as to why this should be so, because it emphasizes that everything we remember automatically links sensory input with the appropriate muscular movements. We see something in our environment as being *out there* because our eye movements are consistent with the appearance of the object at the distance it lies from us. There are, in fact, strong indications that we remember things in terms of the requisite muscular movements. But before considering the evidence, we ought to make another digression.

An organism devoid of consciousness, because of its lack of sufficient neural sophistication, is driven by environmental non-uniformities and also by its internal needs, the former being coupled directly to the latter. If motion caused by these dual agencies gives rise to a significant response from the environment, *significant* here meaning that the response is causally related to the movement, the creature will not be able to appreciate the fact. The provoked correlation will merely be added to the passive conditioning that is the creature's sole mechanism of learning. The animal will naturally be able to respond immediately to a stimulus, but it will do so only in a manner dictated by that conditioning.

Humans, and similarly endowed species, also generate movements of the above unconscious type. Indeed, given the greater range of their motivations, their scope for such generation must be much wider. But when the resulting movement produces a significant response from the environment, the implied correlation will be instantly detected, because of consciousness, and the organism will immediately be able to incorporate the correlation's perceived consequences into its repertoire of possible reactions. Movement may also be a result of prior thought, of course, but the above scenario is important because it shows that no exploration of the environment has to be burdened by the need to invoke a spontaneous and dualist act of will.

How does the system capture stimulus–response correlations? I believe that such capture is made possible by two vital factors: the anatomical provision of what is known as *efference copy*, and the presence of the appropriate receptor molecules on certain neurons. Let us look at these factors in turn.

bb / A colleague of mine, who is the curator of a zoological museum, has told me that he knows of no example, amongst animals possessing vision, in which the eyes are not located rather close to the mouth. Visually guided navigation toward what an animal might consume clearly enjoys high priority.

It is a common attribute of many neurons that their axons branch quite soon after emerging from the soma, so as to produce two or more efferent, signal-carrying, routes. In the case of neurons whose primary axons provide a conduit for instructions to muscles, the secondary routes (known as *axon collaterals*) are said to carry efference copy. A classic example of the usefulness of such efference copy is seen in the monitoring of (and correction of) retinal slip in the visual system. The situation was analysed by Erich von Holst,[132] who showed how the existence of a mechanism for comparing efference copy of the instructions sent to the eye muscles with the resulting signals being returned from the retina could obviate such slip. In this system, therefore, we see the eye functioning so as to hold an image steady on the retina. As Roger Carpenter[133] remarked: '... a saccade is a question directed at a particular part of the world'. I suggested above that one can generalize this idea, by noting that *every muscular movement asks questions of the environment, and our only means of asking such questions is through muscular movement.*

Turning to the second vital factor, namely the presence of appropriate receptor molecules, it would appear that the NMDA-type glutamate receptor is a particularly good candidate, because it is ideally suited to promote capture of correlations between different stimuli arriving in the vicinity of a given synapse.[cc,134] And because its activation produces excitatory postsynaptic potentials which are unusually long-lasting,[135] about 150–200 milliseconds, this receptor also mediates further capture of correlations by the receiving neuron. Returning to the underlying anatomy, we see that the premotor area is the source both of instructions to the motor cortex (and thence to the muscles) and of the efference copy passed back toward those regions of the sensory cortex that receive input from the thalamus. In effect, the signals transmitted via the latter routes would be 'saying' to their target areas: *BE ON THE ALERT! – the premotor area has just given the orders for muscular movements which could lead to a response from the environment!*

There is more to consciousness than the mere capturing of correlations, however; there would also have to be a means of gauging the significance for the body of any given action. Given the importance of the muscles, it would not be surprising if their role extended to such evaluation. There is circumstantial evidence which suggests that this could be the case, even though the underlying mechanism is somewhat surreptitious. Indeed, one might have assumed that it lay below the level of consciousness, but this is apparently not the case.

Consider, for example, a fascinating investigation carried out by Guy Goodwin and his colleagues.[136] They created a proprioceptive illusion in a blindfolded subject who was asked to hold his forearms mutually parallel, his upper arms lying parallel and horizontal on a support. When vibration at 100 cycles per second was applied to the biceps of one of the arms, the subject

cc / An analogous capture of correlations observed by Benjamin Libet and his colleagues involved tandem functioning of dopaminergic and cholinergic synapses.

actually held the forearms at an offset angle of about forty degrees, while still believing them to be parallel. The conclusion was that the muscles were thus indeed contributing to conscious sensation. Similarly, Goodwin and his colleagues[137] reported that estimates, by human subjects, of the magnitudes of weights and tensions are influenced by the state of the muscles, fatigued muscles leading to overestimates. Finally, it has been reported[138] that patients recovering from paralysis tend to have sensations of undue effort and heaviness when attempting to move, the illusion disappearing only when recuperation is complete.

It is tempting to take this line one step farther and to identify the normal versions of these muscle-related sensations with qualia. In other words, I am suggesting that qualia arise naturally from the need to monitor the significance for the body of the environment's response to a volition-provoked stimulus. It is equally tempting to suppose that the system can mimic the functioning of the stimulus–response loop that closes via the environment, simply by employing the alternative route provided for the efference copy. This would seem to be a viable alternative, because the signals carried by that part of the circuit must possess all the attributes of the main route. If that were not the case, it is difficult to see how they could adequately perform their task in the scheme. I submit that such mimicry is in fact the process of thought, and am thus advocating the reinstatement of what Roger Sperry feared was defunct, namely *the motor theory of thought*.

A difficulty facing this theory stems from the ambiguity of connection between sensory stimulus and motor response. Francis Crick[139] stressed this problem when he noted that whereas the receptive field of a neuron in the early visual cortex is fully specified and readily determined, that of a neuron in the motor cortex is not well-defined; in principle, it can be influenced by any of the senses, by combinations of them, and also by the emotions. Moreover, whereas tactile perception is intimately linked to muscular movement and proprioception, the connection to other senses seems more tenuous. But we could recall the experiments of Robert Wurtz, Michael Goldberg and David Lee Robinson,[140] which revealed the saccadic scanning of visual objects. This too involved muscular movements, and although saccades may appear to be unconscious, we should bear in mind the experiment by Goodwin, cited above, which revealed a conscious component of muscle function.

Even for objects entirely encompassable within the fovea, and thus not requiring saccades, recognition might be possible only if there has been prior saccadic scanning of a similar object, identification then invoking a size-invariance capability.[dd] Results reported by David Noton and Lawrence Stark[141] appear to endorse this idea because they find that perception and recall of a visual image is mediated by recognition of the movements made

dd / Size-invariance is achievable in feed-forward neural networks if their units have appropriate receptive fields, and are arranged in layers. The hierarchically deployed visual areas could mediate a similar attribute.

during the scanning of it, rather than of the image as a whole. This has been a long digression, but it has ended with the memory-of-movement argument establishing its credibility.

We could take this argument to its logical conclusion and say that because the primary motor area (M1) is the governor of movement (except that of the eyes), it must serve as the fiducial structure, against which we can rationalize the positions of all other areas in the cortex (see figure 9.11). One would then expect that the more remote from M1 a particular sensory modality makes its entry, into the cortex, the more sophisticated is its connection with movement. Touch and taste are relatively rudimentary, in this respect, since the environmental feedback resulting from movements of the underlying muscles (finger muscles, for example, in touch, and mouth and tongue muscles for taste) is detected by bodily regions lying very close to these muscles. Those two senses make their entry into the cortex very close to M1. As one can see from figure 3.2, the motor and somatosensory maps are essentially in registry. And the left-hand part of figure 9.11 reveals that gustatory input enters the cortex very close to the muscles that control mouth and tongue movements. It is also worth noting that the four basic tastes (sweet, sour, bitter and salt – see the small inset to figure 9.11) are served by different regions of the tongue, so the gustatory and tactile senses are very similar and, I suggest, rather primitive.

Vision, surely the most developed sense in the primates,[142] enters the cortex at the extreme caudal position, despite the fact that the eyes are at the front of the head, and the connection to muscular movements in this sensory modality is particularly indirect. Hearing and smell lie between these extremes, and they enter the cortex at an intermediate position.[143] One can nevertheless think of these other senses as being more sophisticated versions of their tactile counterparts. Regarded in this manner, smell becomes chemically mediated touching at a distance, hearing is remote touching mediated by the agency of (mechanical) sound waves, and sight can be likened to remote touching of objects by the retina, mediated by light rays.

Another consequence of all this is that we can now identify the anatomy underlying Ronald Melzack's theory of the *phantom limb effect*, discussed earlier in this chapter. We recall that this invoked what he referred to as the neuromatrix, which contrived to monitor the state of all the body's components, through a comprehensive network of neural circuits. In view of what we have just been discussing, we can say that the monitoring exploits those efference copy routes, which will still be in place despite the loss of the corresponding body part. In other words, the limb may have gone but all the schemata related to it survive. There is, indeed, an intimate relationship between what Melzack calls the neurosignature and what we have been referring to as schemata.

Likewise, our emphasis on the role of motor planning suggests a particularly direct solution of the *binding problem*. I noted earlier that gamma-band oscillations might serve binding of features being perceived through a given sensory channel, as Peter Milner and Christoph von der Malsburg suggested. But Antonio Damasio has stressed that there is an even more severe problem of binding between elements belonging to *different* sensory modalities; different senses must also know what goes with what. Damasio's solution was the postulated existence of neural systems with the special task of integrating functional regions of the brain (or, to be more precise, the telencephalon).[144] In fact, his idea bears more than a superficial resemblance to Melzack's neuromatrix. I would like to suggest that the system overcomes the multimodal binding problem in a rather more direct manner. Let us see what is involved. Once again, we will have to start with a digression.

A passage in a book by John Eccles[145] expressed his astonishment that a structure so heterogeneous as the human nervous system could salvage unity for the consciousness it mediates. He (and others) may have viewed the issue from the wrong direction. *I believe that the problem confronted during evolution of complex organisms like ourselves was not to unify conscious experience but rather to avoid destroying the unity that Nature provided.* The single-celled organisms from which we sprang were not burdened by a unity problem. Devoid of volition, each was driven by its environment. Non-uniformities of illumination, chemical concentration and the like were directly coupled to its mobility mechanism.[146] In a very real sense, the environment of such an organism and the *map* of its environment were one and the same.

One of the milepost achievements of the last decade was the confirmation by Apostolos Georgopoulos and his colleagues[147] that muscular movements (in the monkey) are coded for, vector fashion, by populations of the appropriate neurons. This followed earlier observations and discussion along these lines.[148] R. P. Erickson had coined the term *neuronal response function* to describe the discharge rate of cells in the relevant cortical area, and Georgopoulos and his colleagues verified that this function is broadly tuned. They demonstrated, moreover, that despite the complexity of muscular movement, the relationship between movement direction and the rate of motor cortical discharge is both simple and robust.

The ultimate targets of those motor cortical discharges are, of course, the appropriate muscles, which receive their instructions via the neuronal axons known as alpha efferents. And it is through these muscles alone that the body acts upon the physical surroundings. A conspicuous attribute of this environment is its lack of ambiguity, provided we confine our attention to length and time scales that are comfortably removed from the quantum domain; our world is comprised of objects that have uniquely defined positions at any given time. Muscles themselves are physical, of course, and they too are subject to the same constraint of uniqueness; a given muscle has, at any instant, a given length and a given tension. And if it is undergoing change at that instant, then the rates of change of those parameters are themselves characterized by a similar lack of ambiguity.

Although the validity of these statements might seem trivially obvious, I believe that they reveal something which could easily be missed: irrespective of which brain processes precede them, muscle activations must be the unique products of decision mechanisms carried out in the underlying

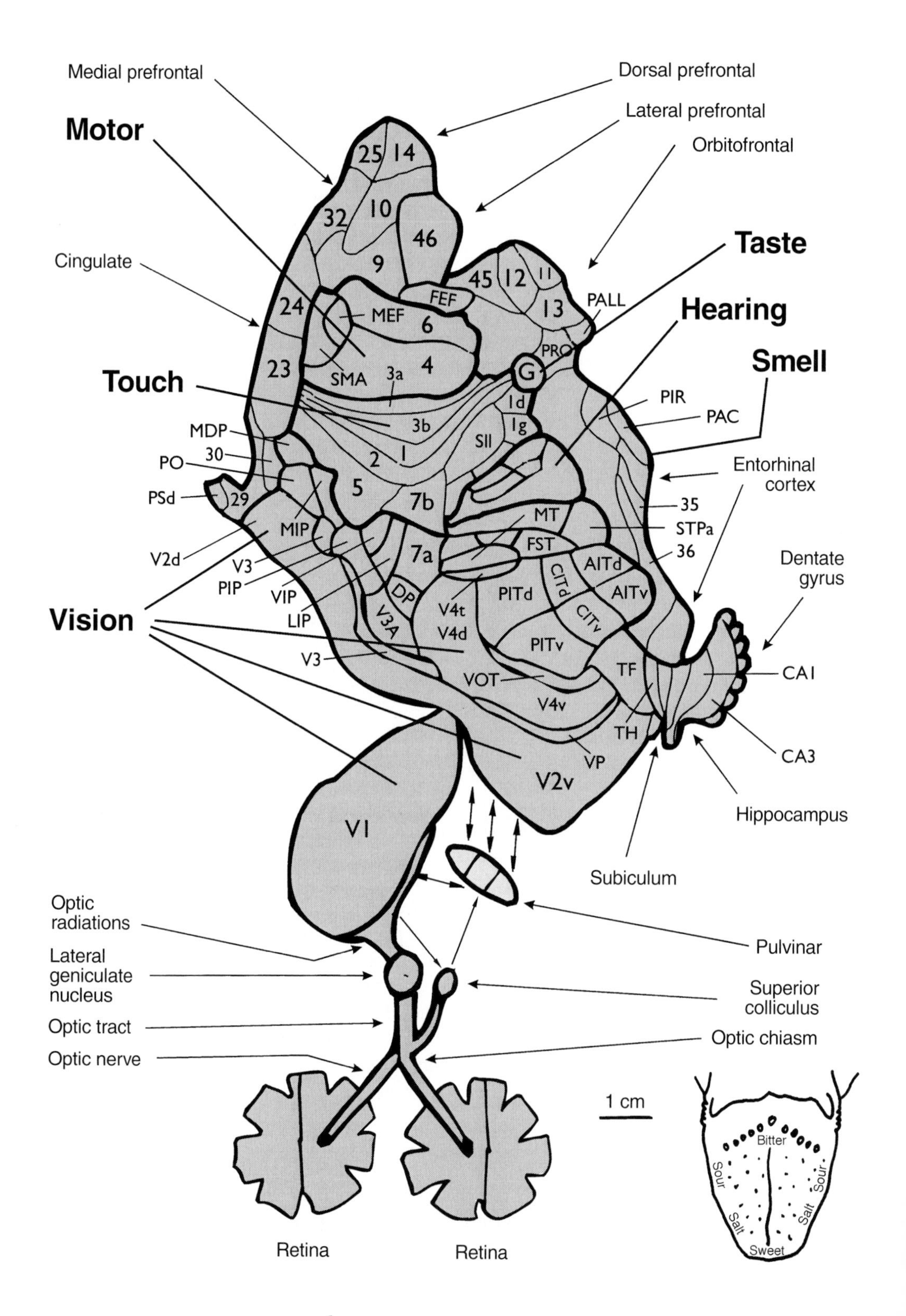
Medial prefrontal
Dorsal prefrontal
Lateral prefrontal
Orbitofrontal
Motor
Taste
Hearing
Smell
Cingulate
Touch
Vision
25
14
10
32
46
9
45
12
11
13
24
23
FEF
MEF
6
4
SMA
3a
3b
2
1
5
7b
7a
PALL
PRO
G
1d
1g
SII
PIR
PAC
Entorhinal cortex
35
STPa
36
MDP
30
PO
PSd
29
MIP
V2d
V3
PIP
VIP
LIP
DP
V3A
V3
MT
FST
AITd
AITv
CITd
CITv
PITd
PITv
V4t
V4d
VOT
V4v
VP
TF
TH
V2v
V1
Dentate gyrus
CA1
CA3
Hippocampus
Subiculum
Pulvinar
Optic radiations
Lateral geniculate nucleus
Superior colliculus
Optic tract
Optic chiasm
Optic nerve
1 cm
Bitter
Sour
Sour
Salt
Salt
Sweet
Retina
Retina

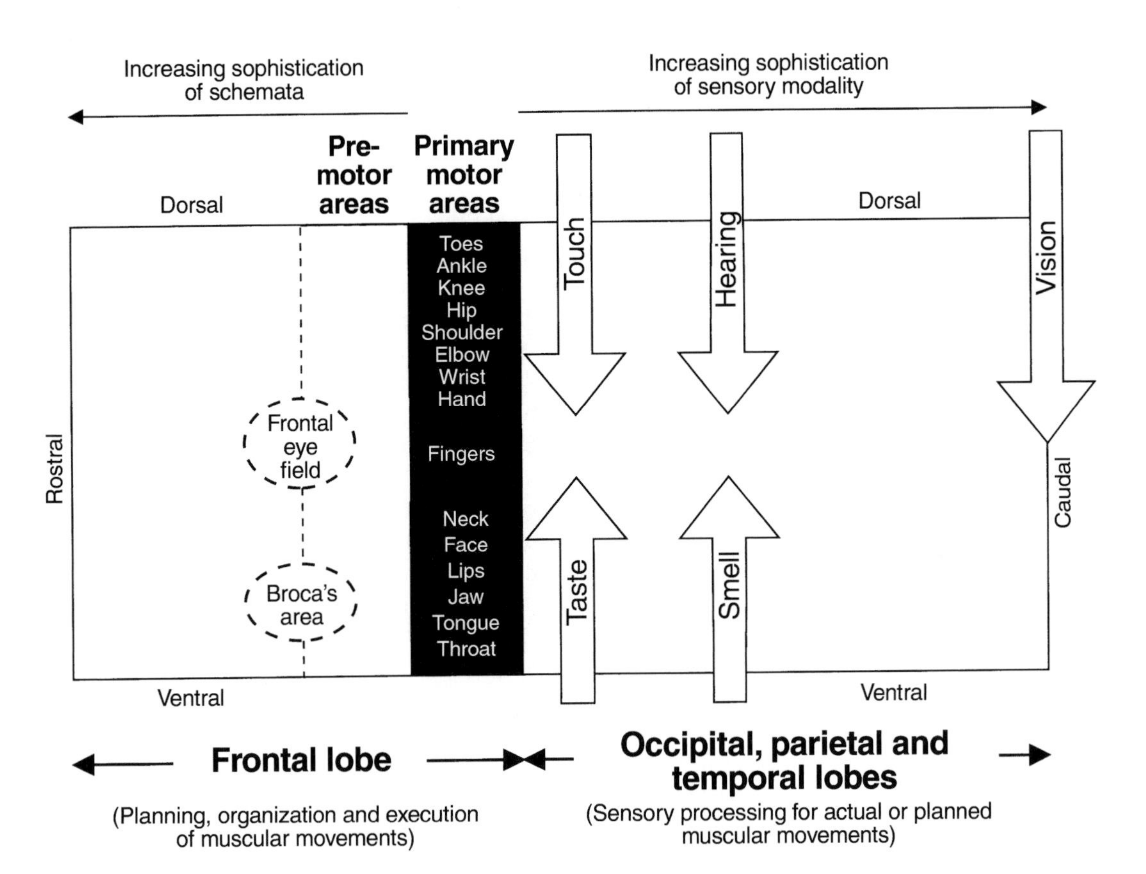

Figure 9.11
The primacy of motor planning postulated by the theory of consciousness presented here suggests that the positions of all cortical regions should be rationalized with respect to the motor and premotor areas. The more remote from the latter a particular sensory modality makes its entry into the cortex (right-hand diagram), the more sophisticated is its relationship to movement. A particularly revealing feature of the picture of the spread-out right hemisphere of the monkey brain (left-hand diagram), due to Daniel Felleman and David Van Essen, is the fact that the gustatory (taste) area, denoted by the letter G, is located as a continuation of the somatosensory region, and it is positioned in close proximity to the motor area corresponding to movement of the tongue (see figure 3.2). Because the different primary tastes (salt, sweet, sour and bitter) are served by different regions of the tongue (see the small inset figure), one sees that taste can be regarded as a chemically mediated tactile sense. Hearing, which is served by mechanoreceptors in the cochlea, can be regarded as a tactile sense that can detect things remotely, via (mechanical) sound waves, and smell can similarly be thought of as a chemically mediated form of tactile sense that can act remotely because certain molecules can be carried by the air. Finally, even sight can be looked on as a type of touching at a distance, acting through the agency of light rays, the presence in the retina of various types of receptor cell giving this sense its sophistication. Likewise, and because the time scale of neuronal persistence increases in the rostral direction, schemata become progressively more sophisticated as increasingly anterior regions of the frontal lobe are recruited. The theory also predicts that Broca's area might not exclusively serve speech; it suggests that this area will be activated by all well-rehearsed mouth movements (including those of the lips and tongue).

neuronal networks. Even in the absence of actual volition, there will be neuronal-level decisions which produce what could be termed the unity of unmeditated action. Although the preceding neural processes are distributed,[149] there must ultimately be a focussing down to a unique vector in the hyperdimensional space that is required to describe the collective state of the body's total musculature.

What can we say about those preceding neural events? The ones immediately prior to the making of that hyperdimensional decision, at least, also appear to be vectorial, because David Redish and David Touretzky[150] found evidence of vector arithmetic being carried out in the motor system during their investigation of a reaching task. But what about the processing that lies farther removed from that penultimate stage? Does the vector concept lose its validity as one approaches areas closer to sensory input? Let us consider the case of vision, which we know to be served by a multiplicity of cortical areas.[151]

Every visually perceived object has both position (its *where*) and shape (its *what*), and it is known that the former is handled in the parietal region, while shape is served by the temporal region.[152] Position can certainly be described vectorially, and frequently is indeed, but is its case special and thus not reliable as evidence in the issue? Jacques Paillard[153] has argued that it is *not* special. Regarding shape, he noted that: '... an object contains its own network of spatial relationships. Thus each channel has to compute a topographic representation of spatial features but differs in its basic encoding modes. One concerns the space of shape and the other the space of locality, thus reflecting two different processing modes'.

The channels referred to here are those lying in the parietal and temporal routes, of course: the ones that respectively handle the where and the what. And Paillard's first sentence emphasizes that shape too is defined by a set of vectors. One sees something akin to these vectors in the work of Robert Wurtz,[154] who recorded the saccadic movements of the eye as it scans an object. The centre of gaze darts about, most frequently visiting points of greatest interest. I suggested above that this permits the system to build up an overall impression of the object, by a process I referred to as *aperture synthesis*. A prerequisite would be that the system maintains a record of the most recent saccades, and automatically shifts the image fragments to their correct positions in the overall picture. Recent work[155] suggests that the system does indeed possess this capability.[ee] It thus appears that the parietal and temporal regions both use vector representations, but in different ways.

Perhaps this similarity of representation is not surprising, given the above-mentioned link between such vectors and the neuronal response functions that code for them. And although the vector concept might seem to have no applicability in the other four senses, there is no reason why population coding should not be equally efficacious in those modalities too; one might

ee / I am much indebted to Roger Carpenter for bringing the investigations of Sparks, and those of Noton and Stark, to my attention.

profitably think in terms of *pseudo*vectors in those faculties. This attractive universality of representation, across all senses, would be convenient for the system, given the analogous universality of output emphasized here.[ff]

From what has been stated about the premotor area's decision-mediating role, one could easily get the impression that it would act as a sort of gate for signals passing on their way to the muscles; one might picture it permitting passage of some signals while blocking others. Such a view would be unnecessarily clear-cut, however. It seems more likely that the governing region would have the form of an extended piece of neural circuitry arranged in parallel with what could be called the direct signalling route. It might then operate by modifying the signals in the latter. I believe that this is the case, and that the direct and indirect paths go through the motor and premotor areas, respectively.

Although analysis of the overall dynamics of the resulting dual circuit would probably be difficult, there would nevertheless be the underlying simplification that the state of each and every muscle would have to be uniquely defined; as mentioned earlier, a muscle cannot simultaneously be in two different states. A good example of the governing principles is seen in the recent analyses of the swimming motion of the lamprey by Sten Grillner and his colleagues.[156]

Returning to our theme, we should note that as further cells were added, during the evolution of more complicated organisms, tissue differentiation was accompanied by an information problem: which tissues were to be coupled to which factors in the environment, and within the organism, what was to be coupled to what? I believe that evolution selected those developmental patterns that, in effect, simply worked backwards from the desired result of unambiguously coupling the various inputs to the mobility mechanism, notwithstanding the fact that the latter was evolving its own sophistication. As is often remarked, Nature adopts the tinker-toy approach, adding to what is already present without jeopardizing overall function.[157]

In the present context, this amounted to rejecting evolutionary developments that compromised temporal integrity; if stimuli emanating from a single environmental feature or event, but passing through different sensory modalities, did not *simultaneously*[gg] arrive at the region responsible for activating the appropriate muscles, the organism was not viable. One could call this a pusillanimous solution to the inter-modal binding problem, but it was an eminently effective one. In advocating this view, I have had to part

ff / Muscular reactions to tactile stimulation are rather obvious. Those to sound are particularly conspicuous in animals that respond by moving their ears, in addition to the more familiar moving of the entire head. More subtly, an odour can cause flaring of the nostrils, and possibly movement of the diaphragm, while tasted substances can provoke movements of the tongue and mouth.

gg / Simultaneity is here to be interpreted as meaning that the various signals merely arrive within such a short interval that the system reacts as if they are concurrent.

company with Eccles; he saw the nervous system as a democracy, I am suggesting that it is really a dictatorship.

The neural circuitry that permits muscular output to follow sensory input is structured hierarchically. We must nevertheless bear in mind that the nervous system just serves a conglomerate of cells, furthering the latter's twin goals of nourishment and perpetuation. The hierarchy's existence implies no spiritual pinnacle; as I mentioned in an earlier chapter, it just provides the organism with a more sensitive antenna to the environment. The division of labour accompanying tissue differentiation led to the need for, and to the formation of, *multiple* maps of the environment, and of the body's relationship to the latter.[158] The unity of our consciousness experience merely demonstrates that these maps have been created without endangering the operational needs of our musculature.

Attention and binocular rivalry

The ideas expressed in the preceding sections might hold the key to a continuing puzzle regarding attention. Stefan Treue and John Maunsell[159] measured attentional modulation of neuronal signals related to visual motion. They found that the responses of direction-selective neurons in the monkey visual cortex are markedly influenced by attention, and that this modulation is detectable surprisingly early in the visual hierarchy. From the experimental set-up in such studies, it is clear that the attentional control is being driven by top-down (or exogenous) cues. As Kenneth Britten[160] remarked, when commenting on these results, '... how these signals reach back into the early stages of processing to select certain synaptic inputs over others remains a profound mystery'. According to the theory being presented here, the ultimate source of those top-down signals must be the frontal lobes, and they reach the early visual areas via the thalamus.

Further endorsement of these principles is provided by another type of visual experiment, namely that involving binocular rivalry.[161] It is not difficult to arrange for the two eyes to be shown different scenes, and the question arises as to how the brain will interpret the situation. When the scenes are both stationary, and with practice, it is possible to perceive them simultaneously, for a brief period at least. When they are both moving, but in opposing directions, the two percepts alternate at irregular intervals; the system follows one for a few seconds, and then switches over to the other.

David Leopold and Nikos Logothetis[162] monitored the activities of neurons in various visual areas of macaque monkeys subjected to binocular rivalry involving two counter-moving sets of alternating light and dark lines. They found evidence of modification of the firing patterns of some of the neurons in each of three areas, namely MT (the midtemporal cortex), V4 (which mediates colour perception), and V1/V2 (near the border between the primary and secondary visual areas). As in the case of the Treue–Maunsell study, however, the firing patterns of the majority of the neurons in these areas remained unaltered. This work is particularly valuable because the psychophysical investigations can be followed up[163] by post-mortem studies which pinpoint the location of the neurons displaying the firing-pattern modification: the deep layers 5 and 6, both of which are known to project out of the cortex, the latter to the thalamus.

An alternative form of rivalry pits a stationary scene against a moving one. What is observed has come to be known as the Cheshire Cat effect (see chapter 8), because a typical choice for the former picture is a smiling cat, whereas the other eye is initially presented with a blank expanse of white.[164] If one then moves one's hand across the latter, those parts of the cat that are briefly occluded disappear. With practice, one can contrive to leave only the cat's smile intact. Interestingly, this illusion involves similar time constants to those observed in the other version; the eclipse of the rest of the cat lasts a few seconds. But there is a very suggestive difference between the two cases: in the latter version, the movement always dominates over the fixed image. (This has obvious survival value for the animal, since moving objects usually pose a greater threat.)

As Crick[165] has noted, work on binocular rivalry may well play a decisive role in resolving the consciousness issue, so it is important to ask whether the ideas I have been expressing might explain the observations. I believe that they can. By postulating that something will not be perceived unless it has penetrated through to the premotor area, and set up an appropriate schema which it can 'latch on to', the model shows why the rival pictures cannot be simultaneously accommodated by the system; the schemata for two counter-directed movements must be mutually exclusive. Moreover, since schemata serve movement, it is not surprising that this attribute enjoys easier access to the available schemata in the Cheshire Cat version of the phenomenon. (The shapes of fixed objects presumably require preliminary translation into the movements that would be required to grasp and handle them, which would inevitably take more time.) The few seconds persistence of the phenomenon might be related to the duration of neuronal inhibitory effects, and the alternation between percepts in the former (conflicting-motions) version might be provoked by habituation, each switch involving a process of what is referred to as *self-organization*. (This process is a specific facet of the wider phenomenon of *synergetics*, the study of which was pioneered by Hermann Haken.[166] Haken[167] and Scott Kelso[168] have employed the self-organization principle to great effect in their analyses of brain function.)

The notion of the system latching on to a schema is so important that it deserves further illustration. A particularly fascinating example is seen in the so-called autostereogram, which became a popular diversion in the 1990s. A typical specimen is shown in figure 9.12, which initially looks like a mere collection of random dots. If one deliberately focusses one's gaze beyond the plane of the page, however, an intelligible shape suddenly emerges; in this case it is a set of concentric undulations, rather reminiscent of the wave pattern set up when a stone is thrown into water. This percept does not come gradually; either it is there or it is not. My interpretation is that one can see the percept because it is something that one could handle, and

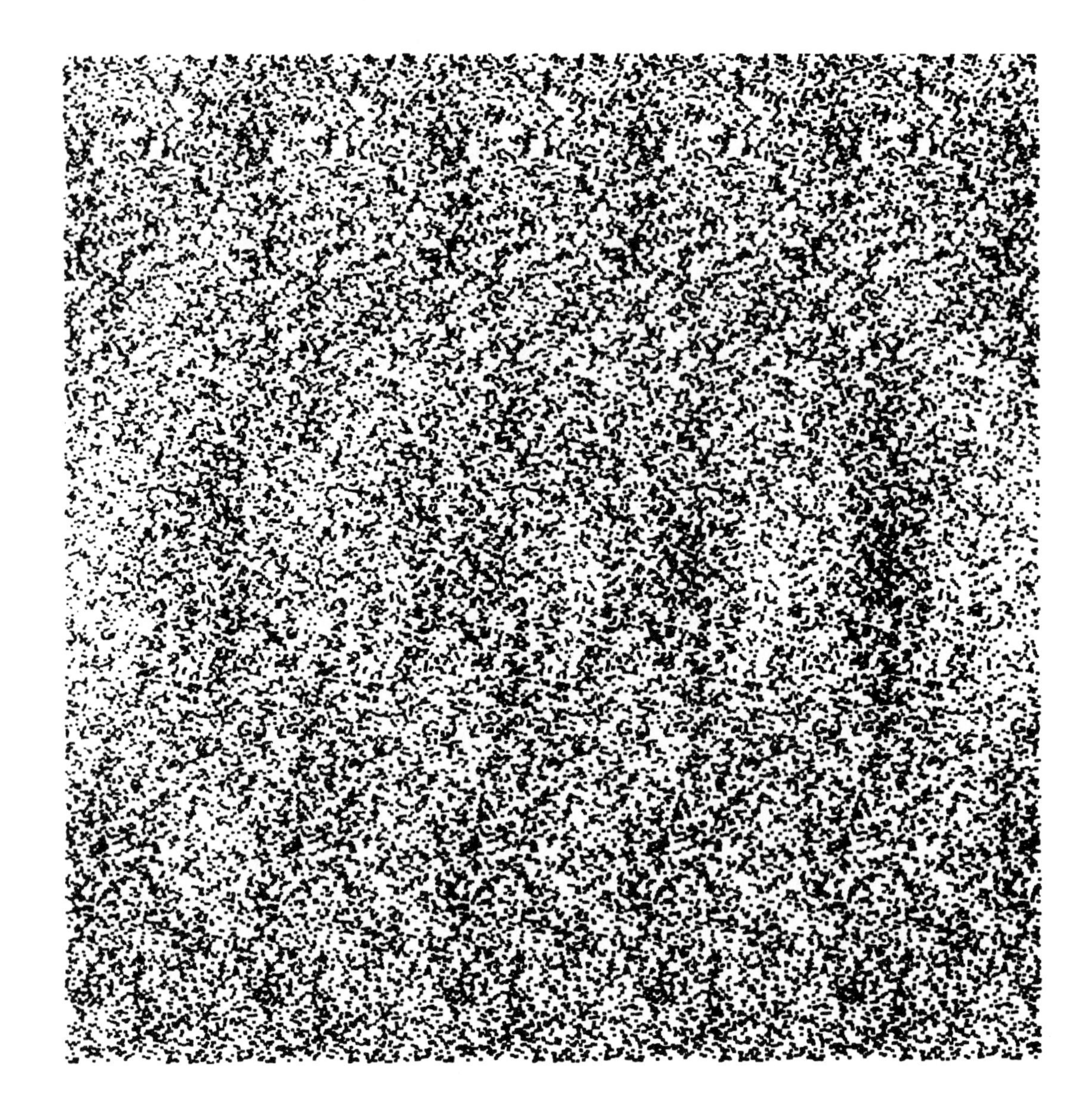

Figure 9.12
If one is not experienced in interpreting autostereograms, like the one reproduced here, it can be quite difficult to discover what is being conveyed. One trick is to position the picture right in front of the nose, so that focusing on the dots is impossible, and then gradually move it away, without changing the position of gaze. When the picture is about 40 centimetres from the tip of the nose, a beautiful three-dimensional shape should suddenly be perceived. The final result is well worth the effort. It is very significant that the entire three-dimensional shape emerges simultaneously, this being a consequence of a self-organizing process of the type described by Hermann Haken. In the present author's view, this process has to implicate movement-preparation events in the premotor cortex.

therefore something for which one possesses a relevant schema. The time required for self-organization may actually have been measured by recent studies of the firing of a neuron in the frontal eye field, when the animal (a monkey) is discriminating between a target and distractors,[169] both falling within the neuron's receptive field. If the firing frequency of the neuron is measured as a function of time from stimulus onset, there is initially no discernable difference between the responses to a target and to a distractor, but after about 120 milliseconds, the responses clearly begin to diverge, and they are totally separated by the 200 millisecond mark. It seems, then, that self-organization takes about 120 milliseconds to establish itself.

There is another spectacular illustration which cannot be reproduced here because it involves motion. One is shown a stationary pattern of what appear to be randomly drawn dots. One is unable to explain what they represent. The dots are then set in motion, but although it is clear that each dot is tending to oscillate within a reasonably constrained region of the picture, it is still impossible to say what is being shown. The observer is then shown the original stationary pattern once more, and asked if it can be given explanation when it is then rotated by 180 degrees; the answer is still no. The spots are set in motion once more, and again it is clear that the each individual spot is moving in a confined region. Then, quite suddenly, the interpretation seems to leap out of the picture – one is observing two people dancing!

Let me quickly explain how this movie was made. Two people were dressed from top to toe in black tights, and small lights were attached to all their joints (ankles, knees, hips, wrists, elbows, and so on). They were then filmed in a darkened room, as they did a sort of aerobics dance. In this example too, the solution of the mystery comes very suddenly; there are no half measures in one's perception of the scene. And once again I submit that such perception is mediated by one's possession of a suitable schema: one can, so to speak, project one's own imagined motions into the moving dots.

Let me close with just one more example, of a rather different, and humorous, type. In one of my familiar domestic situations, hardly unique, my wife accuses me of not listening while she recounts her latest interaction with a group of cronies. When I assure her that she has had my full attention, she challenges me to repeat her most recent sentence. My inability to do so unmasks me for the fibber that I am. I feel that these little scenarios are quite revealing; if one equates concentrating on something with being conscious of it, they indicate that consciousness is not involved unless the thing one is supposed to be attending to is penetrating right through to the premotor area, and being set up as a series of motor acts, which can be put in motion if one so decides.

I have seen a music-hall turn in which the artist can repeat, within a very small fraction of a second, everything another person is saying. Volunteers from the audience were astonished by this 'instant echo', because it seemed to usurp their power over their own voices. The ultimate upshot was invariably embarrassed and uneasy laughter on the part of each 'victim'. The performer was clearly concentrating intensely on the task, and lip-reading may have augmented listening in his remarkable technique. His impressive feat must have been related to a particularly efficient coupling to and from the premotor area.

Qualia – the feelings of consciousness

The most difficult aspect of consciousness concerns *qualia*, which are the raw feelings associated with the phenomenon. Because of the paramount role played by muscles, either in actual movements or merely in the proxied versions involved in thought, they would be expected to play a role in these feelings. As Peter Matthews[170] has discussed, muscle components come in two varieties: the *extrafusal fibres*, which are activated by the *alpha nerve axons* (stemming from the *alpha neurons*), and which provide the means of movement, and the *spindles*, which are composite structures consisting of centrally located sensors linked at either end to *intrafusal fibres*. Both the extrafusal and the intrafusal fibres are joined to the relevant part of the skeleton via tendons, which themselves contain the sensors known as *Golgi tendon organs*. Charles Bell appears to have been the first to realize that we are conscious of muscular exertion, and he linked this together with our being apprised of joint position and movement, in what he referred to as the *sixth sense*.[171]

The latter could be said to function as stress gauges (measuring the load on the muscle), while the central parts of the spindles serve as static and dynamic strain gauges (measuring the current length of the spindle, relative to that of the extrafusal fibres, and the rate of change of that length). As was discovered by Lars Leksell[172] in 1945, the intrafusal fibres are activated by the *gamma nerve axons* (stemming from the *gamma neurons*). Half of all the spindles in the human body are located in the neck, where they mediate fine control of head position, thereby stabilizing the eyes and balance apparatus in space.

It is not the case that muscle activation is merely handled by the alpha axons, these receiving their instructions from the motor cortex. The gamma axons initiate the process, by inducing the intrafusal fibres to change their length, thereby probing their current length-wise relationship to the extrafusal fibres. The signals resulting from this probing are transmitted by a route which ultimately involves messages being given to the alpha neurons, whereafter the alpha neurons do indeed dispatch their commands to the extrafusal muscle fibres, causing the entire muscle to contract.

In the lower vertebrates, collaterals from the alpha nerve fibres are the only source of activation of the intrafusal fibres, and even in more advanced species the alpha and gamma neurons often function in tandem, producing stereotyped sequences of movements. In 1951, Carlton Hunt and Stephen Kuffler[173] showed that the gamma neurons are activated by the central nervous system, in order to maintain the tension of the intrafusal fibres, and Ragnar Granit[174] subsequently showed that electrical stimulation of the motor cortex leads to what he called *alpha–gamma coactivation*, that is to say simultaneous galvanizing into action of the alpha and gamma neurons. Of particular interest to our story here is the fact that the mammalian nervous system has acquired an additional embellishment, namely *independent gamma neurons*, thereby decoupling the control of the spindles from the control of the muscles. I believe that this was a *highly* significant development. It permitted not only greater flexibility of motor operation, but also the exploitation of the spindle in the navigational processes described above. It provided, I believe, the basis of qualia themselves. It would have to follow that only mammals possess qualia.

The innervation of the intrafusal muscle fibres, and the resulting return of signals to the higher command centres of the nervous system, bears a striking resemblance to the eliciting of reafferent signals from the external environment, by the system's overt muscular movements. The difference in the case of the intrafusal fibres lies simply in the fact that it is the *internal* environment that is being probed. In other words, there is probing through the use of what could be called *internal reafference*.

It might seem rather far-fetched to attribute a role in *all* qualia to a process which merely reports on the muscular state of play, as it were, but it must be borne in mind that these monitoring processes will inevitably be correlated with signals being transmitted from the various internal sensory receptors, such as those that serve proprioception, or those that sense pain.

Charles Sherrington,[175] in his Linacre Lecture delivered at St John's College, Cambridge, on 6 May 1924, expressed the opinion that muscles are not merely at the passive service of the nervous system. That this system is also informed, by the muscles, about the *internal* world is, in his own words '... shown by their possession of receptors of their own. On their own behalf they send messages into the central exchanges. This must mean they have some voice in their own conditions of service, perhaps ring themselves up and ring themselves off'. Later on in his text, he adds: '... Following the functional scheme of all receptors, we may be sure that the central reactions provoked by the receptors of muscle will be divisible into, on one hand, the purely reflex and on the other hand, those which subserve mental experience'.

It is instructive to consider the feelings one experiences when the degree of attention is varied. We have considered before the successive concentration on parts of the hand from finger to wrinkles. As one concentrates on a more limited portion of the finger the rest of the finger becomes blurred. One can ultimately focus right down to a pinhead-sized spot on the finger, at the expense of making everything else unclear. When one makes these field-diminishing adjustments, one has the vague feeling of making an effort, the exertion increasing as the focus becomes narrower; if one relaxes, the width of attention reverts to its original size. But where is that effort being applied? It does not seem too far-fetched to ascribe all such feelings to events proceeding in the body's muscles.

Unfortunately, it is not possible to muster experimental evidence supporting this idea, though there is something that has a tenuous connection with the issue. Once again, we can cite Sherrington. Discussing depth perception, he noted that this is served, in part, by the sensation of movement possessed by the eye, and that an eye rotation as small as about one minute of arc can be detected by the system. He was thus suggesting that the discernment of muscular movement indeed does underlie one important aspect of visual performance.

That there is a muscle sense now appears to be well established,[176] but it remains to be shown which of the associated nerve systems actually generates the feelings. Meanwhile, it is interesting to note that Vahe Amassian, Roger Cracco and Paul Maccabee[177] have found that a single transcranial pulse to the human frontal lobe, generated by an external magnetic coil, can elicit a sense of limb movement, even when that limb has been deliberately immobilized by ischemic blockage.

The idea that qualia are inextricably related to the body's musculature receives support from observations on victims of multiple personality disorder (MPD). Margaret Boden[178] examined the clinicians' film of Eve White/Eve Black, a famous MPD patient,[179] and wrote of two separate streams of consciousness. She noted that these were associated with two distinctive sets of physical attitudes, involving not just the gait and body-language but even the lines of the face. The changes from the one to the other were instantaneous and compelling; Eve White's posture and locomotion were recognizably different from Eve Black's. The influence of mental state on the body's musculature is familiar to the physiotherapist, of course, and the goal of the masseur is as much a relaxed mind as it is relaxed muscles.

A well-known philosophical argument establishes that there is something that it is like to see the colour red, and it is of obvious interest to ask whether the ideas developed here could shed any light on this issue. According to the theory, a feature of the external environment will not be perceived until the afferent signals it gives rise to have set up a response in the premotor area, in the form of possible motor acts, and this response will, in turn, put previously established associations on the alert. One's perception of red will thus inevitably strike up possible connections to fire-engines, ripe strawberries, blood and the like. But most significantly, it will inevitably generate activity in those parts of the motor-planning region that would, if permitted, activate the vocal apparatus required to articulate the word ***red***. (In the neonate, the colour red will naturally activate the appropriate sensory receptors, but the experience of the colour will be quite impoverished; the appropriate contexts will not yet have been established in the brain.)

It is important to note that a given pattern of motor-planning activity unambiguously determines the resulting muscular movements; one could say that the activity pattern can thus serve as a *proxy* for a (set of) motor act(s). It must be emphasized, however, that the obligatory involvement of muscular movements (actual or merely proxied) in the laying down of memories is *not* the same as saying that all memories are merely memories of movements. It is worth noting that the envisaged mechanism of perception would not preclude perceptive acts by individuals who are paralysed. Such perception is still possible because it can be mediated by the loop from the sensory processing areas to the premotor area, and thence back to the sensory processing areas via the thalamic intralaminar nuclei. It remains a moot point, however, as to whether the qualia experienced by such unfortunate individuals could be as full as those of normal people, given that the spindle system would presumably also be out of action. Perhaps the system is able to capture and store, somewhere in the cortex, correlations between efference copies of the signals sent to the gamma motor neurons and the returning copies of the signals sent by the spindles to the alpha neurons. This would make the upshot of such correlations available for remote use, and it could provide the paralysed person with reasonably normal qualia.

As Anton Lethin[180] has noted, in some private correspondence with me, the body expresses intentionality by preparing a motor interaction with the environment. This is an important concept, because it permits us to bring into the compass of our spindle-mediated qualia sensations that might not otherwise seem to have any relationship to muscles, such as nausea, hunger, thirst, and pain, and, for that matter, seeing colour. Thus the sensation of nausea is mediated by muscular preparations to vomit, hunger by the muscular preparations to bring food to the mouth and masticate it, and so on. In this respect, suffocation, which I earlier classified together with thirst, hunger and pain, is particularly enlightening; we can contemplate those other discomforts, while immobilized, whereas suffocation demands *immediate* muscular response to give access to air.

Apropos nausea, John Skoyles, in other private correspondence, has noted that the seemingly arbitrary connection between being on a pitching boat and wanting to throw up can be explained in neural terms. Detection of the boat's motion, by the vestibular apparatus, happens to produce the same aberrant signalling as results when certain noxious substances have been inadvertently consumed. In such latter situations, the evolved response of the system clearly makes good biological sense. The trouble is that evolution could not anticipate that Nature's highest ape would return to the seas!

Let us close this section with a couple of clinical notes. A particularly familiar way of losing consciousness is by being knocked out. This causes disruption of the functioning of certain midbrain neurons and the associated axons, especially those of the reticular activating system,[181] which, as we saw earlier, controls the nRt. The truly dreadful *locked-in syndrome*,[hh,182] caused by a stroke in the upper medulla and lower pons, leaves all the muscles permanently paralysed, except for those that control the eyes. The latter are therefore still available for signalling. The victim retains consciousness,[183] and this should not surprise us, given that the mediating loops discussed above remain intact. One would imagine that the patient's spectrum of qualia would be rather truncated, however, since the normal functioning of the muscle spindles would be curtailed. It is a moot point as to whether even their signals have proxied counterparts, acquired through past experience. One can hardly

hh / Related, but milder, conditions have been discussed by Jonathan Cole and Jacques Paillard.

expect sufferers of this terrible fate to enter into discussions about such matters, however.[ii]

The content of consciousness

According to the picture being advocated here, conscious experience would be mediated by the flow of signals around the triangle comprising the sensory cortex, the premotor cortex and the thalamic intralaminar nuclei, with the anterior cingulate and the nucleus reticularis thalami adding the important factor of veto-like control. The actual content of consciousness would be additionally dependent upon processes occurring at the 'apices' of this triangle. Indeed, these structures are interconnected with the various components of the limbic system and the basal ganglia. Analysis of the parts played by these other brain regions is hampered by the scarcity of relevant data. The information that is available endorses the view that there is consolidation, in these regions, of signals emanating from diverse posterior areas.

The processing of information in the posterior cortical regions, and the resulting dispatch of signals to the frontal lobe occurs in a distributed and parallel fashion,[184] and it was earlier thought that the object and spatial processing domains remained distinct and separate in the frontal lobe.[185] This has recently been challenged by S. Chenchal Rao, Gregor Rainer and Earl Miller,[186] however. Monitoring of the activity of many individual neurons in the prefrontal cortex of two monkeys revealed numerous examples which responded to signals from both the parietal (spatial-serving) and temporal (object-serving) regions. Possibly the most difficult aspect to incorporate into the present theory is the manner in which the limbic structures exert their influence over the time scale of short-term memory. And this brings us to the question of language. The emotional aspect of speech must act over a longer time scale than that which applies to the mere stringing together of phonemes.

Although this might be an oversimplification, it is tempting to link the emotional and non-emotional aspects of speech, respectively, with the declarative and procedural forms of memory.[187] By studying language impairments in subjects with well-defined pathological conditions, which are known to be related to lesions in specific brain regions, Michael Ullman and his colleagues[188] and also William Marslen-Wilson and Lorraine Tyler[189] have been able to show that the lexical aspect of speech is served by declarative memory, which is handled in the temporal–parietal/medial–temporal regions (which latter includes the hippocampus), while the grammatical rules are served by the procedural memory, which is looked after by the frontal and basal ganglia regions. From the line being taken here, it is not difficult to believe that it is the grammatical aspects of language which are mediated by

ii / A remarkable book was published in 1997 'written' by a French editor suffering from this syndrome: *The Diving-bell and the Butterfly* by J.-D. Bauby (Fourth Estate).

the frontal and basal ganglion regions. Despite the above reservations, these results are most encouraging for the picture being advocated here.

The updating of one's personal lexicon requires an ability to detect novelty, and it is interesting to note that this facility has recently been identified with the hippocampus.[190] It must be borne in mind, however, that one's lexicon is no mere collection of snapshots; words are concatenations of phonemes, so the hippocampus is being called upon to store temporal sequences.[191] Just how this fits in with the massively documented role of the hippocampus in spatial memory (see, for example, the recent work on so-called *knockout mice* by Joe Tsien, Patricio Huerta and Susumu Tonegawa,[192] and Thomas McHugh, Kenneth Blum and Matthew Wilson, together with Tsien and Tonegawa[193]) is not yet clear. It is probably the case, however, that mice need to store sequenced information when they learn to run mazes. Novelty detection has recently been found to implicate the premotor and anterior cingulate regions, and that responsiveness to novelty can even occur in the absence of awareness. We will return to the important issue of language in the next chapter.

It hardly needs emphasizing that the sensory cortex is a major shaper of the signals passing around the triangular route, and it is important to bear in mind that there are both forward and reverse projections in this part of the brain. The arrival at the sensory cortex of what could be called primary signals, that is to say those that emanate from the various sensory receptors, has been shown by Hubert Dinse and Katharina Krüger to lead to activity which proceeds simultaneously in essentially the entire sensory hierarchy.[194] It is clearly not the case that these primary signals give rise to secondary activity that merely progresses through the system tidal-wave fashion; there are signals running simultaneously in both the forward and reverse directions. What are they doing? Wolf Singer has put forward a well-documented case for the reverse projections mediating the type of memory that would store useful associations,[195] and it seems probable that the simultaneous forward-and-backward signalling permits the picking up of these significant memories.

Perhaps this is what underlies the *preconscious* processing that concerned Max Velmans,[196] when he asked whether human information processing is conscious. To illustrate his fascinating epiphenomenalist argument, let us consider one of his examples. You should pause a while after *silently* reading the following sentence:

The forest ranger did not permit us to enter the national park without a permit.

There is a two-syllable word which occurs twice, but you will have silently pronounced it in two different ways, putting the stress on the *second* syllable at the word's first occurrence, and on the *first* syllable at the second occurrence. How did you know which pronunciation to use, in each case? As Velmans points out, the syntactic and semantic analysis required to make this choice must have preceded the allocation of the stress pattern. And this, in turn, must have preceded the phonemic image's entry into awareness. We must conclude, therefore, that an entire sentence can be processed unconsciously, despite the fact that its message enters consciousness.

This might seem paradoxical, were it not for the fact that this is just another example of the navigation that was discussed earlier; when we read, we navigate our way through a text, unconsciously making articulatory decisions, while consciously absorbing the overall meaning. We can happily agree with Velmans, therefore, when he suggests that human information processing is unconscious. This is not the same as saying that it is unsophisticated, however. On the contrary, and as had been suggested earlier,[197] preconscious analysis must activate memory traces of input stimuli and traces of semantically related stimuli, while failing to excite unrelated stimuli. It seems likely that Singer's reverse-projection memory processes would make an important contribution to the overall mechanism. As Semir Zeki[198] archly put it: 'Ask not what perception does for the mind, ask what the mind does to perception'. In fact, William Shakespeare took pretty much the same line about four hundred years earlier when, in his one hundred and thirteenth sonnet, he wrote:

> *The most sweet favour, or deform'dst creature,*
> *It shapes them to your feature,*
> *Incapable of more, replete with you,*
> *My most true mind thus makes mine eye untrue.*

The reader might like to bear these words in mind, while turning back to illusions such as the Kanizsa triangle, reproduced in chapter 7. Roger Carpenter[199] added an interesting new dimension to this idea when he wrote: 'It can be argued that much artistic endeavour consists in trying to recapture the novelty of primary perceptions which our internal model continually discounts'.

Bottom-up meets top-down?

We have reached the point at which we should check to see whether the criteria identified by other authors, some of which were listed earlier in this chapter, have been fulfilled, and whether the theory that has been described passes muster.

The subjectivity of our private experiences is really tied to the issue of qualia, which have been accounted for by the interplay between internal and external reafference. The unity of conscious experience stems from the fact that the premotor area acts as a bottleneck, policing planned movements so as to prevent them from bringing the body's muscles into mutual conflict; the unity therefore stems from the fact that a muscle can adopt only one state at any given time, and this forces the system to follow only a single muscular path, which essentially determines the direction taken by the stream of consciousness. These same constraints are seen by the theory as automatically solving the multi-modal binding problem.

John Searle linked the issue of intentionality to that of choice, and this played a decisive role in the discussion of what I have been calling navigation, in the general sense. I argued that such choice, even in the absence of free will, was the pivotal advantage, and that this paradoxically required conscious awareness in order to function; this was the surprising mechanism that enabled the higher species to ratchet themselves up a notch on the evolutionary ladder.

Searle's distinction between the centre and the periphery of conscious awareness really comes down to our ability to do more than one thing at once, and the critical division in this case is between the serial nature of events attended to and the parallel nature of the unconscious remainder. The crux of this issue lies in the mechanism by which one is (automatically) switched from the one to the other. As we have seen, the anterior cingulate and the nucleus reticularis thalami are intimately involved in this facility, although the details still await final elucidation.

Searle's issue of familiarity, exemplified by our recognition of things seen from an unusual angle, continues to be a difficult one. I have couched a central part of my own argument in terms of schemata, but just how a particular schema can be exploited to produce related schemata remains a mystery. It seems likely, however, that population coding is involved. We have seen that there is strong evidence for such coding dictating our muscular movements. Perhaps there is a more general form of coding, which is amenable to manipulation, permitting one schema to be smoothly transformed into another. On the other hand, Searle's boundary conditions, that is to say our general awareness of things like time, date and season, do not seem to pose a major problem; these are surely mediated by our memories.

Turning to the three main categories of internal representation identified by Gerd Sommerhoff, the items falling within his Category A (a comprehensive representation of the current structure and properties of the environment) will be present in our conscious awareness if, and only if, the corresponding afferent signals are actively giving rise to motor-planning activities. His Category B representations (serving hypothetical objects, events and situations) will, of course, be one product of thought, and their very existence will depend upon the adequate functioning of those parts of the brain which handle prediction, including the cerebellum and the anterior parts of the frontal lobes.

Sommerhoff's Category C representations (representing the fact that internal representations or individual stimuli are part of the current state of the organism) are, according to the ideas presented here, intimately connected with the way in which schemata are related to the body's musculature, through the internal reafference mechanism; they are, in other words, inseparable from the qualia that accompany them. In effect, this means that Sommerhoff's integrated global representation (IGR), must be strongly related to the output side of the system, which I have argued provides our consciousness with its unity.

Sommerhoff saw the IGR as serving acquired states of conditional expectancy or their derivatives, which can indeed be couched in terms of schemata. The question really comes down to the actual mechanics of combination – the process of integration in the IGR. To take a very simple example, we can walk, we can chew gum, but we can also simultaneously walk *and* chew gum. To some extent at least, schemata thus appear to be additive. Whether this is true of all schemata is quite another matter, however, because the dynamic nature of these prescriptions for movement must lead to considerable complexity. We need what could be called a general algebra of schemata, not the least because working memory can apparently handle several of these simultaneously.[200] Schemata could be said to be the coinage of thought, and the latter is nothing more than a perpetual cascading of schemata, which, at every instant, are vying with each other for the nervous system's centre stage. The bottleneck present on the output side provides a simple means of selecting from among the many things that simultaneously bombard our senses, and of arranging them into an orderly sequence of perceptual events – the 'story line' we know as the stream of consciousness.

Sommerhoff's IGR bears a certain resemblance to what Bernard Baars had earlier called the *global workspace*.[201] This is a central information exchange that allows many individual specialized processors to interact, those that gain access to the workspace being permitted to broadcast their message to the entire system. Some of the things that would be mediated by this workspace are attended percepts, clear mental images, deliberate inner speech, and things deliberately retrieved from memory. The reader will not be surprised to learn that I believe this workspace to be anchored to the motor output, the permission to broadcast being granted by the above bottleneck. And when Daniel Dennett[202] writes of a choice being made from between *multiple drafts* of possible action, the agency mediating the choice is, I believe, that same arbiter.

Let us close this section by briefly returning to the question of why exploitation of the unconscious veto mechanism paradoxically requires us to be conscious. That paradox has emerged in this book, so earlier authors cannot be taken to task for not addressing it directly. Several have nevertheless been struggling with the essence of the issue, however. For example, Naomi Eilan[203] asks what the *consciousness* of conscious perception actually is. Gareth Evans[204] suggested that a key requirement is that perceptual information entering the system must be input to a concept-using mechanism. In the model I have been advocating in this book, that concept-using mechanism relates to the anchoring of schemata to what I have called the vital triangle, with planned movement being the *sine qua non*.

It is extremely important to bear in mind, however, that consciousness will not be present unless a schema, or a set of schemata, are *actually being activated*, this being my main point of departure from the views of Crick and Koch. Shaun Gallagher[205] expressed much the same attitude when he wrote that: '... a body schema involves an extraintentional operation carried out prior to or outside of intentional awareness; although it has an effect on conscious awareness, it may be best to characterize it as a sub-conscious system ...'. Finally, it is worth noting the connection between these planned movements and what Ulric Neisser[206] calls *affordances* for action, these requiring the subject to be an actual agent, rather than a mere mover. The ideas put forward in this chapter should provide a fitting capstone for these previous efforts.

The apparent freedom of volition

My working assumption, in daily life, is that I function as a free agent. As discussed earlier, I have it in common with the vast majority of my fellow beings in feeling that I am being carried around in my head, and that I make decisions that are exclusively my own; I feel, in other words, that I possess *free will*. My impression that this is the case is, if anything, reinforced by my knowledge that such decisions can, if I am so inclined, be made capriciously; nearly all my decisions are predictable, of course, because I generally prefer to function within the constraints of society, but if I chose, I can behave out of character. What could be more free than that? This begs the question, however, because such whimsical decisions might nevertheless reflect certain inner workings of my mind that even my conscious I is not party to. As Willard Quine[207] has noted, freedom of will might be nothing more than the freedom to do as one will.

In this chapter, I have been putting the case for a consciousness mechanism that would confer great evolutionary advantage despite its determinism and lack of what could be called dualistic freedom. A persistent worry for those seeking the mechanism of consciousness is the possibility that they have inadequately defined their goal, and are thus using an inevitably flawed approach. This spectre was raised by Brian Josephson and his colleagues.[208] And for the committed physicalist, a major anxiety arises from suggestions of mind components beyond the immediate confines of the nervous system, the sophisticated and resourceful theory conceived by Karl Popper and John Eccles[209] probably being the best known current example.

The dualist model of Descartes identified the pineal body as the site at which mind and body commune, on what is now seen as the flimsy evidence of its lying in the brain's median plane. In the recent theory of Eccles,[210] followed up with Freidrich Beck,[211] the point of dualist interaction is the supplementary motor area, and the mechanism owes its exquisite sensitivity to quantum influences on vesicle discharge at presynaptic membranes. The choice, by Eccles, of the supplementary motor area was made on the basis of electrode studies of monkeys by Robert Porter and Corbie Brinkman, and of radio-tracer studies of humans by Niels Lassen and Per Roland.[212]

The monkeys studied by Porter and Brinkman initiated voluntary movements by pulling levers, in order to be rewarded with food, and it was found that this was associated with the firing of many neurons in the supplementary motor area, which became active before cells in the primary motor area. The investigations by Lassen and Roland included a motor sequence test, in which the thumb must touch the fingers in quick succession: finger-1 twice, finger-2 once, finger-3 three times, and finger-4 twice. A subsequent reverse sequence, following a brief pause, started with finger-4 twice, and so on. Completion of such a sequence demands continuous voluntary attention, and never becomes automatic. Lassen and Roland found that this type of performance is accompanied by increased activity in the supplementary motor area. Eccles was particularly excited about the discovery that a merely imagined repetition of such a sequence still produced activity in the supplementary motor area, whereas the other cortical regions remained quiescent. (Chris Frith and his colleagues later found evidence for increased activity in the dorsolateral prefrontal cortex during willed actions.[213]) This, he felt, provided strong evidence of an external influence on synaptic transmission in the active area.

One could argue that the system would be robust against such delicate perturbations down at the level of single vesicles. We have seen that there is overwhelming evidence that motor responses employ population coding, which is a sort of safety-in-numbers principle at the neuronal level, and this makes it even less likely that single vesicle release could have the critical influence that Eccles envisaged. What is still lacking, of course, is identification of the source of the supplementary motor cortex's activation. I feel that there is little doubt that one will find that it derives from a number of brain regions, acting in unison. I say this because activation by a single area would probably have already been exposed by currently available imaging techniques. I suspect that improvements to the sensitivity of these methods will one day reveal the identities of the contributing brain centres, and I would not be surprised if there are at least half a dozen of them.

In a night's sleep

If only for the sake of completeness, we ought to take a brief look at the interesting issue of sleep. It is, after all, the state in which we spend about one third of our lives. Only the higher animals appear to sleep, lower creatures being awake all the time. On evolution's ladder, the rungs in the vicinity of our own level are occupied, in descending order, by the mammals (of which we are an example, of course), the birds, the reptiles, then the amphibians, and finally, because we need not consider creatures that are even lower down, the fish. Sleep is seen in mammals, birds and most reptiles, whereas there is no real evidence of it in amphibians and fish. Dreaming in adults has been clearly identified only in mammals, although there appear to be brief dreaming periods in recently hatched birds.

Despite the commendable efforts of researchers in this important area, science still does not have a consensus as to why sleep evolved.[214] Seemingly common-sense explanations, such as the one which has it that we sleep to conserve energy, are not supported by the available evidence. Neither do we sleep to permit recuperation of tired limbs. In evolutionary terms, all that matters is that sleep is advantageous to those animals which possess it. But such pragmatism comes no closer to explaining how the advantage accrues.

One explanation states that sleep keeps us out of harm's way during the hours in which our senses would leave us at a competitive disadvantage.[215]

There are mammals which see better in the dark than we do, and some of them sleep only in very brief stretches, so as to minimize the danger from predators. The giraffe, for example, rests its head on the ground for only a few minutes at a time. But the fact that these animals *do* nevertheless sleep suggests that this state must serve some positive purpose. After all, the lion, which enjoys a position at the summit of its food chain, sleeps almost all the time. It is difficult to believe that it does so out of sheer boredom.

That sleep itself serves 'housekeeping' functions was demonstrated by J. Allan Hobson,[216] who found evidence of influences on anabolic hormone release, thermoregulation and immune responsiveness. The latter effect indicates that sleep may indeed protect the sleeper from predators, namely the microorganisms that prey from within.

Our main concern should be with dreaming, however, because it seems to represent a unique coupling of sleep with something that appears akin to consciousness. Recent advances in the subject have owed much to EEG studies, so we ought to briefly remind ourselves of the basic facts. As was originally demonstrated by Hans Berger,[217] in the early 1930s, the most prominent characteristic of the EEG for a resting brain is the periodicity of about ten cycles per second, this being replaced by a spectrum of higher frequencies, but lower amplitudes, when the subject becomes mentally active. In 1952, Eugene Aserinsky and Nathaniel Kleitman[218] discovered that the EEG trace of the sleeping brain displays a similar period of higher-frequency oscillations about once every ninety minutes. It became known as the paradoxical period, because this activity was more typical of the waking brain.

It was subsequently found that the irregular EEG is accompanied by an almost frenzied motion of the eyes beneath their lids, this too being detected by electrodes taped to the latter. Whence the term rapid eye movement sleep (REM sleep) now used to describe the paradoxical period. A year after the discovery of the paradoxical EEG pattern, Aserinsky and Nathaniel Kleitman established that a sleeping person is actually quite difficult to awaken during that period. This is now known to result from the immobilization of a large fraction of the motor system. And two years later, Kleitman and William Dement went on to discover something still more intriguing. If nevertheless awaked during the REM phase, a subject will usually be able to remember having dreamt, and will often be able to recall the details of the dream in considerable detail.[219] Although the distinction is not a very sharp one, it does seem that the paradoxical EEG, the REM, dreaming, and the decoupling of some motor functions are all part of the same phenomenon.

As Dement went on to establish, people persistently deprived of REM sleep, through being awakened every time their EEG pattern enters the paradoxical phase, display impaired mental function when awake.[220] Their powers of concentration are markedly diminished, and their capacity for logical thinking appears to evaporate. And as if in an attempt to make up for lost dreaming time, the sleeping brain then enters REM periods more frequently. Prolonged REM deprivation often leads to hallucinations, which are essentially waking dreams.

In 1979, Michel Jouvet destroyed the locus ceruleus alpha (which is located at the top of the brainstem, near the brainstem reticular formation) of an experimental cat.[221] During subsequent sleep, the animal periodically displayed a number of stereotyped behavioural patterns. It licked and washed itself, and attacked imaginary prey. It even stood up and explored its immediate surroundings, in a feline counterpart of human sleep-walking. But all these actions were performed in a desultory and disconnected manner. Jouvet's conclusion was that the observed motoric fragments, which were known to be directed by structures in the midbrain rather than the cortex, need the locus ceruleus to knit them into a useful fabric of behavioural events.

In humans, likewise, the dream is characterized by its grotesque qualities, the most prominent of which is the disjointed nature of the typical dream's plot. Although the senses of smell and taste are not much in evidence, we do seem to see and hear normally during dreaming. There is something decidedly odd about the sense of touch, however, when we are in the REM period. Our movements appear normal enough, but we feel no pain, despite all the physical ills which are a common feature of dream sequences; we do not come to harm, even in those terrifying falls, which seem never to end with a bump. It is possible that this absence of tactile sensation is related to the decommissioning of those motor functions.

Sigmund Freud saw the dream as serving to expunge neuroses.[222] The fact which militates most strongly against this idea concerns the observation that infant children actually have more REM sleep than adults. A newborn baby spends as much as eight hours a day in the dreaming state, and this aggregate duration is actually exceeded by the foetus in the last three months prior to birth. It is difficult to see how the neonate could have acquired neuroses enough to warrant so much time spent in REM sleep. A more mundane viewpoint started to emerge in 1968, when Ernest Hartmann noted that the sleep–dream cycle closely parallels the metabolism-linked cycles of the body.[223] A higher body temperature reduces the time lapse between REM periods, for example, and vice-versa. This is in accord with the view that dreaming is related to enzymatic processes, since these are known to proceed more rapidly at elevated temperatures. Moreover, Hartmann was able to demonstrate an inverse relationship between metabolic rate and the sleep–dream cycle time in a number of mammals, including the mouse, rat, rabbit, cat, human and elephant; the higher the metabolic rate, the shorter is the inter-dream period. Thus the mouse has a brief spell of REM sleep every five minutes or so, which is approximately eighteen times more frequently than for humans, and it has a metabolic rate that exceeds our own by a factor of about twenty. Conversely, the metabolic rate of an elephant is about 40 per cent lower than ours, and it dreams roughly every two hours, compared with our ninety minutes.

But what purpose does REM sleep actually serve? Avi Karni and colleagues[224] showed that a subject's cognitive performance gradually improves during the eight to ten hours following a specific task, both during wakefulness and sleep. But the improvement is suspended by REM deprivation. As J. Allan Hobson and Robert Stickgold[225] recently remarked, this merely shows that sleep is a permissive state, one that is not necessary for learning, but nevertheless sufficient. The stronger hypotheses that REM sleep alone can mediate memory-error correction (put forward by Francis Crick and Graeme Mitchison[226]), must await definitive confirmation or rejection.

A powerful point in favour of the Crick–Mitchison theory is the particular prevalence of REM sleep in the neonate (and in the third-trimester foetus), when the brain is undergoing rapid development. It is difficult to believe that newly formed synapses will automatically adopt the strengths required for harmonizing with the memory traces they inadvertently impinge upon, and the Crick–Mitchison prescription offers a mechanism for their adjustment. But what aspect of the system's performance is it which requires correction?

Two facts, considered in conjunction, give a strong hint: during REM sleep, many of the muscles are more relaxed than in non-REM sleep; and REM dreams are invariably bizarre whereas non-REM dreams tend to have a thought-like character.[227] These aspects of REM sleep suggest that the bizarre nature of its dreams stems from inactivation of normal control mechanisms during that phase.[228] Pedro Calderón de la Barca put the idea more poetically, in 1635, in his play *Life as a Dream*: 'Dreams are rough copies of the waking soul yet uncorrected of the higher will, so that men sometimes in their dreams confess an unsuspected, or forgotten, self; – since dreaming, madness, passion, are akin in missing each that salutary rein of reason, and the grinding will of man'.

The muscular relaxation that accompanies the REM phase occurs because the final common-path motor neurons of the spinal cord are rendered ineffective by postsynaptic inhibition. This was established by Ottavio Pompeiano,[229] who also showed that sensory input is blocked during REM sleep, the brain stem producing signals which depolarize thalamic sensory-relay nuclei. But what about the neuromuscular spindles? There is no mention of them being inactivated, and in any case it is difficult to imagine the system being able to compensate for the loss of their appropriate input to the cerebellum, appropriate here meaning that which would be expected to be returned from the now immobilized muscles. As John Eccles[230] stressed, the unity of experience nevertheless extends to the dreaming state; this corruption of information from the spindles may be the source of that state's bizarreness.

As the very use of the term REM reminds us, the eyes remain active. The system presumably 'knows' that the eyelids are closed and, parsimonious with its resources, thus has no need to inhibit the eye muscles. Hermann von Helmholtz[231] believed that the latter are devoid of receptors, but Charles Sherrington[232] showed that they too possess spindles. Indeed, it has been demonstrated[233] that these muscles are particularly well endowed with those monitoring gauges. With both muscles and spindles still on duty, the visual aspects of REM dreams might be expected to be less bizarre, and this seems to be case; it is not with respect to its visual imagery that the typical REM dream is grotesque, but rather with respect to its *plot*, as in the fall down a staircase that involves no touching of steps or walls.

Before we leave this subject, mention should be made of the so-called PGO waves, which take their name from the structures involved, namely the pons, the geniculate nuclei and the occipital cortex. J. Allan Hobson and Robert McCarley have shown that there is a strong three-way correlation between the phasic firing of certain large neurons in the reticular formation, the phasic PGO waves, and the phasic movements of the eyes from which the REM state derives its name.[234] When each wave or eye movement starts, moreover, it is found to be immediately preceded by the firing of one of those large cells. The conclusion drawn by Hobson and McCarley was that it is the large cells which are directing these linked processes.

It might seem, however, that eye movements are only a minor aspect of dreaming, and that other in-phase processes remain to be revealed. That may well be so, but we ought to bear in mind how important vision is to waking consciousness; it may be equally important to sleeping pseudo-consciousness. Two points must be borne in mind in this context. Firstly, eye movements are directed by the superior colliculus, as has been discussed earlier in this book, but this sub-cortical structure is itself under the control of the primary visual area, which is the most prominent member of the above-cited occipital lobe. Secondly, we have Lawrence Weiskrantz's work on blindsight.[235] Damage to the primary visual area produces blindness. But as Weiskrantz demonstrated, patients with lesions in that area retain some residual ability to distinguish the shape and colour of objects, although they do so without conscious awareness of the fact. This indicates that the primary visual cortex is involved in consciousness of perception. Seen in this light, the PGO waves take on added significance.

One could ask why dreams *have* plots, given that the senses are switched off. That they do is suggestive of automation, and this is reminiscent of what was described earlier in this chapter, in connection with filling in. Once excited, items in memory may excite other items by automatic association. But because there is no sensory input to keep the train of thought on the rails, as it were, dream sequences fly off at tangents, and thus become bizarre. If there is a memory-consolidating process, provoked by the PGO waves, it might be that the latter simply inject activity into the various neural networks in a haphazard fashion, and that recently stored memories are preferentially aroused. It is a common experience that recent experiences provide the favoured themes for our dreams. Those recent events which require the participation of conscious awareness (to qualify as candidate memories) are temporarily stored in the hippocampus. But we have also seen that timing is of the essence in that prominent limbic structure. Perhaps this is why the possibly random activation provided by the PGO waves is sufficient to excite recent memories into providing the starting point for dream sequences, but insufficient to place the sequences themselves in memory.

What I am suggesting, therefore, is that dreams are the incidental consequence of a process which is both fundamental and vital, in that it protects recently acquired memories from erasure. Bernard Davis has put forward a similar hypothesis, and he has suggested that the short-term memory refreshment process might also serve to transfer items to long-term memory.[236] This too would tie the hippocampus into the dreaming state.

Other brain structures have been identified with the REM phase through PET studies of healthy human subjects while they were dreaming. Pierre Maquet and his colleagues[237] found increased activity in the pontine tegmentum, the amygdaloid complex, the right parietal operculum and the anterior cingulate. They observed a simultaneous decrease of activity in the dorsolateral prefrontal cortex, the supramarginal gyrus and the posterior cingulate. As we will be discussing in the next chapter, the amygdaloid complexes play a central role in the acquisition of explicit memories, their function probably being related to what we have here been calling evaluation and choice. The participation of the anterior cingulate should not surprise us; we have seen that it has to be on duty when pain might be in the offing, which essentially means *all the time*. And we have also seen that it is thereby ideally suited as the system's grand adjudicator. The absence of activity anterior to the premotor cortex, which appears to support normal activity during the REM phase, indicates that dreaming activates simple schemata but does not involve their consolidation into more complex scenarios. This gives a further glimpse of the dream's bizarreness.

This chapter has been devoted to the origins of conscious awareness and thought, while the subjects of the next chapter are emotion, intelligence and reason. Straddled between the two, this last section of the former has taken a brief look at what happens when neither consciousness nor reason are in operation. Like Hobson,[jj] I have been struck by the relevance of the following delightful little couplet in John Dryden's *The Cock and the Fox*, published almost three hundred years ago:

Dreams are but interludes which fancy makes
When monarch reason sleeps, this mimic wakes.

jj / See reference 216.

10 The depth of reason

Know thyself.
Thales of Miletus

Intelligence is the capacity to guess right by discovering new order.
Horace Barlow (*Nature*, 1983[1])

Our organ of thought may be superior, and we may play it better, but it is surely vain to believe that other possessors of similar instruments leave them quite untouched.
Stephen Walker (*Animal Thought*[2])

Language, introspection and Helen Keller

It is frequently conjectured that *Homo sapiens* acquired language and conscious awareness around the same time, and that the former was a prerequisite for the latter. This idea comes in two versions, one maintaining that all conscious awareness is predicated on the spoken word, while the second merely postulates that language gave us the power of introspection. Because it requires knowledge of one's thoughts, the latter capacity presumably requires a deeper brand of cogitation.

It seems justifiable to equate language with the spoken word, in this context, because the verbal form of language surely preceded the written variety. And few would claim that conscious awareness emerged somewhere *en route* between speaking and writing. This is an important point, because speaking involves audition whereas writing requires vision. We have seen that vision is the sense which is best represented in terms of cortical area; I have referred to it as the senior sense. But this confronts us with a paradox: it indicates that conscious awareness, despite its primacy amongst our faculties, was not originally connected with our primary sense.

I find this suggestion quite extraordinary. And the language-based theory of conscious awareness requires the precursor, that is to say language, to have emerged without the benefit of conscious awareness. For the sake of argument, however, let us accept the idea and see where it leads. I am going to assume that the reader believes that the evolutionary ideas of Darwin and Wallace are essentially correct, and that we descended from the apes. Various scenarios suggest themselves. Firstly, there is the possibility that apes possessed conscious awareness before we arrived on the scene, and that we merely inherited it from them. Secondly, we have the version in which we acquired conscious awareness the moment we emerged as a separate species. Finally, there is the idea that we learned to speak later, and that conscious awareness followed on the heels of that auspicious event.

If the first of these alternatives is the correct one, it merely pushes the problem back to our simian predecessors, with the decision still having to be made as to whether language or conscious awareness came first. It might be countered that this would be different, because monkeys make sounds rather than words. The second scenario encounters a different sort of trouble, namely that science still does not have a plausible theory of speciation, and it is hard to see how we could continue a meaningful debate without one. The final option is the one that appears to be favoured by those who support the language theory of conscious awareness. Let us take a closer look at it.

Julian Jaynes put forward an extreme form of this theory.[3] His hypothesis is that consciousness arose through social pressures, the triggering events being catastrophes. He sees the preconscious human as having communicated more directly with gods than with peers, and he postulates that consciousness began to develop as the authority of the deities started to wane. This forced an individual's thoughts to turn inward. There are several points one could make about these ideas. For a start, one could wonder whether there really has ever been a sharp diminution of the belief in gods, since the idea of divine beings first impressed itself. After all, a large fraction of the world's population still believes in one or another supreme being. Just as puzzling is the fact that Jaynes happily describes the cities of our supposedly preconscious forebears, without questioning those ancestors' ability to accomplish such impressive feats of engineering without the benefit of consciousness.

Daniel Dennett puts forward a similar thesis, and he even goes so far as to identify the scene in which the great event took place.[4] One of our predecessors, out there on the veldt, attempted to call to her mates one day, not realizing that she was alone. There was no response, of course, and she found herself listening to her own verbalizations. This got her talking to herself, and consciousness gradually followed.

The Jaynes and Dennett theories are more relevant to introspection than they are to conscious awareness. Since words are a product of the mind, both theories see the mind as being able to talk to itself. I have difficulties with these ideas, however. To begin with, words are not the only sounds we humans make. We emit all manner of noises which can come to the attention of our fellow beings. At various times, we belch, cough, groan, grunt, howl, moan, scream, screech, shout, shriek, sigh, sneeze, sniffle, sob, squawk, wail, weep, wheeze, whimper and yell, just to name a mentionable fraction of the full repertoire. Every one of these utterances carries the potential for communication. Indeed, as every newborn baby rapidly discovers, sounds initially made by reflex are just as effective as words, and babies have been drawing attention to themselves with their cries as long as there have been babies. A baby is born – it cries, the parent reacts, and communication begins. This has been going on at least since our species appeared on the scene, and one observes the same thing in other species.

This is only the first of my misgivings, however, because visual communication can surely be just as effective. And as I noted earlier, vision is at least as developed as audition, in our species, and probably more so. Let us return to those individuals in Dennett's little masque, the disappointed caller and the absent remainder. The fact of the call indicates that they were members of a group, and this has to imply that they were used to cooperating. What would be the point of their being together if this were not the case? But how were they cooperating if they were not communicating? Moreover, that communication could just as well have been based on gestures. Think of the group collectively partaking of some food, for example. Etiquette was probably not at a premium in those remote times, and there was probably a lot of pushing and shoving. But the push is a close cousin of the raised arm, and a raised arm is a gesture. What is more, a gesture is pointless unless it can be heeded, and how can one heed a gesture if one is not consciously aware of it? I find it hard to accept that deliberate communication between humans is not just as old as humans themselves; and I find it even harder to believe that conscious awareness was not a requirement for communication rather than a consequence of it.

As noted earlier, however, the Jaynes and Dennett theories seem more suited to introspection. This is presumably more demanding of the nervous system since it requires conscious awareness of one's own thoughts. But here again we could ask whether the two theories do not err on the side of sophistication. The primitive human must have had less use for abstract thought than his modern counterpart, so what introspection he indulged in must have sprung from more primitive drives. Going back to that communal food supply, surely a major factor in life, there would occasionally have been individuals who were denied a proper share of the day's takings. If we assume that the emotions played at least as prominent a role then as they do today, would this not have provided motivation enough for the occasional bit of private aggravation? And there is not such a long hop from aggravation to introspection.

This is all speculation, however, and it is difficult to see how the issue is going to be resolved; it is one thing to excavate ancient remains, it is quite another to draw conclusions about the mental states of antiquity. Let us close this section, instead, by turning to a well-documented case which appears to strike at the heart of the matter. It is the famous one of Helen Keller, who became deaf, dumb and blind at the age of two, in 1882, following an attack of what was diagnosed as an acute congestion of the stomach and the brain. That we have her written record of her experiences is due to the magnificent efforts of her teacher, Anne Sullivan. This gifted person had the patience and ingenuity to teach Helen sign language.

The first word Helen learned was *doll*. Anne gave her a doll one day, and after the child had played with it for a while, she spelled out d-o-l-l in the palm of Helen's hand. The blind, deaf mute immediately showed interest in this finger play, and tried to imitate it. In her biography, Helen emphasizes that she did not know that she was spelling out a word, although she quickly grasped the significance of sequence.[5] Nouns like *pin*, *hat* and *cup* soon followed, as did simple verbs such as *sit*, *stand* and *walk*. These are all words of one syllable, but that is probably less significant than the fact that they are also rather short. As Helen has noted, many weeks were still to pass before it dawned on her that everything has a name. There were subsequent difficulties over the words *mug* and *water*, with Helen tending to confound them, possibly because they were often experienced in conjunction. This problem was resolved when Anne held Helen's hand under water that was issuing from a spout, and then spelled out w-a-t-e-r on her palm. Another difficulty arose when Anne gave her pupil a different type of doll; Helen was slow to grasp that two distinct items could go by the same name.

What interests us here, of course, was the state of Helen's mind, particularly before Anne came into her life. Fortunately, we have her own words to go by, as recorded in a later volume.[6] In a chapter entitled *Before the Soul Dawn*, she states

> *Before my teacher came to me, I did not know that I am. I lived in a world that was a no-world. I cannot hope to describe adequately that unconscious, yet conscious time of nothingness. I did not know that I knew aught, or that I lived or acted or desired. I had neither will nor intellect. I was carried along to objects and acts by a certain blind natural impetus. I had a mind which caused me to feel anger, satisfaction, desire. These two facts led those about me to suppose that I willed and thought. I can remember all this, not because I knew that it was so, but because I have tactual memory. It enables me to remember that I never contracted my forehead in the act of thinking. I never viewed anything beforehand or chose it. I also recall tactually the fact that never in a start of the body or a heart-beat did I feel that I loved or cared for anything. My inner life, then, was a blank without past, present, or future, without hope or anticipation, without wonder or joy or faith.*

Later, in the same chapter, Helen writes: '*I remember, also through touch, that I had the power of association*', and further on again she states: '*Since I had no power of thought, I did not compare one mental state with another*'. Let me quote just one more passage from this moving and enlightening document. It runs

> *When I learned the meaning of 'I' and 'me' and found that I was something, I began to think. Then consciousness first existed in me. Thus it was not the sense of touch that brought me knowledge. It was the awakening of my soul that first rendered my senses their value, their cognizance of objects, names, qualities, and properties.*

From what was stated earlier, one should probably interpret the word *consciousness* in this last passage as actually meaning *introspection*. Seen in that light, the rest of the paragraph harmonizes with the use of the word *value*.

Even more revealing is Helen's reference to her lack of thought, and inability to compare mental states, before she had developed a sense of personal identity. Similarly, her use of the word *unconscious*, in the first long quotation, is probably best taken to indicate a lack of introspection.

Our main concern is with conscious awareness, however, and it is particularly interesting to note Helen's allusion to the associative power of touch. Denied sight and hearing, she was an individual for whom touch had been promoted to the role of senior sense. And despite the poverty of her experiences, she was nevertheless able to observe correlations between cause and effect. Moreover, and this is the real litmus test of the attribute, she was able to hold these observations in short-term memory long enough to draw further inferences from them. From what was discussed in the preceding chapter, we would have no hesitation in saying that Helen therefore possessed conscious awareness, even in those meager days before Anne came into her life.

Let us close this discussion by returning once more to the question of language, to the point at which humans acquired it, and to its possible connection with introspection. There is a growing body of evidence which suggests that no new neural systems evolved to exclusively serve language,[7] and that there was no discontinuity of language from other cognitive systems.[8] Instead, language appears to be a new mechanism that Nature constructed out of old parts,[9] these being cortical maps of sensorimotor origin.[10] Since such maps have a close connection with the generation of an animal's sense of its own local geography, these ideas are very much in line with what was stated earlier regarding personal space, identity, and introspection.

As William Calvin has emphasized, we do well to use language as an example when discussing intelligence.[11] Before we continue on this theme, however, it would be a good idea to review the main thrust of the preceding chapter.

A recapitulation

The crux of my argument has been that exploitation of an unconscious veto mechanism, mediated by a specific cortical circuit, and made possible by an adequate short-term memory span, paradoxically requires conscious awareness. And I see the same circuit, interacting with other brain components, as permitting internal simulations of the body's interactions with the environment, thereby providing the basis of mind.

I have stressed that the evolutionary advantage of consciousness is the power of real-time choice. Feed-forward neural networks also choose, and we saw them doing so in a previous chapter. But these networks are merely fed with information, and their outputs are read off by intelligent agencies; they usually do not acquire experience of their own 'volition'. As Andy Clark has stressed, animals *do* gain information in this fashion.[12] Similarly, I have argued that the environment is just as much a black box as is the brain of the creature exploring it.[13] This might seem strange to a person with normal sight, but not to one who is blind. It follows that acquisition of information is always an active process, even when merely the upshot of thought. Such accumulation is not *necessarily* conscious, however. For example, each walking stride is effectively a question: *is the ground still there?* Only when we encounter an irregularity, a hole perhaps, do our motor explorations suddenly become conscious.

We learned that one must master six technical terms: afference, efference, exafference, exefference, reafference and efference copy. In fact, there is even a seventh term: *internal reafference*, which is important in the generation of the qualia (or raw feelings) associated with consciousness. *Afference* refers to those nerve signals that travel from the sensory receptors toward the brain, while *efferent* nerve signals travel from the brain to the muscles, muscular movement being the nervous system's only external product. But the brain is an extended structure, through which nerve signals pass, so this raises the question of where afferent signals terminate and where efferent signals begin.

Exafference and *exefference* refer, respectively, to incoming signals from the environment which impinge upon the sensory receptors (thereby producing on-going afference signals) and to signals sent out to the environment (as a consequence of prior efference signals). Exafference comes in five different classes, corresponding to the five human senses, whereas all exefference is generated as a consequence of the body's motor output. Examples of exefference are the sounds of human speech and facial expressions, both of which are used for communication. Social intercourse is also mediated by the muscles we use to gesticulate. In speech, the exefference is in the form of sound waves, while the exefferent messages sent to the environment by facial expressions and gesticulations are communicated by light.

The traditional view of consciousness sees exafference producing afferent nerve signals which travel to the brain, where they may make a conscious impression, provided there is the necessary attention. The result can be production of efferent nerve signals and ultimately exefferent signals, if there is the necessary motivation. In the case of a reflex, the exefferent response follows automatically from the exafferent stimulus, and volition is not involved.

Reafference is that portion of exefference that re-enters the system, by way of the sensory receptors; one could loosely call it *environmental feedback*. Reafference is fundamentally different from exafference, *because the nervous system is aware that it is the origin of these feedback signals*. There is an internal signalling system that alerts the relevant sensory-processing regions of the brain to the possible arrival of such reafferent messages, and it involves the nerve signals known by the term *efference copy*. These signals are copies of the corresponding efference, and are thus intimately related to the resulting exefference, and thereby to the reafference. The equivalence between the efferent and efferent copy signals stems from their common origin, namely the premotor cortex or the anterior cingulate, or a combination of both those regions. (I have referred to them collectively by the term *master module*, in view of the processing bottleneck that they represent.[14]) Efference copy signals keep the nervous system apprised of when and how it is probing the environment.

Failure in the efference copy route has been conjectured to be the cause of the seemingly extraneous voices reported by some schizophrenics.[15] The voices are in fact the victim's own, but are not recognized as such because the warning efference copy signals fail to arrive, causing reafference to be mistaken for exafference. Indeed, *any* deficits in the mechanisms whereby the various forms of reafference are detected will impair a person's ability to probe the environment. Putting it in simple terms, one could say that this is malfunction of one's personal radar.

The key aspect of conscious choice, I have conjectured, is that it permits what I have called *veto-on-the-fly*. An insect also explores and makes choices. But when about to make a move, it never suddenly 'realizes' the significance of its intended action, because of the current feedback from the environment and the creature's own currently evoked memories of danger, and then vetoes its own intended action, even though the vetoing might perforce be mechanical and deterministic. This ability could sometimes make the difference between life and death.

I suggested that functioning of the veto mechanism is predicated on the existence of efference copy loops, and that there is also the requirement of an adequate short-term memory span. When both of these conditions are satisfied, there can be internally simulated muscular navigations of the environment (analogous to Antonio Damasio's *as if* processes[16]), and these can invoke memories of related situations. They can also draw upon the predictions provided by the cerebellum (the *what if* processes discussed by Chris Miall and his colleagues[17]), and exploit the planning capabilities provided by the prefrontal cortex, while automatically notifying the body of their significance.[18]

The real-time veto mechanism provided the great stride forward, in evolutionary terms. But because this unconscious process required conscious awareness, making us in effect a captive audience of our own environmental probings, it brought other developments in its wake. And although these were not, in my opinion at least, the critical factors some have made them out to be, there was nothing to prevent them from being very useful in their own right. I believe that language was a case in point. The veto-imbued navigation I wrote of in the preceding chapter must initially have been advantageous in the avoidance of predators, but its dependence on keeping its possessor apprised of current muscular events later provided the basis for other things. Prominent amongst these was language. We would not have this faculty if we could not avail ourselves of short-term memory; speaking a sentence would be impossible if we could not hold in mind what we had just said, and simultaneously be preparing what we were just about to say. The point has come, therefore, at which we should take a closer look at the kinematics of speech. And a necessary preparation for that will be a look at the dynamics of consciousness itself, at the neuronal level.

The kinematics of language

I have suggested that there is, in the brain's neural circuitry, what could be called a *vital triangle*, and that this mediates thought. The apices of this triangle lie in the sensory cortex (that is to say, in the occipital, parietal and temporal lobes), in the premotor area, and in the thalamic intralaminar nuclei.[19] The flow of signals around this three-sided loop, in the direction in which the three apices have just been identified, was taken to lead to the selection of a sequence of relevant schemata. Indeed, the sequential activation of such schemata was taken to be the basis for the familiar *stream of consciousness*. We should now check whether the available experimental data support this idea and, in the event of this producing affirmation, what one might then conclude regarding the kinematics of consciousness.

To begin with, it *does* seem sensible to think in terms of a flow of signals, because reflexes are undeniably vectorial; there is a well-defined directionality, stretching from sensory stimulus to muscular response. The alternative is what could be called scalar behaviour, with groups of neurons being excited, but with no discernable directionality in the underlying signalling. The activity would then take the form of a sort of resonance excitation. Although this might seem to be in obvious conflict with the ultimate need, mentioned above, for there to be a well-defined input and an equally unambiguous output, it is a fact that certain excitations associated with consciousness do not display a clear directionality. Reconciling this with the information flow postulated above must therefore be our immediate concern.

The observations in question were reported by Daniel Barth and Kurt MacDonald,[20] and they concerned detection of gamma-band oscillations (the actual frequency being 28 Hz) in some of the thalamic intralaminar nuclei during wakefulness (and spasmodically during sleep). Given that these nuclei lie at one apex of the above-mentioned vital triangle, these observations are clearly of relevance to the present discussion. Mircea Steriade[21] has argued that these nuclei coordinate the oscillations, rather than drive them. Now the interesting thing about these observations, and those made by Steriade and his colleagues,[22] is that the thalamic intralaminar nucleus in question (the centrolateral in the latter work, and the posterior in the Barth–MacDonald study) and the patch of cortex being probed (the association suprasylvian cortex in the latter work, and the auditory cortex in the Barth–MacDonald study) are found to oscillate *in phase*, or very nearly so.

This could be taken to indicate that the above-mentioned non-directional resonance behaviour is the one that prevails, and that there is no actual flow of signals from one brain region to another. If this were really so, it would obviously undermine the role of what is here being called the vital triangle.

Before jumping to that conclusion, however, we ought to consider two other important pieces of evidence. The first concerns the recent theoretical work of Roger Traub, Miles Whittington, Ian Stanford and John Jefferys,[23] which demonstrated how long-range synchronization of gamma-band oscillations can be achieved. They used a model comprising excitatory pyramidal cells and inhibitory interneurons, and found that synchronization prevails when each of the latter cells is producing pairs of spikes in rapid succession.

The inter-spike intervals in these doublets was found to have an average value of 5.25 milliseconds, and the long-range synchrony was observed to be in the millisecond time scale. Strong endorsement of these theoretical findings was provided by the same research team, in the form of experimental observations on hippocampal slices, in which the inhibitory interneurons were indeed seen to spike in doublets, under conditions in which the gamma oscillations are synchronized over distances of several millimetres (whereas they fire only in single spikes under other conditions). Their mechanism would be capable of establishing zero phase-lag in the oscillations of neurons spaced over several millimetres, in agreement with the experimental observations of Charles Gray and his colleagues.[24] In a somewhat different vein, Diego Contreras, Alain Destexhe, Terrence Sejnowski and Mircea Steriade[25] studied barbiturate anaesthetized cats, before and after removal of the cortex, and were thereby able to establish that spatiotemporal coherence is dependent upon corticothalamic feedback.

The second important piece of evidence comes from the recent work of Michael Wehr and Gilles Laurent.[26] They demonstrated the presence of rhythmic synchronization of roughly 20 Hz oscillations in the locust nervous system, as a consequence of the detection of odours, the synchrony appearing transiently. Moreover, a given neuron in an oscillating assembly of cells was observed to emit spikes for a period of about 200 milliseconds, whereafter it fell silent for a while, only to become active again at a still later epoch. Another cell in the same assembly would, in general, be spiking during a period that was not precisely coincident with that of any other of its brethren. The overall firing pattern was found to be stimulus-specific. As Wehr and Laurent themselves noted, '... neural coding with oscillations thus allows combinatorial representations in time as well as in space'.

These observations are mutually consistent, and they can be reconciled with the vital triangle model described above. This can be achieved if it is assumed that: (*a*) Gamma-band oscillations provide the system with a *carrier wave*; (*b*) This carrier wave is *stationary*, so the oscillations are in widespread spatial synchrony; (*c*) The carrier wave is *amplitude modulated*, the modulation pattern being *non*-stationary; and (*d*) This enables the system to transmit information from one region to another in spatiotemporal patterns.

Let us consider how these moving activity patterns might be used by the system. In the experienced animal, the premotor area serves perception by finding spatiotemporal motor-generating patterns of activation which match those being transmitted to it by the above amplitude-modulated waves. How does it accomplish this? The obvious candidate is the variable temporal delays inherent in the transmission of signals along the dendrites of its constituent neurons.[27] The axons via which the neurons in the premotor area receive signals from the sensory-processing areas will clearly make a host of random synaptic connections on these dendritic arborizations.[28] If these are sufficiently numerous, connections will always exist which produce just the right timing of spikes from the premotor area's neurons, these signals being available for transmission both to the muscles and via the efference copy routes. It is in this way, I believe, that the dynamic links between the sensory cortex and the premotor area are set up.

In fact, of course, they are originally set up in the reverse order, since it is the animal's own exploration of its environment that permits the correlations between the signals in the reafference and efference copy routes to be captured. This, then, is the manner in which the system is conjectured to acquire its schemata. We will soon be ready to turn to the specific example of language, but let us first consider the flow and content of consciousness in more general terms.

Oli Lounasmaa and his colleagues[29] used whole-head functional magnetic resonance imaging during a study of picture naming by humans, and monitored the progression of signals from the occipital lobe toward the temporal lobe and, ultimately, the frontal lobe. They found that the first signals detectable in the frontal lobe had already arrived after about 200 milliseconds, although the strongest signal in this region was not detected until about the 500 millisecond epoch. Simon Thorpe and his colleagues,[30] who used electrode recordings in their study of macaque monkeys, found an even shorter latency for the transit of signals from the retina to the orbitofrontal cortex, namely 120 milliseconds.

The vital triangle, as it is being called here, would be an impoverished structure were it not for the fact that each of its apices is, in itself, an extended neuronal network. In the case of the apex that constitutes the sensory cortex, for example, we have seen that this involves both forward and reverse projections, that is to say both up and down the processing hierarchy. And we also saw that Wolf Singer[31] argued for the presence of two complementary strategies to cope with the immense number of combinatorial possibilities when sensory and motor patterns are to be mutually linked. The first strategy, handled by the forward projections, looks after the analysis and representation of frequently occurring and behaviourally relevant relations by groups of neurons with fixed but broadly tuned response properties. The second strategy, which is the responsibility of the reverse projections, serves the dynamic association of these neurons into functionally coherent assemblies.

With signals simultaneously passing up and down the hierarchy, the spatiotemporal signals being dispatched to the premotor area would also be expected to evolve. It would be as if the priorities of the system were perpetually being updated by top-down influences, mediated by the signals conveyed by the reverse projections. I suggested earlier that this could be the mechanism whereby the stream of consciousness is always moving on to fresh pastures, as it were.

Now, finally, we can turn to language itself. Ghislaine Dehaene-Lambertz and Stanislas Dehaene[32] used the high-density event-related potential technique to investigate the processing of simple syllables by two-month-old babies. They found evidence for the existence of three temporally and spatially separated responses, and the most parsimonious interpretation of their data indicates that these are connected with the perception of sound intensity (in the temporal lobe), phonetic processing (in the vicinity of Wernicke's area) and prosodic changes (in the premotor region, at or near

Broca's area). The epochs at which the signals reached these three regions were, respectively, 220, 390 and 700 milliseconds.

If one makes reasonable assumptions regarding the number of synapses that the advancing signals in the Thorpe and company work must traverse, in order to reach the frontal lobe, it can readily be shown that 20–30 milliseconds of neuronal activity in any given cortical area is sufficient for the discrimination of even the most complicated visual stimuli.[33] There is only sufficient time for two or three action potentials in such a brief span. Moreover, in the above-mentioned locust studies of Wehr and Laurent, each envelope of oscillations comprised no more than five or six peaks around the maximum intensity, and this too is suggestive of very rapid information processing.

The credibility of these arguments is admittedly undermined by the fact that we are citing data from widely differing species; it is to be hoped that analogous information drawn exclusively from the higher primates will soon be at hand. Meanwhile, it is tempting to take the argument one stage farther, by considering the question of individual phonemes.

Let us assume that those 20 Hz oscillations in the locust are the counterparts of the gamma-band oscillations observed in cats and monkeys. The indication would then be that the spatiotemporal coded signals required by the premotor area, as argued above, are indeed being produced by the sensory cortex. In the case of human language, this would mean that spatiotemporal coded signals are dispatched from Wernicke's area, to be subsequently received by Broca's area, and there converted into patterns of sequenced signals that are sent to the appropriate units in the motor cortex. It would be these latter signals which actually lead to the utterance of the required phonemes.

Meanwhile, efference copy signals would be dispatched via the thalamic intralaminar nuclei back toward Wernicke's area. And in the event that the thereby-transported information was found to harmonize with what had just been sent to Broca's area, there would be no remnant mismatch signal and the system would proceed to the next point on the agenda, as it were.

Just how this comparator mechanism functions is still a matter of speculation. I earlier speculated that it involves interplay between excitation and inhibition, but just how these would conspire to provide the basis for the comparator mechanism is unclear. It is worth noting, however, that Toshiyuki Sawaguchi[34] has found evidence of modular connections between excitatory and inhibitory neuronal groups in the primate neocortex, using optical imaging of the exposed cortical surface. As mentioned earlier, the comparator function could involve the NMDA version of the glutamate receptor.

The words of which speech is composed are themselves made up of syllables, each of which comprises a number of phonemes. A given syllable is not exclusively associated with a particular word, of course, and neither is a specific phoneme exclusively identified with a single syllable. On the contrary, the flexibility of language stems from this lack of exclusivity, its underlying modularity being the reason why there are very many more words in a dictionary than there are phonemes.

This is common knowledge, but there is something about the modularity of language which could be overlooked: whereas sentences, and in some cases even single words, can have an emotional undertone, there is usually no emotion associated with an individual syllable, let alone a single phoneme. This surely reveals an important byproduct of consciousness: it imbues its possessor with the ability to string together emotion-free elemental movements, so as to produce emotion-charged *sequences* of elemental movements. And in order to do this, the system needs a certain span of short-term memory, for how otherwise, in our example here, would one know which phoneme appropriately comes next in such a series of phonemes. This is navigation again, in the verbal sense.

The concatenation of phonemes involves sequencing the elemental muscular movements of which speech is composed, the muscles in question being those of the tongue, the lips, the jaw, and those serving the vocal cords.[a] According to the line being taken here, those muscular movements must be set up in Broca's area, by signals arriving from the relevant part of the sensory cortex, namely Wernicke's area. And the flow of spoken words is the product of a corresponding flow of signals around what I am calling the vital triangle (see figure 9.10*a*).

It is the interaction of these signals with the associated nuclei of the limbic system which is believed to bring the emotional dimension into play,[35] in a manner that would presumably be impossible during the brief time it takes to articulate a single phoneme.

An estimate of the time it takes for such an elementary articulation can be obtained from the world record for rapid speech,[36] namely the 327 words per minute attained by President John F. Kennedy during a political speech. Allowing an average of two syllables per word, and assuming an average of two phonemes per syllable, this gives a phoneme-articulation frequency of just over 20 Hz, which is about half that of the so-called gamma band. The time taken in uttering the typical phoneme is thus slightly less than 50 milliseconds, an estimate which is in excellent agreement with the recent experimental findings.[37]

As just noted, none of us is so fickle that our emotions can fluctuate at this high rate, so the system must work with two rather different time constants. And as we saw earlier, this duality of temporal scales makes possible that unconscious veto mechanism that, paradoxically, requires consciousness. It remains to note that there is excellent agreement between the 50 milliseconds that we have just calculated for the average time of phoneme duration and the durations of the individual pulses in the oscillations we considered above; the gamma band appears to be ideally suited to mediate human speech.

If all these speculations prove to be of value, they would vindicate the

a / Speech also involves the muscles of the diaphragm, of course, but these are not necessarily moved at the same high rate as those of the other speech muscles.

ideas of Francis Crick and Christof Koch,[38] which linked consciousness to gamma-band oscillations. One final point should be made, namely that the above-discussed stationarity might have to apply simultaneously to all three apices of the vital triangle, if the system is to be able to properly link the spatiotemporal events in the sensory cortex, premotor area and thalamic intralaminar nucleus. If these three regions fall out of mutual synchrony, the corresponding conscious impressions might be lost. Putting it in terms of another concept that was prominent in the preceding chapter, one could say that the generation of an appropriate schema might depend on such global synchrony.

Theories win their spurs by making predictions, and also by explaining things that have previously seemed mysterious. In the next section, I shall be making a couple of predictions that this theory could be said to force upon me. And in the section after that, I shall be asking whether the ideas presented here might have any relevance to a mental infirmity which, until now, has defied explanation. If that proved to be the case, it too would lead to predictions, possibly even to suggestions for the infirmity's amelioration.

Why is language cortically lateralized?

The above ideas suggest a revision of the accepted view of the frontal lobe. If the role of a given patch of the premotor area is merely to set up schemata involving those muscles served by the nearby part of the primary motor cortex, we would have to predict that Broca's area serves *all* well-rehearsed sets of lip and tongue movements, not merely those used in speech. (The specificity for speech would presumably then be embodied in the synaptic connections *to* Broca's area.) Likewise, one would expect to find an analogous area, lying dorsal to that of Broca (see figure 9.11*b*), which serves manual dexterity.

A complicating factor arises from the presence of the frontal eye field,[39] close to the premotor area,[40] but even the placing of this area is suggestive: close to the finger region of primary motor area, presumably in the interest of finger–eye coordination. It is possible that the influence of the frontal eye field is seen in the apparent distinction between externally cued and internally cued movements that have been ascribed to the lateral and dorsal regions of the premotor area, respectively[41] (even though this distinction is not a sharp one); the frontal eye field is inevitably related to external influences.

Let us turn to another prediction. As is well known, speech is usually mediated by only one of the two cortical hemispheres. In ninety-six per cent of right-handers, this faculty is served exclusively by the left hemisphere, and that same hemisphere serves speech in seventy per cent of left-handers. Of the remaining thirty per cent of left-handers, about half have their speech centres located exclusively in the right hemisphere, while the remainder display evidence of bilateral control, which is not seen at all in right-handers.[42]

Why is speech thus lateralized? I would like to offer a remarkably simple rationalization of this fact: one related to the role played for the system by the premotor area, which, as we have been discussing, permits the setting up of possible sequences of motor movements, in such a way as to hinder the various muscles from coming into mutual conflict.

The avoidance of such conflict is not necessarily a difficult task in the case of the arms, say, because one can easily move one's left arm while the right arm is inactive, and vice-versa. Only when both arms are moved, and/or one is trying to make them occupy the same region of external space, does a clash occur which the premotor area must resolve. A different situation prevails in the case of the tongue, however, because it is not possible to move its left side without the right side being moved simultaneously.

Now it is an interesting fact that the muscles involved in speech all serve structures which lie on the body's median plane, and the two sides of each of these structures tend to be constrained to move in unison. This is true of the tongue, as we have just noted, and it is also true of the lips, when these are making the puckering movements encountered in the articulation of phonemes. It is almost certainly the case for the muscles of the larynx as well.

This must place severe restrictions on the possible modes of movement of these muscles, and avoidance of conflict is best achieved if one side of the cortex simply takes on the job of dictating the patterns of movements of the muscles on *both* sides of the median plane. Activation of the relevant regions of the motor cortex on the subordinate side would presumably then be accomplished by signals passing through the corpus callosum. As has been argued recently,[43] temporal sequencing of complex movements requires high activity in the premotor area, which must be the source of those trans-callosal signals.

That the left cortical hemisphere serves speech in the majority of people might be related to the fact that most people are right-handed, because the hand we most tend to manipulate our mouths with is that same right hand, as in feeding. This, I would like to submit, is the reason for the cortical *uni*lateralization of speech.

That the right hand has come to dominate over the left may originally have been the product of chance, the dominance thereafter being passed down through the generations. In any event, the cortex has come to display considerable structural asymmetry, the planum temporale on the left side being larger than that on the right in sixty-five per cent of people, the opposite being the case in a mere ten per cent of people.[44] In some cases, the planum temporale on the left side is five times larger than that on the right.

There is a way in which this hypothesis might be tested, albeit in a rather indirect manner. One could study the activation of other sets of muscles that serve structures lying in the body's median plane, that is to say muscles not implicated in speech. Obvious candidates are those used when flexing the abdomen or flaring the nostrils. (A particularly suitable group of subjects, drawn from the entertainment profession, might be those performers specializing in rhythmic undulation of the abdominal surface, that is to say *belly dancers*.) Another possibility is the sphincter muscle at the nether extremity of the alimentary canal.

My suggestion is that if brain activity monitoring, with positron emission tomography, say, or functional magnetic resonance imaging, were used to investigate the cortex during the flexing of one of these sets of muscles, or even the mere *imagination* of that act, it might transpire that activity could be detected in only one cortical hemisphere. This is admittedly a long shot, but it would be worth trying.

But so much for predictions. Let us now turn to the above-mentioned infirmity, and ask whether the arguments developed in this chapter might have anything useful to say. This will occupy us in the following two sections, whereafter we will be ready to turn to the entire question of intelligence itself.

An invisible shell

In scientific investigation, understanding of the general is often furthered by contemplation of the particular. When a phenomenon is the product of several contributing agencies, the complex interplay between the latter may be illuminated by observing what happens when one of them malfunctions; as the saying goes, *treasure your exceptions*. In this book, we have noted many advances that came through considering the consequences of accidental or deliberate brain lesions, this approach being especially useful when closely related to sensory input, such as the visual perception of form, colour or motion. When a faculty is more general, it is reasonable to assume that a greater number of brain regions will be participating. This makes it more difficult to identify the neural circuits involved. Science has been getting to grips with identifying the cortical regions implicated in vision, in particular, but where is the seat of personality? What, for that matter, are the real determinants of intelligence? Answers to those two fundamental questions might come through consideration of one particular exception to the norm, namely *autism*.[45]

This syndrome affects about 4 out of every 10,000 people, apparently from birth, there being a preponderance amongst males.[b] The main characteristic is severely impaired social interaction, and this has led some to liken the typical autistic person's existence to living in a invisible shell. The syndrome was first properly described by Leo Kanner in 1943,[46] and independently by Hans Asperger a year later.[47] The term comes from the Greek word *autos*, meaning self, a reference to the social withdrawal that is the affliction's hallmark. It was coined in 1911 by Ernst Bleuler, who is better known for his pioneering identification of schizophrenia, with which autism was initially coupled. Kanner pinpointed two aspects of autism which he took to be its defining features, namely social isolation and resistance to novelty; the autistic person is profoundly aloof to the human environment and obsessively desires constancy. But he also noted the frequent presence of isolated abili-

b / The actual proportion appears to be about four males to each female.

ties, particularly those related to memory. Asperger's list of characteristics included such additional items as a paucity of facial expressions and gestures, lack of eye contact, and meagerness or absence of language.

These inventories of deficits have been augmented over the intervening years, and there are now roughly twenty recognized autistic characteristics. Following a survey by Lorna Wing and Judith Gould,[48] carried out in the London borough of Camberwell, one can highlight a *triad of handicaps:*[c] truncation or absence of verbal and non-verbal communication; severe impairment of (reciprocal) social interaction, particularly with peers; and absence of imaginative pursuits, including pretending, coupled with the presence of stereotyped behaviour.

I tend to divide autistic characteristics in two broad classes, rather than three. The first comprises emotional detachment, aversion to affection, aversion to physical contact, lack of eye contact, lack of personal identity, lack of facial expression, disturbed communication and lack of imaginative play. The second comprises that remarkable resistance to change, fixation on certain objects, bizarre ritualistic behaviours, abnormal anxiety patterns, impaired motility,[d] disturbed perception and attention, language deficits, lack of initiative, and retarded intellectual development. I would have to add that, with very few exceptions, creativity is not much in evidence amongst autistic people.

There is more to communication than language. The diminished facility with language prominent in autism deserves consideration, however, because of the increasing understanding of language's neuronal basis. Nouns and verbs have now been identified with specific cortical regions, whereas cortical areas serving adjectives, prepositions and other types of word have not yet been pinpointed; it seems that their production is a more nebulous affair. When an autistic person possesses some language, it is usually nouns and verbs that are present, whereas there is scant evidence of adjectives, prepositions, adverbs or pronouns.

Left to their own devices, autistic people appear content to continue whatever they are doing, often for hours on end; they seem to singularly lack initiative. Continuity is broken only by the calls of basic bodily needs. A concomitant feature appears to be a lack of curiosity, though this might not be so pronounced as some have taken it to be. Autistic people can nevertheless be found to have considerable stores of passive knowledge. It is not that they are furtive about their acquisition of facts. Rather, it is the lack of communication which allows the learning process to pass unnoticed. I personally suspect that autistic memory is predominantly of the implicit type.

Given the woefully disadvantaged state with respect to language, as

c / The Camberwell study screened 35,000 children, up to 14 years of age, who lived in the borough on 31 December 1970. Of the 914 known to be physically or mentally handicapped, 7 had the symptoms described by Kanner, an incidence of 2 in 10,000.

d / The word motile merely means capable of movement.

well as the reticence in taking the initiative, it is hardly surprising that the average autistic person is intellectually backward. The autodidactic processes that is so important an aspect in normal learning obviously labours under difficult conditions. Moreover, it takes great patience on the part of a parent or teacher to continue attempts at communication when feedback is sparse or totally absent. Another issue that is not clear-cut relates to mental retardation; there are autistic people with normal IQs, and there are people with low IQs who are not autistic. However, there is a clear trend which makes autism more common as IQ decreases.

Holes in the mind

It has been suggested that autistic people do not possess a *theory of mind*, and that they are thus unaware that other people think (see figure 10.1). (Whether they are also unaware of their own thoughts is a more difficult question.) Such an inference is properly viewed as a theory, because the state in question is not directly observable, and its existence can be used to make predictions about

Figure 10.1
A normal person encounters little difficulty in getting the point of this piece of humour by the cartoonist Petit. When the soldier on the right pops the paper bag, the officer is going to think that the rifle he is inspecting was loaded, and that it has gone off in his face. This is not the case for autistic people with lower IQs, however. They appear to be unable to impute thought processes to others, and they may even be unaware of their own consciousness.

the behaviour of others. The term is now widely adopted, also, in connection with the mental powers of other animals. Later in this chapter, I will be considering whether chimpanzees, for example, have a theory of mind.

The ingenious *Sally-Anne test*, sometimes attributed to Heinz Wimmer and Josef Perner,[49] but originally devised by D. Lewis in 1969,[50] involves a scenario with two dolls, Sally and Anne, acted out in front of the subject being tested. Sally has a basket, and initially also a marble, while Anne has only an empty box. Sally puts her marble into her basket, covers it with a cloth, and retires from the stage, leaving the basket behind. During Sally's absence, Anne transfers the marble from the basket to her box, closes the lid, and then she too departs. Sally returns, and the question is where she will look for her marble: in the basket or in the box? A normal person, not having been present when the marble was transferred, will obviously look in the basket. Simon Baron-Cohen, Alan Leslie and Uta Frith applied this test to a number of children, all with a mental age above three, some autistic, some normal, and some mentally retarded (Down's syndrome) but not autistic.[51] The majority of the children in the latter two groups answered correctly: Sally would look in the basket because this is where she believed the marble to be. The majority of the autistic children opted for the box. This was where the marble was located, and this is where they believed that Sally would look. They were unable to attribute belief to her, and the logical conclusion was that they thus lacked a theory of mind. This experiment was as simple as it was elegant.

Given the absence of imaginative play in autistic children, the choice of dolls for this little drama was not particularly fortunate. Leslie and Frith therefore repeated the procedure with a different group of autistic children, replacing Sally and Anne by themselves, and using a coin rather than a marble.[52] The outcome was roughly as before, with 15 out of 21 going for the inappropriate option.[e] Leslie and Frith also looked for the appreciation of beliefs in others by subjecting the children to a test in which a (tubular) Smarties (small candy-covered chocolate drops) box contained a pencil stump rather than the usual sweets. The children were tested individually and in sequence. Each, after discovering the inedible content, was asked what the next person in line would say was in the box. Only 4 out of 20 correctly realized that the answer would be *Smarties*.

One could be concerned about this test's dependence on verbal

e / To put this result in perspective, *random choice* would have given 10.5 failures (i.e. 50%). The *standard deviation*, which is used in statistic tests of significance, is a measure of the amount of scatter of a series of numbers or measurements about their mean value. It is defined as the square root of the average value of the squares of the deviations from their mean value. If the deviation from the random-choice score is within the standard deviation, the result has dubious significance. The standard deviation for 21 measurements, all having equally probable outcomes, would be the square root of $21 \times 0.5 \times 0.5$, namely 2.3, so 15 out of 21 *is* significant.

reports by the autistic subjects, who typically have language deficits, particularly with prepositions. Although this would have made the trials more difficult to carry out, it would have been preferable to use protocols that remotely checked for surprise, independent of language. This is now routinely achieved by monitoring the size of the pupil; when an individual is surprised, the pupil dilates, and this is of course a totally involuntary process.

Lack of a theory of mind does appear to account for the symptoms I referred to above as belonging to the first class. The Wing–Gould triad of impairments are all in this class, if one takes lack of verbal communication *not* to embrace language deficits. Frith calls this triad the core symptoms of autism, while she relegates the items in the second class to peripheral roles.[53] I feel that this is counter-productive. Lack of a theory of mind does not obviously account for resistance to change, and it is difficult to see what relevance it has to the questions of fixation and ritualistic behaviour. The same could be said of autistic language deficits. It is also hard to see how not having a theory of mind could lead to delayed crawling and walking in autistic people. Thus although one must admire the ingenuity that went into what is surely a significant piece of progress, it would be unfortunate if this milepost came to be regarded as the end station. I feel that one should delve deeper.

In what respect are high-IQ autistic people, who usually *do* have a theory of mind, nevertheless autistic? Lack of a theory of mind is clearly not a *necessary* condition for having autism, even though it might be a *sufficient* one. Different tasks vary in the demands they make on our mental powers, and we readily accept that *certain things are beyond us*. In this respect, I happily agree with Colin McGinn's concept of *cognitive closure*.[54,f] The neonate's abilities are obviously circumscribed; just after birth, there are numerous *holes in the mind*. As learning proceeds, these holes are filled in, there being an environmental contribution to this process; except for the occasional flash of insight, we tend to know what we have been told. Although it is possible for an individual to work things out for himself, holes usually exist where information has been lacking.

It is uncommon for autistic people to write books. There have been a few notable exceptions, however, and their cases are clearly significant. Two autobiographies were produced by Donna Williams[55] and Temple Grandin.[56] Donna was initially labelled deaf, abnormal, retarded, spastic, crazy and insane, amongst other things. By her own admission, she existed in a state of dream-like recession, but she yearned to become normal. She had reached the age of 25 before hearing the word autism for the first time, and she had the insight to appreciate that the description of its symptoms fitted her own case. Against the odds, Donna achieved a place at university, lived independently, and ultimately wrote about her experiences and, importantly, her thoughts.

f / McGinn's formal definition of *cognitive closure* states that a mind *M* is cognitively closed with respect to a property *P* (or theory *T*) if and only if the concept-forming procedures at *M*'s disposal cannot extend to a grasp of *P* (or an understanding of *T*).

Temple had a similar uphill struggle, and she too acquired academic qualifications. With a vengeance, indeed, for she has a Ph.D. in veterinary science, and holds a professorship in animal husbandry.

There is no mention in either of these books of the author being subjected to a theory of mind test. And this is hardly necessary, because their texts reveal that they *are* aware of other people's thoughts. Donna makes the first reference to this on her second page, when she recounts that her parents thought that she had leukaemia. Temple's first reference to the thoughts of others is on her third page, and it refers to her mother's belief that her daughter was merely somewhat slow in developing.

It appears, then, that one must look beyond appreciation of thought in others, in order to get at the real origins of this syndrome. One is looking for something which can lead to a lowering of the IQ, even though this is not an essential aspect, and for something which is not in itself a consequence of low IQ. One is looking for an explanation of the triad of autistic deficits in general, and of the theory of mind deficit in particular. But one ought also to be looking for an explanation of those things in the second category.

To make any further progress, one has to do what we have done so many times in these pages: turn to the bottom-up approach and ask what anatomy and physiology are able to expose. There are, for a start, strong indications of a genetic origin of autism, as can be seen in the relevant statistics gathered by Susan Folstein and Michael Rutter.[57] An ultimate product of genetic endowment is the individual brain structure, and there is ample evidence of malformation in autistic brains. Margaret Bauman,[58] following up an earlier post-mortem investigation by herself and Thomas Kemper,[59] described abnormalities in the hippocampus, subiculum, entorhinal cortex, septal nuclei, mammillary body, and selected nuclei of the amygdala. These are all components of the limbic system, and the aberrations point to impaired emotional function in autism.

Deformities were also discovered in the cerebellum (specifically, the neocerebellar cortex and the roof nuclei). Because there was no observable loss of neurons in the inferior olive, which is the last port of call for signals before they impinge on the cerebellum, Bauman was able to conclude that the observed abnormalities began at, or occurred before, birth. Working with Deborah Arin, Bauman and Kemper established that there is a reduction in the number of Purkinje cells in the autistic cerebellum, the deficit lying in the range of 50 to 95 per cent.[60] Evidence of Purkinje cell depletion in autism has also been reported by Edward Ritvo and his colleagues.[61] Given the regulatory role in movement that the cerebellum plays, this loss could underlie the motility deficits in autism.

The results obtained by computed tomography (of X-ray images – also known as CT scanning), magnetic resonance imaging (MRI), and positron emission tomography (PET) reported by Eric Courchesne and his colleagues corroborate the evidence for reduction of cerebellar tissue.[62] Some parts of the cerebellum were found to be only half the normal size. There were indications that the reduction of tissue was a result of developmental abnormalities rather than damage following full development. The statistical significance of this work was beyond reproach; the MRI study alone involved 283 autistic patients. Not all parts of the cerebellum are affected, the regions most depleted being the cerebellar hemispheres and the vermis.[63]

There are a number of other observations which are of particular relevance to the things discussed earlier in this chapter, and in the preceding one. For example, autistic communication deficits, and lack of verbal spontaneity, were found to bear a striking resemblance to those seen in patients with frontal lobe lesions, particularly in the premotor and supplementary motor areas.[64] Edward Ornitz has also noted[65] that a number of autistic symptoms point to dysfunction of the gating mechanism mediated by the nucleus reticularis thalami. Just as intriguing is his suggestion, again based on a review of autistic symptoms, that there might actually be confusion between the exafferent and reafferent sensation in this syndrome. He suggests that the autistic nervous system might not receive the instruction, via efference copy routes, to ignore the sensory input associated with the person's own movement. Finally, turning to the biochemical level, there is evidence of malfunction in the interaction between the frontal lobe and the thalamus.[66]

Figure 10.2 reproduces data published by Francesca Happé,[67] which shows how the probability of passing the theory of mind test steadily increases with increasing verbal mental age. This is the case for both normal and autistic children, but the latter are clearly seen to lag behind their normal counterparts; whereas virtually all normal children pass the test by the time they have acquired a verbal mental age of about six years, that same point is reached by autistic children only when they have a verbal mental age of almost twenty.

Putting these widely differing pieces of information together, and in the light of what this and the preceding chapter have discussed, I cannot help speculating whether a major factor in autism is an innate inability to manipulate schemata: a lack of proficiency in exploiting the modularity of muscular

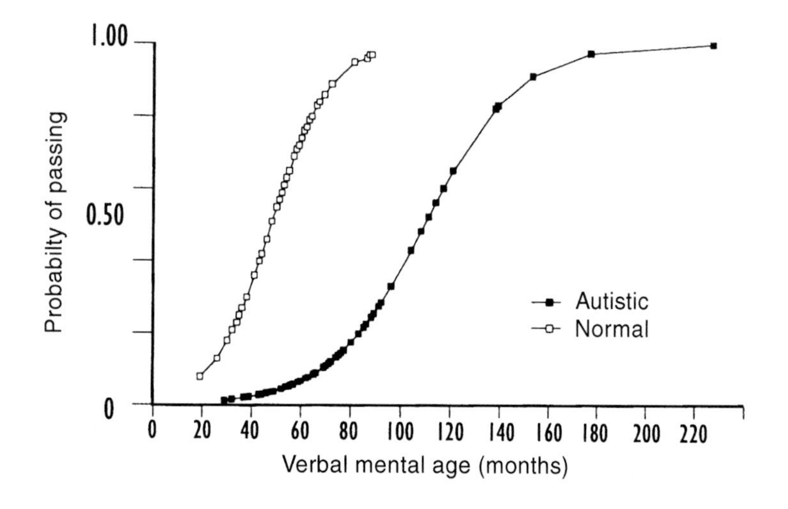

Figure 10.2
There is a marked difference between the aptitudes of normal-ability and autistic people for grasping the fact that other people think. Francesca Happé has found that the ability varies systematically with verbal mental age, as shown here. Very few autistic people have such a 'theory of mind' when their verbal mental age is that at which virtually all normal-ability people possess it.

actions. This could explain the difficulties with language and those related to motility. Moreover, and this would be particularly exciting, because the symptom is especially hard to understand, this idea might even be capable of explaining that peculiar insistence on sameness; there might be an emotional impairment that disrupts the process whereby simple schemata are consolidated into more complicated schemata. Indeed, it is tempting to speculate whether such manipulation and consolidation might underlie intelligence in general. Let us now turn to that very important issue.

Roots of intelligence and creativity

There are quite pronounced differences between the EEG patterns of autistic and normal people. The normal pattern, as we have seen, continues out to about 500 milliseconds, and the latter part of chapter 8 was devoted to explaining this in terms of what Benjamin Libet and his colleagues termed neuronal adequacy. The latter was seen to be related to conscious awareness. In the EEG trace of the typical autistic person, the amplitudes of the various peaks are diminished – more markedly so as time progresses over those 500 milliseconds. The indication is thus that whatever causes the various swings in signal is truncated in the case of the autistic person.[68] It is as if signals are having a harder time getting through to their destinations, and they therefore decay away faster than normal.

This result is interesting in its own right, but it takes on added significance when we note that a relationship has been observed between EEG characteristics and intelligence. This was established by John Ertl and Edward Schafer in 1969.[69] If one compares the evoked potentials of people with high IQ to those recorded from people with low IQ, the former are seen to have more troughs and valleys, for a given time interval (of about 250 milliseconds). This indicates that people with low IQs lack certain high-frequency components in their response characteristics, the missing frequencies lying in the gamma band.[g] This result is rather suggestive, because that was the frequency range of those oscillations observed to be related to coherence in visual scenes, which were conjectured to relate to feature binding (see chapter 8). If gamma oscillations are a ubiquitous feature in conscious awareness, the suggestion could be that people with low IQ may have a harder time capturing correlations in sensory input.

In 1956, George Miller wrote a fascinating paper in which he referred to a magical number of information processing. This was actually not so much a single number as a *range* of numbers, stretching from 5 to 9; he called it *the magical number seven plus or minus two*. He had discovered that normal people can repeat between five and nine digits that have just been read out to them, and this number is referred to as a person's *digit span*.[70] Digit

g / The actual data are for 10 people with IQs in the range 120 to 142, compared with 10 people having IQs in the range 62 to 89.

span was the first thing that Donald Broadbent employed when he was carrying out the investigations on attention that led to his filter theory, as discussed in chapter 8. One can feel the circle of evidence beginning to close, because digit span has been found to correlate with the evoked potential. John Polich, Lawrence Howard and Arnold Starr have established this, by showing that the normal 300-millisecond peak in the EEG trace tends to be delayed and diminished in people with lower spans.[71]

With IQ being related to EEG characteristics, and EEG characteristics being related to digit span, one might guess that digit span would have to be related to IQ. This is indeed the case.[72] I have measured digit span on a number of autistic people. The results were highly consistent, for the five probands studied, and they contrasted sharply with what Miller observed for normal people. For five digits, virtually all normal people have a 100 per cent score. For autistic people, I found that the chance of recalling five digits is virtually zero; the average proband has a magical number of three and a half, plus or minus a half.

What are these numbers and EEG characteristics a measure of, however? Let us first note that there is more to mental power than the mere gathering of information, not the least because such acquisition can also occur unconsciously. Information must be used for making predictions, because in default of forecasts there can be no surprises. This surely is a prerequisite for what we usually think of as common sense. Recalling the suggestion by Christopher Miall and his colleagues[73] that the cerebellum may function as a predictor, and also that the cerebellum of the autistic person is often rather small, we could conclude that the typical autistic person is a poor forecaster.

Individual predictions will not be useful, however, unless they can be incorporated into scenarios, and these are put together in the frontal lobe, probably with the aid of the basal ganglia; as we have seen, the combining of individual schemata into more complex schemata implicates these brain regions. Let us return to that issue of language. As William Calvin has noted, wild chimpanzees use about three dozen different vocalizations, to convey about the same number of different meanings.[74] But they do not string these sounds together, in order to augment their repertoires.

Humans use about the same number of phonemes. But as I emphasized earlier, these do *not* convey meaning; they acquire significance only when strung together to make words, for it is only at the word and sentence levels that emotion comes in. Normal children are rapidly acquiring single words around the end of the first year, two-word phrases by the end of the second year, and multi-word sentences by the end of the third year (see figure 10.3). I shall now try to bring in that vital advantage provided by the veto-on-the-fly mechanism, using examples at the single-word level. Although there might be the possibility of unconscious acquisition through repeated passive exposure – brainwashing, as it could be called – let us confine our consideration to the conscious appropriation of new words. I have suggested that consciousness is perforce an *active* process, and that the *sine qua non* of the

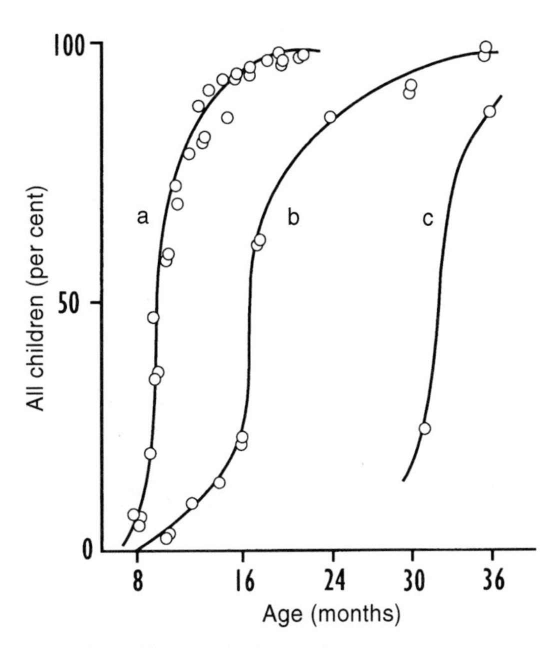

Figure 10.3
Normal-ability children acquire language at an impressive rate, virtually all of them managing single words at a year and a half, and two-word phrases by the age of three. As this diagram redrawn from the work of William Calvin demonstrates, the majority of children are mastering primitive sentences not long after that.

phenomenon is that *planning of movement must be actually proceeding, rather than merely intended.*

The child can learn the significance of words like *salt* and *sweets* (the noun forms, for simplicity) by tasting these things for herself (it could also, of course, be himself), while silently adding appropriate phonemes to the initial S-sound, as she hears someone else pronounce each name aloud. If she subsequently sees jars labelled **SALT** and **SWEETS**, she will automatically know what sensations their contents would give her. But what will happen as she reaches for a third container, while mutely following the S-sound with an I-sound, an AN-sound, another I-sound, and finally a D? Her nervous system will unconsciously recall her being told that the word *cyanide* has bad connotations, and her hand will stop in its tracks. In fact, those negative associations may well start to emerge even before her silent intonation of the word is complete. (If she has *not* heard the word, but nevertheless encounters the substance, she will of course be in grave danger; this is why we are advised to keep such things out of reach of children.)

I admit that this example is rather contrived, but it does reveal the great advantage language furnishes us. There are many things that we cannot afford to learn about by direct experience; they are simply too dangerous. Fortunately, we are able to learn about them second hand, and nevertheless incorporate them into schemata that can be activated by reflex, despite their sophistication. But this takes time, and the young of our species require careful supervision. Individuals with below-normal IQ are at even greater risk.

Let us bring in the work of Margaret Boden, at this point. Her analysis of creativity led her to suggest that the key attribute is *the ability to vary the variable*, which she illustrated with reference to jazz music, amongst other things.[75] Indeed, one could say that the first hurdle is recognition *of* variability. Albert Einstein's path to the *theory of relativity* was through mentally letting the speed of an object *vary*. It does not seem too far-fetched to suggest that Alan Leslie's ideas about the origin of a *theory of mind*[76] could likewise be perceived as an ability to imagine *variability*. Isn't this what we do when we pretend that one thing represents another? And wouldn't insistence on sameness, as displayed by the autistic person, be detrimental to this facility?

When pondering the concept of theory of mind, I am often reminded of a scene in Peter Ustinov's *Romanov and Juliet* (a play written during the Cold War era), in which an obsequious character in an imaginary country has a series of meetings with the American and Soviet ambassadors, alternating between them. Informing the American that the Soviets have a secret weapon, he is told that the Americans already *know* this. He then tells the Soviet ambassador that the Americans know about the weapon, only to be informed that the Soviets *know* that the Americans know. At his next meeting with the American ambassador, again attempting to ingratiate himself by exposing something he believes to be valuable, he is dismayed to discover that the Americans already *know* that the Soviets know that the Americans know. The procedure continues, the story getting more complicated each time, and it ends with one of the ambassadors exclaiming: '*Oh! We didn't* ***know****, that they know that we know that they know that we know ...*'. The trouble is that once the story contains more than a few links, it becomes impossible to keep track of the significance of what is being revealed. My point in relating this fascinating piece of theatre is that once things exceed a certain logical depth, we cannot fathom them. The psychologist speaks of *embedded inference*, and human limits on embedding are not impressive; beyond about four steps, our species becomes uncomfortable. I appears that autistic people's ability to embed is seriously compromised.

In fact, limitation of this ability might be a good indicator of impaired intelligence in general. We have noted, however, that one cannot merely equate autism with low intelligence. This underlines the fact that intelligence has many facets; a person might be normally endowed in most respects, but display deficits (those holes in the mind, as we called them earlier) in certain domains. That autistic people are particularly disadvantaged is attested to by the results of a genetic study, carried out by Andrew Pickles and a number of colleagues,[77] which indicates that several different genes may be implicated in the syndrome.

Emotional intelligence?

It has recently been implied that intelligence does not merely possess a variety of facets, but that it even comes in different basic *forms*, the particularly important type being what Daniel Goleman refers to as *emotional intelligence*.[78] Goleman expresses the opinion that the latter, designated by the letters EQ, is more important than the traditional IQ. His assumption appears to be that IQ does not involve the emotions. Let us consider whether that is a viable hypothesis.

Goleman credits Peter Salovey with the actual concept of emotional intelligence, and one could speculate as to why they do not use the letters EIQ, which would have been more logical. EQ literally suggests *emotional quotient*, which might well be a useful parameter, but which need not necessarily have anything to do with intelligence as such. Indeed, Goleman's book seems to be more preoccupied with the way in which possession of an adequate EQ can help an individual to get on in the world. His message is thus to be compared with those earlier prescriptions for success that give prominence to empathy with one's fellow beings, the how-to-make-friends-and-influence-people maxim that is, indeed, difficult to gainsay. Our argument here is whether the parameter EIQ, which they eschewed, would really have been distinguishable from the traditional IQ. I strongly suspect that it would *not*.

Goleman argues that IQ, as usually measured, does not seem to be a particularly reliable measure of an individual's prospects in many walks of life. Here too, he seems to have made a valid point, because the type of test often used to assess IQ actually checks only a limited set of what I earlier called intelligence's facets. Goleman is right to point out the inadequacies of such tests. But he moves into less reliable territory when he applauds the holding in check of impulses, hinting that failure of this is a recipe for social truncation, if not downright calamity. As Stuart Sutherland pointed out:[79] '... neither Hitler nor Stalin were notable for their empathy or self-restraint, but by their own lights both were successful'.

It is not with Goleman's social science that we should be concerned here, however; it is with the possible neuroscientific underpinnings, or the lack of them. It was suggested above that the crux of intelligence lies in the ability to manipulate schemata. And I indicated that such orchestration would make demands on working memory, as well as on the mechanism by which possible consolidations of schemata are adjudicated. The actual possibility for juxtaposing existing schemata, and transforming their combinations into new schemata, was seen as being provided by the core circuit that mediates consciousness itself, a prerequisite being an adequate short-term memory span. And I earlier stressed that exploitation of modularity depends upon the fact that a given emotional mind-set persists for a considerably longer time than the span of iconic memory, my illustration being taken from the important case of language. The later change to a different emotional mind-set results from the process that I have been calling adjudication, and the moment has come for us to delve deeper into its underlying mechanism.

It will already be clear, however, that intelligence as here defined cannot be divorced from emotion. Rather direct endorsement of this attitude is provided by studies reported by Antoine Bechara, Daniel Tranel, Hanna Damasio and Antonio Damasio.[80] They measured the skin (electrical) conductance responses (SCRs), during the making of decisions, of patients with damage to the prefrontal cortex, comparing their observations with what is seen in normal control subjects. The trials included rewards and penalties. After a number of trials, the normal subjects began to display SCRs in anticipation of the results of further trials, whereas this effect was absent in the patients. As discussed earlier in this chapter, the prefrontal cortex figures centrally in working memory, and thus in the manipulation of schemata, while the generation of an SCR is a clear sign of emotional activity.

In the previous chapter, I cited Edgar Adrian's use of the word *chief*, in describing the muscular output as a product of activity in the nervous system, and noted that this reminds us that there is a purely internal product of such processes, namely that which controls the endocrine system. Joseph LeDoux[81] has been prominent in the charting of the neuronal routes in that vital part of the overall system, particularly with respect to the amygdala, a composite structure comprising several nuclei, located near the tip of the temporal lobe (on each side of the brain). These nuclei dispatch signals via the *stria terminalis* to various parts of the hypothalamus, which, in turn, exercise control over the endocrine system. This is mediated by the connections between the latter and the pituitary, and both these structures interface with the vascular system, the former monitoring the chemical composition of the blood and the latter directly controlling it. The upshot of all this circuitry is the control that the amygdala has on the involuntary behaviour related to arousal. But it is important to bear in mind that the amygdala is also connected (albeit indirectly) to motor output.

The strategy favoured by LeDoux and his colleagues has been to lesion various parts of the system, and observe what impairment ensues. They classically conditioned rats, pairing sounds or light signals with mild electric shocks to the feet, while observing the impact on blood pressure, pulse, and the susceptibility to being startled. Bruce Kapp and his colleagues[82] demonstrated that it is the central nucleus of the amygdala that is responsible for that structure's response-controlling output, although Simon Killcross, Trevor Robbins and Barry Everitt[83] later showed that other amygdaloid nuclei also influence the integrated emotional response.

It is the inputs to the amygdala which will be our main concern, however, for it is they that can reveal the ultimate origins of the control. The lateral nucleus[84] sends signals to the above-mentioned central nucleus, and LeDoux, Claudia Farb and Lizabeth Romanski[85] showed that the former receives inputs from both the thalamus and the cortex, the latter finding confirming that of Norman Weinberger and his colleagues.[86] The former route, which is clearly the more direct, and therefore the faster, was found to involve glutamate,[87] and long-term potentiation (LTP) in this part of the circuit was demonstrated by Marie-Christine Clugnet and LeDoux.[88] Jonathan Gewirtz and Michael Davis, studying the more complicated scenario known as second-order fear conditioning, have found evidence that the receptor implicated is of the NMDA type.[89] Analogous LTP in the route from the cortex to the amygdala was demonstrated by Thomas Brown and Paul Chapman.[90]

The plot thickened when Michael Fanselow and Jeansok Kim discovered that lesioning the *hippocampus*, following fear conditioning, blocked the expression of responses to the environment.[91] As has been discussed earlier in this book, the hippocampus appears to be vital in the laying down of what could be called contextual memories, a sub-class of these being those that help rats to run mazes, for example; whence the concept of hippocampal place cells. Russell Phillips and LeDoux showed that it is the part of the hippocampus known as the *subiculum* which projects to the amygdala's lateral nucleus. As one would expect, however, it is not all hippocampus-mediated memories that require the involvement of the amygdala, only those implicating emotion. This distinction was underlined by the work of Larry Squire and his colleagues.[92] The hippocampus has also been shown to contribute to novelty detection,[93] and in view of what was discussed above regarding the kinematics of language, it might be significant that Minoru Tsukada, Takeshi Aihara, Makoto Mizuno, Hiroshi Kato and Ken-ichi Ito have discovered that LTP in this part of the brain is sensitive to the temporal pattern of signals impinging upon it.[94] Returning briefly to the question of autism, Robert DeLong has suggested that hippocampal dysfunction might be the underlying cause, this impairing the modulation of emotion, amongst other things.[95]

Let us briefly take stock of all these results. We have seen that there are at least three distinct routes into the amygdala (which also has its own *internal* pathways). The most direct leads from the thalamus, and although its signalling does not have the benefit of the cortex's discrimination (LeDoux calls it the quick and dirty path), it is adequate to give the immediate type of reaction that could make the difference between life and death. Then there is the route from the cortex, which would provide a more nuanced cognitive response. Finally, and most sophisticatedly, there is the route that stems from the hippocampus, this being the one that triggers the amygdala only if the context is right (see figure 10.4).

One would think that this should be a sufficiently comprehensive bag of tricks to cover all contingencies. It might come as something of a surprise, therefore, to learn that yet another part of the brain gets into the act; surprise, that is, unless one recalls what was stated above regarding the circuitry that serves intelligence itself. To appreciate this last piece of the story, we will have to consider something known as *extinction*. If, after a period of classical conditioning, a rat (say) is repeatedly presented with the conditioned stimulus, but *without* the accompanying electric shock, its fear response is observed to gradually diminish; this is called extinction. Edmund Rolls, in studies of primates, found that damage to the prefrontal cortex makes extinction of an emotional memory far more difficult.[96] Michael Davis and his colleagues went on to show that extinction can be precluded by blocking the action of NMDA receptors in the amygdala,[97] and this indicated that extinction is an active learning process, not just the passive decline of a previously acquired memory.

LeDoux rationalized these findings by suggesting that the prefrontal areas normally control the expression of emotional memory, and that they can actually suppress emotional responses when these are no longer useful to the system. It might seem that this is related to the issue of context, which, as we have seen, implicates the hippocampus. There are clear indications, however, that the prefrontal and hippocampal regions do not function in tandem.[98] PET studies showed that the hippocampus is involved when overall novelty of presented stimuli must be registered, as we saw earlier, whereas the prefrontal regions come into play when previously formed connections have to be changed. Such changes are frequent when the elements of language are manipulated.

It is tempting to draw wider comparisons, here, and recall that three of the autistic symptoms *not* explainable on the basis of a theory-of-mind deficit are the difficulty with language, the insistence on sameness, and the tendency toward stereotyped movements. I would like to suggest that all of these features may be related to prefrontal dysfunction. Using the terms employed earlier in this chapter, and in the preceding chapter, one could say

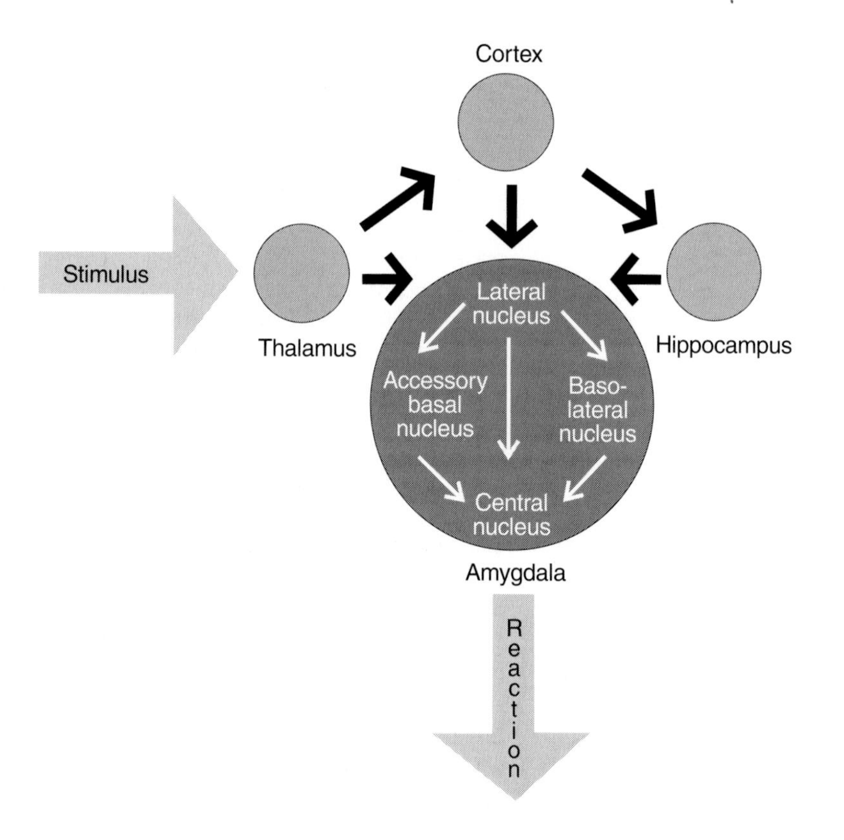

Figure 10.4
The amygdala can be regarded as the brain's central adjudicator of the emotional values of sensory input. As this figure inspired by the work of Joseph LeDoux shows, there are three different routes into this component, which is located near the base of the brain (on each side) – see figure 7.10. The quickest path leads directly from the thalamus, and thus does not pass through the cortex at all; it serves reflexes that do not require conscious awareness. The second path is from the cortex, and requires bare cognition, while the most circuitous route, through the hippocampus, produces activation only in the case of cognition in the appropriate context.

that the autistic person encounters particular difficulty with the manipulation of schemata. So putting phonemes together in novel sequences, to make new words, or words together in novel order to make new sentences, gives these people problems. Likewise, and recalling the cognitive theory of emotions put forward by Philip Johnson-Laird and Keith Oatley, discussed in the previous chapter, we see that the emotional malfunction that manifests itself in the autistic resistance to change may be traced to the prefrontal regions, and their interactions with the amygdala. Finally, the repetitive muscular movements characteristically displayed by autistic people might simply indicate that the system tends to get locked into a particular schema.

John Skoyles[99] has suggested that cognitive processes in general have become part of the human heritage because of the felicitous coming together of no less than four factors: neural plasticity; large functionally uncommitted prefrontal, temporal and parietal areas; the ability of their neural circuits (because of plasticity) to acquire novel symbolic and non-symbolic skills; and a large prefrontal cortex, which mediates working memory. It is this happy combination of endowments, he suggests, which enabled humans to develop mental skills that were not present immediately after the emergence of our species. Similarly, both John Duncan[100] and I, writing separately, have suggested that such factors provide the real basis of intelligence.

Recall, recognition and the detection of novelty have been the concern of Endel Tulving and his colleagues.[101] Together with Roberto Cabeza, Shitij Kapur, Fergus Craik, Anthony McIntosh and Sylvain Houle, he has demonstrated that both recall and recognition require activation of the prefrontal and anterior cingulate regions, recall additionally activating the globus pallidus, thalamus and cerebellum, whereas recognition needs participation of the inferior parietal cortex. (The activation of cerebello-frontal pathways in recall is interesting, in view of the diminished size of cerebellum seen in many autistic people.)

It is also noteworthy that this work brings us back to the anterior cingulate. Richard Lane has specialized in bilateral cingulotomies, for the relief of chronic pain, and he has suggested, in collaboration with Geoffrey Ahern, Gary Schwartz and Alfred Kaszniak, that functional disconnection of incoming emotional information to the anterior cingulate might underlie a condition known as *alexithymia*.[102] People suffering from this impairment have difficulty putting emotions into words, and Lane and his colleagues suggested that the situation is analogous to blindsight; indeed, they referred to it as emotional blindness. That the affliction is serious is attested to by Schwartz's earlier suggestion that conscious awareness of emotions promotes both psychological and physiological health.[103] (It is interesting to note that Jacques Paillard and his colleagues have discovered a tactile equivalent of blindsight.[104])

As we saw earlier, the anterior cingulate is intimately involved in the perception of pain, and this makes it eminently suitable as the king-pin of what I have called the veto-on-the-fly mechanism. It has also been emphasized that the significance of *any* sensory input only emerges as a consequence of the animal's own motor explorations, because only then will it be reliable in the navigation that is the be all and end all of behaviour. The best illustration of this principle that I have encountered was described in Donald Hebb's *Essay on Mind*.[105] The observations were made on Scottish terriers, reared in two contrasting ways. Some were given the degree of freedom normally accorded pets, whereas the others were restricted to small cages during the entire period of growth. The latter group could hear and smell other dogs, and humans, but could neither see nor touch them, or indeed be touched by them.[106]

When fully grown, the dogs reared in the severely restricted environment were found to be clearly inferior to the pet group when it came to running mazes. They were also found to be untrainable; as one of the investigators put it:'I never saw a dog like that before'. But the most dramatic finding concerned pain; these dogs were unmindful of it! Their reflex responses to painful stimuli were normal, in fact, but they evinced no emotional or motivational reaction to them. On the strength of these important observations, pain is thus seen to have two distinct components: an automatic sensory one and another that is charged with emotion. The experimenters remarked that the restricted dogs acted as if they had been lobotomized. These observations were made four decades ago, however, and we would now be more inclined to compare them with Richard Lane's cingulotomy patients. One of the dogs noticed a lighted cigar fall to the floor, and probed the glowing ash with its nose. It pulled back, reflexively, but then repeated its action twice more. Subsequent tests with electric shocks exposed the same lack of appreciation of danger.

We have come far, in this analysis of emotion and intelligence, and the number of brain components invoked has been considerable. This is in keeping with ideas that have been around for at least sixty years,[107] and which have recently resurfaced in Jeffrey Gray's conjectures regarding the content of consciousness,[108] though he admitted that his ideas were unable to explain why the brain should generate conscious experience in the first place. This is what I attempted to do in the preceding chapter, of course, when I was particularly concerned with showing why consciousness is a prerequisite for experiencing subjective emotional states. I agree with LeDoux when he says that: '... emotions or feelings are conscious products of unconscious processes; it is crucial to remember that the subjective experiences we call feelings are *not* [my italics] the primary business of the system that generates them'.

To close this section, then, my answer to the question, do we have emotional intelligence must be: yes, indeed, because *emotional intelligence is the only type there is!*

The scope of the simian mind

The concept of a theory of mind did not arise from consideration of its possible absence in autistic individuals. It was introduced in a famous article by

David Premack and Guy Woodruff, in 1978.[109] If any other animal is able to impute mental states to its peers, it is likely to be one of our evolutionarily close relations.[110] It is now possible, using genetic data, to work out the evolutionary distance between different species. These techniques are so sensitive that one is even able to differentiate between the various members of the *Hominidae* family, in terms of their genetic proximity to ourselves. The result can be stated in terms of an arbitrarily defined distance scale, in which importance is to be attached to relative rather than absolute magnitudes. The list is as follows: chimpanzee 1.8, gorilla 2.4, orangutan 3.6, gibbon 5.2, and monkey 7.7. Premack and Woodruff specifically addressed the issue of whether chimpanzees have a theory of mind, since members of that species ought to have the best chance of possessing this attribute.

They carried out a number of different types of investigation, in an effort to settle the question either way. In one set of tests, they showed subjects four different video sequences, each depicting a person (let us call him an actor) who was encountering a simple problem. Typical tasks were the retrieval of something that was out of arm's reach, when there were adequately lengthed sticks in the vicinity, and reaching a suspended item when there were convenient boxes at hand. On each occasion, the sequence was put on hold just before the actor solved the problem. The subject was then required to choose from between a number of alternative photographs that picture which illustrated the correct solution. The important thing about these sequences was that they each showed the person being confronted with a decision. The question thus arose as to whether the subject would be able to *put himself in the place of* the person. By and large, the chimpanzees were able to make the correct selections, but there were lingering doubts as to whether they were really imputing thought to the actor.

The most interesting results were obtained from tests which involved two trainers, only one of them being 'on stage' at any time. One of these plays a *benevolent* role, in a conspicuous green hat, while the other is *villainous*, and his hat is an equally bright red. These hats served to clearly identify the trainer who was present for each repetition. The scene played out in the presence of the subject ran as follows. A large number of empty containers were separated from the subject by a wire screen, and the subject was permitted to watch while food was put into one of them. On some occasions, this was done in the presence of the benevolent trainer, while at other times the villain was in attendance. Both made a show of having seen the food put in place. Following each placement, the participating trainer was to offer the subject one of the containers, the benevolent trainer making the correct choice and the villain always deliberately making a mistake. The chimpanzee naturally became frustrated in those latter episodes. Premack and Woodruff noted that its anger was so intense that continuation of the experiment was actually deemed to be too dangerous.

As impressive as these tests were, they do not yield an unequivocal answer as to whether a chimpanzee can attribute states of knowledge to others. They demonstrate that these animals can ascribe purposes and affective attitudes, but that is not the same thing as being conscious of the thoughts of other individuals. There is something that does appear to establish this in the chimpanzee, however, and that is deceit. Emil Menzel has verified this through an experiment in which he revealed the location of food to a chimpanzee who had only a modest ranking in the local pecking order of its group. Before the creature had time to reach it, Menzel admitted the rest of the group to the enclosure, and was intrigued to observe that the animal with the knowledge of the food's whereabouts deliberately misled its senior companions in order not to lose it to them.

Hans Kummer and his colleagues have observed an even higher order of deception in hamadryas baboons.[111] A female took over twenty minutes to gradually work her way over to a rock just a couple of metres distant, in an apparent attempt to give the impression that her short journey was capricious and without ulterior motive. In fact, however, the rock was concealing a young male from her harem master's view. She then proceeded to groom the young fellow, carefully adopting a posture in which all of her body, except her grooming hands, was visible to the older male (see figure 10.5). Richard Byrne and Andrew Whiten have described another case of such deceit.[112] A youngster being bullied by one of the pack's adolescents made such a din that it attracted the attention of several adults, including the victim's mother. The offender was soon being chased, with the adults in hot pursuit. Suddenly, the adoles-

Figure 10.5
Until recently, it was believed that humans are the only animals which think, but detailed observations on several other species has now refuted this oversimplification. In some cases, it has even transpired that non-human animals are able to attribute thought to their fellow creatures, a process known as embedding. The sketch shows a female hamadryas baboon grooming a young male that is hidden from her harem master's view by a rock. She took twenty minutes to work her way casually over to the rock, so as not to arouse suspicion, and it was clear that she could calculate that the young male would be invisible to the senior baboon.

cent stopped and stood on its hind legs, scanning the horizon. This is the standard reaction of baboons when they sense danger from predator. His would-be punishers stopped in their tracks, and joined in the reconnaissance, leaving the bully to slink away unpunished.

I mentioned earlier that human performance with embedded logic is not impressive. We encounter great difficulty with tasks that require a depth of more than four steps. Franz Plooij has reported evidence of a depth of two in chimpanzees, because they can apparently deceive deceivers.[113] Let me first describe something simpler that Jane Goodall reported. She saw a young animal suddenly notice a banana that she had hidden in a tree, while the other chimpanzees' attention was elsewhere. A larger chimpanzee happened to be sitting closer to the tree than the observant youngster, and Goodall was intrigued to notice that the youngster moved away so as to lead its potential competitor off the scent. When the larger animal moved away, the young one dashed back to claim its prize. Plooij reported a case that was rather more complex. A male who was on the point of eating some bananas that only he knew about was interrupted by the sudden appearance of another chimpanzee. He promptly acted just as Jane Goodall's animal had done: moved away and made an ostentatious show of innocence. The intruder then left the scene, but only to position himself behind a tree. When the first animal then attempted to recover the bananas, the unseen observer rushed out and grabbed them for himself. As Byrne and Whiten have remarked, this complex of deceptions indicates that a chimpanzee is capable of saying to himself: 'The other fellow thinks that I think that he doesn't know where to find the bananas...but I know perfectly well that *he* knows where the bananas are'. This is embedding of the second order, and it seems to remove any doubt surrounding a chimpanzee's theory of mind.

Deep Blue and Deep Thought

Chimpanzees, then, appear to be capable of second-order embedding, while we can manage about fourth-order. This is hardly surprising, because the anatomy and physiology of the simian brain are not very different from those of our own. Autistic people in the lower IQ range seem to have even more limited embedding than chimpanzees. All these numbers must be taken for what they are, however: convenient short-hand descriptions of what is probably a rather more complex issue. Let us nevertheless go along with the simple view and ask how deep a machine might be able to reason. Those who write chess programs for computers appear convinced that machine reasoning can be deep, because that word is used in their programs' names.

In 1988, Feng-hsiung Hsu, Thomas Anantharaman, Murray Campbell and Andreas Nowatzyk produced such a program, and they called it *Deep Thought*. In November of that year, the program tied with Grandmaster Anthony Miles for first place in a chess tournament in Long Beach, California, beating Danish Grandmaster Bent Larsen on the way. It could analyse 750 thousand positions per second, and its rating performance lay around 2600 on the international scale. This put the program in the bottom half of the grandmaster range. An average tournament player lies around 1500. The successor to Deep Thought, which its inventors named *Deep Blue*, was analysing 8 million chess positions per second in 1991, and it was up to 1 billion positions per second by the following year. Australian chess champion Darryl Johansen, who drew 1–1 with Deep Thought in 1991, refused to be impressed by the program's rate of analysis. Echoing something originally attributed to Richard Réti, in the 1920s, he said that he could boast of seeing only one position ahead, in about one second. *But it's the right one*, he quipped.

It was Claude Shannon, around the middle of the twentieth century, who first appreciated the significance of computer chess. He noted that it addresses a problem which is sharply defined, because the allowed operations and the final goal are known. He argued that chess programs are a good demonstration of the analytical power of a computer, because their performance can be checked directly against that of a human. In this respect, he was echoing Goethe, who called chess the touchstone of the intellect. Shannon was the founder of *information theory*, and his analysis of machine chess built upon the pioneering efforts of John von Neuman and Oskar Morgenstern, the originators of the *theory of games*. One product of the latter had been the so-called minimax algorithm, with which one can derive the best move, in any situation.

The computer chess program could be said to adopt the brute force approach. For a given position, there are a certain number of permissible moves that a player can make, and each of these could give rise to a number of responses by the opponent. Those responses, in turn, could give rise to a certain number of next moves by the player, and so on. The trouble is, of course, that this branching structure produces a tree of mammoth proportions. Indeed, the number of different games can be calculated to lie around a number that is 10 raised to the power of 120; to write this huge number, one would put a 1 followed by 120 zeros. *Tree structures* are a common feature of programs in what is now referred to as traditional artificial intelligence, the word *traditional* being required in order to distinguish it from artificial intelligence based on neural networks, of the type we considered in chapter 6.

When computers entered the world of chess, in the 1950s, there were two broad schools of thought. There were those who advocated a pseudo-human approach, in which moves were to be chosen through logical reasoning, while the opposing faction regarded the challenge as an exercise in trial and error. The latter group, it must be admitted, had grasped the true strength of the computer, which stems from its enormous appetite for tedious calculation. But one can sympathize with those who opted for what could be called the more elegant route. In 1964, eight years before he became World Champion, I had the pleasure of playing Bobby Fischer in a simultaneous tournament held in Chicago. It was an intriguing experience to observe the speed with which he made his moves. There were fifty of us, our boards arranged in a large circle, and he rapidly shuffled from board to board, seldom taking more than 15 seconds to come to a decision. He lost one game, drew five others, and won the rest. Contemplating the beauty of his performance, all those years ago, it would have been difficult to imagine anything further removed from brute force. And it was the inspiration of such gifted play which held sway in the computer chess of that era.

By the time Fischer had become champion, however, the pragmatism of trial and error had firmly established itself, and it has been in the driver's seat ever since. This is not to say that progress has only been dependent on improvements to computer speed. On the contrary, the trial and error method has given rise to strategies which are elegant in their own right. The basic problem is to minimize the waste of time that comes from investigating branches of the tree which a human player would instantly recognize as futile. But this is easier said than done. For a given computer speed, and a given amount of time available per move, there will always have to be a cut-off, beyond which there simply isn't sufficient time to probe. This might be seven of eight moves, for example. The trouble is that this temporal horizon must occasionally cut through an exchange of pieces. As Hsu and his colleagues point out,[114] a situation can arise in which the computer can only see as far as the exchange of one of its minor pieces for one of the opponent's major pieces, the machine remaining oblivious of the certain loss of one of its own major pieces the very next move after the horizon.

Such problems are now circumvented by the addition of special sequences which follow such exchanges through to their local conclusion. That strategy, in turn, has spawned something known as the *singular extension algorithm*. It probes with particular diligence those routes of the tree in which there is only a single beneficial response. And in this respect, the trial and error method could be said to be getting closer to what the good human player does. Singular extension has proved to be a very powerful device. On one occasion, a shocked chess master was informed by the machine that he would be check-mated within 19 moves.

A common feature of computer simulations is the difficulty of assigning appropriate values to the various adjustable parameters. It is these parameters which determine how much significance each factor is to be accorded, and small changes in them can lead to large changes in the ultimate outcome. It is such sensitivity that makes weather forecasting a hazardous undertaking, for example. It is thus particularly impressive that Deep Blue's parameters are actually adjusted by the program itself. Naturally, the program must have a model on which to base its modifications, and the obvious choice was actual games played by experts. This is not as unfair as it might seem. After all, human players become experts by playing with those who are better than themselves; why should a machine not be allowed to do the same thing?

The thing that I find fascinating about this is its similarity to the training of artificial neural networks. In that case too, the ultimate performance is determined by the quality of the data that the network is trained on. There have been suggestions that a neural network might indeed be a viable alternative for computer chess, but this has not led to anything that even approaches Deep Blue. The latter has long since had special hardware designed for it, and this has led to even faster search procedures. The machine of Hsu and his colleagues goes on improving its rating, as computer speed increases and the algorithms become even more efficient.

The show-down took place during the early summer of 1997, Deep Blue finally getting the better over a somewhat disgruntled World Champion Gary Kasparov, under normal tournament conditions. I do not know how one would go about calculating the program's effective degree of embedding, but it seems clear that the word deep, in its name, is not misplaced. And yet I find this accomplishment not so terribly impressive. After all, we would hesitate to award the World weight-lifting championship to a fork-lift truck, would we not, so why should being beaten by a chess-playing computer program be so traumatic. I submit that those who tend to be overawed at such technical feats have not fully appreciated the great difficulties that confronted nervous systems when they made that vital step upward, and acquired the faculty of real-time-choice.

The circuit had to be just right, of course, and Paolo Nichelli and his colleagues, using PET scanning, have shown just how many different brain regions chess playing calls into action.[115] Nature also had to tune the system's temporal parameters so as to manoeuvre them into the correct range. And as we saw above, developing a sufficiently large prefrontal region, in order to provide us with that erasable blackboard we call working memory, was no mean feat on Nature's part either.[116] It is a *highly* impressive thing that we can embed as deeply as we do, and after all, it was that ability which enabled some of our program-writing kinsmen to produce those chess-playing strategies.

11 The message and the medium

Had I been present at the birth of this planet, I would probably not have believed on the word of the Archangel that the blazing mass, the incandescent whirlpool there before our eyes at a temperature of fifty million degrees would presently set about the establishment of empires and civilizations, that it was on its way to produce Greek art and Italian painting.
W. Macneile Dixon (*Gifford Lectures, Glasgow* (1935–37))

There is no long-distance target, no final perfection to serve as a criterion for selection, although human vanity cherishes the absurd notion that our species is the final goal of evolution.
Richard Dawkins (*The Blind Watchmaker*[1])

Of all possible worlds

When the automobile superseded the horse-drawn carriage, about a hundred years ago, the event was not the abrupt change that one might have imagined it to be. It is true that the horse had disappeared from the front of the vehicle, but the rest of the contraption was pretty much as it had been before. In the very earliest horseless conveyances, there was a raised seat at the front for the driver, and the passengers entered through doors which bore a striking resemblance to those used in the horse-drawn variety. And the various accoutrements, such as the brass lamps, looked as if they might have merely been transferred from their predecessors. Similarly, when the first skyscrapers began to appear, around the same time, many of their structural details were identical to buildings of two or three storeys. I recall flying above New York several decades ago, and being surprised to see old skyscrapers topped by gabled roofs which were tiled and replete with chimney stacks. But today's automobiles and skyscrapers are a quite different story. The potential for change which they really represented has long since been realized.

My reason for starting this final chapter with those historical snippets has been to emphasize that radical modifications in the underlying design of an object do not necessarily imply immediate changes in its overall structure. Given time, however, the changes will come, and they may ultimately be quite sweeping ones. In the long run, as Marshall McLuhan pointed out, a change of medium inevitably leads to a change in the message.[2] As his famous sentence assured us, the medium *becomes* the message. A case in point that is particularly relevant to this book concerned the early days of electronic digital computing. Noting that these new devices were emerging, the British government appointed a committee to look into the question of how many such machines would be required. After due deliberation, the committee's learned members made their recommendations: it would be sufficient, they felt, if there were one computer in each of the four university cities, Cambridge, Oxford, London and Manchester, and a fifth machine at the National Physical Laboratory. Like the time-honoured generals who are always fighting the last war, these experts had overlooked McLuhan's dictum. As we now know, the computer generated its own new outlets, and it revolutionized not only scientific research but also accounting, banking, all manner of ticket offices, as well as the daily life of just about every secretary on Earth. In the case of the computer, the medium certainly did become the message.

A couple of decades ago, it would have been tantamount to treason in biological circles to suggest that the same could be true in that domain. The message was the supreme entity, and it was contained in the genetic material, be that DNA or RNA. According to what was called the central dogma of molecular biology, it was these molecules which determined everything about organisms. It is true that the thorny problem of tissue differentiation had not been solved, but no one doubted that the only route (in eukaryotes, say) went from DNA to messenger-RNA, to protein, and thence to cellular component, to cell, and finally to organism. Certainty began to crumble with the discovery of reverse transcriptase, and the existence of backward journeys. And when Barbara McClintock's long claimed jumping of genes finally became respectable, Pandora's box was suddenly seen to be wide open. Given that substances can penetrate the membranes of cells, the possibility arose that genetic events might be influenced by external agencies. The spectre of Lamarckism began to raise its head once more. Here too, at the microscopic level, the medium was threatening to become the message.

The agents of change at the molecular level are not like those feats of engineering cited earlier. They are random, and Nature evolves through a process of trial and error. But the changes in the message that are wrought by mutations have it in common with those early automobiles, skyscrapers and computers that their significance is first felt long after the new course has been set. Only in retrospect, can one appreciate that a bold new bough has been added to evolution's tree. It is through such branching that Nature explores the multi-dimensional space of possible organisms, but it manages to sample only a minute fraction. And because there is a requirement of continuity, in that new steps must always be taken from an established starting point, the available space is not probed haphazardly. This *cladistic* constraint means that the possibilities, in any era, are dictated by what went before, and usually *long* before. A new branch thus develops a life of its own. In this respect too, the medium becomes the message.

It was through the generation of such new branches that our Earth became populated by the various animal phyla and plant divisions that we know today. We share our own sub-phylum, vertebrata, with the fish, amphibians, reptiles, birds and other mammals, all of which are inexorably ratchetting themselves up to ever-higher levels of sophistication. Exploring their corner of that multi-dimensional space, they never know what lies ahead.

Signalling was an early arriver in this evolutionary maelstrom. But in the beginning, it merely took the primitive form of excretion, from the cellular membrane, of unwanted chemicals that happened to have a beneficial influence on the surroundings. Beneficial, that is, for the cell itself. Aquatic monocells have been discovered which eject a poison, and then absorb the dissolving carcasses of more advanced organisms that they have paralysed.[3] When the more benign forms of signalling arrived on the scene, they permitted cells to cooperate rather than compete. And the local passing of messages ultimately gave place to wholesale signalling in nervous systems. As with all else in evolution, there was no guiding forethought. It was only subsequently that systems of communicating neurons would be inventing those automobiles, building those skyscrapers, and designing those computers. Only then was the medium of nervous systems becoming the message of civilization. And only then were some nervous systems becoming the investigators of nervous systems themselves.

One frustration of the latter derives from the fact that brains do not fossilize. If they did, it would be an easier job to trace those branches back to the bifurcations which gave rise to them. In default of such information, progress has been slow in charting the brain's contribution to those searches of multi-dimensional space. That we mammals have a more advantageous ratio of brain weight to body weight than the birds, which are our closest competitors, is not in doubt (see figure 11.1). But just how our developments diverged is still pretty much a mystery. As R. Glenn Northcutt has emphasized, the lack of fossilized brains means that phylogenetic sequences for central nervous system characters have to be based on the patterns of variation of those characters in living organisms.[4] Phylogenetically, the pathway from the cortex to the nerve fibres in the spinal region emerged with the mammals. In the beginning, these projections served merely to modify sensory input to spinal interneurons, but for more advanced mammals one sees the more complicated arrangements that have been discussed many times in this book. A interesting intermediate example is provided by the hedgehog, because its somatosensory and motor cortices actually overlap to some extent.

If the logarithm of brain weight is plotted as a function of the logarithm of body weight, for the mammals, the result is an approximate straight line with a slope of two-thirds (see figure 11.2). In other words, the brain weights of mammals are proportional to the two-third powers of the body weights. Now body weight to the two-thirds power is proportional to body area, so brain weight is proportional to the surface area of the body. What is more, because the largest part of the brain is the cortex, which is a sheet of roughly constant thickness, brain weight is *very roughly* proportional

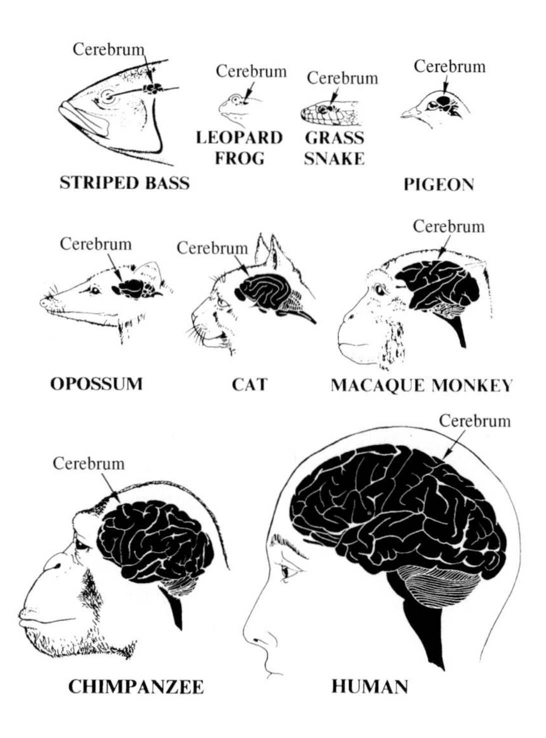

Figure 11.1
The evolutionary progression from bony fish to amphibians, reptiles, birds and finally mammals is accompanied by a gradual increase in the ratio of brain size to body weight. The increase in the relative size of the cerebrum is even more pronounced, as shown here. Within the mammals, there is a systematic increase in the relative size of the prefrontal region, and also of the cerebellar hemispheres, the closer the species is to our own.

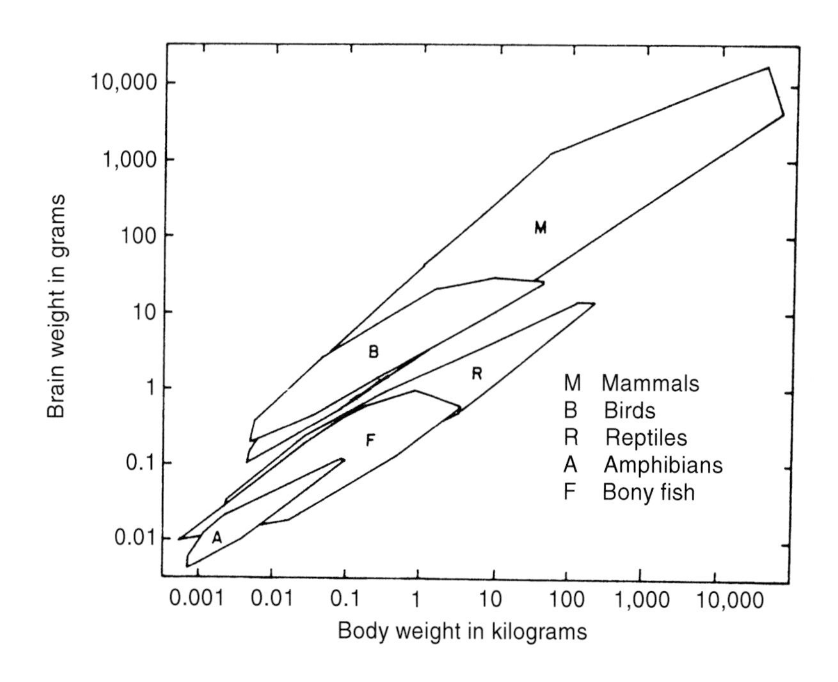

Figure 11.2
A plot of the logarithm of brain weight as a function of the logarithm of body weight is an approximate straight line with a slope of about two-thirds. Brain weight is thus proportional to the two-thirds power of body weight, which means that it is proportional to the body's surface area. This suggests that the brain is concerned with the interaction of the body's surface with the surrounding environment.

to cortical area.[a] So although this is a rather crude approximation, we see that cortical area is roughly proportional to the surface area of the body. In view of this, it is not surprising that the cortex comprises so many different maps of the body's surface, as do many sub-cortical structures.

This reminds us, once again, that personal space is important. In the phylogeny of species close to ourselves, the number of corticospinal neurons progressively increases from prosimians to monkeys, and then to anthropoid apes and finally to us. Likewise, the number of fibres conveying information to the cortex systematically increases. It is as if the cortex were providing a progressively larger antenna with which to detect correlations, and a progressively larger control panel with which to direct movement. In addition, and I would agree with Gerald Edelman on this point, there was increasing scope for the looping flow of information between the different cortical regions and areas. Indeed, the main message of this book has been that such re-entrant circuitry provided its owners with what might seem to have been initially only a marginal advantage. Perhaps it was just that, but by paradoxically requiring its possessor to be consciously aware, Nature was nudging evolution along a particularly promising branch of that multi-dimensional space: one that would ultimately produce human culture and technology.

John Searle's Chinese Room

The emergence of a *proof* that machine-based consciousness and intelligence is impossible would be important, if for no other reason than that it would release resources for other causes. An attempt at such a demonstration was made by John Searle, and it involves what has come to be known as the *Chinese Room*.[5] The gist of Searle's case hinges on the word *understanding*. He perceives a fundamental difference between what the artificial intelligence advocate means by the word, and thus demands of his machine, and what a human being believes understanding to be. Machines, Searle maintains, merely behave *as if* they understand, whereas we understand in an altogether deeper sense. A line is thus being drawn between our depth of reasoning and machine superficiality, and the Chinese Room was enlisted to lay bare the schism.

Searle's motivation for conceiving the Chinese Room test sprang from the development of clever computer programs that are apparently able to understand pieces of text, to the extent that they are able to draw inferences about things that are not explicitly stated. A typical example would be a little tale about a person going into a restaurant, ordering a hamburger, but subsequently leaving the premises without paying because the thing was served to him in a badly charred state. The program is able to conclude that the man did

a / About half the neurons in the human brain are located in the cerebellum, and this is almost exclusively composed of its cortex, so regarding the overall brain as a thin sheet *is* a reasonable first approximation.

not eat the hamburger, not because this fact is stated in the text, but rather because the software includes a dossier of typical aspects of such a restaurant scenario. The question thus arises as to whether the computer program really *understands* the story. Searle believes that its ability to handle questions on the subject does not constitute a proof of such comprehension.

Searle's refutation of such machine intelligence invokes the Chinese Room. This is a cubicle containing a comprehensive selection of Chinese symbols, a complete set of instructions, in English, for manipulating and combining Chinese characters, and a person who is able to communicate with the outside world only by passing combinations of such symbols through a slot in the room's wall, and receiving other combinations via the same route.[b] Searle's points are firstly that such a person will be in the same situation as the programmed computer, the instructions being the program, and secondly that such manipulations, however extensive, will be insufficient to give the person a true *understanding* of Chinese. To put it more succinctly, Searle emphasizes the distinction between manipulating the syntactical elements of a language and actually understanding the semantic concepts that this manipulation serves; it is a question of syntax *vis-à-vis* semantics.[c]

A computer program must, perforce, be composed of syntax. It cannot reach the depth of meaning that we are inclined to take for granted. This is why, in Searle's view, one cannot equate the mind to computer software any more than one can liken the brain to computer hardware. It is important not to misinterpret his statements. He is not saying that one cannot implement brain-like functions in a computer program, or on a silicon chip. His point is more profound, because it invokes the lack of causal powers in artificial devices. The significance of Searle's ideas should not be underestimated; they are not merely philosophical wrangling. Some have attempted to refute his argument, however. Let us consider a challenge mounted by Richard Gregory.[6]

One aim of artificial intelligence is understanding the intelligence of real brains through the concepts of engineering. The task is obviously difficult, for even relatively primitive things can involve physical phenomena that are far from trivial. When it comes to understanding, which was the crux of Searle's refutation of what he referred to as *strong AI*, we must cope with some-

b / Care is required concerning the interpretation of the Chinese Room's walls. They are clearly not comparable to the interface between the brain and rest of the body, since this would make a homunculus of the manipulator of those Chinese symbols. Neither can they be seen as being equivalent to the person–environment interface, as depicted in figure 9.7, for much the same reason. The walls are best regarded as boundaries in space–time, which thus enable them to circumscribe *events*.

c / We recall that *syntax* is the admissible combinations of words in phrases and sentences; it is what is commonly referred to as grammar. *Semantics* is concerned with the meanings that correspond to all words in a given language, and to all possible sentences.

thing which is quite beyond anything that technology can reliably measure. If Searle's pessimistic view is to be countered, therefore, it must be met on its own ground, namely that of analysis. This is what Richard Gregory has attempted to do.

Gregory's prescription was remarkably simple: *put a baby in the Chinese Room*. His point is that the infant would have an impossible task trying to learn the significance of the symbols, so it would fare no better than a machine. In other words, the Chinese Room is too restricted an environment to serve as a criterion for the acquisition of intelligence. We must not underestimate the significance of all the feedback that the growing infant receives by way of comment, encouragement, discouragement, reaction, exclamation, and the host of other human interactions which a large portion of its fumbling meets with every day. This is the stuff that human intelligence springs from, and I myself have stressed the importance of such self-initiated probing of the environment. Learning from one's own mistakes is a vital aspect of intelligence, and there seems no reason why such a facility could not be built into a computer program. It seems likely that a sufficiently well-constructed example would actually fare rather admirably in a Chinese Room.

Unless such a machine also had the same structure as ourselves, however, it would not mimic all aspects of our behaviour. And it is easy to overlook the importance *to us* of things that the computer engineer would never dream of trying to emulate. As I have stressed, our bodies are an indispensable part of our consciousness, as is our personal space. Designing a computer to simulate pain, hunger, anxiety and even elation, does not appear to be too tall an order. But such things are relatively primitive compared with the nuances of experience we take as commonplace. It is difficult to imagine a computer program being able to appreciate a good limerick, let alone *write* one. And when will there be devices capable of giving an appropriate answer to the excellent question composed by Martin Fischler and Oscar Firschein:[7] *if a young man of 20 can gather 10 pounds of blackberries in one day, and a young woman of 18 can gather 9, how many will they gather if they go out in the woods together?*

The sub-title of this book promises that the subject of consciousness in computers will be at least addressed. And it would not be unreasonable of the reader to anticipate that I would even make some predictions as to when such things will appear on the scene. I intend to do precisely that, after I have given a bird's-eye view of what the preceding chapters have told us.

The brain – the overall picture

The vital evolutionary step was, according to the view I have been advocating, the provision of a real-time choice mechanism. By this, I mean the ability to gauge the possible consequences of one's actions as events actually unfold, and take evasive action if need be. Let us begin by reiterating the difference between this ability and what the non-conscious animal can draw upon. Such lower animals also make choices, of course, but these are of the type made by those feed-forward neural networks we encountered in chapter 6. Likewise, and using the term favoured by Larry Squire,[8] amongst others, these creatures possess only the *implicit*, or *procedural*, variety of memory. Their sole type of reaction to an environmental stimulus is equivalent to our own unconscious reactions. This is not to say that these reactions are unimpressive. Our own procedural memories allow us to traverse unfamiliar musculo-sensory territory, by which I mean novel couplings of movement with resulting environmental feedback, and the unconscious creature can match us in this. Moreover, for both us and the unconscious creature, such unconscious probings lead to memory traces that are acquired gradually, over many repetitions; in our own case, there is no 'one shot learning' unless the probing leads to a breakthrough to consciousness.

We enjoy the tremendous advantage of being able to veto an action that has already been given the system's permission to proceed, even though that veto is exercised deterministically and unconsciously. I must emphasize that this is *not* the same as merely modifying one's muscular course as a response to changing external influences, such as when one of an action's immediate results is alteration of the sensory input, and the consequent sudden presentation of danger. The unconscious creature can obviously manage *that*. The real-time-choice I am referring to includes the vital extra ingredient of being able to react (deterministically) to stimuli that are generated *internally*, through the agency of short-term memory.

Now in order for such memories to be evoked, the system requires some form of priming, and what could be more appropriate than an arrangement whereby that priming is supplied by a combination of the intended motor act and the current sensory input. This will guarantee that the unconscious veto mechanism gets the best possible instantaneous information, under the prevailing circumstances. I made what I consider to be a strong case for the recall of relevant memories being provoked by planned motor movements (figure 11.3 is of interest, in this respect). Moreover, we noted that it is particularly efficient for the activation of relevant memories to recruit the same areas that are used for perception, this being supported by PET imaging.

I argued that such memory evocation requires motor planning to be *actually proceeding*, and that this will give rise to signals being passed around what I referred to as the vital triangle. The latter comprises the sensory cortex (that is, the occipital, parietal and temporal lobes), the premotor area of the frontal lobe, and the thalamic intralaminar nuclei. I also argued that the actual veto mechanism requires the simultaneous circulation of signals around a second triangle comprising the sensory cortex, the premotor area and the anterior cingulate, the latter region being known to be intimately associated with the perception of pain (see figure 9.10 – and most importantly, figure 11.4). Its inclusion in that second circuit is hardly surprising, given the great importance to the animal of any prospects of pain.

The final link in this circuitry, I argued, is the one via which the anterior cingulate exercises control over the nucleus reticularis thalami, the latter, in turn, providing the gating control of what is circulating around the vital triangle, and thus of what is at the centre of attention. We also noted that the memories invoked in the overall mechanism will be able to avail themselves of the added information provided by the reverse projections present in the sensory cortex, as discussed by Wolf Singer.

The biggest surprise came when we saw why this deterministic and unconscious choice (veto) mechanism paradoxically requires the animal to be consciously aware. Let us go over the argument again, couching it in terms of *navigation*, as I did in chapter 9. As Edgar Adrian,[9] Charles Sherrington[10] and Roger Sperry[11] all reminded us, the only thing an animal can do is move muscles. When it does so, and if it is to function as anything other than an automaton, its concern must also be with the here and now, and not exclusively with how the environment *usually* reacts to its muscular probings. Its need is access to the specific, rather than to the general.[d] The unconscious parallel processing that implicates procedural memory is not enough; the creature must also be able to tap into the sequential processing that implicates *explicit*, or *episodic*, memory.

This, in turn, requires the animal to retain, briefly, a record of what movements it has just performed, as well as a record of the resulting reactions of both the environment and of the creature's own body. A very important reporting system of those latter reactions, I have argued, is the muscular system's spindle apparatus, which provides its information via the internal reafference routes. Indeed, I have suggested that this apparatus plays a key role in the generation of the feelings we group under the name *qualia*. What I am saying, therefore, is that the so-called *hard problem of consciousness* identified by David Chalmers[12] is no more difficult to explain than any other aspect of the phenomenon, for the fact is that navigation in the above sense requires that consciousness and qualia are inextricably tied.

Almost ironically, this means that an adequate definition of consciousness, which Francis Crick and Christof Koch[13] fear is still not available, can be found in almost any good dictionary. When the *Concise Oxford Dictionary of Current English* tells us that consciousness is '[the] totality of a person's thoughts and feelings', it serves us handsomely. What the dictionary does not tell us, however, is how and why consciousness arose. And I suspect that the difficulties most people have with the subject really stem from that issue rather than from the obvious first-person products of the phenomenon. In this book, I have offered explanations of the how and the why, and the biggest surprise has been the indication that the brain's information processing is carried out unconsciously, as Max Velmans[14] had suggested. This means, in turn, that Roger Carpenter[15] had perceived the truth when he concluded that each of us is a captive audience of his or her own motor acts. It also strengthens my belief that the system is indeed a lowerarchy, as I conjectured in chapter 3.

It is interesting to note that the tight coupling between the alpha and gamma motor neurons is replaced by what one could call a negotiable

d / The addition of the specific to the general is indeed the *hallmark* of consciousness, according to this view.

Figure 11.3
To emphasize one of the key points made in this book, one should read this sentence and then pass directly to the footnote at the bottom of the page.

FINISHED FILES ARE THE RE-
SULT OF YEARS OF SCIENTIF-
IC STUDY COMBINED WITH
THE EXPERIENCE OF YEARS

interaction when one reaches the mammals. It is first in these that one has the spindles reporting to higher brain centres. And, as if on cue, it is in these animals that one sees the linking of the anterior cingulate with the nucleus reticularis thalami; the link is clearly visible in the rat, for example. The tentative indications are, therefore, that at least a primitive form of consciousness may be present in all mammals.

I continue to follow Roger Sperry in believing that the system's output tells us (indeed, *has* told us) more about its internal workings than has the input. And I shall continue to think of the real-time-choice, or veto-on-the-fly, mechanism as being under the control of what I have called the *master module*.[16] This module, which comprises the components shown in figure 9.10, has dictatorial dominion over our muscular movements; it is the system's best bet as an agency for keeping us out of harm's way. The brain is often referred to as a democracy, and mention of a dictatorship will no doubt prove distasteful to many. In retrospect, however, I do not feel that hypothesizing such a module was particularly daring, and it was emphatically not eccentric. After all, we already know that the brain *does* possess a master module, namely that which exercises higher control over the supply of those of its products destined for internal use: the hypothalamus. Why should there not also be a master module to control the system's external products, namely its muscular movements? I submit that if Nature unwittingly took that particular route through the hyperspace of possibilities, it merely happened to follow good engineering practice.

The master module concept offers a particularly transparent explanation of inter-modal binding, and of the unity of conscious experience. It also provides a straightforward explanation of the limitations Donald Broadbent[17] showed were present in the system, and one of the module's

Now read the sentence a second time, *from start to finish and not repeating any of the lines*, counting the number of times the letter F occurs. Then pass immediately to the footnote at the bottom of the following page.

ramifications is a neuronal interpretation of the breakthrough phenomena studied by Anne Treisman.[18] Moreover, when the interactions of the module with the prefrontal cortex are taken into account, one glimpses an anatomical and physiological basis for the working memory investigated by Allan Baddeley.[19] I feel that one of its particular attractions is the relatively simple manner in which it accounts for the temporal phenomena discovered by Benjamin Libet and his colleagues;[20] the linking of planned motor movements to what these would lead the sensory cortex to expect, by way of reafferent input, is one of Hermann Haken's self-organizing processes, and such processes inevitably consume a certain amount of time.[21] Likewise, the master module concept shows how the rivalry phenomena studied by David Leopold and Nikos Logothetis,[22] amongst others, fits directly into the Broadbent picture.

The elegance of the overall system is seen in its ability to permit different sets of motor patterns to be executed simultaneously, as when we hold a conversation while driving a car. It is fortunate for us that some of these patterns can safely be handled by the autopilot of familiarized procedures, with the valuable participation of the cerebellum. Indeed, the very word procedure was linked above to one form of memory. And when I say that conscious awareness requires motor planning to be actually proceeding, the link is strengthened further, because only if this condition is satisfied will expectations be generated which can commandeer attention if they are not fulfilled. The trained system husbands its resources, and when our unconscious mental processes detect no threat, automation reigns and any inconsequential detail that is lacking is merely compensated for, as in the process of filling in.

The system can cope with a great variety of situations, of course, and this sophistication demands the presence of numerous embellishments. The adjudication provided by that real-time choice would not be able to operate if it were not backed up by devices for starting and stopping movements, on the basis of detected advantages and disadvantages; hence the need for the amygdala, the limbic system and the basal ganglia (which might collectively function like a more sophisticated counterpart of an automobile clutch). And because some advantages and disadvantages can only be recognized as such if they are sensed in context, participation of the hippocampus confers its own special benefits.

The nuances of behaviour that are being rationalized by Philip Johnson-Laird and Keith Oatley[23] require fine orchestration of the couplings between motor planning and sensory reafference, and this is no doubt mediated by the impressive number of different neurotransmitters that the system has at its command. A contributing factor must also arise from the multiplicity

Letter F occurs six times. The reader should not feel embarrassed if half of the target letters were missed; the majority of people make the same error, which arises because the three overlooked letters are pronounced as if they were Vs. This example is important, because it shows that the searching process draws on sequenced muscular movements (in this case, the ones used in articulation) rather than on visual images.

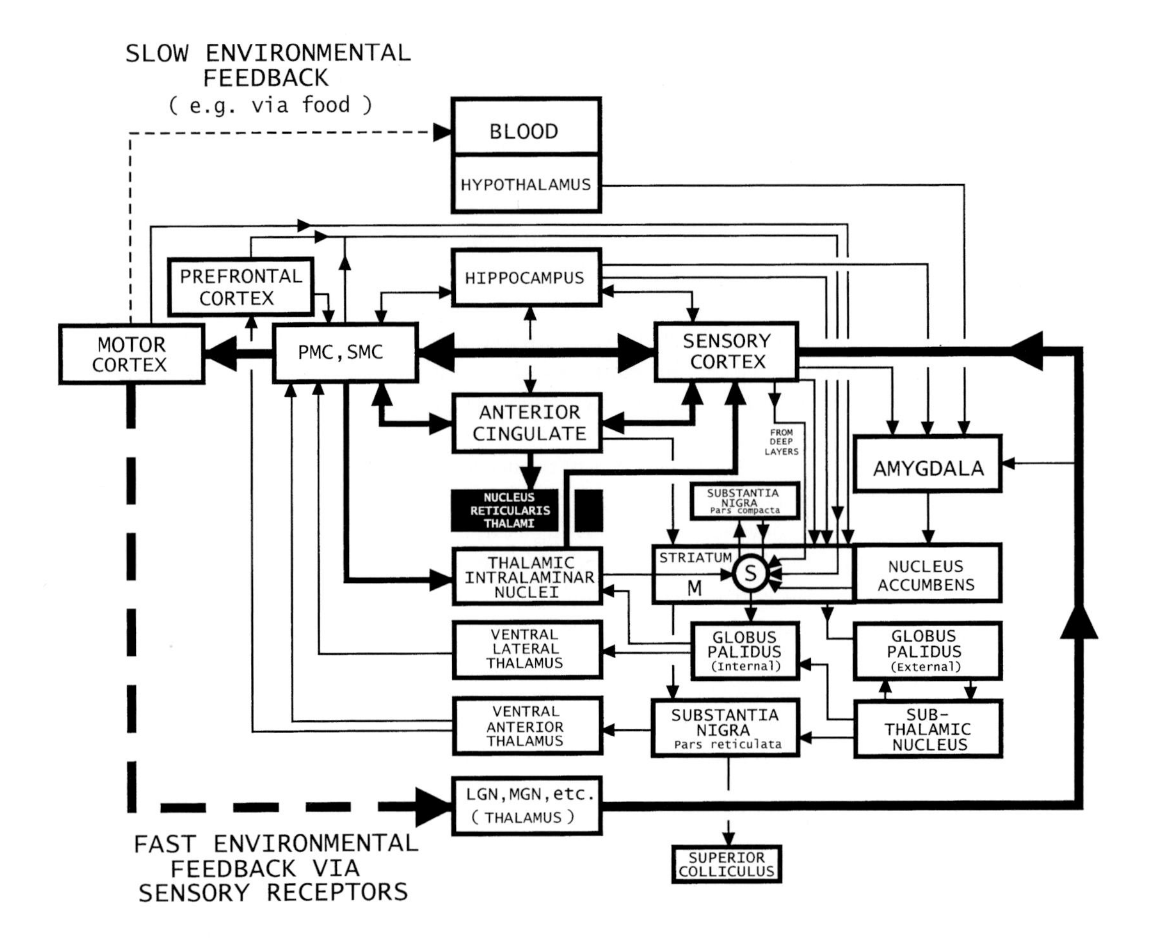

Figure 11.4
The locations of some of the components of the cerebral cortex, the basal ganglia and the limbic system, in this highly schematic diagram, do not reflect their actual positions in the human brain. They have been placed in this manner so as to minimize the amount of overlap in the inter-component connections. Moreover, several such connections have been omitted, in order to emphasize the interactions believed to be most intimately associated with conscious and unconscious mental processes. These omissions include the connections through which the amygdala influences the hypothalamus, and the related ones via which the latter regulates the endocrine system, by way of the pituitary gland. The reverse projections that run from the sensory cortex back to the thalamus have also been left out. Finally, no attempt has been made to include the reflex routes, which mediate direct coupling of sensory stimulus to motor response.

As stressed in this book, we acquire knowledge of the environment only through our own motor explorations. These produce feedback that is returned rapidly through the various senses, and also more slowly through the consequences of those explorations, such as the effect on blood sugar of our ingestion of food. The inter-component connections involved in the rapid feedback are indicated by the thickest lines, which are seen to form a closed loop (LGN denoting the lateral geniculate nucleus and MGN its medial counterpart). Every motor act can be regarded as a question put to the environment, such as – *is the ground still there?* – for every stride that we take. But only when we encounter something that the system did not predict – a hole, perhaps – does the question-and-answer event break through to consciousness. The *un*conscious capture of correlations between sequences of movements and the resulting feedback is mediated by the reverse connections from the premotor cortex (PMC) and the supplementary cortex (SMC) to the sensory cortex (see the centrally lying double-arrowed portion of the thick-line circuit). This produces an unconscious form of short-term memory, which can ultimately contribute to long-term procedural (implicit) memory. (An even more direct form of correlation capture, which links elemental motor acts to routine sensory feedback, strengthens connections that link the motor cortex to the relevant sensory areas; an example is the mutual touching of fingers and palm, when clenching the fist.)

Conscious mental processing is conjectured to additionally involve participation of the circuitry indicated by the lines having intermediate thickness (see also figure 9.10), with the loop through the thalamic intralaminar nuclei being an essential participant. Part of this circuit is under the control of the nucleus reticularis thalami, which, in turn, is influenced by the brainstem reticular formation (not shown) and the anterior cingulate. The latter component is known to be implicated in pain perception, and is thus well suited to influence conscious attention. But retention of consciousness by bilateral cingulotomy patients shows that the anterior cingulate is not an obligatory participant in the phenomenon. Neither is the hippocampus, according to clinical observations, but it is strategically placed to capture correlations between activity in the PMC/SMC and the sensory cortex. Such capture would parallel the *un*conscious variety mentioned above, and it might be conditional on the presence of theta oscillations generated in the septum, and ultimately provoked by the limbic system (see figure 5.9 and the nearby text). A single theta oscillation lasts about as long as an individual cognitive event, and the difference between unconsciously and consciously acquired memories may lie in the contextual dimension of the latter. Subsequent (and possibly periodic) activation of the hippocampus may contribute to the mechanism whereby such consciously captured correlations are consolidated into long-term declarative (explicit) memory. (This may happen preferentially during REM sleep.) But the hippocampus is not required for their later recall. The recorded linking of a specific sequence of muscular movements to the corresponding sensory input is believed to be the basis of each elemental schema, and the sequential activation of context-related schemata may underlie thought. This does *not* imply that all we ever think about is movement, but it strongly suggests that a *proxied* version of a motor sequence *is* associated with every thought.

We can think without acting, act without thinking, act and simultaneously think about that act, and act while thinking about something else. Our acts can be composite affairs, in which we concurrently achieve several things, though we appear not to be able to simultaneously maintain two streams of thought. Such impressive flexibility would seem to demand that the system possesses a sophisticated counterpart of the familiar clutch mechanism found in some automobiles: one that comprises two circuits which are nevertheless able to cooperate, and thereby divide mental and motor tasks between them. It is possible that an important contribution is made by the striatum. This has a heterogeneous structure, in which the continuous matrix (M) is interdigitated with the isolated patches known as striosomes (S), here represented by a single member. The input to the striosomes seems to be the more special, since it is fed by the thalamic intralaminar nuclei, invoked above in the discussion of consciousness, the emotion-related amygdala, the PMA/SMA and prefrontal cortex, and the *deep* layers of the sensory cortex; of the two components, the striosomes may be more intimately linked to consciousness. The matrix, conversely, appears to be more directly related to raw movement and its sensory consequences. The periodic shifting of attention, as when simultaneously holding a conversation (or merely thinking) and driving in a busy thoroughfare, would make considerable demands on the underlying differential control mechanism, and this could be the duty of the substantia nigra pars compacta and the sub-thalamic nucleus, which appear to serve as gain controls for the striosome-related and matrix-related routes, respectively.

of cortical areas serving most sensory modalities, the extraordinary diversification uncovered by Semir Zeki,[24] in the visual system, being the prize example.

Those frontal-lobe areas lying rostral to the premotor area serve the system by supplying what has been called the blackboard of working memory. It is there that the predictions made by other parts of the system are briefly deposited, thereby being made available for the modification of those motor-to-reafference couplings, and the consequent acquisition of new schemata. But the system is perpetually fighting against the fading of these short-term memory traces, and individuals with impairments of this part of the brain must live with below average intelligence. Turning once more to the *Concise Oxford Dictionary of Current English*, we read that the definition of this attribute is 'intellect, understanding', and here too it would be difficult to find an objection. But as with consciousness, it is not the description of the phenomenon that gives us trouble, only its origins and determinants. In this book, I have attempted to identify those origins and determinants.

The cerebellar hemispheres, which do Trojan service for the above-mentioned auto-pilot functions, and the prefrontal region, which exploits, in the cause of higher things, the resources thereby released, are particularly prominent in our lucky species; they are the great contributors to intelligence. It is they which make possible the embedding in inter-human transactions that are so wonderfully present in the writings of Jane Austen, say, and so woefully absent in the average soap opera. But every member of *Homo sapiens sapiens* who has acquired sufficient proficiency with mental manipulations has had to pay the price of knowledge of his own impending death. So John Steinbeck had perceived a sad truth when he wrote of consciousness being a tragic miracle.

Finally, in this brief survey of the whole picture, as it appears to the author, let us turn to a particularly venerable issue. The flow of signals around that core circuit, and their modification through interactions with other brain components, has been conjectured to underlie thought itself, this being perceived to be nothing other than a sort of internally generated, but *out-there-experienced*, masque of the body's transactions with the surroundings, including the latter's human components. Seen in this light, there is no longer a *problem* of body/mind, only a *situation* of body/ability-to-simulate-body's-interactions-with-environment.

Consciousness will be seen in computers

From what has been stated regarding the phenomenon, it will be clear that there could be no disembodied consciousness. This makes it rather unlikely that computers as such could be given consciousness merely through the use of a specific type of software. There would have to be something that could be likened to a body, equipped with counterparts of our muscle-monitoring apparatus. Moreover, the state of the body would have to have relevance for the overall system, and there would have to be some form of internal adjudication of the significance of currently impinging stimuli.

These statements might seem to clash with this final section's title. There is no conflict, however, because a computer could also be used to *simulate* consciousness. *Computer simulation* has become a familiar aspect of modern science, and one could identify it as a *bone fide* third branch of the scientific endeavour, to be reckoned on a par with the experimental and theoretical enterprises; computer simulation could be called *theoretical experimentation*. For those who are not conversant with this technique, let me use weather forecasting as an example. Around the mid-1980s, it became apparent that meteorological experts could now reliably predict weather changes as much as five days in advance. A major contribution to this progress was provided by computer simulation. The approach involves the setting up of a model of the Earth's atmosphere, and investigation of the temporal evolution of the various parameters. The technique appears to have reached its limit, in fact, because of the inevitable intrusion of chaotic factors.

To say that one is simulating weather in a computer is obviously not the same as saying that the weather is actually *in* the computer. In precisely the same manner, one could *simulate* consciousness in a computer, without maintaining that the computer itself was conscious; whence this section's carefully worded title.

What, then, is my prescription for simulating consciousness in a computer. That is one of the things this entire book has been leading up to, because the only reliable way to achieve this goal would be to emulate the anatomy and physiology of the only structure we can be sure is conscious: the human brain. Given the complexity of the latter, this prescription might seem to dodge the issue; how could one ever hope to build in so much detail, given that such a brain comprises a hundred billion neurons. I am not convinced that the task has to be all that daunting, however, because one should bear in mind how many different tasks the human brain serves. One would not have to simulate a complete human being in order to mimic basic consciousness; something far more primitive would probably suffice, as long as the essential components were present.

What would these include? I suspect that a couple of primitive senses, and a couple of internal monitors, would be a good starting point, and the simulated creature could probe a relatively simple environment. One would naturally build in the possibility of real-time-choice, that veto-on-the-fly process. The sensing of the external and internal environments (including the states of the simulated body's muscles) would have also been built into the model, of course, as would the necessary spans of iconic and short-term memory. There would also have to be provision for the consolidation of selected short-term memories into long-term memory. Finally, there would have to be a suitable mechanism for adjudication of the significance of sensory inputs, as well as a mechanism for working memory.

These would be only the preliminaries, however, because one would now have to investigate whether the simulated creature was capable of thought. Given that thought is essentially simulation of the body's interactions with the environment, as I have said, this means that the computer would be simulating simulation. Ultimately, one would be looking for evidence of intelligence. One would be impressed, I am sure, by behaviour that smacked of an ability to draw non-obvious conclusions about the environment.

The creature would naturally acquire its expertise through its own probings of the surrounds, and it is interesting to contemplate how many of these we humans can manage in a lifetime. On the basis of a generous allowance of fifty formative years, and assuming one meaningful probe of the environment every ten seconds, which is certainly a prodigious rate, throughout a sixteen-hour day, one arrives at a figure of about a hundred million probings in a lifetime. And one must bear in mind all the diverse things that our senses must take in. This is not a particularly impressive number.

With the speed of modern computers, it would be a relatively simple matter to give one's simulated creature far more opportunities for environmental probings. If one had also built reverse projections into one's modelled brain, to permit the sort of correlation-capturing mechanisms variously discussed by Gerald Edelman[25] and Wolf Singer,[26] the creature would have the possibility of detecting features of its environment which might even have escaped the attention of its human master. One could then find that the creature was, of its own volition, doing what Margaret Boden has suggested lies at the root of creativity: *varying the variable.*[27] I can give the reader a watertight guarantee that computer simulators will be trying to do all this, because this is precisely what my students and I are already attempting. The coming years promise to be exciting ones, for us and for all the other research teams trying to accomplish the same sort of thing. But I must emphasize that our goals are nevertheless rather modest. All we will be looking for is evidence of intelligent behaviour, and success in this venture will be a sufficient reward.

We humans appear to be mesmerized by the prospect of artificially producing copies of ourselves, and in this book's introduction I put this down to the reinforcing factor of our culture, not the least literature, and latterly the film. We now know enough about genetics to be able to size up the magnitude of such a task, and it is daunting to say the least. One has only to contemplate the huge international project that will soon have mapped the entire human genome, and this is merely an exercise in biological cartography. Surely the more sensible goal is to continue refining artificial neural networks so that they can relieve us of the routine and drudgery of everyday life, not to mention the taking over of all sorts of tasks that are dangerous to humans, in one way or another.

I believe that computer programs will be written which will have counterparts of at least the basic human emotions, if this is required to let them function more effectively. This would be a most interesting development, but it is difficult to see why it should also be looked upon as a threat.

There is, in any event, a much easier way of making a brain.

Bibliography

Chapter 1

1 / Sherrington, C. S. (1940). *Man on his Nature* (Cambridge: Cambridge University Press). This is a transcript of the Gifford Lectures delivered in Edinburgh, 1937–38.

2 / Eysenck, H. J. (1979). *The Structure and Measurement of Intelligence* (Berlin: Springer-Verlag); Kail, R. & Pellegrino, J. W. (1985). *Human Intelligence – Perspectives and Prospects* (New York: Freeman); Sternberg, R. J. & Detterman, D. K. eds. (1986). *What is Intelligence? – Contemporary Viewpoints on Its Nature and Definition* (Norwood, NJ: Ablex); Anderson; M. (1992). *Intelligence and Development – A Cognitive Theory* (Oxford: Blackwell); Khalfa, J. ed. (1994). *What is Intelligence?* (Cambridge: Cambridge University Press).

3 / Binet, A. (1909). *Les Idées Modernes sur les Enfants* (Paris: Ernest Flammarion); Binet, A. & Simon, T. (1908). Le développement de l'intelligence chez les enfants. *Année Psychologie*, **14**, 1–94.

4 / Turing, A. (1950). Computing machinery and intelligence. *Mind*, **59**, 433–60.

5 / Kulli, J. & Koch, C. (1991). Does anesthesia cause loss of consciousness? *Trends in Neurosciences*, **14**, 6–10.

6 / Sutherland, N. S. (1989). In N. S. Sutherland ed. *The Macmillan Dictionary of Psychology* (London: Macmillan).

7 / Dennett, D. C. (1987). In R. L. Gregory ed. *The Oxford Companion to The Mind* (Oxford: Oxford University Press).

Chapter 2

1 / Eccles, J. C. & Gibson, W. C. (1979). *Sherrington – His Life and Thought* (Berlin: Springer International). This includes an account of Sherrington's Linacre Lecture, delivered at St John's College, Cambridge, on 6 May 1924.

2 / Koestler, A. (1967). *The Ghost in the Machine* (New York: Random House); Ryle, G. (1949). *The Concept of Mind* (New York: Barnes & Noble).

3 / Singer, C. (1956). *Galen on Anatomical Procedures* (London: Oxford University Press).

4 / Woollam, D. H. M. (1958). In F. N. L. Poynter ed. *The History and Philosophy of the Brain and its Functions* (Oxford: Blackwell).

5 / McMurrich, J. P. (1930). *Leonardo da Vinci the Anatomist (1452–1519)* (Baltimore: Williams & Wilkins).

6 / Debus, A. G. (1965). *The English Paracelsians* (London: Oldbourne).

7 / Singer, C. (1952). *Vesalius on the Human Brain* (London: Oxford University Press).

8 / Kosslyn, S. M. (1975). Information representation in visual images. *Cognitive Psychology*, **7**, 341–70; Kosslyn, S. M. (1980). *Image and Mind* (Cambridge, MA: Harvard University Press).

9 / Perky, C. W. (1910). An experimental study of imagination. *American Journal of Psychology*, **21**, 422–52.

10 / Hamlyn, D. (1990). *In and Out of the Black Box* (Oxford: Basil Blackwell).

11 / Swade, D. D. (1993). Redeeming Charles Babbage's Mechanical Computer. *Scientific American*, **268(2)**, 62–7.

12 / Mackintosh, A. R. (1988). Dr. Atanasoff's Computer. *Scientific American*, **259(2)**, 90–6.

13 / Boring, E. G. (1930). A new ambiguous figure. *American Journal of Psychology*, **42**, 444–5.

Chapter 3

1 / Breasted, J. H. (1930). *The Edwin Smith Surgical Papyrus* (Chicago: University of Chicago Press).

2 / Descartes, R. (1664). *L'Homme* (Paris: Charles Angot).

3 / Steensen, N. (1965). Trans. G. Scherz *Lecture on the*

Anatomy of the Brain (1669) (Copenhagen: Arnold Busck).

4 / Offray de la Mettrie, J. (1966). *L'homme-machine* (Paris: Pauvert).

5 / Harlow, J. (1868). Recovery after severe injury to the head. *Massachusetts Medical Society Publications*, **2**, 327–47.

6 / Head, H. (1926). *Aphasia and Kindred Disorders of Speech* (Cambridge: Cambridge University Press).

7 / von Bonin, G. (1960). *Some Papers on the Cerebral Cortex* (Springfield, ILL: Thomas); Wilkins, H. (1963). Neurosurgical classics XII. *Journal of Neurosurgery*, **20**, 904–16; Ferrier, D. (1878). *The Localization of Cerebral Disease* (London: Smith, Elder).

8 / Adrian, E. D. (1943). Sensory areas of the brain. *The Lancet*, **245**, 33–5. The transcript of his 1943 Sharpey-Schafer memorial lecture delivered to the faculty of Medicine, Edinburgh University.

9 / Wernicke, C. (1900). *Grundriss der Psychiatrie in klinischen Vorlesungen* (Leipzig: Thieme).

10 / Bianchi, L. (1922). Trans. J. H. Macdonald *The Mechanisms of the Brain and the Function of the Frontal Lobes* (Edinburgh: Livingstone).

11 / Dejerine, J. (1900). *Semiologie des Affections du Systeme Nerveux* (Paris: Masson).

12 / Sperry, R. (1982). Some effects of disconnecting the cerebral hemispheres. *Science*, **217**, 1223–6.

13 / Kandel, E. R., Schwartz, J. H. & Jessel, T. M. (1991). *Principles of Neural Science* (New York: Elsevier).

14 / Bernard, C. (1857). *Lecons sur les Effets des Substances Toxiques et Medicamenteuses* (Paris: Bailliere et Fils).

15 / Sherrington, C. S. (1897). On reciprocal innervation of antagonistic muscles. *Proceedings of the Royal Society*, **60**, 414–7; Sherrington, C. S. & Laslett, E. E. (1903). Observations on some spinal reflexes and the interconnection of spinal segments. *Journal of Physiology*, **29**, 58–96; Sherrington, C. S., Creed, R. S., Denny-Brown, D. E., Eccles, J. C. & Liddell, E. G. T. (1932). *Reflex Activity of the Spinal Cord* (London: Oxford University Press). Sherrington's early studies of the reflex arc extended the work of Jiri Prochaska, around 1780, and Ivan Sechenov (described by Pavlov as the father of Russian physiology), around 1865.

16 / Houk, J. C. & Rymer, W. Z. (1981). Neural control of muscle length and tension. In V. B. Brooks ed. *Handbook of Physiology, Section 1: The Nervous System, Volume II: Motor Control, Part 1* (Bethesda, MD: American Physiological Society).

17 / Eckmiller, R. (1987). Neural control of pursuit eye movements. *Physiological Reviews*, **67**, 797–857; Guitton, D. (1992). Control of eye–head coordination during orienting gaze shifts. *Trends in Neurosciences*, **15**, 174–9; Cheron, G., Mettens, P., & Godaux, E. (1992). Gaze holding defect induced by injections of ketamine in the cat brainstem. *NeuroReport*, **3**, 97–100.

18 / Brodmann, K. (1909). *Vergleichende Lokalisationslehre der Grosshirnrinde in ihren Prinzipien dargestellt auf Grund des Zellenbaues* (Leipzig: Barth); Haymaker, W. & Schiller, F. (1970). *The Founders of Neurology* (Springfield, ILL: Thomas).

19 / Hubel, D. H. & Wiesel, T. N. (1962). Receptive fields, binocular interaction and functional architecture in the cat's visual cortex. *Journal of Physiology*, **160**, 106–154; Hubel, D. H. (1988). *Eye, Brain and Vision* (Oxford: Freeman).

20 / Zeki, S. (1992). The visual image of mind and brain. *Scientific American*, **267(3)**, 42–51.

21 / Livingstone, M. S. & Hubel, D. H. (1984). Anatomy and physiology of a colour system in the primate visual cortex. *Journal of Neuroscience*, **4**, 309–56.

22 / Damasio, A. R. & Damasio, H. (1992). Brain and Language. *Scientific American*, **267(3)**, 62–71.

23 / Martin, K. A. C. (1992). Visual cortex: parallel pathways converge. *Current Biology*, **2**, 555–7.

24 / Sherrington, C. S. (1947). *The Integrative Action of the Nervous System* (New Haven, CN: Yale University Press).

25 / Lassen, N. A., Ingvar, D. H. & Skinhøj, E. (1978). Brain function and blood flow. *Scientific American*, **239(4)**, 50–59; Petersen,, S. E., Fox, P. T., Posner, M. I., Mintun, M. & Raichle, M. E. (1988). Positron emission tomographic studies of the cortical anatomy of single-word processing. *Nature*, **331**, 585–9.

26 / Berger, H. (1929). Uber das Elektroenkephalogramm des Menschen. *Archiv für Psychiatrie und Nervenkrankeiten*, **87**, 527–70; Basar, E. (1980). *EEG-Brain Dynamics* (Amsterdam: Elsevier/North-Holland); Moore, E. J. (1983). *Bases of*

Auditory Brain-Stem Evoked Responses (New York: Grune & Stratton).

27 / Tank, D. W., Ogawa, S. & Ugurbil, K. (1992). Mapping the brain with MRI. *Current Biology*, **2**, 525–8; Ogawa, S., Tank, D. W., Menon, R., Ellermann, J. M., Kim, S-G., Merkle, H. & Ugurbil, K. (1992). Intrinsic signal changes accompanying sensory stimulation: functional brain mapping with magnetic resonance imaging. *Proceedings of the National Academy of Sciences, USA*, **89**, 5951–5.

28 / Reitboeck, H. J., Pabst, M. & Eckhorn, R. (1988). Texture description in the time domain. In R. M. J. Cotterill ed. *Computer Simulation in Brain Science* (Cambridge: Cambridge University Press).

29 / Darwin, C. (1872). *The Expression of the Emotions in Man and the Animals* (London: John Murray).

30 / Oatley, K. & Johnson-Laird, P. N. (1987). Towards a cognitive theory of emotion. *Cognition and Emotion*, **1**, 29–50.

31 / Cotterill, R. M. J. (1990). Is the brain a lowerarchy? In H. Bohr ed. *Characterizing Complex Systems* (Singapore: World Scientific).

Chapter 4

1 / van Leeuwenhoek, A. (1693). *Philosophical Transactions of the Royal Society*, **17**, 949–60.

2 / Schwann, T. (1847). Trans. H. Smith *Microscopical Researches* (London: Sydenham Society).

3 / Golgi, C. (1908). La doctrine du neurone. In *Les Prix Nobel en 1906* (Stockholm: Norstedt & Sons); Ramón y Cajal, S. (1908). Structure et connexions des neurones. In *Les Prix Nobel en 1906* (Stockholm: Norstedt & Sons).

4 / von Waldeyer-Hartz, H. (1920). *Lebenserinnerungen* (Bonn: Cohn).

5 / Du Bois-Reymond, E. (1884). *Untersuchungen uber thierische Elektrizitat* (Berlin: Reimer).

6 / Galvani, L. (1791). *De viribus electricitatis in motu musculari commentarius* (Ex typographia Instituti Scientiarum) – English trans. by M. G. Foley *Commentary on the Effects of Electricity on Muscular Motion* (Norwalk, CN: Burndy Library, 1953).

7 / von Helmholtz, H. (1852). Messungen uber Fortpflanzungsgeschwindigkeit der Reizung in der Nerven, zweite Reihe. *Archives of Anatomy and Physiology*, 199–216.

8 / Adrian, E. (1913–1914). The all-or-none principle in nerve. *Journal of Physiology, London*, **47**, 460–74.

9 / Adrian, E. (1947). *The Physical Background of Perception* (Oxford: Clarendon Press).

10 / Hodgkin, A. L. & Huxley, A. F. (1952). A quantitative description of membrane current and its application to conduction and excitation in nerve. *Journal of Physiology (London)*, **117**, 500–44.

11 / Sigworth, F. J. & Neher, E. (1980). Single Na^+ channel currents observed in cultured rat muscle cells. *Nature*, **287**, 447–9.

12 / Skou, J. C. (1957). The influence of some cations on an adenosine triphosphate from peripheral nerves. *Biochimica et Biophysica Acta*, **23**, 394–401.

13 / Sherrington, C. S. (1951). *Man on His Nature* (London: Cambridge University Press).

14 / Bernard, C. (1857). *Lecons sur les Effets des Substances Toxiques et Medicamenteuses* (Paris: Bailliere et Fils).

15 / Loewi, O. (1935). On problems connected with the principle of humoral transmission of nerve impulses. *Proceedings of the Royal Society, Series B*, **118**, 299–316. This was Loewi's Ferrier Lecture.

16 / Katz, B. & Miledi, R. (1963). A study of spontaneous miniature potentials in spinal motoneurones. *Journal of Physiology (London)*, **168**, 389–422; Fatt, P. & Katz, B. (1952). Spontaneous subthreshold activity at motor nerve endings. *Journal of Physiology (London)*, **117**, 109–28; Katz, B. (1966). *Nerve, Muscle and Synapse* (New York: McGraw-Hill) – a pocket-sized gem.

17 / Dale, H. (1953). *Adventures in Physiology* (Oxford: Pergamon Press).

18 / Collingridge, G. L., Kehl, S. L. & McLennan, H. (1983). *Journal of Physiology*, **334**, 34–46; Dingledine, R. (1983). *Journal of Physiology*, **334**, 385–405; Lynch, G., Larson, J., Kelso, S., Barrionuevo, G. & Schottler, F. (1983). *Nature*, **305**, 719–21; Müller, W. & Connor, J. A. (1991). Dendritic spines as individual neuronal compartments for Ca responses. *Nature*, **354**, 73–6; Guthrie, P. B., Segal, M. & Kater, S. B. (1991). Independent regulation of calcium revealed by imaging dendritic spines. *Nature*, **354**, 76–80.

19 / McCulloch, W. S. & Pitts, W. (1943). A logical calculus of the ideas immanent in nervous activity. *Bulletin of Mathematical Biophysics*, **5**, 115–33.

20 / Caianiello, E. (1961). Outline of a theory of thought processes and thinking machines. *Journal of Theoretical Biology*, **1**, 204–35.

21 / Lashley, K. S. (1950). In search of the engram. *Society of Experimental Biology Symposium, No.4: Psychological Mechanisms in Animal Behavior* (Cambridge: Cambridge University Press), pages 454–5; 468–73; 477–80.

22 / Barlow, H. (1971). Single units and sensation: a neuron doctrine for perceptual psychology. *Perception*, **1**, 371–94. This is where Barlow's famous *grandmother cell* first saw the light of day.

23 / Feldman, J. A. & Ballard, D. H. (1982). Connectionist models and their properties. *Cognitive Science*, **6**, 205–54.

24 / Nicoll, R. A. (1982). Neurotransmitters can say more than just 'yes' and 'no'. *Trends in Neurosciences*, **5**, 369–74.

25 / Cooper, J. R., Bloom, F. E. & Roth, R. H. (1991). *The Biochemical Basis of Neuropharmacology* (New York: Oxford University Press).

Chapter 5

1 / Hebb, D. O. (1949). *The Organization of Behavior* (New York: Wiley).

2 / Kandel, E. R. (1979). Small systems of neurons. *Scientific American*, **241(3)**, 66–76; Kandel, E. R. & Hawkins, R. D. (1992). The biological basis of learning and individuality. *Scientific American*, **267(3)**, 52–60.

3 / Hawkins, R. D., Abrams, T. W., Carew, T. J. & Kandel, E. R. (1983). A cellular mechanism of classical conditioning in *Aplysia*: activity-dependent amplification of presynaptic facilitation. *Science*, **219**, 400–5.

4 / Pavlov, I. (1941). Trans. W. H. Gannt. *Lectures on Conditioned Reflexes* (London: Lawrence).

5 / Wigström, H., Gustafsson, B., Huang, Y. Y. & Abraham, W. C. (1986). Hippocampal long-lasting potentiation is induced by pairing single volleys with intracellularly injected depolarizing current pulses. *Acta Physiologica Scandinavia*, **126**, 317–19.

6 / Kelso, S. R., Ganong, A. H. & Brown, T. H. (1986). *Proceedings of the National Academy of Sciences USA*, 83, 5326–30.

7 / Artola, A. & Singer, W. (1987). *Nature*, **330**, 649–52.

8 / Bliss, T. V. P. & Lømo, T. (1973). Long-lasting potentiation of synaptic transmission in the dentate area of the anaesthetized rabbit following stimulation of the perforant path. *Journal of Physiology*, **232**, 331–56.

9 / Watkins, J. C. & Collingridge, G. L. eds. (1989). *The NMDA Receptor* (Oxford: IRL Press).

10 / Lynch, G. & Baudry, M. (1991). Reevaluating the constraints on hypotheses regarding LTP expression. *Hippocampus*, **1**, 9–14.

11 / Bekkers, J. M. & Stevens, C. F. (1990). Presynaptic mechanism for long-term potentiation in the hippocampus. *Nature*, **346**, 724–9; Malinow, R. & Tsien, R. W. (1990). Presynaptic enhancement shown by whole-cell recordings of long-term potentiation in hippocampal slices. *Nature*, **346**, 177–80; Bliss, T. V. P. (1990). Maintainance is presynaptic. *Nature*, **346**, 698–9.

12 / Snyder, S. H. & Bredt, D. S. (1992). Biological roles of nitric oxide. *Scientific American*, **266**, 28–35.

13 / Kandel, E. R. & Hawkins, R. D. (1992). The biological basis of learning and individuality. *Scientific American*, **267(3)**, 79–86; Zhuo, M, Small, S. A., Kandel, E. R. & Hawkins, R. D. (1993). Nitric-oxide and carbon-monoxide produce activity-dependent long-term synaptic enhancement in hippocampus. *Science*, **260**, 1946–50.

14 / Morris, R. G. M. (1984). Developments of a water-maze procedure for studying spatial learning in the rat. *Journal of Neuroscience Methods*, **11**, 47–60.

15 / Silva, A. J., Stevens, C. F., Tonegawa, S. & Wang, Y. (1992). Deficient hippocampal long-term potentiation in alpha-calcium-calmodulin kinase II mutant mice. *Science*, **257**, 201–6; Silva, A. J., Paylor, R., Wehner, J. M. & Tonegawa, S (1992). Impaired spatial learning in alpha-calcium-calmodulin kinase II mutant mice. *Science*, **257**, 206–11.

16 / Changeux, J. P. & Danchin, A. (1976). Selective stabilization of developing synapses as a mechanism for the specification of neuronal networks. *Nature*, **264**, 705–12.

17 / Rose, S. R. (1989). Can memory be the brain's Rosetta Stone? In R. M. J. Cotterill ed. *Models of Brain*

Function (Cambridge: Cambridge University Press).

18 / Rose, S. (1992). *The Making of Memory* (London: Bantam).

19 / Merzenich, M. M. (1989). Representational plasticity in somatosensory and motor cortical fields. *Biomedical Research*, **10**, 85–6.

20 / Gilbert, C. D. & Wiesel, T. N. (1992). Receptive-filed dynamics in adult primary visual cortex. *Nature*, **356**, 150–2.

21 / Perkel, D. H. & Perkel, D. J. (1985). Dendritic spines: role of active membrane in modulating synaptic efficacy. *Brain Research*, **325**, 331–5; Pongrácz, F. (1985). The function of dendritic spines: a theoretical study. *Neuroscience*, **15**, 933–46.

22 / Standing, L. (1973). Learning 10,000 pictures. *Quarterly Journal of Experimental Psychology*, **25**, 207–22.

23 / Steinbuch, K. (1961). Die Lernmatrix. *Kybernetik*, **1**, 36–45; Taylor, W. K. (1959). Pattern recognition by means of analogous automatic apparatus. *Proceedings of the IEE, London*, **106B**, 168–72; Taylor, W. K. (1985). Electrical simulation of some nervous system functional activities. In C. Cherry ed. *Information Theory* (London: Butterworths); Taylor, W. K. (1968). Machines that learn. *Science Journal*, **102**, part 6.

24 / Willshaw, D., Buneman, O. P. & Longuet-Higgins, C. (1969). Non-holographic associative memory. *Nature*, **222**, 960–2; Willshaw, D. (1972). A simple model capable of inductive generalization. *Proceedings of the Royal Society, Series B*, **182**, 233–47.

25 / Kohonen, T. (1987). *Content-Addressable Memories* (Berlin: Springer); Kohonen, T. (1972). Correlation matrix memories. *IEEE Transactions on Computers*, **C-21**, 353–9; Anderson, J. A. (1972). A simple neural network generating an interactive memory. *Mathematical Biosciences*, **14**, 197–220; Hopfield, J. J. (1982). Neural networks and physical systems with emergent collective computational abilities. *Proceedings of the National Academy of Sciences, USA*, **79**, 2554–8.

26 / Penfield, W. & Rasmussen, T. (1950). *The Cerebral Cortex of Man* (New York: Macmillan).

27 / Milner, B. (1966). Amnesia following operation on the temporal lobes. In C. W. M. Whitty & O. L. Zangwill eds. *Amnesia: Clinical, Psychological and Medicolegal Aspects* (London: Butterworths).

28 / Warrington, E. K. & Weiskrantz, L. (1982). Amnesia: a disconnection syndrome?' *Neuropsychologia*, **20**, 233–48.

29 / Squire, L. R. (1992). Memory and the hippocampus: a synthesis from findings with rats, monkeys and humans. *Psychological Review*, **99**, 195–231.

30 / Squire, L. R., Shimamura, A. P. & Amaral, D. G. (1989). Memory and the hippocampus. In J. H. Byrne & W. O. Berry eds. *Neural Models of Plasticity* (New York: Academic Press).

31 / Ramón y Cajal, S. (1911). *Histologie du Système Nerveux* (Paris: A. Maloine).

32 / Rolls, E. (1989). Functions of neuronal networks in the hippocampus and neocortex in memory. In J. H. Byrne & W. O. Berry eds. *Neural Models of Plasticity* (New York: Academic Press).

33 / Brown, T. H. & Zador, A. M. (1990). Hippocampus. In G. M. Shepherd ed. *The Synaptic Organization of the Brain* (Oxford: Oxford University Press); Deadwyler, S. A., Hampson, R. E., Foster, T. C. & Marlow, G. (1988). The functional significance of long-term potentiation: relation to sensory processing by hippocampal circuits. In P. W. Landfield & S. A. Deadwyler eds. *Long-Term Potentiation – From Biophysics to Behavior* (New York: Alan R. Liss).

34 / Just, M. A. & Carpenter, P. A. (1992). A capacity theory of comprehension: individual differences in working memory. *Psychological Review*, **99**, 122–49.

35 / Goldman-Rakic, P. S. (1992). Working memory and the mind. *Scientific American*, **267(3)**, 72–9.

36 / Funahashi, S., Bruce, C. J. & Goldman-Rakic, P. S. (1989). Mnemonic coding of visual space in the monkey's dorsolateral prefrontal cortex. *Journal of Neurophysiology*, **61**, 331–49.

37 / Wilson, F. & Skelly, J. (1992). Unpublished work cited by Goldman-Rakic, P. S. (1992). Working memory and the mind. *Scientific American*, **267(3)**, 73–9.

38 / Knight, R. T. (1987). Aging decreases auditory event-related potentials to unexpected stimuli in humans. *Neurobiology of Aging*, **8**, 109–13; Robertson, L. C., Lamb, M. R. & Knight, R. T. (1988). Effects of lesions of temporal-parietal junction on perceptual and attentional processing in humans. *Journal of Neuroscience*, **8**, 3757–69.

39 / Park, S. & Holzman, P. S. (1992). Schizophrenics

show spatial working memory deficits. *Archives of General Psychiatry*, **49**, 975–82.

40 / Bruce, C. J. & Macavoy, M. G. (1990). Response field biases in parietal, temporal, and frontal-lobe visual areas. *Behavioral and Brain Sciences*, **13**, 546–9.

41 / Goldman-Rakic, P. S. (1991). Prefrontal cortical dysfunction in schizophrenia: the relevance of working memory. In B. J. Carroll & J. E. Barrett eds. *Psychology and the Brain* (New York: Raven Press).

42 / Squire, L. R. (1987). *Memory and Brain* (Oxford: Oxford University Press).

43 / Kohonen, T. (1984). *Self-Organization and Associative Memory* (New York: Springer-Verlag). Together with Shun-ichi Amari and Stephen Grossberg, Kohonen established the International Neural Network Society in 1987, and its official journal, *Neural Networks*, was first published early in 1988.

Chapter 6

1 / Wiener, N. & Schadé, J. P. eds. (1963). *Nerve, Brain, and Memory Models* (Amsterdam: Elsevier).

2 / von Neumann, J. (1958). *The Computer and the Brain* (New Haven, CN: Yale University Press); Johnson-Laird, P. N. (1988). *The Computer and the Mind* (London: Fontana Press).

3 / Boden, M. A. (1988). *Computer Models of the Mind* (Cambridge: Cambridge University Press).

4 / Amongst the journals now wholly or partly devoted to articles on neural networks are: *Neural Networks* (Pergamon Press) edited by S. Grossberg, M. Kawato and J. Taylor; *Connection Science – Journal of Neural Computing, Artificial Intelligence and Cognitive Research* (Carfax Publishing Company) edited by N. E. Sharkey; *Network – Computation in Neural Systems* (Institute of Physics, UK) edited by D. J. Amit; *Neural Computation* (MIT Press) edited by T. Sejnowski; *International Journal of Neural Systems* (World Scientific) edited by B. Lautrup; *The International Journal of Neural Networks: Research & Applications* (Learned Information) edited by K. N. Karna and I. F. Croall; *Neurocomputing – An International Journal* (Elsevier) edited by V. D. Sánchez; *The Journal of Computational Neuroscience* (Kluwer Academic) edited by J. M. Bower, J. Miller, J. Rinzel, I. Segev and C. Wilson; *IEEE Transactions on Neural Networks* (IEEE) edited by R. J. Marks; *Neural Computing & Applications* (Springer International) edited by D. Bounds, H. James and N. Beard; *Biological Cybernetics* (Springer International) edited by G. Hauske; *International Journal of Pattern Recognition and Artificial Intelligence* (World Scientific) edited by H. Bunke and P. S. Wang; *Neural, Parallel and Scientific Computations* (Dynamic Publishers, Inc) edited by W. Sambandham. The field is still expanding, and there may well be other periodicals which have not come to the author's attention.

5 / There are now so many books available on the subject of neural networks that the following list is merely intended to be representative. The great classic of the field is M. L. Minsky and S. A. Papert's *Perceptrons – An Introduction to Computational Geometry* (Cambridge, MA: MIT Press, 1988), which is now available in an expanded edition of the 1969 original. Another absolute classic is T. Kohonen's *Self-Organization and Associative Memory* (Berlin: Springer-Verlag, 1989). Two books by S. Grossberg have been referred to as the old and new testaments of neural networks: they are *Studies of Mind and Brain* (Boston: Reidel Press, 1982) and *The Adaptive Brain* (Amsterdam: North-Holland, 1987). Excellent historical reviews of the subject are J. A. Anderson and E. Rosenfeld's *Neurocomputing – Foundations of Research* (Cambridge, MA: MIT Press, 1988) and J. A. Andersen, A. Pellionisz and E. Rosenfeld's *Neurocomputing 2 – Directions for Research* (Cambridge, MA: MIT Press, 1990). The all-time best sellers of the field are D. E. Rumelhart, J. L. McClelland and the PDP Research Group's *Parallel Distributed Processing – Explorations in the Microstructure of Cognition: Volume 1 – Foundations* and J. L. McClelland, D. E. Rumelhart and the PDP Research Group's *Parallel Distributed Processing – Explorations in the Microstructure of Cognition: Volume 2 – Psychological and Biological Models* (Cambridge, MA: Bradford Books, 1986), together with J. L. McClelland and D. E. Rumelhart's *Explorations in Parallel Distributed Processing – A Handbook of Models, Programs, and Exercises* (including software on a diskette) (Cambridge, MA: Bradford Books, 1988).

6 / A brief introduction to neural networks is R. Beale and T. Jackson's *Neural Computing – An Introduction* (Bristol: Adam Hilger, 1990), while J. Hertz, A. Krogh and R. G. Palmer's *Introduction to the Theory of Neural*

Computation (Redwood City, CAL: Addison-Wesley, 1991) is more extensive and mathematical. Another excellent introduction is I. Aleksander and H. Morton's *An Introduction to Neural Computing* (London: Chapman and Hall, 1990).

7 / Much useful information on neural networks is to be found in P. D. Wasserman's *Neural Computation – Theory and Practice* (New York: Van Nostrand Reinhold, 1989); P. Peretto's *An Introduction to the Modeling of Neural Networks* (Cambridge: Cambridge University Press, 1992); I. Aleksander and P. Burnett's *Thinking Machines – The Search for Artificial Intelligence* (Oxford: Oxford University Press, 1987); D. J. Amit's *Modelling Brain Function – The World of Attractor Neural Networks* (Cambridge: Cambridge University Press, 1989); H. Ritter, T. Martinetz and K. Schulten's *Neural Computation and Self-Organizing Maps – An Introduction* (Reading, MA: Addison-Wesley, 1992); P. S. Churchland and T. J. Sejnowski's *The Computational Brain* (Cambridge, MA: Bradford Books, 1992).

8 / The number of edited volumes on neural networks continues to increase rapidly. A necessarily incomplete list is I. Aleksander's *Neural Computing Architectures – The Design of Brain-like Machines* (Oxford: North Oxford Academic, 1989); W. Richards's *Natural Computation* (Cambridge, MA: Bradford Books, 1988); R. Durbin, C. Miall and G. Mitchison's *The Computing Neuron* (Wokingham: Addison-Wesley, 1989); R. Eckmiller and C. von der Malsburg's *Neural Computers* (Berlin: Springer-Verlag, 1989); R. Eckmiller's *Advanced Neural Computers* (Amsterdam: North-Holland, 1990); C. Koch and I. Segev's *Methods in Neuronal Modeling* (Cambridge, MA: Bradford Books, 1989); R. M. J. Cotterill's *Computer Simulation in Brain Science* (Cambridge: Cambridge University Press, 1988) and *Models of Brain Function* (Cambridge: Cambridge University Press, 1989); S. J. Hanson and C. R. Olson's *Connectionist Modeling and Brain Function – The Developing Interface* (Cambridge, MA: Bradford Books, 1990); D. Z. Anderson's *Neural Information Processing Systems* (New York: American Institute of Physics, 1988); D. S. Touretzky's *Advances in Neural Information Processing Systems 2* (San Mateo, CAL: Morgan Kaufmann, 1990); R. P. Lippmann, J. E. Moody and D. S. Touretzky's *Advances in Neural Information Processing Systems 3* (San Mateo, CAL: Morgan Kaufmann, 1990); L. Personnaz and G. Dreyfus's *Neural Networks from Models to Applications* (Paris: I.D.S.E.T., 1989); I. Aleksander and J. G. Taylor's *Artificial Neural Networks* (Amsterdam: North-Holland, 1992); and J. G. Taylor, E. R. Caianiello, R. M. J. Cotterill and J. W. Clark's *Neural Network Dynamics* (Berlin: Springer-Verlag, 1992).

9 / Rosenblatt, F. (1958). The perceptron: a probabilistic model for information storage and organization in the brain. *Psychological Review*, **65**, 386–408.

10 / Minsky, M. L. & Papert, S. A. (1969). *Perceptrons – An Introduction to Computational Geometry* (Cambridge, MA: MIT Press).

11 / Widrow, B. & Hoff, M. E. (1960). Adaptive switching circuits. *1960 WESCON Convention Record* (New York: IRE Press), **part 4**, 96–104.

12 / Werbos, P. (1974). *Beyond Regression – New Tools for Prediction and Analysis in the Behavioral Sciences.* (Harvard University: Ph.D. Thesis).

13 / Parker, D. B. (1985). Learning logic. *Technical Report TR-47* (Cambridge, MA: Center for Computational Research in Economics and Management Science, MIT).

14 / Le Cun, Y. (1985). Une Procédure d'Apprentissage pour Réseau à Seuil Assymétrique. *Cognitiva 85 – A la Frontière de l'Intelligence Artificielle des Sciences de la Connaissance des Neurosciences* (Paris: CESTA), 599–604.

15 / Rumelhart, D. E., Hinton, G. E. & Williams, R. J. (1986). Learning internal representations by error propagation. In D. E. Rumelhart & J. L. McClelland eds. *Parallel Distributed Processing: Explorations in the Microstructure of Cognition, Vol 1* (Cambridge, MA: MIT Press).

16 / Lehky, S. R. & Sejnowski, T. J. (1988). Network model of shape-from-shading: neural function arises from both receptive and projective fields. *Nature*, **333**, 452–4.

17 / Hartline, H. (1940). The receptive fields of optic nerve fibers. *American Journal of Physiology*, **130**, 690–9.

18 / Barlow, H. (1953). Summation and inhibition in the frog's retina. *Journal of Physiology*, **119**, 69–88.

19 / Kuffler, S. W. (1953). Discharge patterns and functional organization of the mammalian retina. *Journal of Neurophysiology*, **16**, 37–68.

20 / See reference 15.

21 / Sejnowski, T. J. & Rosenberg, C. R. (1987). NETtalk: a parallel network that learns to read aloud. *Complex Systems*, **1**, 145–60.

22 / Minsky, M. L. & Papert, S. A. (1988). *Perceptrons* (Cambridge, MA: MIT Press).

23 / Crick, F. H. C. (1989). The recent excitement about neural networks. *Nature*, **337**, 129–32.

24 / Fodor, J. A. & Pylyshyn, Z. W. (1988). Connectionism and cognitive architecture: a critical analysis. *Cognition*, **28**, 3–72.

25 / Sorabji, R. trans. (1972). *Aristotle on Memory* (Providence, RI: Brown University Press). Aristotle's views, as recorded in his scroll writings *De memoria et reminiscentia*.

26 / Clark, A. (1992). Representation, development and situated connectionism. *Connection Science*, **4(3&4)**, 171–4.

27 / Nguyen, D. & Widrow, B. (1990). The truck backer-upper: an example of self-learning in neural networks. In R. Eckmiller ed. *Advanced Neural Computers* (Amsterdam: North-Holland).

28 / See reference 9.

29 / See reference 11.

30 / Aleksander, I., Thomas, W. V. & Bowden, P. A. (1984). WISARD, a radical step forward in image recognition. *Sensor Review*, **4**, 120–4.

31 / Hubel, D. H. & Wiesel, T. N. (1962). Receptive fields, binocular interaction and functional architecture in the cat's visual cortex. *Journal of Physiology*, **160**, 106–54.

Chapter 7

1 / Steensen, N. (1965). Trans. G. Scherz *Lecture on the Anatomy of the Brain (1669)* (Copenhagen: Arnold Busck).

2 / Sperry, R. W. (1963). Chemoaffinity in the orderly growth of nerve fiber patterns and connections. *Proceedings of the National Academy of Sciences USA*, **50**, 703–10.

3 / Levi-Montalcini, R. (1975). NGF: An uncharted route. In F. G. Worden ed. *The Neurosciences – Paths of Discovery* (Cambridge, MA: MIT Press); Thaller, C. & Eichele, G. (1987). Identification and spatial distribution of retinoids in the developing chick limb bud. *Nature*, **327**, 625–8; Gierer, A. (1988). Spatial organization and genetic information in brain development. *Biological Cybernetics*, **59**, 13–21; Palka, J. & Schubiger, M. (1988). Genes for neural differentiation. *Trends in Neurosciences*, **11**, 515–17.

4 / Dowling, J. E. (1987). *The Retina – An Approachable Part of the Brain* (Cambridge, MA: Belknap Press); Nauta, W. J. H. & Feirtag, M. (1986). *Fundamental Neuroanatomy* (New York: Freeman).

5 / Knudsen, E. I. & Konishi, M. (1978). A neural map of auditory space in the owl. *Science*, **200**, 795–7.

6 / Hartline, H. (1938). The response of single optic nerve fibres of the vertebrate eye to illumination of the retina. *American Journal of Physiology*, **121**, 400–15.

7 / Barlow, H. (1953). Summation and inhibition in the frog's retina. *Journal of Physiology*, **119**, 69–88.

8 / Kuffler, S. (1953). Discharge patterns and functional organization of mammalian retina. *Journal of Neurophysiology*, **16**, 37–68.

9 / Hermann, L. (1870). Eine Erscheinung simultanen Kontrastes. *Plugers Arch. geselschaft Physiologi*, **3**, 13–45; Spillman, L., Ransom-Hogg, A. & Oehler, R. (1987). A comparison of perceptive and receptive fields in man and monkey. *Human Neurobiology*, **6**, 51–62.

10 / Bullock, T. H., Orkand, R. & Grinnell, A. (1977). *Introduction to Nervous Systems* (San Francisco: Freeman).

11 / Held, R. & Hein, A. (1963). Movement-produced stimulation in the development of visually guided behavior. *Journal of Comparative and Physiological Psychology*, **56**, 872–6.

12 / Blakemore, C. & Cooper, G. F. (1970). Development of the brain depends on the visual environment. *Nature*, **228**, 477–8; Hirsch, H. V. B. & Spinelli, D. N. (1970). Visual experience modifies distribution of horizontally and vertically oriented receptive fields in cats. *Science*, **168**, 869–71; Rauschecker, J. P & Singer, W. (1981). The effects of early visual experience on the cat's visual cortex and their possible explanation by Hebb synapses. *Journal of Physiology*, **310**, 215–39.

13 / Creutzfeldt, O. D. (1978). The neocortical link: thoughts on the generality of structure and function of the neocortex. In M. A. B. Brazier & H. Petsche eds.

Architectonics of the Cerebral Cortex (New York: Raven Press); White, E. L. (1989). *Cortical Circuits – Synaptic Organization of the Cerebral Cortex – Structure, Function, and Theory* (Basel: Birkhäuser); Shepherd, G. M. (1990) *The Synaptic Organization of the Brain* (Oxford: Oxford University Press); Peters, A. & Jones, E. G. eds. *Cerebral Cortex* (New York: Plenum Press) – multiple volumes; Goldman-Rakic, P. S. & Rakic, P. eds. *Cerebral Cortex* (Cary, NC: Oxford University Press) – multiple volumes.

14 / Barlow, H. B. (1972). Single units and sensation: a neuron doctrine for perceptual psychology. *Perception*, **1**, 371–94.

15 / Gross, C. G., Bender, D. B. & Rocha-Miranda, C. E. (1969). Visual receptive fields of neurons in inferotemporal cortex of the monkey. *Science*, **166**, 1303–6; Perrett, D. I., Rolls, E. T. & Caan, W. (1982). Visual neurones responsive to faces in the monkey temporal cortex. *Experimental Brain Research*, **47**, 329–42; Heit, G., Smith, M. E. & Halgren, E. (1988). Neural encoding of individual words and faces by the human hippocampus and amygdala. *Nature*, **333**, 773–5.

16 / Crick, F. H. C. & Jones, E. G. (1993). Backwardness of human neuroanatomy. *Nature*, **361**, 109–10.

17 / Rockland, K. S. & Pandya, D. N. (1979). Laminar origins and terminations of cortical connections of the occipital lobe in the rhesus monkey. *Brain Research*, **179**, 3–20.

18 / Zeki, S. M. & Shipp, S. (1988). The functional logic of cortical connections. *Nature*, **335**, 311–17.

19 / Van Essen, D. C., Anderson, C. H. & Felleman, D. J. (1992). Information processing in the primate visual system: an integrated systems perspective. *Science*, **255**, 419–23; Felleman, D. J. & Van Essen, D. C. (1991). Distributed hierarchical processing in the primate cerebral cortex. *Cerebral Cortex*, **1**, 1–47.

20 / Scannell, J. W. & Young, M. P. (1993). The connectional organization of neural systems in the cat cerebral cortex. *Current Biology*, **3**, 191–200. This paper includes an impressive colour-coded map showing 1134 inter-area groups of connections in the cat cerebral cortex.

21 / Gilbert, C. D. & Wiesel, T. N. (1985). Intrinsic connectivity and receptive-field properties in visual cortex. *Vision Research*, **25**, 365–74; Gilbert, C. D. (1993). Circuitry, architecture, and functional dynamics of visual cortex. *Cerebral Cortex*, **3**, 373–86.

22 / Cotterill, R. M. J. & Nielsen, C. (1991). A model for cortical 40 Hz oscillations invokes inter-area interactions. *NeuroReport*, **2**, 289–92.

23 / Mountcastle, V. B. (1979). An organizing principle for cerebral function: the unit module and the distributed system. In *Neurosciences Fourth Study Program* (Boston: MIT Press); Szentagothai, J. (1983). The modular architectonic principle of neural centers. *Reviews of Physiology, Biochemistry and Pharmacology*, **98**, 11–61.

24 / Fujita, I., Tanaka, K., Ito, M. & Cheng, K. (1992). Columns for visual features of objects in monkey inferotemporal cortex. *Nature*, **360**, 343–6.

25 / Livingstone, M. & Hubel, D. (1988). Segregation of form, color, movement, and depth: anatomy, physiology, and perception. *Science*, **240**, 740–9.

26 / Martin, K. A. C. (1992). Parallel pathways converge. *Current Biology*, **2**, 555–7.

27 / Ferrera, V. P., Nealey, T. A. & Maunsell, J. H. R. (1992). Mixed parvocellular and magnocellular geniculate signals in visual area V4. *Nature*, **358**, 756–8.

28 / Olds, J. & Milner, P. (1954). Positive reinforcement produced by electrical stimulation of the septal area and other regions of rat brain. *Journal of Comparative Physiology and Psychology*, **47**, 419–27.

29 / Routtenberg, A. & Kuznesof, A. W. (1967). Self-starvation of rats living in activity wheels on a restricted feeding schedule. *Journal of Comparative Physiology*, **64**, 414–21; Routtenberg, A., Gardner, E. L. & Huang, Y. H. (1971). Self-stimulation pathways in monkey macaca-mulatta. *Experimental Neurology*, **33**, 213–24; Wauquier, A. & Rolls, E. T. (1976). *Brain Stimulation Reward* (Amsterdam: North-Holland).

30 / O'Keefe, J. & Nadel, L. (1978). *The Hippocampus as a Cognitive Map* (Oxford: Clarendon Press).

31 / Eccles, J. C. (1973). The cerebellum as a computer: patterns in space and time. *Journal of Physiology*, **229**, 1–32; Eccles, J. C. (1977). An instruction-selection theory of learning in the cerebellar cortex. *Brain Research*, **127**, 327–52.

32 / Eccles, J. C., Ito, M. & Szentagothai, J. (1967). *The Cerebellum as a Neuronal Machine* (Berlin: Springer-Verlag).

33 / Miall, R. C., Weir, D. J., Wolpert, D. M. & Stein,

J. F. (1993). Is the cerebellum a Smith Predictor? *Journal of Motor Behavior*, **25**, 203–16.

34 / Smith, O. J. M. (1959). A controller to overcome dead time. *Instrument Society of America Journal*, **6**, 28–33.

35 / Albus, J. S. (1971). A theory of cerebellar function. *Mathematical Bioscience*, **10**, 25–61; Marr, D. (1969). A theory of cerebellar cortex. *Journal of Physiology, London*, **202**, 437–70.

36 / Gilbert, P. F. C. & Thach, W. C. (1977). Purkinje cell activity during motor learning. *Brain Research*, **128**, 309–28.

37 / Georgopoulos, A. P., Schwartz, A. B. & Kettner, R. E. (1986). Neuronal population coding of movement direction. *Science*, **233**, 1416–19.

38 / Lee, C., Rohrer, W. H. & Sparks, D. L. (1988). Population coding of saccadic eye movements by neurons in the superior colliculus. *Nature*, **332**, 357–60.

39 / Shepard, R. N. & Metzler, J. (1971). The mental rotation of three-dimensional objects. *Science*, **171**, 701–3; Cooper, L. A. & Shepard, R. N. (1973). In W. G. Chace ed. *Visual Information Processing* (New York: Academic Press).

40 / Georgopoulos, A. P., Lurito, J. T., Petrides, M., Schwartz, A. B. & Massey, J. T. (1989). Mental rotation of the neuronal population vector. *Science*, **243**, 234–7.

41 / Ivry, R. B. & Baldo, J. V. (1992). Is the cerebellum involved in learning and cognition? *Current Opinion in Neurobiology*, **2**, 212–16.

42 / Thompson, R. F. (1990). The neural mechanisms of classical responses. *Philosophical Transactions of the Royal Society, Series B*, **329**, 161–70.

43 / Graybiel, A. M. & Ragsdale, C. W. (1978). Histochemically distinct compartments in the striatum of human, monkeys, and cat demonstrated by acetylcholinesterase staining. *Proceedings of the National Academy of Sciences, USA*, **75**, 5723–6.

44 / Yahr, M. D. & Bergmann, K. J. eds. (1987). Parkinson's disease. *Advances in Neurology*, **45**, 277–93; Tanner, C. M. (1989). The role of environmental toxins in the etiology of Parkinson's disease. *Trends in Neurosciences*, **12**, 49–54.

45 / Carlsson, A. (1959). The occurrence, distribution and physiological role of catacholamines in the nervous system. *Pharmacological Review*, **11**, 490–3.

46 / Hayden, M. R. (1981). *Huntington's Chorea* (New York: Springer); Wasmuth, J. J., Hewitt, J., Smith, B., Allard, D., Haines, J. L., Skarecky, D., Partlow, E. & Hayden, M. R. (1988). A Highly polymorphic locus very tightly linked to the Huntington's disease gene. *Nature*, **332**, 734–6.

47 / Cannon, W. B. & Britton, S. W. (1925). Studies on the conditions of activity in endocrine glands. XV: Pseudoaffective medulliadrenal secretion. *American Journal of Physiology*, **72**, 283–94.

48 / Bard, P. (1928). A diencephalic mechanism for the expression of rage with special reference to the sympathetic nervous system. *American Journal of Physiology*, **84**, 490–515.

49 / Klüver, H. & Bucy, P. C. (1939). Preliminary analysis of functions of the temporal lobes in monkeys. *Archives of Neurology and Psychiatry*, **42**, 979–1000.

50 / Ransom, S. W. (1934). The hypothalamus: Its significance for visual innervation and emotional expression. *Transactions of the College of Physicians Philadelphia*, **2**, 222–4; Hess, W. (1954). *Diencephalon – Autonomic and Extrapyramidal Functions* (New York: Grune & Stratton).

51 / Papez, J. W. (1937). A proposed mechanism of emotion. *Archives of Neurology and Psychiatry*, **38**, 725–43.

52 / Weiskrantz, L. (1956). Behavioral changes associated with ablation of the amygdaloid complex in monkeys. *Journal of Comparative and Physiological Psychology*, **49**, 381–91.

53 / Bandler, R. (1987). Brain mechanisms of aggression as revealed by electrical stimulation: suggestion of a central role for the midbrain periaqueductal grey region. In A. Epstein & A. R. Morrison eds. *Progress in Psychobiology and Physiological Psychology Volume 13* (New York: Academic Press).

54 / Rolls, E. T. (1990). A theory of emotion, and its application to understanding the neural basis of emotion. *Cognition and Emotion*, **4**, 161–90.

55 / Moruzzi, G. & Magoun, H. (1949). Brainstem reticular formation and activation of the EEG. *Electroencephalographic and Clinical Neurophysiology*, **1**, 455–73.

56 / Bremer, F. (1935). Cerveau isolé et physiologie du sommeil. *Comptes Rondue Societé Biologie*, **118**, 1235–41.

57 / Singer, W. (1977). Control of thalamic transmission by corticofugal and ascending reticular pathways in the visual system. *Physiological Reviews*, **57**, 386–420.

58 / Scheibel, A. B. (1984). The brain stem reticular core and sensory function. In I. Darian-Smith ed. *Handbook of Physiology – The Nervous System III*, **1**, 213–56.

59 / Fox, K. & Daw, N. W. (1993). Do NMDA receptors have a critical function in visual cortical plasticity? *Trends in Neurosciences*, **16**, 116–22; Bekkers, J. M. & Stevens, C. F. (1989). NMDA and non-NMDA receptors are co-localized at individual excitatory synapses in cultured rat hippocampus. *Nature*, **341**, 230–3.

60 / Singer, W. (1995). Development of plasticity of cortical processing architectures. *Science*, **270**, 758–64.

61 / Tononi, G., Sporns, O. & Edelman, G. M. (1992). Reentry and the problem of integrating multiple cortical areas: simulation of dynamic integration in the visual system. *Cerebral Cortex*, **2**, 310–35.

62 / Edelman, G. M. (1987). *Neural Darwinism – The Theory of Neuronal Group Selection* (New York: Basic Books); Edelman, G. M. (1989). *The Remembered Present – A Biological Theory of Consciousness* (New York: Basic Books); Edelman, G. M. (1992). *Bright Air, Brilliant Fire – On the Matter of the Mind* (London: Allen Lane, Penguin).

63 / Carman, G. J. & Welch, L. (1992). Three-dimensional illusory contours and surfaces. *Nature*, **360**, 585–7.

64 / von der Heydt, R., Peterhans, E. & Baumgartner, G. (1984). Illusory contours and cortical neuron responses. *Science*, **224**, 1260–2.

65 / Jackson, J. H. (1958). Evolution and dissolution of the nervous system. In J. Taylor ed. *Selected Writings of John Hughlings Jackson* (New York: Basic Books). This collection includes Hughlings Jackson's 1884 Croonian Lectures.

Chapter 8

1 / Shibuya, H. & Bundesen, C. (1988). Visual selection from multielement displays: measuring and modeling effects of exposure duration. *Journal of Experimental Psychology*, **14**, 591–600.

2 / Gross, C. G., Bender, D. B. & Rocha-Miranda, C. E. (1969). Visual receptive fields of neurons in inferotemporal cortex of the monkey. *Science*, **166**, 1303–6; Perrett, D. I., Rolls, E. T. & Caan, W. (1982). Visual neurones responsive to faces in the monkey temporal cortex. *Experimental Brain Research*, **47**, 329–42.

3 / Rolls, E. T. (1992). Neurophysiological mechanisms underlying face processing within and beyond the temporal cortical visual areas. *Philosophical Transactions of the Royal Society, London, Series B*, **335**, 11–21.

4 / Koffka, K. (1950). *Principles of Gestalt Psychology* (London: Routledge & Kegan Paul).

5 / Rock, I. & Palmer, S. (1990). The legacy of gestalt psychology. *Scientific American*, **263(6)**, 84–91.

6 / Wertheimer, M. (1945). *Productive Thinking* (New York: Harper & Row).

7 / Sutherland, N. S. (1968). Outlines of a theory of visual pattern recognition in animals and man. *Proceedings of the Royal Society, Series B*, **171**, 297–317.

8 / von der Malsburg, C. (1981). The correlation theory of brain function. *Internal Report Number 81-2, of the Max-Planck Institute for Biophysical Chemistry, Göttingen*.

9 / Milner, P. M. (1974). A model for visual shape recognition. *Psychological Review*, **81**, 521–35.

10 / Penrose, L. S. & Penrose, R. (1958). Impossible objects: a spacial type of illusion. *British Journal of Psychology*, **49**, 31.

11 / Gregory, R. L. (1981). *Mind in Science* (London: Weidenfeld & Nicolson).

12 / Gregory, R. L. & Gombrich, E. H. eds. (1973). *Illusion in Nature and Art* (London: Gerald Duckworth).

13 / Gray, C. M. & Singer, W., (1989). Stimulus-specific neuronal oscillations in orientation columns of cat visual cortex. *Proceedings of the National Academy of Sciences, USA*, **86**, 1698–702.

14 / Gray, C. M., König, P., Engel, A. K. & Singer, W. (1989). Oscillatory responses in cat visual cortex exhibit inter-columnar synchronization which reflects global stimulus properties. *Nature*, **338**, 334–7.

15 / Eckhorn, R., Bauer, R., Jordan, W., Brosch, M., Kruse, W., Munk, M. & Reitboeck, H. J. (1988). Coherent oscillations: a mechanism of feature

linking in the visual cortex. *Biological Cybernetics*, **60**, 121–30.

16 / Singer, W. (1990). The formation of cooperative cell assemblies in the visual cortex. *Journal of Experimental Biology*, **153**, 177–97.

17 / Gilbert, C. D. & Wiesel, T. N. (1985). Intrinsic connectivity and receptive field properties in visual cortex. *Vision Research*, **25**, 365–74.

18 / Rockland, K. S. & Lund, J. S. (1982). Widespread periodic intrinsic connections in the tree shrew visual cortex. *Science*, **215**, 1532–4.

19 / Cotterill, R. M. J. (1988). A possible role for coherence in neural networks. In R. M. J. Cotterill ed. *Computer Simulation in Brain Science* (Cambridge: Cambridge University Press).

20 / Freeman, W. J. (1979). Nonlinear dynamics of paleocortex manifested in the olfactory EEG. *Biological Cybernetics*, **35**, 21–37; Freeman, W. J. & Viana Di Prisco, G. (1986). Relation of olfactory EEG to behavior: time series analysis. *Behavioral Neuroscience*, **100**, 753–63; Freeman, W. J. & Skarda, C. A. (1987). How brains make chaos in order to make sense of the world. *Behavioral and Brain Sciences*, **10**, 161–95; Freeman, W. J. (1995). *Societies of Brains – A Study in the Neuroscience of Love and Hate* (Hillsdale and Hove: Laurence Erlbaum Associates).

21 / Schöner, G., Kopecz, K., Spengler, F. & Dinse, H. R. (1992). Evoked oscillatory cortical responses are dynamically coupled to peripheral stimuli. *NeuroReport*, **3**, 579–82.

22 / Pfurtscheller, G. & Neuper, C. (1992). Simultaneous EEG 10 Hz desynchronization and 40 Hz synchronization during finger movements. *NeuroReport*, **3**, 1057–60.

23 / Steriade, M., Curró Dossi, R., Paré, D. & Oakson, G. (1991). Fast oscillations (20–40 Hz) in thalamocortical systems and their potentiation by mesopontine cholinergic nuclei in the cat. *Proceedings of the National Academy of Sciences, USA*, **88**, 4396–400.

24 / Soong, A. C. K. & and Stuart, C. I. J. M. (1989). Evidence of chaotic dynamics underlying the human alpha-rhythm electroencephalogram. *Biological Cybernetics*, **62**, 55–62.

25 / von der Malsburg, C. & Buhmann, J. (1992). Sensory segmentation with coupled neural oscillators. *Biological Cybernetics*, **67**, 233–42.

26 / Kleinfeld, D., Raccuia-Behling, F. & Chiel, H. J. (1990). Circuits constructed from *Aplysia* neurons exhibit multiple patterns of persistent activity. *Biophysical Journal*, **57**, 697–715.

27 / Sporns, O., Gally, J. A., Reeke , G. N. Jr. & Edelman, G. M. (1989). Reentrant signaling among simulated neuronal groups leads to coherency in their oscillatory activity. *Proceedings of the National Academy of Sciences, USA*, **86**, 7265–9.

28 / Cotterill, R. M. J. & Nielsen, C. (1991). A model for cortical 40 Hz oscillations invokes inter-area interactions. *NeuroReport*, **2**, 289–92.

29 / Nakamura, K., Mikami, A. & Kubota, K. (1992). Oscillatory neuronal activity related to visual short-term memory in monkey temporal pole. *NeuroReport*, **3**, 117–20.

30 / Julesz, B. (1975). Experiments in the visual perception of texture. *Scientific American*, **232(4)**, 34–43.

31 / Julesz, B. (1981). Textons, the elements of texture perception, and their interactions. *Nature*, **290**, 91–7.

32 / Hubel, D. H. & Wiesel, T. N. (1968). Receptive fields and functional architecture of monkey striate cortex. *Journal of Physiology*, **195**, 215–43.

33 / Marr, D. (1982). *Vision – A Computational Investigation into the Human Representation and Processing of Visual Information* (San Francisco: Freeman).

34 / Treisman, A. M. (1993). The perception of features and objects. In A. Baddeley & L. Weiskrantz eds. *Attention: Selection, Awareness, and Control – A Tribute to Donald Broadbent* (Oxford: Clarendon Press); Treisman, A. M. & Gelade, G. (1980). A feature-integration theory of attention. *Cognitive Psychology*, **12**, 97–136.

35 / Broadbent, D. E. (1958). *Perception and Communication* (Oxford: Oxford University Press); Broadbent, D. E. (1982). Task combination and selective intake of information. *Acta Psychologica*, **50**, 253–90.

36 / Sakrison, D. J. (1968). *Communication Theory – Transmission of Waveforms and Digital Information* (New York: Wiley).

37 / Moray, N. (1969). *Listening and Attention* (Harmondsworth: Penguin).

38 / Treisman, A. M. (1960). Contextual cues in

selective listening. *Quarterly Journal of Experimental Psychology*, **12**, 242–8.

39 / Crick, F. H. C. & Koch, C. (1992). The problem of consciousness. *Scientific American*, **267(9)**, 110–17.

40 / Enroth-Cugell, C. & Robson, J. G. (1966). The contrast sensitivity of retinal ganglion cells of the cat. *Journal of Physiology*, **187**, 517–52.

41 / McWhirter, N. & McWhirter, R. (1963). *The Guinness Book of World Records* (New York: Bantam Books). This lists President John F. Kennedy's record speaking rate.

42 / Yarbus, A. L. (1967). *Eye-Movements and Vision*. (New York: Wiley).

43 / Wurtz, R. H., Goldberg, M. E. & Robinson, D. L. (1980). Behavioral modulation of visual responses in the monkey: stimulus selection for attention and movement. *Progress in Psychobiology and Physiological Psychology*, **9**, 43–83; Wurtz, R. H., Goldberg, M. E. & Robinson, D. L. (1982). Brain mechanisms of visual attention. *Scientific American*, **246**, 124–32.

44 / Ferrier, D. (1878). *The Localization of Cerebral Disease* (London: Smith, Elder).

45 / Mountcastle, V. B. (1978). In G. M. Edelman & V. B. Mountcastle eds. *The Mindful Brain* (Boston: MIT Press).

46 / Bushnell, M. C., Goldberg, M. E. & Robinson, D. L (1981). Behavioral enhancement of visual responses in the monkey cerebral cortex, I: modulation in posterior parietal cortex related to selective visual attention. *Journal of Neurophysiology*, **46**, 755–72.

47 / Robinson, D. L. & Petersen, S. E. (1992). The pulvinar and visual salience. *Trends in Neurosciences*, **15**, 127–32.

48 / Noton, D. & Stark, L. (1971). Scanpaths in saccadic eye movements while viewing and recognizing patterns. *Vision Research*, **11**, 929–41; Sparks, D. L. & Mays, L. E. (1981). The role of the monkey superior colliculus in the control of saccadic eye movements: a current perspective. In A. Fuchs & W. Becker eds. *Progress in Oculomotor Research* (New York: Elsevier); Sparks, D. L. & Mays, L. E. (1983). The role of the monkey superior colliculus in the spatial localization of saccade targets. In A. Hein & M. Jeannerod eds. *Spatially Oriented Behavior* (New York: Springer-Verlag); Schlag, J. & Schlag-Rey, M. (1990). Colliding saccades may reveal the secret of their marching orders. *Trends in Neurosciences*, **13**, 410–14; Sparks, D. L. (1991). The neural encoding of the location of targets for saccadic eye movements. In J. Paillard ed. *Brain and Space* (New York: Oxford University Press); Glimcher, P. W. & Sparks, D. L. (1992). Movement selection in advance of action in the superior colliculus. *Nature*, **355**, 542–4.

49 / Paillard, J. (1991). Knowing where and knowing how to get there. In J. Paillard ed. *Brain and Space* (New York: Oxford University Press).

50 / Guitton, D., Munoz, D. P. & Pélisson, D. (1991). Spatio-temporal patterns of activity on the motor map of cat superior colliculus. In J. Paillard ed. *Brain and Space* (New York: Oxford University Press); Guitton, D. (1992). Control of saccadic eye and gaze movements by the superior colliculus and basal ganglia. In R. H. S. Carpenter ed. *Eye Movements* (London: MacMillan).

51 / Posner, M. I., Walker, J. A., Friedrich, F. J. & Rafal, R. D. (1984). Effects of parietal injury on covert orienting of attention. *Journal of Neuroscience*, **4**, 1863–74; Posner, M. I. & Driver, J. (1992). The neurobiology of selective attention. *Current Opinion in Neurobiology*, **2**, 165–9.

52 / Mountcastle, V. B. (1978). Brain mechanisms for directed attention. *Journal of the Royal Society of Medicine*, **71**, 14–28. This is a transcript of Mountcastle's 1977 Sherrington Memorial Lecture.

53 / Singer, W. (1977). Control of thalamic transmission by corticofugal and ascending reticular pathways in the visual system. *Physiological Review*, **57**, 386–420.

54 / Crick, F. H. C. (1984). The function of the thalamic reticular complex: the searchlight hypothesis. *Proceedings of the National Academy of Sciences, USA*, **81**, 4586–90.

55 / Mitrofanis, J. & Guillery, R. W. (1993). New views of the thalamic reticular nucleus in the adult and the developing brain. *Trends in Neurosciences*, **16**, 240–5.

56 / Troscianko, T. (1987). Colour vision: brain mechanisms. In R. L. Gregory ed. *The Oxford Companion to the Mind* (Oxford: Oxford University Press).

57 / Fukurotani, K. (1982). Color information coding of horizontal-cell responses in fish retina. *Color Research and Applications*, **7**, 146–8.

58 / Shallice, T. & Warrington, E. K. (1970). Independent functioning of verbal memory stores: a neuropsychological study. *Quarterly Journal of Experimental Psychology*, **22**, 261–73.

59 / Baddeley, A. D. & Hitch, G. J. (1974). Working memory. In G. H. Bower ed. *The Psychology of Learning and Motivation, Volume 8* (New York: Academic Press).

60 / Baddeley, A. D. (1992). Working Memory. *Science*, **255**, 556–9.

61 / Daneman, M. & Carpenter, P. A. (1980). Individual differences in working memory and reading. *Journal of Verbal Learning and Verbal Behavior*, **19**, 450–66; Kyllonen, P. C. & Christal, R. E. (1990). Reasoning ability is (little more than) working-memory capacity?! *Intelligence*, **14**, 389–433.

62 / Baddeley, A. D. (1986). *Working Memory* (Oxford: Oxford University Press); Baddeley, A. D. (1990). *Human Memory – Theory and Practice* (Needham Heights, MA: Allyn & Bacon).

63 / Becker, J. T. (1987). Rates of forgetting of verbal and non-verbal material in Alzheimers disease. In R. J. Wurtman, S. H. Corkin & J. H. Growdon eds. *Alzheimer's Disease – Advances in Basic Research and Therapies* (Cambridge: Center for Brain Sciences and Metabolism Charitable Trust); Baddeley, A. D., Logie, R., Bressi, S., Della Sala, S. & Spinnler, H. (1986). Dementia and working memory. *Quarterly Journal of Experimental Psychology*, **38A**, 603–18.

64 / Farah, M. J., Hammond, K. M., Levine, D. N. & Calvanio, R. (1988). Visual and spatial imagery: dissociable systems of representation. *Cognitive Psychology*, **20**, 439–62; Farah, M. J. (1988). Is visual imagery really visual? Overlooked evidence from neuropsychology. *Psychological Review*, **95(3)**, 307–17.

65 / Libet, B., Wright Jr, E. W., Feinstein, B. & Pearl, D. K. (1979). Subjective referral of the timing for a conscious sensory experience: a functional role for the somatosensory specific projection system in man. *Brain*, **102**, 193–224; Libet, B. (1982). Brain stimulation in the study of neuronal functions for conscious sensory experience. *Human Neurobiology*, **1**, 235–42.

66 / Honderich, T. (1984). The time of a conscious sensory experience and mind-brain theories. *Journal of Theoretical Biology*, **110**, 115–29.

67 / Libet, B., Wright, E. W., Feinstein, B. & Pearl, D. K. (1992). Retroactive enhancement of a skin sensation by a delayed cortical stimulus in man: evidence for delay of a conscious sensory experience. *Consciousness and Cognition*, **1**, 367–78.

68 / Libet, B. (1989). Conscious subjective experience vs. unconscious mental functions: a theory of the cerebral processes involved. In R. M. J. Cotterill ed. *Models of Brain Function* (Cambridge: Cambridge University Press).

69 / Kornhuber, H. H., Becker, W., Taumer, R., Hoehne, O. & Iwase, K. (1969). Cerebral potentials accompanying voluntary movements in man: readiness potential and reafferent potentials. *Electroencephalography and Clinical Neurophysiology*, **26**, 439; Deecke, L., Scheid, P. & Kornhuber, H. H. (1969). Distribution of readiness potential, pre-motion positivity, and motor potential of the human cerebral cortex preceding voluntary finger movements. *Experimental Brain Research*, **7**, 158–68.

70 / Libet, B., Wright, E. W. & Gleason, C. (1982). Readiness-potentials preceding unrestricted 'spontaneous' vs. pre-planned voluntary acts. *Electroencephalography and Clinical Neurophysiology*, **54**, 322–35.

71 / Libet, B., Gleason, C. A., Wright, E. W. & Pearl, D. K. (1983). Time of conscious intention to act in relation to onset of cerebral activities (readiness-potential); the unconscious initiation of a freely voluntary act. *Brain*, **106**, 623–42.

72 / Libet, B., Wright, E. W. & Gleason, C. A. (1983). Preparation- or intention-to-act, in relation to pre-event potentials recorded at the vertex. *Electroencephalography and Clinical Neurophysiology*, **56**, 367–72; Libet, B. (1985). Unconscious cerebral initiative and the role of conscious will in voluntary action. *Behavioral and Brain Sciences*, **8**, 529–39; Libet, B. (1985). Theory and evidence relating cerebral processes to conscious will. *Behavioral and Brain Sciences*, **8**, 558–66.

73 / Watt, H. (1905). Experimentelle Beiträge zu einer Theorie des Denkens. *Archiv für die Gesamte Psychologie*, **4**, 389–436.

Chapter 9

1 / James, W. (1890). *The Principles of Psychology* (New York: Henry Holt & Co.).

2 / Gregory, R. L. (1978). Consciousness. In R. Duncan & M. Weston-Smith eds. *The Encyclopedia of Ignorance* (Oxford: Pergamon).

3 / James, W. (1890). *The Principles of Psychology* (New York: Henry Holt & Co.).

4 / Freud, S. (1955). *The Interpretation of Dreams* (New York: Basic Books).

5 / Ryle, G. (1949). *The Concept of Mind* (New York: Barnes & Noble).

6 / Penfield, W. (1975). *The Mystery of the Mind – a Critical Study of Consciousness and the Human Brain* (New York: Princeton University Press).

7 / Rorschach, H. (1942). *Psychodiagnostics* (New York: Grune & Stratton).

8 / Weiskrantz, L. (1986). *Blindsight – A Case Study and Implications* (Oxford: Oxford University Press); Cowey, A. & Stoerig, P. (1995). Blindsight in monkeys. *Nature*, **373**, 247–9.

9 / Kihlstrom, J. F. (1996). Perception without awareness of what is perceived, learning without awareness of what is learned. In M. Velmans ed. *The Science of Consciousness* (London: Routledge).

10 / Cowey, A. & Stoerig, P. (1991). The neurobiology of blindsight. *Trends in Neurosciences*, **14**, 140–5; Braddick, O., Atkinson, J., Hood, B., Harkness, W., Jackson, G. & Vargha-Khadem, F. (1992). Possible blindsight in infants lacking one cerebral hemisphere. *Nature*, **360**, 461–3; Cowey, A. & Stoerig. P. (1993). Insight into blindsight? *Current Biology*, **3**, 236–8.

11 / Nagel, T. (1974). What is it like to be a bat? *Philosophical Review*, **83**, 435–50.

12 / Dennett, D. C. (1991). *Consciousness Explained* (New York: Little, Brown & Co.).

13 / Rushton, W. A. H. (1979). Charles II and the blind spot. *Vision Research*, **19**, 225.

14 / Gilbert, C. D. & Wiesel, T. N. (1992). Receptive field dynamics in adult primary visual cortex. *Nature*, **256**, 150–2.

15 / Gattass, R., Fiorani, M., Rosa, M. G. P., Pirion, M. C. G., Sousa, A. P. B. & Soares, J. G. M. (1992). Changes in receptive field size in V, in relation to perceptual completion. In R. Lent ed. *Visual System from Genesis to Maturity* (Boston: Birkhauser).

16 / Ramachandran, V. S. (1992). Filling in the blind spot. *Nature*, **356**, 115.

17 / Ramachandran, V. S. & Gregory, R. L. (1991). Perceptual filling in of artificially induced scotomas in human vision. *Nature*, **350**, 699–702.

18 / Ramachandran, V. S. (1992). Blind spots. *Scientific American*, **266(5)**, 44–9.

19 / McCauley, R. N. (1993). Why the blind can't lead the blind: Dennett on the blind spot, blindsight, and sensory qualia. *Consciousness and Cognition*, **2**, 155–64.

20 / Ramachandran, V. S. (1993). Filling in gaps in logic: some comments on Dennett. *Consciousness and Cognition*, **2**, 165–8.

21 / Moran, J. & Desimone, R. (1985). Selective attention gates visual processing in the extrastriate cortex. *Science*, **229**, 782–4.

22 / Libet, B. (1989). Conscious subjective experience vs. unconscious mental functions: theory of the cerebral processes involved. In R. M. J. Cotterill ed. *Models of Brain Function* (Cambridge: Cambridge University Press).

23 / Deutsch, D. (1992). Some new pitch paradoxes and their implications. *Philosophical Transactions of the Royal Society of London, Series B*, **336**, 391–7; Deutsch, D. (1992). Paradoxes of musical pitch. *Scientific American*, **267(2)**, 70–5.

24 / Shepard, R. N. (1964). Circularity in judgments of relative pitch. *Journal of the Acoustical Society of America*, **36**, 2346–53.

25 / Geldard, F. A. & Sherrick, C. E. (1972). The cutaneous rabbit: a perceptual illusion. *Science*, **178**, 178–9; Geldard, F. A. & Sherrick, C. E. (1983). The cutaneous saltatory area and its presumed neural base. *Perception and Psychophysics*, **33**, 299–304; Geldard, F. A. & Sherrick, C. E. (1986). Space, time and touch. *Scientific American*, **255(1)**, 85–9.

26 / Melzack, R. (1992). Phantom limbs. *Scientific American*, **266(4)**, 90–6.

27 / Searle, J. (1984). *Minds, Brains and Science*. (London: Penguin) – based on his 1984 BBC Reith Lectures.

28 / Velmans, M. (1991). Consciousness from a first-person perspective. *Behavioral Brain Science*, **14**, 702–26; Velmans, M. (1993). A reflexive science of

consciousness. In G. R. Bock & J. Marsh eds. *Experimental and Theoretical Studies of Consciousness – Ciba Foundation Symposium 174* (Chichester: Wiley).

29 / Arbib, M. A. (1987). Schemas. In R. L. Gregory ed. *The Oxford Companion to the Mind* (Oxford: Oxford University Press).

30 / Crick, F. H. C. & Koch, C. (1990). Towards a neurobiological theory of consciousness. *Seminars in the Neurosciences*, **2**, 263–75; Crick, F. H. C. & Koch, C. (1991). Some reflections on visual awareness. *Cold Spring Harbour Symposia on Quantitative Biology*, **55**, 953–62.

31 / Sommerhoff, G. (1996). Consciousness as an internal integrating system. *Journal of Consciousness Studies*, **3**, 139–57.

32 / Edelman, G. M. (1987). *Neural Darwinism – The Theory of Neuronal Group Selection* (New York: Basic Books); Edelman, G. M. (1989). *The Remembered Present – A Biological Theory of Consciousness* (New York: Basic Books); Edelman, G. M. (1992). *Bright Air, Brilliant Fire – On the Matter of the Mind* (New York: Basic Books).

33 / Carpenter, R. H. S. (1994). Choosing where to look. *Current Biology*, **4**, 341–3.

34 / Llinás, R. & Pellionisz, A. (1979). Brain modeling by tensor network theory and computer simulation. The cerebellum: distributed processor for predictive coordination. *Neuroscience*, **4**, 322–48. Subsequent papers in the series appeared in the same journal: **5**, 1125–36, (1980); **7**, 2949–70, (1982); and **16**, 245–74, (1985).

35. Ackley, D. H., Hinton, G. E. & Sejnowski, T. J. (1985). A learning algorithm for Boltzmann machines. *Cognitive Science*, **9**, 147–69.

36 / Zipser, D. & Andersen, R. A. (1988). A back-propagation programmed network that simulates response properties of a subset of posterior parietal neurons. *Nature*, **331**, 679–84.

37 / Dennett, D. C. (1991). *Consciousness Explained* (London: Penguin).

38 / Fast, J. (1971). *Body Language* (London: Souvenir Press).

39 / Ardrey, R. (1966). *The Territorial Imperative* (New York: Delta); Sack, R. D. (1986). *Human Territoriality* (Cambridge: Cambridge University Press).

40 / Segal, S. J. & Fusella, V. (1970). Influence of imaged pictures and sounds on detection of visual and auditory signals. *Journal of Experimental Psychology*, **83**, 458–64.

41 / Crick, F. H. C. (1988). *What Mad Pursuit – A Personal View of Scientific Discovery* (New York: Basic Books).

42 / James, W. (1890). *The Principles of Psychology* (New York: Henry Holt & Co.).

43 / Crick, F. H. C. (1988). *What Mad Pursuit – A Personal View of Scientific Discovery* (New York: Basic Books).

44 / Carpenter, R. H. S. (1996). *Neurophysiology* (London: Edward Arnold).

45 / Crick, F. H. C. (1989). The recent excitement about neural networks. *Nature*, **337**, 129–32.

46 / Stroop, J. R. (1935). Studies of interference in serial verbal reactions. *Journal of Experimental Psychology*, **18**, 643–62.

47 / Treisman, A. M. (1960). Contextual cues in selective listening. *Quarterly Journal of Experimental Psychology*, **12**, 242–8; Treisman, A. M. (1964). Verbal cues, language, and meaning in selective attention. *American Journal of Psychology*, **77**, 206–19; Cotterill, R. M. J. (1994). Autism, intelligence and consciousness. *Biol.Skr.Dan.Vid.Selsk.*, **45**, 1–93.

48 / Trevarthen, C. B. (1994). Roger W. Sperry (1913–1994). *Trends in Neurosciences*, **17**, 402–4.

49 / Sperry, R. W. (1952). Neurology and the mind–brain problem. *American Scientist*, **40**, 291–312.

50 / Sherrington, C. S. (1940). *Man on his Nature* (Cambridge: Cambridge University Press).

51 / Adrian, E. D. (1932). *The Mechanism of Nervous Action – Electrical Studies of the Neurone* (London: Humphrey Milford).

52 / Ritchie, A. D. (1936). *The Natural History of Mind* (London: Longmans, Green & Co.).

53 / Damasio, A. R. (1994). *Descartes' Error – Emotion, Reason and the Human Brain* (New York: Grosset-Putnam).

54 / Held, R. & Hein, A (1963). Movement-produced stimulation in the development of visually guided behavior. *Journal of Comparative and Physiological Psychology*, **56**, 872–6.

55 / Carpenter, R. H. S. (1988). *Movements of the Eyes* (London: Pion).

56 / Taylor, J. G. & Alavi, F. N. (1993). Mathematical analysis of a competitive network for attention. In J. G. Taylor ed. *Mathematical Approaches to Neural Networks* (Amsterdam: Elsevier); Taylor, J. G. (1998). A competition for consciousness? *Neurocomputing* (in the press).

57 / Hanes, D. P. & Carpenter, R. H. S. (1998). Countermanding saccades in humans: evidence for a race-to-threshold process. (In the press). See also: Carpenter, R. H. S. & Williams, M. L. L. (1995), Neural computation of log likelihood in the control of saccadic eye movements, *Nature*, **377**, 59–62.

58 / Eccles, J. C. & Gibson, W. C. (1979). *Sherrington – His Life and Thought* (Berlin: Springer International).

59 / Hammer, M. (1997). The neural basis of associative reward learning in honeybees. *Trends in Neurosciences*, **20**, 245–52.

60 / Edelman, G. M. (1989). *The Remembered Present – A Biological Theory of Consciousness* (New York: Basic Books).

61 / Humphrey, N. (1992). *A History of the Mind* (London: Chatto & Windus).

62 / Efron, R. (1967). The duration of the present. *Annals of the New York Academy of Science*, **138**, 713–29.

63 / Piaget, J. (1967). G. N. Seagrim trans. *The Mechanisms of Perception* (London: Routledge & Kegan Paul).

64 / Jeannerod, M. (1985). *The Brain Machine – The Development of Neurophysiological Thought* (Cambridge, MA: Harvard University Press); Jeannerod, M. (1994). The representing brain: neural correlates of motor intention and imagery. *Behavioral and Brain Sciences*, **17**, 187–245.

65 / Clark, A. (1997). *Being There – Putting Brain, Body, and World Together Again* (Cambridge, MA: MIT Press).

66 / Gibson, J. J. (1968). *The Senses Considered as Perceptual Systems* (London: Allen & Unwin).

67 / Halpern, B. P. (1983). Tasting and smelling as active exploratory sensory processes. *American Journal of Otolaryngology*, **4**, 246–9.

68 / Iwamura, Y., Tanaka, M. & Hikosaka, O. (1981). Cortical neuronal mechanisms of tactile perception studied in the conscious monkey. In Y. Katsuki, R. Norgren & M. Sato eds. *Brain Mechanisms of Sensation* (New York: Wiley).

69 / Bower, J. M. (1995). Is the cerebellum motor for sensory's sake or sensory for motor's sake? *The Cerebellum – From Structure to Control*. Conference held in Rotterdam, August 31 – September 3, 1995 (The proceedings will be published).

70 / Gao, J.-H., Parsons, L. M., Bower, J. M., Xiong, J., Li, J. & Fox, P. T. (1996). Cerebellum implicated in sensory acquisition and discrimination rather than motor control. *Science*, **272**, 545–7; Barinaga, M. (1996). The cerebellum: movement coordinator or much more? *Science*, **272**, 482–3.

71 / Nicolelis, M. A. L., Baccala, L. A., Lin, R. C. S. & Chapin, J. K. (1995). Sensorimotor encoding by synchronous neural ensemble activity at multiple levels of the somatosensory system. *Science*, **268**, 1353–8.

72 / Larsson, J., Gulyás, B. & Roland, P. E. (1996). Cortical representation of self-paced finger movement. *NeuroReport*, **7**, 463–8.

73 / Darian-Smith, I., Sugitani, M., Heywood, J., Karita, K. & Goodwin, A. (1982). Touching textured surfaces: cells in somatosensory cortex respond both to finger movement and to surface features. *Science*, **218**, 906–9.

74 / Schacter, D. L., Cooper, L. A. & Delaney, S. M. (1990). Implicit memory for unfamiliar objects depends on access to structural descriptions. *Journal of Experimental Psychology: General*, **119**, 5–24; Schacter, D. L. (1995). Implicit memory: a new frontier for cognitive science. In M. S. Gazzaniga ed. *The Cognitive Neurosciences* (Cambridge, MA: MIT Press).

75 / Rubin, E. (1915). *Synsoplevede Figurer – Studier i psykologisk Analyse* (Copenhagen: Nordisk Forlag).

76 / Groenewegen, H. J. & Berendse, H. W. (1994). The specificity of the 'nonspecific' midline and intralaminar thalamic nuclei. *Trends in Neurosciences*, **17**, 52–7.

77 / Crick, F. & Koch, C. (1995). Are we aware of neural activity in primary visual cortex? *Nature*, **375**, 121–3; Pollen, D. A. (1995). Cortical areas in visual awareness. *Nature*, **377**, 293–4; Crick, F. & Koch, C. (1995).

Reply to Pollen. *Nature*, **377**, 294–5; Block, N. (1996). How can we find the neural correlate of consciousness? *Trends in Neurosciences*, **19**, 456–9.

78 / Koch, C. (1995). Visual awareness and the thalamic intralaminar nuclei. *Consciousness and Cognition*, **4**, 163–6.

79 / Bogen, J. E. (1995). The role of the intralaminar nucleus of the thalamus in waking consciousness. *Consciousness and Cognition*, **4**, 137–58.

80 / Kinomura, S., Larsson, J., Gulyás, B. & Roland, P. E. (1996). Activation by attention of the human reticular formation and thalamic intralaminar nuclei. *Science*, **271**, 512–15.

81 / Singer, W. (1977). Control of thalamic transmission by corticofugal and ascending reticular pathways in the visual system. *Physiological Reviews*, **57**, 386–420; Steriade, M. & Glenn, L. L. (1982). Neocortical and caudate projections of intralaminar thalamic neurons and their synaptic excitation from midbrain reticular core. *Journal of Neurophysiology*, **48**, 352–71.

82 / DeVito, J. L. (1969). Projections from cerebral cortex to intralaminar nuclei in monkey. *Journal of Comparative Neurology*, **136**, 193–201.

83 / Jones, E. G. (1985). *The Thalamus* (New York: Plenum); Groenewegen, H. J. & Berendse, H. W. (1994). The specificity of the 'nonspecific' midline and intralaminar thalamic nuclei. *Trends in Neurosciences*, **17**, 52–7.

84 / Baars, B. J. (1995). Surprisingly small subcortical structures are needed for the *state* of waking consciousness, while cortical projection areas seem to provide perceptual *contents* of consciousness. *Consciousness and Cognition* **4**, 159–62.

85 / Steriade, M. & Llinás, R. (1988). The functional states of the thalamus associated neuronal interplay. *Physiological Reviews*, **68**, 649–742.

86 / Cotterill, R. M. J. (1997). On the mechanism of consciousness. *Journal of Consciousness Studies*, **4**, 231–47.

87 / Edelman, G. M. (1989). *The Remembered Present – A Biological Theory of Consciousness* (New York: Basic Books); Cotterill, R. M. J. (1995). On the unity of conscious experience. *Journal of Consciousness Studies*, **2**, 290–312.

88 / Bisiach, E., Luzzatti, C. & Perani, D. (1979). Unilateral neglect, representational schema and consciousness. *Brain*, **102**, 609–18.

89 / Guariglia, C., Padovani, A., Pantano, P. & Pizzamiglio, L. (1993). Unilateral neglect restricted to visual imagery. *Nature*, **364**, 235–7.

90 / Barth, D. S. & MacDonald, K. D. (1996). Thalamic modulation of high-frequency oscillating potentials in auditory cortex. *Nature*, **383**, 78–81.

91 / Munglani, R., Andrade, J., Sapsford, D. J., Baddeley, A. & Jones, J. G. (1993). A measure of consciousness and memory during isoflurane administration: the coherent frequency. *British Journal of Anaesthesia*, **71**, 633–41.

92 / Steriade, M. (1996). Awakening the brain. *Nature*, **383**, 24–5.

93 / Pardo, J. V., Pardo, P. J., Janer, K. W. & Raichle, M. E. (1990). The anterior cingulate cortex mediates processing selection in the Stroop attentional conflict paradigm. *Proceedings of the National Academy of Sciences, USA*, **87**, 256–9.

94 / Posner, M. I. & Raichle, M. E. (1994). *Images of Mind* (New York: Scientific American Library).

95 / Vogt, B. A., Finch, D. M. & Olson, C. R. (1992). Functional heterogeneity in cingulate cortex: the anterior executive and posterior evaluative regions. *Cerebral Cortex*, **2**, 435–43.

96 / Talbot, J. D., Marrett, S., Evans, A. C., Meyer, E., Bushnell, M. C. & Duncan, G. H. (1991). Multiple representations of pain in human cerebral cortex. *Science*, **251**, 1355–8.

97 / Rainville, P., Duncan, G. H., Price, D. D., Carrier, B. & Bushnell, M. C. (1997). Pain affect encoded in human anterior cingulate but not somatosensory cortex. *Science*, **277**, 968–71.

98 / Vogt, B. A. (1987). Cingulate Cortex. In G. Adelman ed. *Encyclopedia of Neuroscience* (Boston: Birkhäuser).

99 / Purpura, K. P. & Schiff, N. D. (1997). The thalamic intralaminar nuclei: a role in visual awareness. *The Neuroscientist*, **3**, 8–15.

100 / Cornwall, J., Cooper, J. D. & Phillipson, O. T. (1990). Projections to the rostral reticular thalamic mucleus of the rat. *Experimental Brain Research*, **80**, 157–71; Lozsádi, D. A. (1994). Organization of cortical afferents to the rostral, limbic sector of the rat

thalamic reticular nucleus. *Journal of Comparative Neurology*, **341**, 520–33.

101 / LaMendola, N. P. & Bever, T. B. (1997). Peripheral and cerebral asymmetries in the rat. *Science*, **278**, 483–6.

102 / Ward, A. W. (1948). The anterior cingulate gyrus and personality. *Research Publications of the Association of Research in Nervous Mental Disorders*, **27**, 438–45.

103 / Merker, B. H. (1971). The nucleus reticularis thalami: a central mechanism of attention? (Unpublished tutorial report, Department of Psychology, Queens College, City University of New York).

104 / Scheibel, M. E. & Scheibel, A. B. (1966). The organization of the nucleus reticularis thalami: a Golgi study. *Brain Research*, **1**, 43–62; Scheibel, M. E. & Scheibel, A. B. (1967). Structural organization of the nonspecific thalamic nuclei and their projection toward the cortex. *Brain Research*, **6**, 60–94.

105 / Crick, F. H. C. (1984). The function of the thalamic reticular complex: the searchlight hypothesis. *Proceedings of the National Academy of Sciences, USA*, **81**, 4586–90.

106 / Luria, A. (1968). *The Mind of a Mnomonist – A Little Book About a Vast Memory* (New York: Basic Books).

107 / Sperling, G. (1960). The information available in brief visual presentations. *Psychological Monographs: General and Applied*, **74**, 1–29.

108 / Coltheart, M. (1980). Iconic memory and visible persistence. *Perception and Psychophysics*, **27**, 183–228.

109 / Peterson, L. R. & Peterson, M. J. (1959). Short-term retention of individual verbal items. *Journal of Experimental Psychology – Learning, Memory, and Cognition*, **58**, 193–8.

110 / Damasio, A. R. (1994). *Descartes' Error – Emotion, Reason and the Human Brain* (New York: Grosset-Putnam).

111 / Miall, R. C., Weir, D. J., Wolpert, D. M. & Stein, J. F. (1993). Is the cerebellum a Smith predictor? *Journal of Motor Behavior*, **25**, 203–16.

112 / Ingvar, D. H. (1985). 'Memory of the future': an essay on the temporal organization of conscious awareness. *Human Neurobiology*, **4**, 127–36.

113 / Bartlett, F. C. (1932). *Remembering – A Study in Experimental and Social Psychology* (Cambridge: Cambridge University Press).

114 / Kant, I. (1933). Trans. N. K. Smith *Critique of Pure Reason* (London: Macmillan).

115 / Johnson-Laird, P. N. (1983). *Mental Models* (Cambridge: Cambridge University Press); Abelson, R. P. & Black, J. B. (1986). Introduction. In J. A. Galambos, R. P. Abelson & J. B. Black eds. *Knowledge Structures* (Hillsdale, NJ: Lawrence Erlbaum Associates); Brewer, W. F. (1987). Schemas versus mental models in human memory. In P. Morris ed. *Modelling Cognition* (Chichester: Wiley); Johnson, M. (1987). *The Body in the Mind – The Bodily Basis of Meaning, Imagination, and Reason* (Chicago: University of Chicago Press).

116 / Neisser, U. (1976). *Cognition and Reality* (San Francisco: Freeman).

117 / Kleinfeld, D. & Sompolinsky, H. (1989). Associative network models for central pattern generators. In C. Koch & I. Segev eds. *Methods in Neuronal Modeling* (Cambridge, MA: Bradford).

118 / Birbaumer, N., Elbert, T., Canavan, A. G. M. & Rockstroh, B. (1990). Slow potentials of the cerebral cortex and behavior. *Physiological Reviews,* **70**, 1–41; Fuster, J. M. (1989). *The Prefrontal Cortex – Anatomy, Physiology, and Neuropsychology of the Frontal Lobe* (New York: Raven).

119 / Cotterill, R. M. J. (1994). Autism, intelligence and consciousness. *Biol.Skr.Dan.Vid.Selsk.*, **45**, 1–93; Duncan, J. (1995). Attention, intelligence, and the frontal lobes. In M. S. Gazzaniga ed. *The Cognitive Neurosciences* (Cambridge, MA: Bradford).

120 / Harlow, J. M. (1868). Recovery after severe injury to the head. *Publications of the Massachusetts Medical Society*, **2**, 327–46.

121 / Damasio, H., Grabowski, T., Frank, R., Galaburda, A. M. & Damasio, A. R. (1994). The return of Phineas Gage: clues about the brain from the skull of a famous patient. *Science*, 264, 1102–5.

122 / Fuster, J. M. (1980). *The Prefrontal Cortex – Anatomy, Physiology, and Neuropsychology of the Frontal Lobe* (New York: Raven Press); Stuss, D. T. & Benson, D. F. (1986). *The Frontal Lobes* (New York: Raven Press); Passingham, R. (1993). *The Frontal*

Lobes and Voluntary Action (Oxford: Oxford University Press); Roberts, A. C., Robbins, T. W. & Weiskrantz, L. eds. (1996). Executive and cognitive functions of the prefrontal cortex. *Philosophical Transactions of the Royal Society, London, Series B*, **351**, 1387–527.

123 / Watanabe, M. (1996). Reward expectancy in primate prefrontal cortex. *Nature*, **382**, 629–32.

124 / Jennings, C. (1996). Tales of the unexpected. *Nature*, **382**, 579.

125 / Gallese, V., Fadiga, L., Fogassi, L. & Rizzolatti, G. (1996). Action recognition in the premotor cortex. *Brain*, **119**, 593–609; Rizzolatti, G., Fadiga, L., Gallese, V. & Fogassi, L. (1996). Premotor cortex and the recognition of motor actions. *Cognitive Brain Research*, **3**, 131–41.

126 / Shadmehr, R. & Holcomb, H. H. (1997). Neural correlates of motor memory consolidation. *Science*, **277**, 821–5.

127 / Berns, G. S., Cohen, J. D. & Mintun, M. A. (1997). Brain regions responsive to novelty in the absence of awareness. *Science*, **276**, 1272–5.

128 / Blakemore, C. (1990). Understanding images in the brain. In H. B. Barlow, C. Blakemore & M. Weston-Smith eds. *Images and Understanding* (Cambridge: Cambridge University Press).

129 / Eccles, J. C. (1965). *The Brain and the Unity of Conscious Experience* (Cambridge: Cambridge University Press).

130 / Harth, E. (1982). *Windows on the Mind – Reflections on the Physical Basis of Consciousness* (Brighton: Harvester Press).

131 / Velmans, M. (1993). A reflexive science of consciousness. In G. R. Bock & J. Marsh eds. *Experimental and Theoretical Studies of Consciousness – Ciba Foundation Symposium 174* (Chichester: Wiley).

132 / von Holst, E. (1957). Aktive leistungen der menschlichen gesichtswahrnehmung. *Studium Generale*, **10**, 231–43.

133 / Carpenter, R. H. S. (1995). R. M. J. Cotterill: Autism, Intelligence and Consciousness [Book review]. *Journal of Consciousness Studies*, **2**, 86–9.

134 / Libet, B., Kobayashi, H. & Tanaka, T. (1975). Synaptic coupling into the production and storage of a neuronal memory trace. *Nature*, **258**, 155–7.

135 / Dale, N. (1989). The role of NMDA receptors in synaptic integration and the organization of complex neural patterns. In J. C. Watkins & G. L. Collingridge eds. *The NMDA Receptor* (Oxford: Oxford University Press).

136 / Goodwin, G. M., McCloskey, D. I. & Matthews, P. B. C (1972). Proprioceptive illusions induced by muscle vibration: contribution by muscle spindles to perception? *Science*, **175**, 1382–4.

137 / McCloskey, D. I., Ebeling, P. & Goodwin, G. M. (1974). Estimation of weights and tensions and apparent involvement of a 'sense of effort'. *Experimental Neurology*, **42**, 220–32.

138 / Gandevia, S. C. (1982). The perception of motor commands or effort during muscular paralysis. *Brain*, **105**, 151–9.

139 / Crick, F. H. C. (1994). *The Astonishing Hypothesis – The Scientific Search for the Soul* (London: Simon & Schuster).

140 / Wurtz, R. H., Goldberg, M. E. & Robinson, D. L. (1980). Behavioral modulation of visual responses in the monkey: stimulus selection for attention and movement. *Progress in Psychobiology and Physiological Psychology*, **9**, 43–83; Wurtz, R. H., Goldberg, M. E. & Robinson, D. L. (1982). Brain mechanisms of visual attention. *Scientific American*, **246(6)**, 124–32.

141 / Noton, D. & Stark, L. (1971). Scanpaths in saccadic eye movements while viewing and recognizing patterns. *Vision Research*, **11**, 929–41.

142 / Zeki, S., Watson, J. D. G., Lueck, C. J., Friston, K. J., Kennard, C. & Frackowiak, R. S. J. (1991). A direct demonstration of functional specialization in human visual cortex. *Journal of Neuroscience*, **11**, 641–9; Howard, R. J., Brammer, M., Wright, I., Woodruff, P. W., Bullmore, E. T. & Zeki, S. (1996). A direct demonstration of functional specialization within motion-related visual and auditory cortex of the human brain. *Current Biology*, **6**, 1015–19.

143 / Felleman, D. J. & Van Essen, D. C. (1991). Distributed hierarchical processing in the primate cerebral cortex. *Cerebral Cortex*, **1**, 1–47.

144 / Damasio, A. R. (1989). The brain binds entities and events by multiregional activation from convergence zones. *Neural Computation*, **1**, 123–32.

145 / Eccles, J. C. (1965). *The Brain and the Unity of*

Conscious Experience (Cambridge: Cambridge University Press).

146 / Carpenter, R. H. S. (1996). *Neurophysiology* (London: Edward Arnold).

147 / Georgopoulos, A. P., Kalaska, J. F., Caminiti, R. & Massey, J. T. (1982). On the relations between the direction of two-dimensional arm movements and cell discharge in primate motor cortex. *Journal of Neuroscience*, **2**, 1527–37; Georgopoulos, A. P., Schwartz, A. B. & Kettner, R. E. (1986). Neuronal population coding of movement direction. *Science*, **233**, 1416–19.

148 / Humphrey, D. R., Schmidt, E. M. & Thompson, L. V. D. (1970). Predicting measures of motor performance from multiple cortical spike trains. *Science*, **179**, 758–62; Erickson, R. P. (1974). Parallel 'population' neural coding in feature extraction. In F. O. Schmitt & F. G. Worden eds. *The Neurosciences – Third Study Program* (Cambridge, MA: MIT Press).

149 / Schwartz, A. B. (1994). Distributed motor processing in cerebral cortex. *Current Opinion in Neurobiology*, **4**, 840–6; Löfqvist, A. & Lindblom, B. (1994). Speech motor control. *Current Opinion in Neurobiology*, **4**, 823–6.

150 / Redish, A. D. & Touretzky, D. S. (1994). The reaching task: evidence for vector arithmetic in the motor system? *Biological Cybernetics*, **71**, 307–17.

151 / Zeki, S. (1992). The visual image in mind and brain. *Scientific American*, **267(3)**, 42–50; Zeki, S. (1993). *A Vision of the Brain* (Oxford: Blackwell Scientific Publications).

152 / Ungerleider, L. G. & Mishkin, M. (1982). Two cortical visual systems. In D. J. Ingle, R. J. W. Mansfield & M. S. Goodale eds. *The Analysis of Visual Behavior* (Cambridge, MA: MIT Press); Shipp, S. (1995). The odd couple. *Current Biology*, **5**, 116–19.

153 / Paillard, J. (1991). Knowing where and knowing how to get there. In J. Paillard ed. *Brain and Space* (New York: Oxford University Press).

154 / Wurtz, R. H., Goldberg, M. E. & Robinson, D. L. (1980). Behavioral modulation of visual responses in the monkey: stimulus selection for attention and movement. *Progress in Psychobiology and Physiological Psychology*, **9**, 43–83; Wurtz, R. H., Goldberg, M. E. & Robinson, D. L. (1982). Brain mechanisms of visual attention. *Scientific American*, **246(6)**, 124–32.

155 / Noton, D. & Stark, L. (1971). Scanpaths in saccadic eye movements while viewing and recognizing patterns. *Vision Research*, **11**, 929–41; Schlag, J. & Schlag-Rey, M. (1990). Colliding saccades may reveal the secret of their marching orders. *Trends in Neurosciences*, **13**, 410–14; Sparks, D. L. (1991). The neural encoding of the location of targets for saccadic eye movements. In J. Paillard ed. *Brain and Space* (New York: Oxford University Press).

156 / Grillner, S., Christenson, J., Brodin, L., Wallén, P., Hill, R. H., Lansner, A. & Ekeberg, Ö. (1989). Locomotor system in lamprey: neural mechanisms controlling spinal rhythm generation. In J. W. Jacklet ed. *Neural and Cellular Oscillators* (New York: Dekker); Grillner, S., Deliagina, T., Ekeberg, Ö., El Manira, A., Hill, R. H., Lansner, A., Orlovsky, G. N. & Wallén, P. (1995). Neural networks that co-ordinate locomotion and body orientation in the lamprey. *Trends in Neurosciences*, **18**, 270–9.

157 / Monod, J. (1972). *Chance and Necessity* (London: Fontana); Jacob, F. (1982). *The Possible and the Actual* (New York: Pantheon); Dawkins, R. (1986). *The Blind Watchmaker* (Harlow: Longman); Dawkins, R. (1996). *Climbing Mount Improbable* (London: Viking).

158 / Gross, C. G. & Graziano, M. S. A. (1995). Multiple representations of space in the brain. *The Neuroscientist*, **1**, 43–50.

159 / Treue, S. & Maunsell, J. H. R. (1996). Attentional modulation of visual motion processing in cortical areas MT and MST. *Nature*, **382**, 539–41.

160 / Britten, K. H. (1996). Attention is everywhere. *Nature*, **382**, 497–8.

161 / Myerson, J., Miezin, F. & Allman, J. (1981). Binocular rivalry in macaque monkeys and humans: a comparative study in perception. *Behavioral Analysis Letters*, **1**, 149–59.

162 / Leopold, D. A. & Logothetis, N. K. (1996). Activity changes in early visual cortex reflect monkeys' percepts during binocular rivalry. *Nature*, **379**, 549–53.

163 / Logothetis, N. K. & Schall, J. D. (1989). Neuronal correlates of subjective visual perception. *Science*, **245**, 761–3.

164 / Crick, F. & Koch, C. (1992). The problem of consciousness. *Scientific American*, **267(3)**, 110–17.

165 / Crick, F. (1996). Visual perception: rivalry and consciousness. *Nature*, **379**, 485–6.

166 / Haken, H. (1977). *Synergetics – An Introduction* (Berlin: Springer); Haken, H. (1987). Synergetic computers for pattern recognition and associative memory. In H. Haken ed. *Computational Systems, Natural and Artificial* (Berlin: Springer).

167 / Haken, H. (1996). *Principles of Brain Functioning – A Synergetic Approach to Brain Activity, Behavior and Cognition* (Berlin: Springer); Ditzinger, T. & Haken, H. (1989). Oscillations in the perception of ambiguous patterns: a model based on synergetics. *Biological Cybernetics*, **61**, 279–87.

168 / Kelso, J. A. S. (1995). *Dynamic Patterns – The Self-Organization of Brain and Behavior* (Boston: MIT Press).

169 / Thompson, K. G., Hanes, D. P., Bichot, N. P. & Schall, J. D. (1996). Perceptual and motor processing stages identified in the activity of macaque frontal eye field neurons during visual search. *Journal of Neurophysiology*, **76**, 4040–55.

170 / Matthews, P. B. C. (1982). Where does Sherrington's 'muscle sense' originate? Muscles, joints, corollary discharges? *Annual Review of Neuroscience*, **5**, 189–218.

171 / Bell, C. (1833). *The Hand – Its Mechanism and Vital Endowments as Evincing Design* (London: Pickering). Reprinted 1979 (Brentwood, Essex: Pilgrims Press).

172 / Leksell, L. (1945). The action potential and excitatory effects of the small ventral root fibres to skeletal muscle. *Acta. Physiologica Scandinavia*, **10 (Suppl. 31)**, 1–84.

173 / Hunt, C. C. & Kuffler, S. W. (1951). Stretch receptor discharges during muscle contraction. *Journal of Physiology (London)*, **113**, 298–315.

174 / Granit, R. (1970). *The Basis of Motor Control* (London: Academic Press).

175 / Eccles, J. C. & Gibson, W. C. (1979). *Sherrington – His Life and Thought* (Berlin: Springer International).

176 / Matthews, P. B. C. (1982). Where does Sherrington's 'muscle sense' originate? Muscles, joints, corollary discharges? *Annual Review of Neuroscience*, **5**, 189–218.

177 / Amassian, V. E., Cracco, R. Q. & Maccabee, P. J. (1989). A sense of movement elicited in paralysed distal arm by focal magnetic coil stimulation of human motor cortex. *Brain Research*, **479**, 355–60; Amassian, V. E., Somasundaram, M., Rothwell, J. C., Britton, T., Cracco, J. B., Cracco, R. Q., Maccabee, P. J. & Day, B. L. (1991). Parasthesias are elicited by single pulse magnetic coil stimulation of motor cortex in susceptible humans. *Brain*, **114**, 2505–20; Amassian, V. E., Hassan, N., Cracco, J. B., Maccabee, P. J., Cracco, R. Q. & Henry, K. (1995). A role of frontal lobe in human visual perception. *Society for Neuroscience*, **21**, 23.

178 / Boden, M. A. (1993). Multiple personality and computational models. *Cognitive Science Research Papers – Number 299* (Brighton: University of Sussex).

179 / Thigpen, C. H. & Cleckley, H. M. (1957). *The Three Faces of Eve* (London: Secker & Warburg).

180 / Lethin, A. (1977). Muscle tone and feeling. *Energy and Character*, **8**, 42–8.

181 / Ommaya, A. K. & Gennarelli, T. A. (1975). Head Injuries. *Second Chicago Symposium on Neural Trauma* (New York: Grune & Stratton).

182 / Cole, J. & Paillard, J. (1995). Living without touch and peripheral information about body position and movement: studies with deafferented subjects. In J. L. Burmúdez, A. Marcel & N. Eilan eds. *The Body and Self* (Cambridge, MA: MIT Press); Cole, J. (1995). *Pride and a Daily Marathon* (Cambridge, MA: MIT Press).

183 / Plum, F. & Posner, J. B. (1980). *The Diagnosis of Stupor and Coma* (Philadelphia: F. A. Davis & Co.).

184 / Miller, E. K. & Desimone, R. (1994). Parallel neuronal mechanisms for short-term memory. *Science*, **263**, 520–2; DeYoe, E. A. Felleman, D. J., Van Essen, D. C. & McClendon, E. (1994). Multiple processing streams in occipitotemporal visual cortex. *Science*, **371**, 151–4.

185 / Wilson, F. A. W., Scalaidhe, S. P. O. & Goldman-Rakic, P. S. (1993). Dissociation of object and spatial processing domains in primate prefrontal cortex. *Science*, **260**, 1955–8; Goodale, M. A. & Milner, A. D. (1992). Separate visual pathways for perception and action. *Trends in Neurosciences*, **15**, 20–5.

186 / Rao, S. C., Rainer, G. & Miller, E. K. (1997). Integration of what and where in the primate prefrontal cortex. *Science*, **276**, 821–4.

187 / Squire, L. R., Knowlton, B. & Musen, G. (1993). The structure and organization of memory. *Annual Review of Psychology*, **44**, 452–95; Parkin, A. J. (1996). Human memory: the hippocampus is the key. *Current Biology*, **6**, 1583–5.

188 / Ullman, M. T., Corkin, S., Coppola, M., Hickok, G., Growdon, J. H., Koroshetz, W. J. & Pinker, S. (1997). A neural dissociation within language: evidence that the mental dictionary is part of declarative memory, and that grammatical rules are processed by the procedural system. *Journal of Cognitive Neuroscience*, **9(2)**, 266–76.

189 / Marslen-Wilson, W. D. & Tyler, L. K., (1997). Dissociating types of mental computation. *Nature*, **387**, 592–4; Pinker, S. (1997). Words and rules in the human brain. *Nature*, **387**, 547–8.

190 / Knight, R. T. (1996). Contribution of human hippocampal region to novelty detection. *Nature*, **383**, 256–9.

191 / Tsukada, M., Aihara, T., Mizuno, M., Kato, H. & Ito, K. (1994). Temporal pattern sensitivity of long-term potentiation in hippocampal CA1 neurons. *Biological Cybernetics*, **70**, 495–503.

192 / Tsien, J. Z., Huerta, P. T. & Tonegawa, S. (1996). The essential role of hippocampal CA1 NMDA receptor-dependent synaptic plasticity in spatial memory. *Cell*, **87**, 1327–38.

193 / McHugh, T. J., Blum, K. I., Tsien, J. Z., Tonegawa, S. & Wilson, M. A. (1996). Impaired hippocampal representation of space in CA1-specific NMDAR1 knockout mice. *Cell*, **87**, 1339–49.

194 / Dinse, H. R. & Krüger, K. (1994). The timing of processing along the visual pathway of the cat. *NeuroReport*, **5**, 893–7.

195 / Singer, W. (1995). Development and plasticity of cortical processing architectures. *Science*, **270**, 758–64.

196 / Velmans, M. (1991). Is human information processing conscious? *Behavioral and Brain Sciences*, **14**, 651–726.

197 / Posner, M. I. & Snyder, C. R. R. (1975). Facilitation and inhibition in the processing of signals. In P. M. A. Rabbitt & S. Dornick eds. *Attention and Performance* (New York: Academic Press); Neeley, J. H. (1977). Semantic priming and retrieval from lexical memory: roles of inhibitionless spreading activation and limited capacity attention. *Journal of Experimental Psychology – General*, **106**, 226–54.

198 / Zeki, S. (1993). *A Vision of the Brain* (Oxford: Blackwell Scientific Publications).

199 / Carpenter, R. H. S. (1995). R. M. J. Cotterill: *Autism, Intelligence and Consciousness* [book review]. *Journal of Consciousness Studies*, **2**, 86–9.

200 / Baddeley, A. D. (1992). Working memory. *Science*, **255**, 556–9.

201 / Baars, B. J. (1988). *A Cognitive Theory of Consciousness* (Cambridge: Cambridge University Press).

202 / Dennett, D. C. (1991). *Consciousness Explained* (New York: Little, Brown & Co.).

203 / Eilan, N. (1995). Consciousness and the self. In J. L. Bermúdez, A. Marcel & N. Eilan eds. *The Body and Self* (Cambridge, MA: MIT Press).

204 / Evans, G. (1983). *The Varieties of Reference* (Oxford: Clarendon Press).

205 / Gallagher, S. (1995). Body schema and intentionality. In J. L. Bermúdez, A. Marcel & N. Eilan eds. *The Body and Self* (Cambridge, MA: MIT Press).

206 / Neisser, U. (1993). The self perceived. In U. Neisser ed. *The Perceived Self – Ecological and Interpersonal Sources of Self Knowledge* (Cambridge: Cambridge University Press).

207 / Quine, W. V. O. (1960). *Word and Object* (Cambridge, MA: MIT Press); see also the dialogue with Bryan Magee in: Magee, B. (1978). *Men of Ideas* (London: BBC Publications); and Quine, W. V. O. (1969). Natural kinds. In *Ontological Relativity and Other Essays* (New York: Columbia University Press).

208 / Josephson, B. D., Rubik, B. A., Fontana, D. & Lorimer, D. (1992). Defining consciousness. *Nature*, **358**, 618; Josephson, B. D. & Rubik, B. A. (1992). The challenge of consciousness research. *Frontier Perspectives*, **3**, 15–19; Josephson, B. D. (1985). Conversation two: with Roger Sperry. In N. Cousins ed. *Nobel Prize Conversations* (San Francisco: Saybrook).

209 / Popper, K. R. & Eccles, J. C. (1977). *The Self and its Brain* (Berlin: Springer-Verlag).

210 / Eccles, J. C. (1986). Do mental events cause neural events analogously to the probability fields of quantum mechanics? *Proceedings of the Royal Society, London, Series B*, **227**, 411–28; Eccles, J. C. (1989). *Evolution of the Brain – Creation of the Self* (New York: Routledge, Chapman & Hall); Eccles, J. C. (1990). A unitary hypothesis of mind-brain interaction in

cerebral cortex. *Proceedings of the Royal Society, London, Series B*, **240**, 433–51.

211 / Beck, F. & Eccles, J. C. (1992). Quantum aspects of brain activity and the role of consciousness. *Proceedings of the National Academy of Sciences USA*, **89**, 11357–61; Beck, F. (1994). Quantum mechanics and consciousness. *Journal of Consciousness Studies*, **1**, 253–5.

212 / Eccles, J. C. (1985). Conversation one: with Roger Sperry. In N. Cousins ed. *Nobel Prize Conversations* (San Francisco: Saybrook). In this conversation, Eccles cites the work of Lassen and Roland.

213 / Frith, C. D., Friston, K., Liddle, P. F. & Frackowiak, R. S. J. (1991). Willed action and the prefrontal cortex in man: a study with PET. *Proceedings of the Royal Society, London, Series B*, **244**, 241–6.

214 / Empson, J (1989). *Sleep and Dreaming* (London: Faber and Faber); Horne, J. (1988). *Why We Sleep* (Oxford: Oxford University Press); Hobson, J. A. (1988) *The Dreaming Brain* (London: Penguin); Hobson, J. A. (1995) *Sleep* (New York: Scientific American Library); Hobson, J. A. (1994) *The Chemistry of Conscious States – How the Brain Changes its Mind* (Boston: Little, Brown & Co.).

215 / Meddis, R. (1975). On the function of sleep. *Animal Behaviour*, **23**, 676–91.

216 / Hobson, J. A. (1990). Sleep and dreaming. *Journal of Neuroscience*, **10**, 371–82.

217 / Berger, H. (1929). Über das Elektrenkephalogramm des Menschen. *Arkiv fur Psychiatrie und Nervenkrankeiten*, **87**, 527–70.

218 / Aserinsky, E. & Kleitman, N. (1953). Regularly occurring periods of eye motility and concomitant phenomena during sleep. *Science*, **118**, 273–4.

219 / Dement, W. C. & Kleitman, N. (1957). Cyclic variations in EEG during sleep and their relations to eye movements, body motility, and dreaming. *EEG Clinical Neurophysiology*, **9**, 673–90.

220 / Dement, W. C. (1960). The effect of dream deprivation. *Science*, **131**, 1705–7.

221 / Jouvet, M. (1967). Neurophysiology of the states of sleep. *Physiological Review*, **47**, 117–77.

222 / Freud, S. (1955). *The Interpretation of Dreams* (New York, Basic Books); Hobson, J. A. (1988). *The Dreaming Brain* (New York: Basic Books).

223 / Hartmann, E. (1968). The 90-Minute Sleep–Dream Cycle. *Archives of General Psychiatry*, **18**, 280–6; Hartmann, E. (1973). *The Functions of Sleep* (New Haven: Yale University Press).

224 / Karni, A., Tanne, D., Rubenstein, B. S., Askenasy, J. J. M. & Sagi, D. (1994). Dependence on REM sleep of overnight improvement of a perceptual skill. *Science*, **265**, 679–82.

225 / Hobson, J. A. & Stickgold, R. (1995). Sleep the beloved teacher? *Current Biology*, **5**, 35–6.

226 / Crick, F. H. C. & Mitchison, G. (1983). The function of dream sleep. *Nature*, **304**, 111–14; Crick, F. H. C. & Mitchison, G. (1986). REM sleep and neural nets. *Journal of Mind and Behaviour*, **7**, 229–49.

227 / Cavallero, C. & Foulkes, D. eds. (1993). *Dreaming as Cognition* (New York: Harvester Wheatsheaf).

228 / Cotterill, R. M. J. (1989). *No Ghost in the Machine – Modern Science and the Brain, the Mind and the Soul* (London: Heinemann).

229 / Pompeiano, O. (1979). Cholinergic activation of reticular and vestibular mechanisms controlling posture and eye movements. In J. A. Hobson & M. A. B. Brazier eds. *The Reticular Formation Revisited* (New York: Raven Press).

230 / Eccles, J. C. (1965). *The Brain and the Unity of Conscious Experience* (Cambridge: Cambridge University Press).

231 / Helmholtz, H. (1867). Handbuch der physiologischen Optik. In G. Karsten ed. *Allgemeine Encyklopädie der Physik*, **9**, 457–529.

232 / Sherrington, C. S. (1918). Observations on the sensual role of the proprioceptive nerve-supply of the extrinsic ocular muscles. *Brain*, **41**, 332–43.

233 / Cooper, S., Daniel, P. M. & Whitteridge, D. (1955). Muscle spindles and other sensory endings in the extrinsic eye muscles: the physiology and anatomy of these receptors and of their connexions with the brain-stem. *Brain*, **78**, 564–83.

234 / McCarley, R. W. & Hobson, J. A. (1977). The neurobiological origins of psychoanalytic dream theory. *American Journal of Psychiatry*, **134**, 1211–21; McCarley, R. W. & Hobson, J. A. (1977). The brain as a dream state generator: an activation-synthesis hypothesis of the dream process. *American Journal of*

Psychiatry, **134**, 1335–48; W. P. Koella & P. Levin eds. (1977). *Sleep 1976* (Basel: Karger).

235 / Weiskrantz, L. (1986). *Blindsight – A Case Study and Implications* (Oxford: Oxford University Press).

236 / Davis, B. D. (1985). Sleep and the maintenance of memory. *Perspectives in Biology and Medicine*, **28**, 457–64.

237 / Maquet, P., Péters, J., Aerts, J., Delfiore, G., Degueldre, C., Luxen, A. & Franck, G. (1996). Functional neuroanatomy of human rapid-eye-movement sleep and dreaming. *Nature*, **383**, 163–6.

Chapter 10

1 / Barlow, H. B. (1983). Intelligence, guesswork, language. *Nature*, **304**, 207–209.

2 / Walker, S. (1983). *Animal Thought* (London:Routledge & Kegan Paul).

3 / Jaynes, J. (1979). *The Origin of Consciousness in the Breakdown of the Bicameral Mind* (New York: Houghton & Mifflin).

4 / Dennett, D. C. (1991). *Consciousness Explained* (New York: Little, Brown & Co.).

5 / Keller, H. A. (1903). *The Story of My Life* (London: Hodder & Stoughton).

6 / Keller, H. A. (1908). *The World I Live In* (London: Hodder & Stoughton).

7 / Deacon, T. (1190). Brain–language coevolution. In J. A. Hawkins & M. Gell-Mann eds. *The Evolution of Human Languages: Proceedings of the Sante Fe Institute Studies in the Sciences of Complexity* (San Francisco: Addison-Wesley).

8 / Deacon, T. (1990). Rethinking mammalian brain evolution. *American Zoologist*, **30**, 629–705.

9 / Bates, E., Thal, D. & Marchman, V. (1991). Symbols and syntax: a Darwinian approach to language development. In N. Krasnegor, D. Rumbaugh, E. Schiefelbusch and M. Studdert-Kennedy eds. *Biological and Behavioral Determinants of Language Development* (Hillsdale, NJ: Erlbaum).

10 / Sereno, M. (1990). *Language and the Primate Brain* (San Diego: California University Center for Research in Language).

11 / Calvin, W. H. (1994). The emergence of intelligence. *Scientific American*, **271(4)**, 79–85.

12 / Clark, A. (1997). *Being There – Putting Brain, Body and World Together Again* (Cambridge, MA: MIT Press).

13 / Cotterill, R. M. J. (1995). On the unity of conscious experience. *Journal of Consciousness Studies*, **2**, 290–312.

14 / Cotterill, R. M. J. (1997). On the mechanism of consciousness. *Journal of Consciousness Studies*, **4**, 231–47.

15 / Frith, C. D. & Done, D. J. (1989). Experiences of alien control in schizophrenia reflect a disorder in the central monitoring of action. *Psychological Medicine*, **19**, 359–63; Frith, C. D. (1992). *The Cognitive Neuropsychology of Schizophrenia* (Hove: Lawrence Erlbaum); Friston, K. J. & Frith, C. D. (1995). Schizophrenia: a disconnection syndrome? *Clinical Neuroscience*, **3**, 89–97; Frith, C. D. (1996). The role of the prefrontal cortex in self-consciousness: the case of auditory hallucinations. *Philosophical Transactions of the Royal Society, Series B*, **351**, 1505–12.

16 / Damasio, A. R. (1994). *Descartes' Error – Emotion, Reason and the Human Brain* (New York: Grosset-Putnam).

17 / Miall, R. C., Weir, D. J., Wolpert, D. M. & Stein, J. F. (1993). Is the cerebellum a Smith predictor? *Journal of Motor Behavior*, **25**, 203–16.

18 / Goodwin, G. M., McCloskey, D. I. & Matthews, P. B. C. (1972). Proprioceptive illusions induced by muscle vibration: contribution by muscle spindles to perception? *Science*, **175**, 1382–4; Lethin, A. (1977). Muscle tone and feeling. *Energy and Character*, **8**, 42–8.

19 / Purpura, K. P. & Schiff, N. D. (1997). The thalamic intralaminar nuclei: a role in visual awareness. *The Neuroscientist*, **3**, 8–15.

20 / Barth, D. S. & MacDonald, K. D. (1996). Thalamic modulation of high-frequency oscillating potentials in auditory cortex. *Nature*, **383**, 78–81.

21 / Steriade, M. (1996). Awakening the brain. *Nature*, **383**, 24–5.

22 / Steriade, M. & Glenn, L. L. (1982). Neocortical and caudate projections of intralaminar thalamic neurons and their synaptic excitation from midbrain reticular core. *Journal of Neurophysiology*, **48**, 352–71; Steriade,

M. & Llinás, R. (1988). The functional states of the thalamus associated neuronal interplay. *Physiological Reviews*, **68**, 649–742.

23 / Traub, R. D., Whittington, M. A., Stanford, I. M. & Jefferys, J. G. R. (1996). A mechanism for generation of long-range synchronous fast oscillations in the cortex. *Nature*, **383**, 621–4.

24 / Gray, C. M., König, P., Engel, A. K. & Singer, W. (1989). Oscillatory responses in cat visual cortex exhibit inter-columnar synchronization which reflects global stimulus properties. *Nature*, **338**, 334–7.

25 / Contreras, D., Destexhe, A., Sejnowski, T. J. & Steriade, M. (1996). Control of spatiotemporal coherence of a thalamic oscillation by corticothalamic feedback. *Science*, **274**, 771–4.

26 / Wehr, M. & Laurent, G. (1996). Odour encoding by temporal sequences of firing in oscillating neural assemblies. *Nature*, **384**, 162–6; Miller, J. P. (1996). Brain waves deciphered. *Nature*, **384**, 115–17.

27 / Koch, C. (1997). Computation and the single neuron. *Nature*, **385**, 207–10.

28 / Mel, B. W. (1992). NMDA-based pattern discrimination in a modeled cortical neuron. *Neural Computation*, **4**, 502–17; Mel, B. W. (1994). Information processing in dendritic trees. *Neural Computation*, **6**, 1031–85.

29 / Salmelin, R., Hari, R., Lounasmaa, O. V. & Sams, M. (1994). Dynamics of brain activation during picture naming. *Nature*, **368**, 463–5.

30 / Thorpe, S. J., Rolls, E. T. & Madison, S. (1983). The orbitofrontal cortex: neuronal activity in the behaving monkey. *Experimental Brain Research*, **49**, 93–115.

31 / Singer, W. (1995). Development and plasticity of cortical processing architectures. *Science*, **270**, 758–64.

32 / Dehaene-Lambertz, G. & Dehaene, S. (1994). Speed and cerebral correlates of syllable discrimination in infants. *Nature*, **370**, 292–5.

33 / Thorpe, S. J. & Imbert, M. (1989). Biological constraints on connectionist modelling. In R. Pfeifer, Z. Schreter, F. Fogelman-Soulié & L. Steels eds. *Connectionism in Perspective* (Amsterdam: Elsevier); Tovée, M. J. (1994). How fast is the speed of thought? *Current Biology*, **4**, 1125–7.

34 / Sawaguchi, T. (1994). Modular activation and suppression of neocortical activity in the monkey revealed by optical imaging. *NeuroReport*, **6**, 185–9.

35 / LeDoux, J. E. (1992). Brain mechanisms of emotion and emotional learning. *Current Opinion in Neurobiology*, **2**, 191–7; LeDoux, J. E. (1994). Emotion, memory and the brain. *Scientific American*, **270(6)**, 32–9.

36 / McWhirter, N. & McWhirter, R. eds. (1963). *The Guinness Book of World Records* (New York: Bantam Books).

37 / Bishop, D. V. M. (1997). Listening out for subtle deficits. *Nature*, **387**, 129–30.

38 / Crick, F. H. C. & Koch, C. (1990). Towards a neurobiological theory of consciousness. *Seminars in the Neurosciences*, **2**, 263–75; Crick, F. H. C. & Koch, C. (1991). Some reflections on visual awareness. *Cold Spring Harbour Symposia on Quantitative Biology*, **55**, 953–62; Crick, F. H. C. (1994). *The Astonishing Hypothesis* (London: Simon & Schuster).

39 / Carpenter, R. H. S. (1994). Choosing where to look. *Current Biology*, 4, 341–3.

40 / Kaas, J. H. (1992). Do humans see what monkeys see? *Trends in Neuroscience*, **15**, 1–3.

41 / Passingham, R. E. (1993). *The Frontal Lobes and Voluntary Action* (Oxford: Oxford University Press).

42 / Rasmussen, T. & Milner, B. (1977). The role of early left-brain injury in determining lateralization of cerebral speech functions. *Annals of the New York Academy of Sciences*, **299**, 355–69.

43 / Tanji, J. & Mushiake, H. (1996). Comparison of neuronal activity in the supplementary motor area and the primary motor cortex. *Cognitive Brain Research*, **3**, 143–50.

44 / Geschwind, N. & Levitsky, W. (1968). Human-brain: left–right asymmetries in temporal speech region. *Science*, **161**, 186–7.

45 / Wing, L. (1971). *Autistic Children – A Guide for Parents* (London: Constable); Schopler, E. & Mesibov, G. B. eds. (1987). *Neurobiological Issues in Autism* (New York: Plenum); Treffert, D. (1989). *Extraordinary People* (London: Bantam); Frith, U. (1989). *Autism – Explaining the Enigma* (Oxford: Basil Blackwell); Gillberg, C. & Coleman, M. (1992). *The Biology of the Autistic Syndromes* (Oxford: Blackwell Scientific); Naruse, H. & Ornitz, E. M. (1992). *Neurobiology of Infantile Autism* (Amsterdam: Excerpta Medica); Hobson, R. P. (1993). *Autism and the Development of*

Mind (Hove: Lawrence Erlbaum); Baron-Cohen, S. & Bolton, P. (1993). *Autism – The Facts* (Oxford: Oxford University Press).

46 / Kanner, L. (1943). Autistic disturbances of affective contact. *Nervous Child*, **2**, 217–50.

47 / Asperger, H. (1944). Die autistichen Psychopathen im Kindesalter. *Arkiv für Psychiatrie und Nervenkrankheiten*, **117**, 76–136.

48 / Wing, L. & Gould, J. (1979). Severe impairments of social interaction and associated abnormalities in children: epidemiology and classification. *Journal of Autism and Developmental Disorders*, 9, 11–30.

49 / Wimmer, H. & Perner, J. (1983). Beliefs about beliefs: representation and constraining function of wrong beliefs in young children's understanding of perception. *Cognition*, **21**, 103–28.

50 / Lewis, D. (1969). *Convention – A Philosophical Study* (Cambridge, MA: Harvard University Press).

51 / Baron-Cohen, S., Leslie, A. M. & Frith, U. (1985). Does the autistic child have a 'theory of mind'? *Cognition*, **21**, 37–46.

52 / Leslie, A. M. & Frith, U. (1988). Autistic children's understanding of seeing, knowing and believing. *British Journal of Developmental Psychology*, **4**, 315–24.

53 / Frith, U. (1989). *Autism – Explaining the Enigma* (Oxford: Basil Blackwell).

54 / McGinn, C. (1991). *The Problem of Consciousness* (Oxford: Basil Blackwell).

55 / Williams, D. (1992). *Nobody Nowhere – The Remarkable Autobiography of an Autistic Girl* (London: Doubleday).

56 / Grandin, T. (1986). *Emergence Labeled Autistic* (New York: Arena Press).

57 / Folstein, S. & Rutter, M. (1977). Infantile autism: a genetic study of 21 twin pairs. *Journal of Child Psychology and Psychiatry*, **18**, 297–321.

58 / Bauman, M. L. (1991). Microscopic neuroanatomic abnormalities in autism. *Pediatrics*, **87**, 791–6.

59 / Bauman, M. L. & Kemper, T. L. (1985). Histoanatomic observations of the brain in early infantile autism. *Neurology*, **35**, 866–74.

60 / Arin, D. M., Bauman, M. L. & Kemper, T. L. (1991). The distribution of Purkinje cell loss in the cerebellum in autism. *Neurology*, **41 (Supplement 1)**, 307.

61 / Ritvo, E. R., Freemen, B. J., Scheibel, A. B., Duong, T., Robinson, H., Guthrie, D. & Ritvo, A. (1986). Lower Purkinje cell counts in the cerebella of four autistic subjects: initial findings of the UCLA–NSAC autopsy research report. *American Journal of Psychiatry*, **143**, 862–6.

62 / Courchesne, E. (1991). Neuroanatomic imaging in autism. *Pediatrics*, **87**, 781–90; Courchesne, E. (1989). Neuroanatomical systems involved in infantile autism: the implications of cerebellar abnormalities. In G. Dawson ed. *Autism* (New York: Guildford Press).

63 / Murakami, J. W., Courchesne, E., Press, G. A., Yeung-Courchesne, R. & Hesselink, J. R. (1989). Reduced cerebellar hemisphere size and its relationship to vermal hypoplasia in autism. *Archives of Neurology*, **46**, 689–94.

64 / Damasio, A. R. & Maurer, R. G. (1978). A neurological model for childhood autism. *Archives of Neurology*, **35**, 778–86.

65 / Ornitz, E. M. (1983). The functional neuroanatomy of infantile autism. *International Journal of Neuroscience*, **19**, 85–124.

66 / Rumsey, J. M. (1992). PET scan studies of autism: review and future directions. In H. Naruse & E. M. Ornitz eds. *Neurobiology of Infantile Autism* (Amsterdam: Excerpta Medica).

67 / Happé, F. (1994). *Autism – An Introduction to Psychological Theory* (London: University College Press).

68 / Dawson, G., Finley, C., Phillips, P. & Galpert, L. (1987). P300 of the auditory evoked potential and the language abilities of autistic children. Cited by Courchesne, E. (1987). A neurophysiological view of autism. In E. Schopler & G. B. Mesibov eds. *Neurobiological Issues in Autism* (New York: Plenum).

69 / Ertl, J. P. & Schafer, E. W. P. (1969). Brain response correlates of psychometric intelligence. *Nature*, **223**, 421–2.

70 / Miller, G. A. (1956). The magical number seven, plus or minus two: some limits on our capacity for processing information. *The Psychological Review*, **63**, 81–97.

71 / Polich, J., Howard, L. & Starr, A. (1983). P300

latency correlates with digit span. *Psychophysiology*, **20**, 665–9.

72 / Ellis, N. R. (1963). The stimulus trace and behavioral inadequacy. In N. R. Ellis ed. *Handbook of mental deficiency* (New York: McGraw-Hill).

73 / Miall, R. C., Weir, D. J., Wolpert, D. M. & Stein, J. F. (1993). Is the cerebellum a Smith predictor? *Journal of Motor Behavior*, **25**, 203–16.

74 / Calvin, W. H. (1994). The emergence of intelligence. *Scientific American*, **271(4)**, 79–85.

75 / Boden, M. A. (1990). *The Creative Mind – Myths and Mechanisms* (London: George Weidenfeld & Nicolson).

76 / Leslie, A. M. (1987). Pretense and representation: the origins of the 'Theory of Mind'. *Psychological Review*, **94**, 412–26.

77 / Pickles, A., Bolton, P., Macdonald, H., Bailey, A., Le Couteur, A., Sim, C-H. & Rutter, M. (1995). Latent-class analysis of recurrence risks for complex phenotypes with selection and measurement error: a twin and family history study of autism. *American Journal of Human Genetics*, **57**, 717–26.

78 / Goleman, D. (1995). *Emotional Intelligence – Why It Can Matter More Than IQ* (New York: Bantam Books).

79 / Sutherland, S. (1996). Sensible to feeling. *Nature*, **379**, 34–5.

80 / Bechara, A., Tranel, D., Damasio, H. & Damasio, A. R. (1996). Failure to respond autonomically to anticipated future outcomes following damage to prefrontal cortex. *Cerebral Cortex*, **6**, 215–25.

81 / LeDoux, J. E. (1992). Brain mechanisms of emotion and emotional learning. *Current Opinion in Neurobiology*, **2**, 191–7; LeDoux, J. E. (1994). Emotion, Memory and the Brain. *Scientific American*, **270(6)**, 32–9.

82 / Kapp, B. S., Pascoe, J. P. & Bixler, M. A. (1984). The amygdala: a neuroanatomical systems approach to its contributions to aversive conditioning. In N. Butler & L. R. Squire eds. *Neuropsychology of Learning* (New York: Guildford); Kapp, B. S., Whalen, P. J., Supple, W. F. & Pascoe, J. P. (1991). Amygdaloid contributions to conditioned arousal and sensory information processing. In J. Aggleton ed. *The Amygdala – Neurobiological Aspects of Emotion, Memory and Mental Dysfunction* (New York: Wiley-Liss).

83 / Killcross, S., Robbins, T. W. & Everitt, B. J. (1997). Different types of fear-conditioned behaviour mediated by separate nuclei within amygdala. *Nature*, **388**, 377–80.

84 / LeDoux, J. E., Cicchetti, P., Xagoraris, A. & Romanski, L. M. (1990). The lateral amygdaloid nucleus: sensory interface of the amygdala in fear conditioning. *Journal of Neuroscience*, **10**, 1062–9.

85 / LeDoux, J. E., Farb, C. R. & Romanski, L. M. (1991). Overlapping projections to the amygdala and striatum from auditory processing areas of the thalamus and cortex. *Neuroscience Letters*, **134**, 139–44; LeDoux, J. E., Farb, C. R. & Milner, T. A. (1991). Ultrastructure and synaptic associations of auditory thalamo-amygdala projections in the rat. *Experimental Brain Research*, **85**, 577–86.

86 / Weinberger, N., Ashe, J., Metherate, R., McKenna, T., Diamond, D., Baking, J., Lennartz, R. & Cassady, J. (1990). Neural adaptive information processing: a preliminary model of receptive-field plasticity in auditory cortex during pavlovian conditioning. In M. Gabriel & J. Moore eds. *Learning and Computational Neuroscience – Foundations of Adaptive Networks* (Cambridge, MA: MIT Press).

87 / LeDoux, J. E. & Farb, C. R. (1991). Neurons of the acoustic thalamus that project to the amygdala contain glutamate. *Neuroscience Letters*, **134**, 145–9.

88 / Clugnet, M. C. & LeDoux, J. E. (1990). Synaptic plasticity in fear conditioning circuits: induction of LTP in the lateral nucleus of the amygdala by stimulation of the medial geniculate body. *Journal of Neuroscience*, **10**, 2818–24.

89 / Gewirtz, J. C. & Davis, M. (1997). Second-order fear conditioning prevented by blocking NMDA receptors in amygdala. *Nature*, **388**, 471–4.

90 / Brown, T. H., Chapman, P. F., Kairiss, E. W. & Keenan, C. L. (1988). Long-term synaptic potentiation. *Science*, **242**, 724–8.

91 / Kim, J. J., DeCola, J. P., Landeira-Fernandez, J. & Fanselow, M. S. (1991). *N*-methyl-D-aspartate receptor antagonist APV blocks acquisition but not expression of fear conditioning. *Behavioral Neuroscience*, **105**, 160–7.

92 / Zola-Morgan, S., Squire, L. R., Alvarez-Royo, P. & Clower, R. P. (1991). Independence of memory functions and emotional behavior: separate contribu-

tions of the hippocampal formation and the amygdala. *Hippocampus*, **1**, 207–20.

93 / Knight, R. T. (1996). Contribution of human hippocampal region to novelty detection. *Nature*, **383**, 256–9.

94 / Tsukada, M., Aihara, T., Mizuno, M., Kato, H. & Ito, K. (1994). Temporal pattern sensitivity of long-term potentiation in hippocampal CA1 neurons. *Biological Cybernetics*, **70**, 495–503.

95 / DeLong, G. R. (1992). Autism, amnesia, hippocampus, and learning. *Neuroscience & Behavioral Reviews*, **16**, 63–70.

96 / Rolls, E. T. (1986). A theory of emotion and its application to understanding the neural basis of emotion. In Y. Oomur ed. *Emotions – Neural and Chemical Control* (Tokyo: Japan Scientific Society Press).

97 / Davis, M., Hitchcock, J. M. & Rosen, J. B. (1987). Anxiety and the amygdala: pharmacological and anatomical analysis of the fear-potentiated startle paradigm. In G. H. Bower ed. *The Psychology of Learning and Motivation* (San Diego: Academic Press).

98 / Dolan, R. J. & Fletcher, P. C. (1997). Dissociating prefrontal and hippocampal function in episodic memory encoding. *Nature*, **388**, 582–5.

99 / Skoyles, J. R. (1997). Evolution's 'missing link': a hypothesis upon neural plasticity, prefrontal working memory and the origins of modern cognition. *Medical Hypotheses*, **48**, 499–501.

100 / Duncan, J. (1995). Attention, intelligence, and the frontal lobes. In M. S. Gazzaniga ed. *The Cognitive Neurosciences* (Cambridge, MA: Bradford).

101 / Cabeza, R., Kapur, S., Craik, F. I. M., McIntosh, A. R., Houle, S. & Tulving, E. (1997). Functional neuroanatomy of recall and recognition: a PET study of episodic memory. *Journal of Cognitive Neuroscience*, **9**, 254–65; Tulving, E., Markowitsch, H. J., Kapur, S., Habib, R. & Houle, S. (1994). Novelty encoding networks in the human brain: positron emission tomography data. *NeuroReport*, **5**, 2525–8.

102 / Lane, R. D., Ahern, G. L., Schwartz, G. E. & Kaszniak, A. W. (1997). Is alexithymia the emotional equivalent of blindsight? *Biological Psychiatry*, **42**, 834–44.

103 / Schwartz, G. E. (1983). Disregulation theory and disease: applications to the repression/cerebral disconnection/ cardiovascular disorder hypothesis. *International Review of Applied Psychology*, **32**, 95–118.

104 / Paillard, J. F., Michel, F. & Stelmach, G. (1983). Localization with content: a tactile analogue of 'blindsight'. *Archives of Neurology*, **40**, 548–51.

105 / Hebb, D. O. (1980). *Essay on Mind* (Hillsdale, NJ: Lawrence Erlbaum Associates).

106 / Thompson, W. R. & Heron, W. (1954). The effects of restricting experience on the problem-solving capacity of dogs. *Canadian Journal of Psychology*, **8**, 17–31; Melzack, R. & Scott, T. H. (1957). The effects of early experience on the response to pain. *Journal of Comparative and Physiological Psychology*, **50**, 155–61.

107 / Papez, J. W. (1937). A proposed mechanism of emotion. *Archives of Neurology and Psychiatry*, **79**, 217–24.

108 / Gray, J. A. (1995). The content of consciousness: a neuropsychological conjecture. *Behavioral and Brain Sciences*, **18**, 659–722.

109 / Premack, D. & Woodruff, G. (1978). Does a chimpanzee have a theory of mind? *Behavioral and Brain Sciences*, **4**, 515–26.

110 / Walker, S. (1983). *Animal Thought* (London: Routledge & Kegan Paul); Radner, D. & Radner, M. (1989). *Animal Consciousness* (Buffalo, NY: Prometheus); Griffin, D. R. (1992). *Animal Minds* (Chicago: University of Chicago Press); Dawkins, M. S. (1993). *Through Our Eyes Only?* (Oxford: Freeman).

111 / Kummer, H. & Goodall, J. (1985). Conditions of innovative behaviour in primates. *Philosophical Transactions of the Royal Society of London, Series B*, **308**, 203–14; Kummer, H., Banaja, A. A., Abokhatwa, A. N. & Ghandour, A. M. (1985). Differences in social behavior between Ethiopian and Arabian hamadryas baboons. *Folia Primatologica*, **45**, 1–8.

112 / Byrne, R. W. & Whiten, A. (1985). Tactical deception of familiar individuals in baboons. *Animal Behaviour*, **33**, 669–73.

113 / Vanderijtplooij, H. H. C. & Plooij, F. X. (1987). Growing independence, conflict and learning in mother–infant relations in free-ranging chimpanzees. *Behaviour*, **101**, 1–86.

114 / Hsu, F., Anantharaman, T., Campbell, M. &

Nowatzyk, A. (1990). A grandmaster chess machine. *Scientific American*, **263(4)**, 18–24.

115 / Nichelli, P., Grafman, J., Pientrini, P., Alway, D., Carton, J. C. & Miletich, R. (1994). Brain activity in chess playing. *Nature*, **369**, 191.

116 / Jonides, J., Smith, E. E., Koeppe, R. A., Awh, E., Minoshima, S. & Mintun, M. A. (1993). Spatial working memory in humans as revealed by PET. *Nature*, **363**, 623–5.

Chapter 11

1 / Dawkins, R. (1986). *The Blind Watchmaker* (Harlow: Longman).

2 / McLuhan, M. (1964). *Understanding Media* (London: Routledge and Kegan Paul).

3 / The discovery of deep-sea mono-cellular organisms which dissolve the fish they paralyse was described in *New Scientist*, July–Sept 1992.

4 / Northcutt, R. G. (1984). Evolution of the vertebrate central nervous system: patterns and processes. *American Zoologist*, **24**, 701–16; Northcutt, R. G. (1981). Evolution of the telencephalon in nonmammals. *Annual Reviews of Neuroscience*, **4**, 301–50.

5 / Searle, J. R. (1992). *The Rediscovery of the Mind* (Cambridge: MIT Press).

6 / Gregory, R. L. (1987). In defence of artificial intelligence – a reply to John Searle. In C. Blakemore & S. Greenfield eds. *Mindwaves – Thoughts on Intelligence, Identity and Consciousness* (Oxford: Basil Blackwell).

7 / Fischler, M. A. & Firschein, O. (1987). *Intelligence – The Eye, the Brain, and the Computer* (Reading, MA: Addison-Wesley).

8 / Squire, L. R. (1987). *Memory and Brain* (Oxford: Oxford University Press).

9 / Adrian, E. D. (1932). *The Mechanism of Nervous Action: Electrical Studies of the Neurone* (London: Humphrey Milford).

10 / Sherrington, C. S. (1940). *Man on his Nature* (Cambridge: Cambridge University Press).

11 / Sperry, R. W. (1952). Neurology and the mind-brain problem. *American Scientist*, **40**, 291–312.

12 / Chalmers, D. J. (1995). Facing up to the problem of consciousness. *Journal of Consciousness Studies*, **2**, 200–19; Chalmers, D. J. (1997). Moving forward on the problem of consciousness. *Journal of Consciousness Studies*, **4**, 3–46; Churchland, P. S. (1996). The Hornswoggle problem. *Journal of Consciousness Studies*, **3**, 402–8; Cotterill, R. M. J. (1997). Navigation, consciousness, and the body/mind 'problem'. *Psyke & Logos*, **18**, 337–41.

13 / Crick, F. H. C. & Koch, C. (1990). Towards a neurobiological theory of consciousness. *Seminars in the Neurosciences*, **2**, 263–75.

14 / Velmans, M. (1991). Is human information processing conscious? *Behavioral and Brain Sciences*, **14**, 651–69.

15 / Carpenter, R. H. S. (1996). *Neurophysiology* (London: Edward Arnold).

16 / Cotterill, R. M. J. (1997). On the mechanism of consciousness. *Journal of Consciousness Studies*, **4(3)**, 231–47; Cotterill, R. M. J. (1997). On the neural correlates of consciousness. *Cognitive Studies: Bulletin of the Japanese Cognitive Science Society*, **4**, 31–44.

17 / Broadbent, D. E. (1958). *Perception and Communication* (Oxford: Oxford University Press).

18 / Treisman, A. M. (1964). Verbal cues, language, and meaning in selective attention. *American Journal of Psychology*, **77**, 206–19.

19 / Baddeley, A. D. & Hitch, G. J. (1974). Working memory. In G. H. Bower ed. *The Psychology of Learning and Motivation, Volume 8* (New York: Academic Press); Baddeley, A. D. (1992). Working Memory. *Science*, **255**, 556–9.

20 / Libet, B., Wright Jr, E. W., Feinstein, B. & Pearl, D. K. (1979). Subjective referral of the timing for a conscious sensory experience: a functional role for the somatosensory specific projection system in man. *Brain*, **102**, 193–224; Libet, B. (1982). Brain stimulation in the study of neuronal functions for conscious sensory experience. *Human Neurobiology*, **1**, 235–42.

21 / Haken, H. (1996). *Principles of Brain Functioning – A Synergetic Approach to Brain Activity, Behavior and Cognition* (Berlin: Springer).

22 / Leopold, D. A. & Logothetis, N. K. (1996). Activity changes in early visual cortex reflect monkeys' percepts during binocular rivalry. *Nature*, **379**, 549–53.

23 / Oatley, K. & Johnson-Laird, P. N. (1987). Towards a cognitive theory of emotion. *Cognition and Emotion*, **1**, 29–50.

24 / Zeki, S. (1993). *A Vision of the Brain* (Oxford: Blackwell Scientific Publications).

25 / Edelman, G. M. (1989). *The Remembered Present – A Biological Theory of Consciousness* (New York: Basic Books).

26 / Singer, W. (1995). Development and plasticity of cortical processing architectures. *Science*, **270**, 758–64.

27 / Boden, M. A. (1990). *The Creative Mind – Myths and Mechanisms* (London: George Weidenfeld & Nicolson).

Glossary

Acetylcholine (ACh) The first neurotransmitter identified, and the standard one at the neuromuscular junction; its excitatory action is blocked by curare.

Action potential The all-or-nothing electrochemical signal (having a standard amplitude and duration – about a millisecond) emitted along the axon by a neuron's soma, if the threshold has been exceeded; also known as a nerve impulse.

Acuity The sharpness with which something can be perceived; usually measured through the ability to resolve two closely spaced points.

Adaptation The progressive diminution of receptor sensitivity during maintenance of stimulation.

Adenosine triphosphate (ATP) A compound produced by metabolism, the small molecules of which store and transport energy, and transfer it to other molecules.

Adenylyl cyclase The enzyme which catalyses conversion of ATP to the second messenger cAMP, (cyclic AMP).

Adrenaline See Epinephrine.

AER See Average evoked response.

Afferent Conducting information inward; said of nerve fibres that conduct signals toward the brain, and of dendrites that conduct signals toward the soma of a nerve cell.

Agnosia A state in which recognition of sensory stimuli is impaired, even though normal acuity is still present; it is a malfunction of perception, often caused by a lesion to the posterior parietal area.

Agonist A compound whose molecules can mimic the promoting action of those of a neurotransmitter, for a given type of receptor.

Algorithm A step-by-step prescription, or method, for solving a problem or accomplishing a task. Algorithms are best regarded as products of human logic; there is no evidence of algorithmic processing at the neuronal level.

Alpha neuron A neuron of the type which directly activates skeletal muscular contraction, by stimulating the extrafusal fibres.

Alzheimer's disease A condition characterized by deterioration of primarily short-term memory, but ultimately all forms of memory; a sort of premature senility, it affects about four per cent of all people over the age of 65.

Amacrine cell A type of retinal neuron in which the processes extend in the plane of the retina; it shares synapses with the bipolar and ganglion cells. It does not generate action potentials.

Amino acids Organic compounds which come in twenty different common types and which, when joined together chain-fashion, form proteins; some, like aspartic acid, glutamic acid and the less-common GABA, also serve as neurotransmitters.

Amnesia See Anterograde amnesia; Retrograde amnesia.

AMPA receptor A glutamate receptor which also responds to the agonist alpha-amino-3-hydroxy-5-methyl-4-isoxazole-propionic acid; this is the chief glutamate receptor in the nervous system, and its simple kinetics contrast with those of the NMDA-type glutamate receptor. (See also NMDA receptor.)

Amygdala An almond-shaped structure in the limbic system; it participates in the integration and control of emotional and autonomic behaviours.

Antagonist The molecule of a compound which inhibits the action of a neurotransmitter, by physically occupying the site on a receptor that is normally the target of the neurotransmitter.

Anterograde amnesia Loss or diminution of the ability to store new memories, following the onset of the disorder.

Aphasia An impairment of the ability to produce or comprehend spoken or written language.

Aplysia A marine mollusc, also known as the sea hare; studies of its simple nervous system (totalling less than twenty thousand neurons) have elucidated molecular mechanisms of memory at the synaptic level.

Arborization The tree-like pattern collectively formed by a neuron's dendrites.

Archicortex The evolutionarily oldest part of the cortex, such as the mammalian hippocampus; its emergence is believed to predate that of the paleocortex.

Association The ability to relate different items of sensory information; it is mediated by neural networks.

ATP See Adenosine triphosphate.

Attention The focussing on specific internal or external stimuli, to the exclusion of others.

Auditory cortex A region of the neocortex which processes auditory frequencies in the range 20–20,000 Hertz; its tonotopic arrangement links specific neurons to specific regions of the inner ear's cochlea. (The primary auditory cortex is designated A1.)

Autism A severe developmental disorder, affecting 4 in every 10,000 children (80 per cent of whom are also retarded); characterized by partial or total inability to communicate and relate to the social environment, the syndrome has been shown to have genetic origins.

Auto-association The ability to reproduce, for further processing, the appropriate full item from memory when presented with only a portion of that item, as in remembering a complete face when shown only a part of it.

Autonomic nervous system An involuntary part of the nervous system which controls such life-sustaining functions as circulation, digestion, excretion and reproduction; divided into the sympathetic, parasympathetic and enteric subsystems.

Average evoked response The average of a number (usually several dozen) of electroencephalograms or magnetoencephalograms which have been recorded following a repeated stimulus.

Axon A single fibre (or process) extending from a neuron's somatic region, along which the latter emits nerve impulses (i.e. action potentials). The axon divides into numerous branches at its extremity, and each of these makes synaptic contacts with other neurons. Closer to the soma, the axon may also give rise to one or more secondary processes known as axon collaterals.

Backward masking Suppression of the perception of one stimulus by a subsequent stimulus.

Backward propagation (of errors) An algorithm for adjusting the synaptic weights (i.e. the signal-transfer efficiencies of the synapses) in a feedforward neural network. There is, at present, no evidence of an analogous mechanism in the brain.

Ballism Uncontrolled ballistic movement (for example, the sudden straightening of an arm), caused by a lesion to the sub-thalamic nucleus.

Bandwidth In the present context, the amount of information which can be processed concurrently by a given component (or set of components) of the brain. The term is also used to characterize the overall processing capacity of the brain, and to emphasize the existence of a processing bottleneck.

Basal ganglia A group of closely connected structures in the forebrain which help to control movement and posture, and which also influence cognitive aspects of behaviour; the group includes the striatum (comprising the putamen and the caudate nucleus), the globus pallidus (which has internal and external segments) the substantia nigra (with its sub-components, the pars compacta and the pars reticulata), and the sub-thalamic nucleus. The inter-component interactions in the basal ganglia are mediated by a variety of neurotransmitters.

Basket cell In the cerebral cortex, a type of inhibitory neuron which makes numerous long-range contacts with certain types of excitatory neuron; in the cerebellar cortex, a type of inhibitory neuron which is activated by the parallel fibres, and which is capable of vetoing the (inhibitory) output of a Purkinje cell.

Behaviourism A movement which enjoyed considerable support in psychological circles earlier in the twentieth century. It focussed exclusively on stimuli and responses, thereby ignoring thought processes and consciousness.

Binary logic A system in which variables can adopt either of only two different values, usually *unity* or *zero* (equivalent to *yes* and *no*).

Binocular rivalry The competition for attention between two conflicting stimuli, one presented to each eye; the visual system alternatingly focusses on one stimulus while ignoring the other, rather than attempting to consolidate both into a single hybrid percept.

Bipolar cell Any neuron possessing only two processes (or neurites). In the retina, a type of neuron located between receptors and ganglions; it also shares synapses with both the horizontal cells and the amacrine cells. Bipolar cells do not generate action potentials.

Bit Short for ***B**inary dig**it***; the unit of binary information, the bit is the coded answer to a yes–no question.

Blindsight A mechanism whereby visual information produces a physical response despite the lack of conscious visual sensation.

Blind spot (optic disc) The region of the retina through which the optic nerve (that is to say, the axons of the ganglion cells) passes, on its route to the thalamus; being devoid of receptor cells, this region gives rise to no visual image.

Blob A group of neurons, primarily located in layer II/III of the primary visual cortex, rich in the enzyme cytochrome oxidase.

Brainstem The structures around the upper part of the spinal cord, below the midbrain, including the medulla oblongata and the pons; also known as the rhombencephalon or hindbrain.

Brain waves Oscillations caused by rhythmic neuronal activity, and detectable by electroencephalography or magnetoencephalography. The oscillations fall within several well-defined frequency ranges: the delta band (about 1–3 Hertz – that is, 1–3 cycles per second); the theta band (about 5–8 Hertz); the alpha band (about 10–14 Hertz); the beta band (about 18–25 Hertz); and the gamma band (about 30–70 Hertz). Theta-band oscillations have been linked to hippocampal function, and their gamma-band counterparts to consciousness.

Breakthrough In the present context, the sudden switch of attention to stimuli that a subject was attempting to ignore.

Broca's area The part of the cerebral cortex that controls speech and possibly, according to the present book, other sequenced movements of the mouth.

Byte A byte is eight bits, that is to say a block of eight 1s and 0s. Eight bits can be arranged in 256 different ways, this number being 2^8. One byte is required to define each of the 256 ASCII (***A**merican **S**tandard **C**ode for **I**nformation **I**nterchange*) characters, so the present author's six-letter first name requires six bytes, or 48 bits.

Catecholamines A group of chemicals that serve as neurotransmitters; including dopamine, epinephrine and norepinephrine.

Caudal In the head, in the direction of the back of the head; in the rest of the body, in the direction of the feet.

Caudate nucleus One of the components of the striatum (in the basal ganglia); in humans, it preferentially processes signals emanating from the association areas of the cerebral cortex. (See also Putamen.)

Causality The doctrine that everything has a physical cause; that every event is completely determined by preceding events.

Cell assembly A group of interconnected neurons which collectively functions as a unit, under given circumstances.

Central nervous system The brain and the spinal cord.

Central pattern generator A neural circuit that provokes rhythmic motor activity, such as that involved in walking.

Central processing unit See CPU.

Central sulcus The groove separating the frontal and parietal lobes, and, more specifically, the motor cortex from the somatosensory cortex.

Cerebellum A structure surrounding the rear of the brainstem which governs motor coordination; its neuronal structure is remarkably regular, with few cell types (e.g. Purkinje, Golgi and granule cells). The cerebellum contains about half of all the neurons in the entire brain. One theory of cerebellar function sees this brain component as being able to remember the nervous system's various temporal delays.

Cerebral cortex The outer (grey-matter) layer of the cerebral hemispheres, comprising regions which receive sensory inputs and others responsible for forming associations; evolution's most recent addition to the nervous system, it is particularly well developed in mammals. It comprises the neocortex, the paleocortex and the archicortex.

Cerebrum The brain proper, in front of and above the cerebellum.

Cheshire Cat effect The occlusion of part of the percept of a static visual stimulus by a competing moving stimulus, in a binocular rivalry situation; such temporary obliteration of most of the image of a cat – a commonly used motif in public displays of the phenomenon – can be contrived to leave only the smile remaining, thereby emulating the event made famous in Lewis Carroll's (Charles Dodgson's) *Alice in Wonderland.*

Chiasm A point at which nerve fibres cross.

Chromosome A strand of genetic material in a cell's nucleus, comprising a large number of genes; visible under a microscope during cell division, a chromosome consists of DNA and the protein histone.

Cingulate gyrus The region of the medial surface of each cerebral hemisphere lying immediately dorsal to the corpus callosum. The anterior cingulate region is implicated in the perception of pain, and it has been linked to the avoidance of danger.

Cladistic Reflecting recent origin from a common ancestor.

Classical conditioning The learning of an association between a response and a neutral stimulus not normally connected with that response.

Climbing fibre One of only two types of input fibre to the cerebellum; emanating from the inferior olive, it makes excitatory synaptic contacts exclusively with the dendrites of Purkinje cells. (See also Mossy fibre.)

Commissure A major sheaf of nerve fibres connecting different regions of the brain.

Cone One of the two types of light-sensitive cells (photoreceptors) present in the retina; different sub-types of cone preferentially react to specific wavelengths, thereby mediating colour vision. The retina's centrally lying foveal region is densely populated with cones. They do not generate action potentials. (See also Rod.)

Consciousness The totality of one's thoughts and feelings; awareness of one's existence, sensations and circumstances. According to the present book's thesis, consciousness arose in evolution because it confers the advantage of being able to anticipate the consequences of muscular acts currently being planned, or actually in the process of being executed. This facility would be particularly useful in the real-time avoidance of danger (veto-on-the-fly), though it probably also has positive counterparts. The ideas put forward here suggest that consciousness requires participation of the premotor (and/or supplementary motor) cortex, the thalamic intralaminar nuclei and (at least parts of) the sensory cortex, and that consciousness will not be present unless all of these brain regions are currently supporting neuronal activity. (Phase locking of gamma-band oscillations in these three regions may be a further prerequisite.) This permits activation of the appropriate schemata, which link motor-planning activity to relevant activity in the sensory cortex. The ability to simultaneously do one thing and think about another, as when driving a car and concurrently holding a conversation, may be related to the heterogeneous structure of the striatum. According to the views expressed here, consciousness is always a muscularly active process, never a passive one, though the muscular processes involved in thought are merely proxied versions of actual movements.

Contralateral On the opposite side of the medial plane.

Convolution See Gyrus.

Coronal section A slice through the brain, made at right angles to the rostral–caudal axis.

Corpus callosum The great commissure, the approximately 200 million fibres of which link the left and right hemispheres of the cerebral cortex.

Cortical areas Specific regions of the cerebral cortex dedicated to sensory, motor or association functions.

Corticofugal Said of efferent nerve fibres that carry feedback signals from the brain to sensory structures.

CPU The **C**entral **P**rocessing **U**nit, such as the *Intel Pentium*, in a computer. One view of the brain likens its anatomy to computer hardware, the mind to software, and consciousness to the functioning of the CPU. Although the analogy is interesting, it should not be taken too seriously; there are many important differences between computers and brains.

Creativity Colloquially, the ability to conceive of new ideas or solutions; in the present book, a facility that is possibly related to the collective functioning of the cerebellum and the reverse projections in the cerebral cortex, the products being fleetingly stored in the frontal lobe. According to this view, novelty of concept would arise from the reverse projections being able to capture correlations not obviously present in the sensory input, real or imagined.

Curare A common chemical blocker of acetylcholine's action at the neuromuscular junction.

Cyclic AMP (cAMP) A cyclic form of adenosine monophosphate which is derived from the energy-storing molecules of ATP (adenosine triphosphate); it serves as a (second messenger) signalling substance in cells.

Declarative memory A cognitive form of memory, which can be acquired through a single exposure to an object or an event; also referred to as explicit memory, it requires the involvement of consciousness, and it may be subdivided into episodic and semantic forms, the former being specific to an individual while the latter refers to memory of objects and events in the public domain. Declarative memory is the complement of procedural memory. (See also Procedural memory.)

Dendrites Highly branched afferent structures (processes) of a neuron; they receive chemical signals from the axon branches of other neurons, via synaptic contacts, and convey the resultant electrochemical signals towards the soma. (See also Spine.)

Dentate gyrus A region of the hippocampus, in which the neurons receive signals from the entorhinal cortex, and send signals to the CA3 region, via mossy fibres.

Depolarization Reduction of the voltage (which, in the quiescent state, is negative inside, to the extent of about 0.1 volts) across a neuron's bounding membrane.

Determinism In general, the theory that a given set of circumstances inevitably produces the same consequences; in particular, the doctrine that human action is not free but determined by external or internal forces acting on the will.

Deterministic chaos The situation in certain systems in which the dynamical development cannot be precisely predicted even though the system itself is deterministic.

Direction selectivity A property of cells in the visual cortex, which respond preferentially to lines lying within a narrow range of orientations, and usually moving in a specific direction (forward or backward, at right angles to the preferred orientation).

DNA See Nucleic acids.

Dopamine A neurotransmitter of the catecholamine class; schizophrenia is possibly caused by an excess of dopamine in the limbic system. Dopamine plays a key role in the functioning of certain members of the basal ganglia.

Dorsal In the head, in the direction of the scalp; in the rest of the body, in the direction of the back.

Dualism The doctrine that mind and body have independent identities.

EEG See Electroencephalogram.

Effectors Neurons at terminal regions of the nervous system, which act on glands or muscles.

Efferent Conducting information outward; said of nerve fibres that carry messages away from the brain, and of the axon of a nerve cell.

Electrode In this book's context, an electrical conductor (metallic or electrolytic) used to measure or influence the activity of a nerve cell, or of a group of such cells.

Electroencephalogram (EEG) A record of the net electrical activity measurable (with attached electrodes) at the scalp, due to the brain's internal workings; the characteristic EEG rhythms reflect general states of arousal, consciousness and the sleep–waking cycle.
Emotions Sensations that arise in connection with the nervous system's evaluation of the significance of internal or external stimuli. In particular, any discrepancy between actual and anticipated results of actions may produce autonomic arousal, accompanied by a state of muscular readiness, which manifests itself in underlying activity in the muscle spindles.
Endocranial Within the skull.
Endocrine Literally, internal secretion; activation of the endocrine system is the nervous system's internal product, just as muscular activation is its external product.
Endorphins A group of small internally produced peptides which function as neurotransmitters, and influence emotional state.
Engram The conjectured physical change in the association cortex's neural network caused by the storage of a memory.
Enkephalins A group of small internally produced peptide neurotransmitters possessing a morphine-like analgesic capacity.
Enteric system The part of the autonomic nervous system that activates the digestive organs.
Epilepsy A disorder of brain function characterized by sporadic recurrence of seizure caused by avalanche discharges of large groups of neurons; known already to the ancient Greeks, it affects about one per cent of humans.
Epinephrine (also known as adrenaline) A neurotransmitter of the catecholamine class.
Episodic memory See Declarative memory.
ERP See Event-related potential.
Event-related potential A discrete pattern of EEG activity reflecting specific perceptual or cognitive processes.
Evoked potential The EEG response measured for a specific stimulus.
Excitation In this book, tending to depolarize a neuron's membrane potential. An excitatory neuron is one whose neurotransmitters tend to depolarize the membrane potentials of its target neurons.
Explicit memory See Declarative memory.
Extrafusal muscle fibres The main fibres of skeletal muscles, which make up the great majority of a muscle's bulk; they supply all the power of muscular contraction.

Feature detector A neuron or small group of neurons which become active only if a particular feature is present in the sensory input.
Feedback The return of nerve signals to a brain region such that the original signal sent from there is reinforced (positive feedback – that is to say, excitatory feedback) or weakened (negative feedback – that is to say, inhibitory feedback).
Feedforward The dispatching of nerve signals to a brain region placed later in the processing sequence. In particular, feedforward inhibition is employed in the mechanism known as winner-take-all.
Feelings See Qualia; Emotions.
Filling-in A process whereby the nervous system compensates for lack of information by reacting to a fictive situation that is most consistent with the available sensory input (that is to say, its best guess). The classic example of filling-in is seen in connection with the blind spot.
Flocculus A minor lobe of the cerebellum, implicated in the vestibular response.
Folia The remarkably regular set of transverse sulci on the surface of the cerebellum, which run in the same direction as the underlying parallel fibres.
Forebrain The evolutionary most recently developed part of the vertebrate brain; it comprises the reticular formation, the limbic system and the cerebral cortex.
Fornix A sheath of fibres which links the hippocampus to the septum and the mammillary bodies.
Fovea A region located axially, at the centre of the retina, possessing a particularly dense packing of receptor cells, which are predominantly of the cone type. The dense packing permits visual observation at high resolution, though the latter probably also

requires participation of higher brain centres.

Free will The feeling that one acts as a spontaneous and independent agent. Because of the inevitable delay in the development of consciousness of sensory input, or of a motor act, this feeling is probably illusory. But the assumption of freedom nevertheless underlies our interactions with our fellow humans, and it is thus an important factor in the self-organizing processes upon which society is based.

Frontal lobe The part of the cerebral cortex that lies within the forehead; it is responsible for various aspects of thought and the planning of actions.

GABA See Gamma-aminobutyric acid.

Gamma-aminobutyric acid (GABA) An amino acid that serves as an inhibitory neurotransmitter. It is widespread in the forebrain.

Gamma neurons Relay cells which exert an indirect influence on skeletal muscles, by activating alpha neurons.

Ganglion cells In general, a group of neurons in the peripheral nervous system; in the eye, the neurons that transmit, in the form of action potentials, the partially processed signals from the retina to the lateral geniculate nucleus of the thalamus; these cells share synapses with the bipolar cells and the amacrine cells. The ganglion axons form the optic nerve.

Gene A section of chromosomal DNA that either codes for a given protein (structural gene) or controls such a structural gene (regulatory gene).

Genetic code The relationships between the sequences of nucleotide bases in DNA and the amino acids that they specify.

Genome The totality of expressible DNA in a cell's nucleus.

Genotype The genetic make-up of a specific member of a species.

Gestalt psychology A movement based on the tenets that form is the primitive unit of perception, and that the form conveyed by any pattern will always depend upon its context. The classic example is provided by figure–ground separation.

Gilles de la Tourette syndrome A mental disorder characterized by a number of bizarre behavioural patterns, it has been linked to over-activity in certain dopamine-active systems; no definite genetic origin has been identified, and about half the (mostly male) patients improve as they approach middle age.

Glial cell A non-signalling type of brain cell that performs nourishment and scavenging duties for the neurons; the most common type of brain cell.

Globus pallidus A component of the basal ganglia, divided into internal (medial) and external (lateral) segments. Both segments receive their main (inhibitory) inputs from the striatum, and the former sends (inhibitory) signals to the thalamus, while the latter interacts with the sub-thalamic nucleus, which appears to function as a gain control for both the pallidal segments.

Golgi cell A type of inhibitory neuron in the cerebellum, which appears to exercise a winner-take-all influence on the mossy fibre inputs; Golgi cells are activated both by the mossy fibres and by the parallel fibres.

Golgi staining A method based on a silver compound, whereby a few entire neurons are rendered visible to the optical microscope while the remaining majority are left invisible.

Golgi tendon organ A *stress*-measuring sensor positioned in series with an extrafusal muscle fibre; it contrasts with the muscle spindle, which is a *strain*-measuring sensor positioned in parallel.

Granule cell The most common type of neuron in the cerebellum, and indeed in the entire brain; granule cells, which are the cerebellum's only excitatory members, receive activation from the mossy fibres, and they activate *all* the other types of cerebellar cell via their systematically arranged axons, the parallel fibres.

Grey matter That part of nervous tissue which contains the somas of the constituent neurons. The outer surfaces of the cerebral and cerebellar cortices are composed of grey matter. White matter, conversely, comprises only nerve cell axons.

Gyrus One of the curved elevations (convex towards the outside) on the surface of the cerebral cortex.

Habituation The reduction of the magnitude of a response following repeated presentation of a stimulus; a form of non-associative learning.

Hardware The physical components of a computer system, such as the electronic processing units, the disc drives, keyboard, monitor, mouse, and so on.

Hebbian learning A mechanism (named after Donald Hebb) whereby the transmission efficiency of a synapse is strengthened if the neurons it connects are both active at roughly the same time.

Hemisphere One side, left or right, of the cerebral or cerebellar cortex.

Hindbrain See Brainstem.

Hippocampal commissure A small sheaf of nerve fibres connecting the cerebral hemispheres near the hippocampus.

Hippocampus A centrally located structure, lying under the cortex, implicated in the formation of memories; an evolutionary development of a more primitive structure found in the reptilian brain. The hippocampus is not necessary for consciousness, and people who have (bilaterally) lost the use of this structure can nevertheless acquire new short-term episodic memories, and recall items from long-term memory. But they are unable to consolidate the former into the latter. According to the ideas expressed in this book, the hippocampus captures correlations between motor planning activity in the frontal lobe and the corresponding activity in the sensory cortex, this capture probably being mediated by the system of recurrent collaterals in its CA3 region. Subsequent activation of the latter feeds temporally correlated signals back to the frontal and sensory regions, thereby permitting strengthening of the appropriate synapses. The involvement of motor planning activity may underlie the fact that certain regions of the hippocampus also appear to code for spatial location.

Hologram Literally, the whole picture; the distributed record of an image, the latter being recovered by illuminating the hologram with one of the two stimuli used during the recording. Certain aspects of associative memory have been likened to the storage and reconstruction of a holographic image.

Homeostasis The automatic control of such vital properties as heart beat, blood temperature and blood pressure, by part of the nervous system.

Homunculus Literally, the little inner man, conjectured by the ancients to reside in the head, observe the environment via the senses, and respond appropriately; his abstract form still survives as the dualist ghost in the machine.

Horizontal cell A type of retinal neuron in which the processes extend in the plane of the retina; it shares synapses with the receptor cells and the bipolar cells. It does not generate action potentials.

Horizontal section This term naturally refers to a cut through the brain when the animal's head is held upright; the plane of the section therefore contains the rostral–caudal and left–right axes.

Hormones Internally secreted organic compounds that control various vital body functions.

Huntington's disease A mental disorder characterized by progressive involuntary dance-like movements (whence the alternative name Huntington's chorea), psychological deterioration and dementia. The victim usually dies of cardiac or pulmonary failure after about two decades.

Hyperpolarization An increase of the voltage across a neuron's bounding membrane, thus decreasing the likelihood of action potential emission.

Hypothalamus A small but very important structure in the limbic system that controls body temperature, eating, emotional tone, drinking, hormonal balance, various metabolic processes and sexual drive. A major part of its influence is exerted through its cooperation with the pituitary gland.

Implicit memory See Procedural memory.

Inferior olive A nucleus located in the medulla oblongata; it is the source of signals entering the cerebellum via the climbing fibres.

Inhibition In this book, tending to hyperpolarize a neuron's membrane potential. An inhibitory neuron is one whose neurotransmitters tend to hyperpolarize the membrane potentials of its target neurons.

Intelligence In general, the ability to learn from experience, and adapt to the prevailing environment; specifically, the capacity for manipulating existing schemata, and for constructing new ones.

Interneuron A neuron which exerts an influence only on other neurons in its immediate vicinity.

Intrafusal muscle fibre See Muscle spindle.

Intralaminar nucleus One of a series of nuclei positioned centrally in the thalamus, with widespread reciprocal connections to the sensory cortex, primarily receptory connections from the premotor cortex, and primarily projectory connections to the striatum; it appears to play a major role in generating consciousness.

Ion An (otherwise electrically neutral) atom which has either gained or lost one or more electrons, thereby acquiring an electrical charge. Ionic movements underlie the action potential, and they are implicated in the storage of memories.

Ion channel A protein present in neural membranes whose structure possesses a voltage-sensitive pore that regulates the passage of ions; channels selective for sodium and potassium provide the basis for the action potential.

Ipsilateral On the same side of the medial plane.

Lateral In the direction (left or right) away from the medial plane.

Lateral geniculate nucleus A group of neurons in the thalamus which receive connections from the retina, and also both send connections to and receive connections from the visual cortex.

Lateralization The separation of mental faculties between the two hemispheres of the cerebral cortex.

Lesion A localized injury; to a part of the brain, in the context of this book.

LGN See Lateral geniculate nucleus.

Limbic system A group of structures intermediate, in both position and evolutionary development, between the brainstem and the cerebral cortex; comprising the amygdala, hippocampus, hypothalamus, thalamus and septum, this region controls emotional behaviour.

Lipid A compound whose molecules are formed by the attachment of hydrocarbon (acyl) chains to a glycerol molecule. Molecules of the important phospholipid sub-group possess two such (hydrophobic) chains and one electrically charged (hydrophilic) head group; this structure favours formation of the phospholipid bilayer that provides the basis for biological membranes.

Lobotomy Surgical disconnection of a cerebral lobe; often the frontal lobe.

Locus ceruleus A structure in the brainstem which exerts its widespread influence through the neurotransmitter norepinephrine; implicated in the sleep-wake cycle.

Long-term memory Memory retained beyond the period of immediate recall.

Long-term potentiation (LTP) An increase in the efficiency of synaptic transmission, lasting for many days, caused by high-frequency stimulation of the presynaptic neuron. Its counterpart, long-term depression (LTD) has also been detected.

Macroscopic Said of things visible to the unaided eye.

Magnetic resonance imaging (MRI) A technique for generating images of body tissue, in general, and of the brain's convolutions in particular. Functional magnetic resonance imaging (fMRI) additionally reveals the brain sites of particularly high activity, under given circumstances.

Magnetoencephalogram (MEG) A recording of the temporal variation of the brain's (very weak) magnetic field, at a particular location, as measured by a SQUID (that is to say, a **S**uperconducting **Qu**antum **I**nterference **D**etector).

Magnocellular A designation used for the larger of the two cell types found in, for example, the lateral geniculate nucleus.

Mammillary bodies Cell groups which integrate

information arriving from the limbic system, via the fornix, and from the midbrain tegmentum.

Masking See Backward masking.

Master module The conjectured constellation of brain components whose collective functioning is believed to underlie the bandwidth limitation in mental processing, and possibly also to be the site of break-through of unattended stimuli. It has not yet been established whether such collective functioning actually takes place, though it does seem possible that the premotor cortex, and/or supplementary motor cortex, might cooperate with the thalamic intralaminar nuclei, the sensory cortex and the striatum to produce the observed limitation.

Matrix Generally applied to the more continuous component of a structurally heterogeneous brain region, such as the striatum.

Mechanistic Explainable in terms of known physical laws.

Medial At or toward the brain's (and body's) plane of symmetry.

Medial geniculate nucleus A group of neurons in the thalamus which receive connections from the ear, and also both send connections to and receive connections from the auditory cortex.

Medulla oblongata A part of the brainstem near the upper part of the spinal cord.

MEG See Magnetoencephalogram.

Membrane In this book's context, the lipid (i.e. fatty) outer skin of all cells, including neurons; its embedded channels, pumps and receptors, all of which are proteins, give the neuron its special signalling capacity.

Memory The mental faculty provided by superimposed changes at the synaptic level; it makes possible the conscious or unconscious recall of past experiences.

MGN See Medial geniculate nucleus.

Microscopic Literally, visible only under a microscope; also applied to such smaller scales as the molecular and the atomic.

Midbrain The region of the brain located between the forebrain and the brainstem.

Mind The seat of consciousness, thoughts, feelings and – perhaps illusorily – volition; the seat of memory.

Mossy fibres 1. One of only two types of input fibre to the cerebellum; the mossy fibres make excitatory synaptic contacts primarily with the granule cells, but also with the Golgi cells. (See also Climbing fibres.) 2. The same term is used for the mossy fibres emanating from the dentate gyrus, which activate the CA3 cells of the hippocampus.

Motor cortex The elongated and centrally located region on each cerebral hemisphere which controls muscle movement via nerve fibres, which it projects to the spinal cord.

Muscle spindle A special form of receptor structure present in all skeletal muscles; it senses muscular tension, and the rate of change of tension, thereby providing a means of controlling muscular state. The spindle consists of a number of centrally located and mutually parallel fibres (divided between bag and chain varieties), the ends of which are attached to intrafusal muscle fibres. The latter are activated by the gamma neurons, and the afferents from the bag and chain fibres activate the alpha neurons (which, in turn, activate the extrafusal muscle fibres, and cause the muscle to contract). Because the gamma neurons are under the control of the higher brain centres, and because the afferents from the bag and chain fibres also send collaterals to those higher centres, the spindle system is involved in supplying what could be called internal reafference; it informs the organism of state of the muscles, and there is evidence that this information enters consciousness. In the present book, therefore, spindles are conjectured to contribute to qualia – the feelings associated with consciousness.

Mutation A chance or deliberate modification of the base sequence in chromosomal DNA; inheritable by subsequent generations if it occurs in the reproductive system.

Myasthenia gravis An autoimmune disorder caused by depletion of the acetylcholine receptors at neuromuscular junctions; formerly often fatal, the condition can now be counteracted by various therapies.

Myelin An electrically insulating lipid (fatty) layer that coats some nerve fibres, thereby enhancing their conductive properties.

Natural selection The reproductive advantage of organisms better adapted to their environment; it provides the driving force for evolution.

Neocortex See Cerebral cortex.

Nerve cell See Neuron.

Nerve fibre A single neural axon or a bundle of them.

Nerve growth factor A chemical substance that promotes the growth of neural processes, in the direction of the substance's gradient.

Nerve impulse See Action potential.

Neural network The brain's actual interconnected mesh of neurons (biological wetware); a theoretical model that attempts to simulate its electrophysical properties (computer software); a brain-inspired computational strategy (computer software); or a brain-inspired computational device (computer hardware).

Neurite See Process.

Neuron (sometimes spelled 'neurone') The nerve cell, whose electrochemically excitable membrane qualifies it as the fundamental signalling unit of the nervous system.

Neuropeptides A group of small molecules whose amino acid chains serve as neurotransmitters.

Neurotoxins A large group of chemical substances that are able to disrupt nerve signal transmission; examples are alcohol, curare, nitrous oxide and tetrodotoxin.

Neurotransmitter A chemical substance whose molecules can physically dock with (protein) receptor molecules, and thereby pass on nerve signals; several dozen such substances are known, and the list is probably not complete.

NMDA receptor A variety of glutamate receptor for which the (artificial) substance *N*-methyl-D-aspartate acts as a particularly effective agonist. The great importance of this receptor lies in its needing both the docking of a glutamate molecule and membrane depolarization in order to have its ion channel opened. It is thus ideally suited to capture correlations between different types of stimulating input.

Node of Ranvier A unmyelinated stretch of an otherwise myelinated axon. (See also Saltatory conduction.)

Noradrenaline See Norepinephrine.

Norepinephrine (also known as noradrenaline). A neurotransmitter of the catecholamine class; widely used in the nervous system (by the locus ceruleus, for example).

Nucleic acids Polymeric chains of nucleotide bases (and the associated sugar and phosphate groups) which store and mediate translation of the genetic message; a gene is a stretch of DNA (deoxyribonucleic acid), that molecule being located in a cell's nucleus, while RNA (ribonucleic acid) transports the messages of the genes out of the nucleus, into the cytoplasm, where they are expressed through formation of proteins.

Nucleotide One of the building blocks of the nucleic acids; there are four types, adenine (A), thymine (T), guanine (G) and cytosine (C), and the fact that the only pairings compatible with double helical DNA are A with T and G with C guarantees perpetuation of the genetic message stored in the chromosomes.

Nucleus Demarcated mass of nerve cell bodies, devoid of the stratification seen in the cerebral cortex. Nuclei are mutually connected by tracts of nerve fibres.

Nucleus accumbens A component of the basal ganglia; also known as the ventral striatum.

Occipital lobe The region of the cerebral cortex at the head's caudal extremity (that is to say, at the back of the head).

Ocular dominance The tendency of a particular nerve cell in the visual system to respond preferentially to stimulus from one eye. Not all visual neurons display such dominance, but those that do contribute to depth perception.

Olfactory bulb Elongated extensions of the forebrain which receive and perform the initial processing of the signals from the odour receptors.

Opiate Chemical substance with molecular structure similar to morphine, which has analgesic properties; several varieties are present in opium.

Optic chiasm The point where a fraction of the optic nerves cross.

Optic nerve The bundle of (approximately a million)

neural axons that connect each retina with the corresponding lateral geniculate nucleus.

Oscillations See Brain waves.

Paleocortex The most ancient region of the cerebral cortex; primarily associated with olfaction.

Parasympathetic Pertaining to that part of the autonomic nervous system that tends to decrease activity; its afferent fibres come from the brainstem.

Parietal lobe The region just rear of centre, and dorsal, on each cerebral hemisphere.

Parkinson's disease A syndrome characterized by tremor and loss of motor control, due to a deficiency in the dopamine system, which normally counterbalances excitatory acetylcholine activity; it usually strikes between the ages of 50 and 65, and it affects almost one per cent of humans.

Parvocellular A designation used for the smaller of the two cell types found in, for example, the lateral geniculate nucleus.

Parallel distributed processing (PDP) 1. A computational brain-inspired strategy which captures input–output correlations, and subsequently permits prediction of the most appropriate output for a given input. 2. A presumably common form of processing in the brain, whereby the various attributes of a stimulus are simultaneously handled by a set of parallelly arranged neural routes.

Parallel fibres The remarkably regular (and transversely deployed) axons of the cerebellar granule cells; they send excitation to the dendrites of all the other types of cerebellar neurons.

Patch-clamping A technique whereby a small area of nerve cell membrane becomes stretched across the end of a capillary, making it possible to investigate the conducting properties of a few ion channels or receptors, or in favourable circumstances just one such molecule.

Peptide More correctly polypeptide; a chain of amino acid residues.

Perceptron A system of two layers of idealized neurons, with interconnecting synapses between the layers, which functions as an input–output (stimulus-response) device; the synaptic strengths are automatically adjusted during a training period in which known input–output matchings are presented to the network, which thereafter is able to recognize, associate and generalize.

Peripheral nervous system The part of the nervous system other than the brain and the spinal cord; it includes the spinal ganglia, the cranial nerves and the autonomic nervous system.

PET See Positron emission tomography.

Phantom limb The paradoxical tactile and pain sensations that seem to emanate from an amputated limb; their occurrence demonstrates the overriding role played by the somatosensory cortex.

Phenotype The collective physical characteristics and properties that an organism develops through the expression of its genes.

Phospholipid See Lipid.

Phosphorylation A process whereby a phosphate group is donated by an ATP molecule, and becomes attached to a target protein. (See also Adenosine triphosphate (ATP).)

Photoreceptors Two light-sensitive types of neuron present in the retina, which convert incident radiant energy into electrochemical signals; the rods primarily sense intensity, while the cones mediate detection of colour, contrast, motion and size.

Phrenology The (now discredited) science which purported to relate the skull's bumps and depressions to mental capacities.

Pituitary gland The supreme endocrine gland, located in the limbic system, which controls the secretions of the other endocrine glands; it, in turn, is controlled by the hypothalamus.

Planum temporale An area on the superior surface of the temporal lobe, located near the primary auditory cortex (A1); in humans, it is frequently much larger in the left hemisphere than in the right.

Plastic In the context of the present book, modifiable by experience.

Pons Part of the brainstem that links the medulla oblongata and midbrain.

Pop-out The sudden capturing of visual attention by one particular object amongst many.

Positron emission tomography (PET) A technique which provides brain images that reveal the currently active areas. It is based on the fact that the sub-atomic particle known as the positron can mutually annihilate with the similarly sub-atomic electron, thereby producing a pair of gamma rays. The latter, being highly penetrating, emerge from the skull, and the angles (with respect to the head) at which they are subsequently detected can be used to work out where the above annihilation occurred. The positron emitter is chemically attached to a non-metabolized form of sugar, which preferentially accumulates at the most active sites in the brain.

Postsynaptic Pertaining to the receiving side of a synapse; the receptor-laden membrane of a dendrite, gland or muscle.

Potential This term is here used interchangeably with the term voltage, and its is conveniently quoted in millivolts.

Prefrontal region The extreme rostral portion of the brain's frontal lobe; believed to be the seat of mental planning, and to make a major contribution to the so-called blackboard of working memory.

Premotor area An area of the frontal lobe concerned with sequences of movements, rather than with activation of individual muscles.

Presynaptic Pertaining to the transmitting side of a synapse; the neurotransmitter laden axon terminal.

Primates A group (taxonomically, an *order*) of mammals comprising apes, humans, lemurs and monkeys.

Procedural memory A form of memory by which the subject learns skills and general motor abilities, not necessarily with accompanying consciousness; also referred to as non-declarative memory or implicit memory, it is acquired slowly, and usually through much repetition. It is the counterpart of declarative memory. (See also Declarative memory.)

Process In this book's context, a filamentous outgrowth from a cell's central (somatic) region; a dendrite or an axon. Also known as a neurite.

Projective field The group of neurons in a subsequent stage of processing to which a given neuron in a prior stage dispatches signals; it is the complement of the receptive field.

Proprioceptive Sensitive to body posture or limb position.

Prosopagnosia A pathological state in which the victim is unable to recognize faces, or certain facial features.

Protein A molecule composed of chain-linked amino acid residues, its structural or enzymatic properties deriving from its three-dimensional (often folded) structure, this being determined by its actual amino-acid sequence; together with nucleic acids, proteins are the major macromolecules of life, and they provide the nervous system with its channels, pumps and receptors.

Psychophysics The science that seeks to quantify sensation, and mathematically model the relationship between brain and mind.

Psychosomatic Related to the mind's influence on bodily processes.

Pulvinar A large thalamic nucleus which serves visual integration, apparently by suppressing processing during the brief period occupied by a saccade.

Pump A protein molecule that transports a specific type of ion through a biological membrane, thereby helping to maintain the ionic imbalance across the latter; the various pumps collectively use about a quarter of all the energy a person consumes in the form of food.

Purkinje cell The main type of neuron in the cerebellum, which provides that brain region's only output route; identified by the dense branching of its dendrites. Its action is inhibitory.

Putamen One of the components of the striatum (in the basal ganglia); it preferentially processes signals emanating from the somatosensory and motor areas of the cerebral cortex. (See also Caudate nucleus.)

Pyramidal cell The principal class of neurons in the cerebral cortex, so named because of their shape; pyramidal cell axons form the white matter.

Qualia (singular: quale) The basic subjective feelings associated with consciousness. It has been said that explaining their existence represents the greatest challenge associated with the phenomenon, but a theory of consciousness that does not include an account of qualia might, in any event, be flawed; feelings are not merely incidental to consciousness, they are part of its *raison d'etre*. The greatest significance of qualia lies in their unifying direct perception and thought; we tend to have the same raw feelings irrespective of whether we are experiencing stimuli first-hand or merely imagining them. Classic examples of qualia include experience of the redness of red, the itchiness of an itch, and the painfulness of pain.

RAM **R**andom **A**ccess **M**emory. The computer memory used to temporarily hold data, while it is being processed. It can be written to and read from, but it involves no permanent record – its contents are lost if the computer is switched off.

Ranvier See Node of Ranvier.

Raphe nuclei Groups of neurons located along the midline of the brainstem, near the medulla oblongata; the neurotransmitter serotonin mediates their widespread influence throughout the central nervous system.

Rapid eye movement sleep See REM sleep.

Reafference That type of afference which stems from the animal's own muscular output, such as the tactile sensation produced by touching something or the audile sensation of hearing one's own voice; the case can be made that all conscious learning necessarily proceeds through detection of reafferent signals, even though such detection is not a sufficient condition for consciousness.

Receptive field The portion of the visual field in which a suitable distribution of light and dark produces the maximum response of a given neuron in the visual system.

Receptor See Receptor cell; Receptor molecule.

Receptor cell A sensory neuron that converts a specific stimulation (light, pressure, odour, etc.) to nerve impulses.

Receptor molecule A membrane-bound molecule capable of recognizing the molecules of a specific substance (neurotransmitter or hormone) and responding by generating a chemical or electrochemical signal.

Recurrent collaterals Axon collaterals which form synapses with the dendrites of the neurons they stem from; a good example is provided by the recurrent collaterals of the CA3 neurons in the hippocampus.

Red nucleus A group of neurons in the midbrain, involved in the control of movement.

Reflex arc A neural route from receptor neuron to motor neuron.

REM sleep Short for rapid eye movement sleep; the portion of sleep devoted to dreaming; also known as paradoxical sleep.

Reptilian complex The evolutionary earliest part of the brain; it governs the faculties related to survival.

Resting potential The voltage difference (about 100 millivolts) across a neuron's membrane, when that cell is in the quiescent state; caused by the (pump mediated) concentration imbalance of sodium and potassium ions between the inside and outside of the cell.

Reticular formation A group of structures stretching from the medulla oblongata to the thalamus; they collectively govern consciousness, and even partial damage can produce coma.

Retina A mesh of receptor and intermediary neurons at the back of the eye which partially processes incident visual information before transmitting it to the lateral geniculate nucleus, via the optic nerve.

Retinotopic Said of the geometrical representation of visual space in a particular (visual) cortical area if its connectivity is identical to that present in the retina; there is usually considerable distortion, however, the deeper one penetrates into the visual hierarchy.

Retrograde amnesia Loss or diminution of the ability to recall items previously stored in memory.

RNA See Nucleic acid.

Rod One of the two types of light-sensitive cell (photoreceptors) present in the retina; rods are specialized for low intensity situations, and they are preferentially located outside the foveal region. (See also Cone.)

Rostral In the head, toward the nose; in the rest of the body, toward the head.

Saccade A sudden change in the direction of gaze; saccades are typically made about four times a second, the actual movements taking about 50 milliseconds (during which the pulvinar appears to suppress visual detection), while the dwell periods last about 200 milliseconds. Given that it usually requires about 500 milliseconds for consciousness to develop, the above durations indicate that saccades are undertaken unconsciously.

Sagittal section A cut through the brain in a plane parallel to, but laterally offset from, the brain's plane of symmetry; a cut with no offset is sometimes referred to as the medial sagittal section.

Saltatory conduction The jumping of axonal conduction from one node of Ranvier to the next, due to the fact that these sites are far more favourable conduction sites than are the axonal stretches that are myelinated.

Schaffer collaterals Those axon collaterals of the CA3 neurons in the hippocampus that activate the nearby CA1 neurons.

Schema (plural: schemata) The reproducible linking of a planned sequence of muscular movements with the corresponding sensory input, the reproducibility stemming from the fact that schemata are laid down in memory. An organism is born with certain default schemata already present in its memory, and the balance of its schematic repertoire is acquired by direct experience. According to the present book's thesis, the ability to manipulate existing schemata, and to thereby construct new ones, is a measure of intelligence.

Schizophrenia A mental syndrome popularly attributed to a splitting of the mind, it manifests different clinical symptoms: withdrawal and emotional dampening (hebephrenia), delusions (paranoia), disturbances in attitudes and movement (catatonia), and impairment of mental faculties (dementia); it is now thought to have a partially genetic origin.

Scotoma A non-responsive patch in one of the processing regions of the visual system, such as the optic disc, or damaged region in one of the visual areas of the cerebral cortex.

Second messenger See Cyclic AMP.

Self-organization A process whereby elementary units rearrange themselves, in space and/or time, so as to produce an overall pattern whose characteristics are not obviously related to the properties of those units. A classic spatial example is the emergence of the structure of a snowflake from clustering water molecules, the latter not possessing the six-fold symmetry of the flake. In the present context, self-organization is believed to underlie the emergence of a suitable schema, to match a sensory input.

Semantic memory See Declaritive memory.

Sensor A structure composed of a number of receptor cells, which converts a stimulus to nerve impulses.

Sensory cortex The region of the cerebral cortex which receives direct sensory input, together with the regions which are involved in associating and processing different inputs or different aspects of a single input; the sensory cortex comprises the occipital, parietal and temporal cortices, and also the somatosensory cortex.

Septum A group of neurons in the limbic system, adjacent to the amygdala; together with the other limbic structures, it is implicated in the emotions. The septum is associated with the generation of theta oscillations.

Serotonin (5-HT) A neurotransmitter (also known as 5-hydroxytryptamine) derived from the amino acid tryptophan; its resemblance to lysergic acid diethylamide (LSD) explains the hallucinogenic properties of the latter.

Short-term memory Memory that cannot be recalled beyond a few hours.

Sodium–potassium pump (Na^+, K^+-ATPase) An enzyme which actively transports sodium out of a cell, and potassium simultaneously into a cell, the energy required for this translocation being provided by ATP.

Software In general, software is a term used to contrast with computer hardware, to refer to all programs which can be run on a given hardware system. A great variety of software is commercially available, and specialized examples have been produced additionally by individual computer users. Types of software include assemblers, compilers, executive programs, operating systems, library routines, generators, and file handling and debugging routines. It is the opinion of the present author that software will ultimately be available which permits simulation of consciousness.

Soma The body; in an individual neuron, the cell body that contains the nucleus and related organelles.

Somatosensory cortex The cortical region that receives sensory information from the body's mechanical (tactile) receptors in the muscles and skin.

Spike A term used loosely as an alternative to action potential or nerve impulse.

Spindle See Muscle spindle.

Spine In this book, a small twig-like protrusion from the branch of a dendrite. When present, spines are the favoured sites of synaptic contact, and the rationale for their existence may lie in the relatively circumscribed environment they provide for the local ionic solution. Spines are commonly found on pyramidal cells and on the excitatory variety of stellate cells.

Stellate cell A type of cortical neuron that influences only its local environment; its axons do not enter the white matter. Spiny stellate cells are excitatory, whereas smooth stellates are inhibitory.

Stretch reflex The automatic tendency for a muscle to contract when it is stretched by an external agency (e.g. the knee jerk).

Striate cortex The primary visual cortex, which receives input from the lateral geniculate nucleus; its name derives from the striped appearance after staining (with, for example, cresyl violet).

Striatum Composed of the putamen and the caudate nucleus, this member of the basal ganglia is located close to the thalamus, around which it could be said to wrap itself. Its structure is heterogeneous, the patch-like striosomes being embedded in the more continuous matrix. Both components appear to comprise complete body maps, the two representations being intimately interdigitated. The input to the matrix, primarily from the neocortex, appears to be less specific than that linking the striosomes to the premotor cortex (and supplementary motor cortex), the thalamic intralaminar nuclei and the amygdala. However, the deep layers (VI and the lower part of V) of the sensory cortex also send signals to the striosomes. It is possible that the heterogeneous structure of the striatum plays an important role in the ability to overtly carry out one motor programme while simultaneously simulating another internally, as when concurrently driving a car and thinking about something else.

Striosome The patchy component of the striatum, which receives specific inputs from the frontal lobe, the thalamic intralaminar nuclei, the amygdala and the deep layers of the sensory cortex.

Substance P One of a group of peptide neurotransmitters (called tachykinins) that are implicated in pain transmission; high concentrations are found in the substantia nigra.

Substantia nigra A two-component structure in the basal ganglia, adjacent to the pons, implicated in the initiation of movement. The *pars compacta* receives signals from the striosomes of the striatum, and returns signals to them, the mediating neurotransmitters being GABA and dopamine, respectively; it may act as a type of gain control for the striosomes. The *pars reticulata* receives signals from the matrix portion of the striatum, and also from the sub-thalamic nucleus, and sends signals to the thalamus and superior colliculus, the neurotransmitters being GABA, glutamate, GABA and GABA, respectively.

Sub-thalamic nucleus A component of the basal ganglia; it appears to function as a sort of gain controller for both regions of the globus pallidus, and also for the pars reticulata of the substantia nigra.

Sulcus A groove on the cortical surface.

Superior colliculus A structure which helps to control eye movement; it mediates blind-sight when the primary visual system is non-functional.

Supplementary motor area An area in the frontal lobe, immediately rostral to the primary motor cortex, which, like the premotor area, appears to be concerned with motor sequences rather than with individual muscular movements.

Suprachiasmatic nucleus A small nucleus of the hypothalamus, located immediately dorsal to the optic chiasm; its position makes it a strategic monitor of the diurnal light–dark cycle, and hence an ideal pace-maker of the circadian rhythm.

Sylvian fissure A major cortical groove (gyrus) running from a point just above the ear, in a forward and downward direction, roughly towards the eye.

Sympathetic Pertaining to that part of the autonomic nervous system that tends to increase activity; its afferent fibres come from the spinal cord.

Synapse The junction of a neuron with another neuron, gland or muscle; message transmission across a synapse is usually chemical (and neurotransmitter mediated), but electrical examples are known.

Synaptic cleft The narrow region (about 20 nanometres wide) which separates the presynaptic and postsynaptic membranes.

Synaptic vesicle See Vesicle.

Temporal lobe The region of the cerebral cortex below and behind the sylvian fissure; it comprises the auditory cortex, the limbic cortex and the temporal association cortex.

Tetanus In this book a period of repetitive stimulation.

Tetrodotoxin A neurotoxin that inhibits the action potential by mechanically blocking the sodium channels.

Thalamus A composite structure in the limbic system that serves as a relay station for nerve fibres from the senses to the neocortex, and the corresponding feedback fibres; the visual system's lateral geniculate nucleus is a prominent component, and the intralaminar nuclei are believed to be implicated in consciousness.

Threshold The minimum membrane depolarization that causes a neuron to emit action potentials.

Thought Internal simulation of an animal's interactions with the environment; probably through activation of schemata; the stream of consciousness appears to invoke kaleidoscopically changing schemata, and it draws freely on memories of past experiences.

Tourette syndrome See Gilles de la Tourette syndrome.

Transverse section A cut through the brain made at right angles to its ventral–dorsal axis.

Triune brain The model, propounded by Paul MacLean, that the brain comprises three evolutionary (and to some extent functionally) independent cognitive systems, viz.: the reptilian cortex, the limbic system and the neocortex.

Uncertainty principle A quantum mechanical rule that limits the precision with which pairs of physical attributes (position and momentum, for example) can simultaneously be measured.

Ventral In the head, toward the jaw; in the rest of the body, toward the front.

Ventricles A series of four interconnected cavities in the brain; filled with cerebrospinal fluid, they were believed by the ancients to harbour animal spirits.

Vergence A term used to cover both convergence and divergence, since each of these two attributes of anatomical connections between cortical layers or cortical areas cannot be present without the other.

Veridical Coincident with, corresponding to, or representing real objects or events.

Vermis The centrally located portion of the cerebellum.

Vesicle A small membrane-bounded container of neurotransmitter present near an axon terminal's presynaptic membrane; spherical for excitatory substances, elongated for inhibitors.

Vestibular system Structures in the inner ear which collectively govern balance.

Virus In the present book, an alien piece of computer code contained in a program or file and capable of destroying these, and indeed of erasing the entire hard disc (that is to say, permanent memory). The analogy with the biological counterpart is not drawn lightly – computer viruses are capable of replicating themselves into other programs and files.

Visual cortex The region of the neocortex that receives (primary) and processes (higher) visual information. It is composed of a number of individual areas (V1, V2, and so on) which serve detection and processing of different attributes of the visual image.

Visual field The amount of the environment that can make a visual impression, for a given direction of gaze.

Visual areas (V1, V2, etc.) See Visual cortex.

Voltage clamp A technique whereby the potential difference across a membrane is maintained at a constant value, while the various components of transmembrane current are measured.

Wernicke's area The cortical region that handles the understanding of language. A lesion in this area causes one type of aphasia.

White matter The layer lying inside the (grey matter) cerebral cortex; it consists of the pyramidal cell axons that connect various cortical regions.

Will The faculty by which one decides, or conceives oneself of deciding, upon and initiating action.

Winner-take-all A mechanism whereby only the strongest among a set of signals impinging upon a group of neurons are able to progress to the next stage of processing. The mechanism depends upon inhibition, preferably of the feedforward type, which sets up what is *effectively* an excitation barrier, through which only the strongest signals can pass. More correctly, the term should be winners-take-all, since several signals may satisfy the penetration criterion. An alternative term is feed-the-fattest.

Working memory The temporary retention of sensory impressions and/or recalled memories, required for the immediate processing of actions or thoughts.

Zygote A fertilized egg cell.

Author index

Subject index